Chambers Materials Science and Technology Dictionary

GENERAL EDITOR

Professor P.M.B. Walker, CBE FRSE

CONSULTANT EDITORS

Dr Nicholas St. J. Braithwaite MInstP, CPhys
Dr Peter R Lewis FIM, CEng
Ken Reynolds FIM, CEng
Dr George W. Weidmann FIM, CEng

Chambers

Published 1993 by Chambers Harrap Publishers Ltd
7 Hopetoun Crescent, Edinburgh EH7 4AY

Reprinted 1997, 1999, 2001, 2003

British Library Cataloguing in Publication Data
A catalogue record for this book is available from
the British Library.

ISBN 0 550 13248 1 Hbk
ISBN 0 550 13249 X Pbk

Cover design by John Marshall

Printed in Great Britain by Clays Ltd, St Ives plc

Contents

How the dictionary was made

Chambers Materials Science and Technology Dictionary was compiled and designed on a COMPAQ 386 personal computer. The original database was made with the INMAGIC library retrieval software from Head Computers Ltd. The text was set using the Xerox VENTURA desktop publishing system and the drawings made with the Micrografx DESIGNER graphics program.

Preface

Chambers Materials Science and Technology Dictionary is designed to provide an up-to-date account of both older materials used in new ways and the more modern materials, like ceramics and high-performance plastics, which are changing the nature of the objects we make and use. It has a number of unique features and was developed from the database of the *Chambers Science and Technology Dictionary* with many new and revised definitions in this important group of subjects which include not only engineering, physics and chemistry but also electronics, plastics, textiles and timber.

A special feature of this book is the series of 65 panels throughout the dictionary which amplify and describe many different topics within the field. Their subjects range from the alloys of the principle metals, bridge design, ceramics processing, magnetic materials and metal fatigue to rusting and tyre technology. Panels are illustrated by over 100 diagrams which help to explain the concepts and techniques which are discussed.

Arrangement

The entries in the dictionary are strictly alphabetical with single letter entries occuring at the beginning of each letter. Panels within the dictionary occur either on the same page as the parent entry or on the pages immediately following with the alphabetical entry stating 'See panel on p 143'. This book is also cross-referenced so that many terms in the dictionary refer to subjects considered at more length in the panels. Such cross-references use the style 'See **fatigue** p 123.' where the page number refers to the first page of the relevant panel. Other terms in **bold** refer directly to a particular entry.

Italic and bold

Italic is used for
(1) alternative forms of, or alternative names for, the headword usually after 'also' at the end of the entry;
(2) the expanded form of an abbreviated headword, provided that the expanded form is not found as a headword elsewhere;
(3) terms derived from the headword, often after 'adj.' or 'pl.';
(4) variables in mathematical expressions;
(5) for emphasis.

Bold is used for
(1) cross-references, either after 'see' and 'cf' or in the **body of the text**;
(2) after 'abbrev. for' when the expanded form can be found as a headword elsewhere;
(3) as vectorial notation in certain mathematical expressions.

Bold italic is used in the panels to highlight a term explained within that article.

Tables

Much information about all the naturally occuring elements will be found in the tables placed in Appendix 2 at the end of the book. The Periodic Table is shown in Appendix 1 and there are also tables of SI and other units and a list of conversion factors in Appendix 3.

Acknowledgements

We gratefully acknowledge permission to make use of material from *Structural Materials*, edited by George Weidmann, Peter Lewis and Nick Reid and *Electronic Materials* (edited by N. Braithwaite and G. Weaver), both published by the Open University and Butterworths. Thanks also to Addison-Wesley for permission to use Fig. 1 (Thermosets panel), from *Material Science for Engineers* by Van Vlach, and to Peter Head of the Maunsell Group for Fig. 2 (Bridges and Materials panel).

Contributors

My special thanks are due to Peter Lewis of the Materials Discipline of the Open University who both suggested the compilation of this dictionary and helped coordinate the writing of the panels, compilation of the tables and revision of many of the entries. He and his three colleagues, Nick Braithwaite, Ken Reynolds and George Weidmann (all of the Materials Discipline, Faculty of Technology, The Open University, Walton Hall, Milton Keynes, Bucks) have written the panels, extensively revised many entries and introduced many new definitions. Their patience and understanding have made this book possible. Nick Braithwaite wishes to acknowledge the assistance of Dr D. A. Cardwell and Dr S. G. Ingram in the preparation of some of the panels on electronic materials.

Extensive use has been made of the entries in *Chambers Science and Technology Dictionary* and I would like to thank all those who contributed to the large number of sections in that dictionary which we have drawn upon.

Nearly all the figures in the book have been drawn in electronic format by myself. Many were freely modified from the sketches and information provided by the contributing editors who were able to base some of them on figures which appeared in *Structural Materials*, acknowledged in the previous section.

P.M.B.W.

A

a- Prefix signifying *on*. Also shortened form of *ab-, ad-, an-, ap-*.

a Symbol for: (1) acceleration; (2) relative activity; (3) linear absorption coefficient; (4) amplitude.

α See under **alpha**. Symbol for: (1) absorption coefficient; (2) attenuation coefficient; (3) acceleration; (4) angular acceleration; (5) fine structure constant; (6) helium nucleus.

α- Symbol for: (1) substitution on the carbon atom of a chain next to the functional group; (2) substitution on a carbon atom next to one common to two condensed aromatic nuclei; (3) substitution on the carbon atom next to the hetero-atom in a hetero-cyclic compound; (4) a stereo-isomer of a sugar.

A Symbol for ampere.

A Symbol for: (1) area; (2) absolute temperature; (3) relative atomic mass (atomic weight); (4) Helmholtz function; (5) magnetic vector potential.

Å Symbol for Ångström.

ablative polymers Materials which degrade controllably in an aggressive environment, esp. on a re-entering spacecraft. Extreme temperatures are reached on heat shield, so it is protected with ablation shield made of eg silicone polymer. Same principle is used in intumescent paints for fire resistance.

abradant A substance, usually in powdered form, used for grinding. See **abrasive**.

abrade Scratch or tear away two surfaces in contact by relative motion.

abrasion Surface damage due to an abrasive or to rubbing contact.

abrasive A hard substance, usually in powdered form, used for the removal of material by scratching and grinding, eg silicon carbide powder (*Carborundum*).

abrasive wear Mechanism of wear due to the presence in one or both surfaces of hard particles (eg carbide in steels), or to hard particles trapped between them.

ABS Abbrev. for *Acrylonitrile-Butadiene-Styrene*. See **copolymer** p. 75.

absolute configuration The arrangement of groups about an asymmetric atom, esp. a tetrahedrally bonded atom with four different substituents. See **chirality**.

absolute temperature A temperature measured with respect to absolute zero, ie the zero of the Kelvin (thermodynamic) scale of temperature. See **kelvin**.

absolute units Those derived directly from the fundamental units of a system and not based on arbitrary numerical definitions. The differences between absolute and international units were small; both are now superseded by the definitions of the SI. See Appendix 3.

absolute viscosity See **viscosity**.

absolute zero The least possible temperature for all matter. At this temperature the atoms, ions or molecules of any substance have no remaining thermal energy. Its generally accepted value is –273.15°C = 0 K. See **kelvin**.

absorbance (1) The logarithm of the ratio of the intensity of light incident on a sample to that transmitted by it. It is usually directly proportional to the concentration of the absorbing substance in a solution. See **Beer's law**, **transmittance**. (2) The capacity of materials such as textile fibres and paper to absorb liquids. See **regain**.

absorbency test Any test method for measuring the capacity of materials such as textile fibres and paper to absorb liquids or fluids. Results are usually expressed as the gain in weight of the test piece, the capillary rise in a test strip in given time, or the time required to reach a predetermined capillary rise.

absorber Any material which converts energy of radiation or particles into another form, usually thermal energy. Energy transmitted is not absorbed. Scattered energy is often classed with absorbed energy.

absorptance A measure of the ability of a body to absorb radiation; the ratio of the radiant flux absorbed by the body to that incident on the body. Formerly *absorptivity*.

absorption band A dark gap in the continuous spectrum of white light transmitted by a substance which exhibits selective absorption.

absorption coefficient (1) At a discontinuity (surface absorption coefficient), (a) the fraction of incident energy which is absorbed, or (b) the reduction in amplitude of a wave incident on a discontinuity in the medium through which it is propagated, or in the path along which it is transmitted. (2) In a medium (linear absorption coefficient), the natural logarithm of the ratio of incident to the emergent energy or amplitude of a beam of radiation passing through unit thickness of the medium. (The mass absorption coefficient is defined in the same way but for a thickness of the medium corresponding to unit mass per unit area.) NB True absorption coefficients exclude scattering losses, total absorption coefficients include them. See **atomic absorption coefficient**.

absorption edge The wavelength at which

there is an abrupt discontinuity in the intensity of an absorption spectrum for electromagnetic waves, giving the appearance of a sharp edge in its photograph. This transition is due to one particular energy-dissipating process becoming possible or impossible at the limiting wavelength. In X-ray spectra of the chemical elements the K-absorption edge for each element occurs at a wavelength slightly less than that for the K-emission spectrum. Also *absorption discontinuity*.

absorption lines Dark lines in a continuous spectrum caused by absorption by a gaseous element. The positions (ie the wavelengths) of the dark absorption lines are identical with those of the bright lines given by the same element in emission.

absorption spectrum The system of absorption bands or lines seen when a selectively absorbing substance is placed between a source of white light and a spectroscope. See **Kirchhoff's laws**.

abura Tropical W African hardwood (*Mitragyna*) with a fine, even texture. Mostly orange-brown to pink sapwood, it is acid-resistant. Used in joinery, furniture, flooring, battery boxes and laboratory fittings. Density 460–690 kg m^{-3} (mean 550 kg m^{-3}).

a.c. Abbrev. for *alternating current*. Also *ac*.

Ac The transformation temperature on heating of the phase changes of iron or steel, subscripts indicating the designated change, eg Ac_1 is the eutectoid (723°C) and Ac_3 the ferrite/austenite phase boundary.

acacia See **myrtle**.

a.c. bias A high-frequency signal applied to a magnetic tape recording head along with the signal to be recorded. This stabilizes magnetic saturation and improves frequency response, at the same time reducing noise and distortion. The bias signal frequency has to be many times the highest recording frequency.

accelerated fatigue test Test which applies a cyclic loading schedule, which can be of varying frequency and/or amplitude, to a machine or component simulating its loading in service, but at a higher rate, to determine its safe fatigue life before it is reached in service. See **fatigue** p. 123.

acceleration The rate of change of velocity, expressed in metres (or feet) per second squared. It is a vector quantity and has both magnitude and direction.

acceleration due to gravity Acceleration with which a body would fall freely under the action of gravity in a vacuum. This varies according to the distance from the Earth's centre, but the internationally adopted value is 9.806 65 m s^{-2} or 32.1740 ft s^{-2}. Abbrev. *g*.

accelerator (1) Any substance which increases the rate of a chemical reaction. See **catalysis**. (2) A substance which catalytically increases the hardening (curing) rate of a thermosetting resin. (3) A substance which increases the rate of **vulcanization** of rubber.

acceptor Impurity atoms introduced in small quantities into a crystalline semi-conductor and having a lower valency than the semiconductor, from which they attract electrons. In this way *holes* are produced, which effectively become positive charge carriers; the phenomenon is known as *p-type conductivity*. See **donor**, **impurity**.

acceptor level See **energy levels**.

accumulator (1) Bottle or other reinforced reservoir for storing pressurized gas or fluid during moulding. Its use helps conserve energy during injection moulding cycle. (2) Voltaic cell which can be charged and discharged. On charge, when an electric current is passed through it into the positive and out of the negative terminals (according to the conventional direction of flow of current), electrical energy is converted into chemical energy. The process is reversed on discharge, the chemical energy, less losses both in potential and current, being converted into useful electrical energy. Accumulators therefore form a useful portable supply of electric power, but have the disadvantages of being heavy and of being at best 70% efficient. More often known as *battery*, also *reversible cell*, *secondary cell*, *storage battery*.

acetal resins See **polyoxymethylene**.

acetate fibres '*Man-made*' *fibres*. Continuous filaments and staple fibres manufactured from cellulose acetate produced from cotton linters or wood pulp. Between 74 and 92% of the hydroxyl groups in the original cellulose are acetylated. For triacetate fibres the cellulose is more highly acetylated.

acetylcelluloses See **cellulose acetates**.

acetylene *Ethyne*. HC≡CH, a colourless, poisonous gas, owing its disagreeable odour to impurities, soluble in ethanol, in acetone (25 times its volume at s.t.p.) and in water. Bp –84°C, rel.d. 0.91. Prepared by the action of water on calcium carbide and catalytically from naphtha. Used for welding, illuminating, acetic acid synthesis and for manufacturing derivatives.

acetyl group Ethanoyl group. The radical of acetic acid, viz., CH_3CO-.

acicular (1) Needle-shaped. (2) The needle-like habit of crystals.

acid Normally, a substance which (a) dissolves in water with the formation of hydrogen ions, (b) dissolves metals with the liberation of hydrogen gas, or (c) reacts with a base to form a salt. More generally, a substance which tends to lose a proton

(Brönsted-Lowry theory) or to accept an electron pair (Lewis theory).

acid brittleness That developed in steel in pickling bath, through evolution of hydrogen. Cf **hydrogen embrittlement**.

acidity (1) The extent to which a solution is acid, normally expressed as its **pH** value. Cf **alkalinity**. (2) The concentration of any species in a solution which is titratable by a strong base.

acid process (1) A steelmaking process in which the furnace is lined with a siliceous refractory, and for which iron low in phosphorus is required, as this element is not removed. See **basic process**. (2) In papermaking, any pulp digestion process utilizing an acid reagent, eg a bisulphite liquor with some free sulphur dioxide.

acid refractory See **silica and silicates** p. 285.

acid resist foils Blocking foils for use in etching metal. The foil is stamped on to paper and the excess foil blocked on to the metal rule or other object which is then exposed to an acidic etching fluid such as ferric chloride.

acid slag Furnace slag in which silica and alumina exceed lime and magnesia.

acid steel Steel made by an **acid process**.

A-class insulation A class of insulating material which will withstand temperatures up to 105°C.

acoustic branch The lattice dynamics of a crystal containing n atoms per unit cell show that the **dispersion curve** (frequency ω against wave number q) has $3n$ branches of which three are acoustic branches. The branches are characterized by different patterns of movement of the atoms. For the acoustic branches, ω is proportional to q for small q. See **optic branch**.

acoustic emission Non destructive testing (*NDT*) method of investigating deformation and failure processes in materials by the signals generated when the elastic waves released by them are detected at the materials' surfaces.

acoustic microscope One based on acoustic waves (longitudinal compressions and rarefactions of density) at microwave frequencies: the interaction of an acoustic wave with a material is sensitive to its elastic properties. Images can be created by modulating a display with the intensity received by a detector/specimen system scanned synchronously (*ultrasonic imaging*). Coupling between electrical signals and acoustic vibrations exploits **piezoelectricity**.

acoustic scattering Irregular and multidirectional reflection and diffraction of sound waves produced by multiple reflecting surfaces the dimensions of which are small compared to the wavelength; or by certain discontinuities in the medium through which the wave is propagated.

Acrilan TN for a synthetic polyacrylonitrile fibre obtained by copolymerizing acrylonitrile (85%) with vinyl acetate (15%).

acro- Prefix from Gk. *akros*, topmost, farthest, terminal.

acrylate rubbers Crosslinkable rubbers based on **copolymers** of ethyl acrylate and about 5% 2-chloroethyl vinyl ether as a site for **vulcanization**. Oil and heat resistant speciality rubbers. Also used to describe elastomers based on acrylate-methacrylate copolymers used in rubber toughening of polyvinyl chloride for example.

acrylic acid *Prop-2-enoic acid.* $CH_2=CH.COOH$, mp 7°C, bp 141°C, of similar odour to acetic acid; a very reactive monomer, forming polyacrylic acid.

acrylic adhesives Wide range of derivatives of acrylic acid, some of which are copolymerized, which gives a variety of adhesive types, eg water soluble, pressure sensitive, anaerobic etc. See **cyanoacrylate**.

acrylic ester An ester of acrylic acid or of a structural derivative of acrylic acid, eg methacrylic acid or its chemical derivatives. Monomers for acrylic polymers.

acrylic fibres Continuous filaments or, more usually, **staple fibres** made from linear polymers which are synthesized from several monomers containing at least 85% by weight of acrylonitrile.

acrylic resins Resins formed by the polymerization of the monomer and derivatives, generally esters or amides, of acrylic acid or α-methylacrylic acid. They include polymethyl methacrylate, acrylic rubbers etc.

acrylonitrile *Vinyl cyanide (1-cyanoethene)*, used as raw material for synthetic acrylic fibres, eg ABS, Acrilan, Courtelle, Orlon, nitrile rubber, SAN etc.

actin A fibrous protein found in muscle in association with myosin.

actinides Radioactive elements after actinium ($Z= 89$). All have similar chemical properties. Cf **lanthanides**.

actinium A radioactive element in the third group of the periodic system. Symbol Ac; half-life 21.7 years. Produced from natural radioactive decay of the ^{235}U isotope or by neutron bombardment of ^{226}Ra. Gives its name to the actinium ($4n+3$) series of radioelements. See Appendix 1, 2.

activated carbon Carbon obtained from vegetable matter by carbonization in the absence of air, preferably in a vacuum. Activated carbon has the property of adsorbing large quantities of gases. Important for gas masks, adsorption of solvent vapours, clarifying liquids and in medicine.

activated charcoal Charcoal treated with acid etc to increase its adsorptive power.

activated sintering Sintering of a **compact** in the presence of a gaseous **reactant**. Also *reaction sintering*. See **ceramics processing** p. 54.

activation (1) An increase in the energy of an atom or molecule, rendering it more reactive. (2) The heating process by which the capacity of carbon to adsorb vapours is increased. (3) Altering the surface of a metal to a chemically active state. Cf **passivation**.

activation energy (1) The excess energy over that of the ground state which an atomic system must acquire to permit a particular process, such as emission or reaction, to occur. (2) The energy required for a thermally activated physical or chemical process. See **Arrhenius's rate equation.**

activator (1) Chemical that promotes **vulcanization** or acceleration of **cross-linking**. (2) An impurity, or displaced atom, which augments luminescence in a material, ie a sensitizer such as copper in zinc sulphide.

active mass Molecular concentration generally expressed as moles dm^{-3}; in the case of gases, active masses are measured by partial pressures. See **activity** (2).

active materials (1) General term for essential materials required for the functioning of a device, eg semiconductor in an integrated circuit, iron or copper in a relay or machine, electrode materials in a primary or secondary cell, emitting surface material in a valve, or photocell, phosphorescent and fluorescent material forming a phosphor in a CRT, or that on the signal plate of a TV camera. (2) Term applied to all types of radioactive isotopes.

activity (1) See **optical activity**. (2) The ideal or thermodynamic concentration of a substance the substitution of which for the true concentration permits the application of the **law of mass action**. (3) The magnitude of the oscillations of a **piezoelectric crystal** relative to the exciting voltage.

activity coefficient The ratio of the activity to the true concentration of a substance.

activity constant The **equilibrium constant** written in terms of activities instead of molar concentrations.

acyclic compound See **aliphatic compound**.

ad- Prefix signifying to, at; from L. *ad*.

adamantine compound Compound with the same tetrahedral, covalently bonded crystal structure as that of **diamond**, eg zinc sulphide.

ADC Abbrev. for **AzoDiCarbonamide**.

adduct The addition product of a chemical reaction between molecules.

adherend A material which is bonded by an adhesive.

adhesion (1) The bonding of materials with adhesives (glues, cements, binders etc), in which the intermolecular forces between adhesive and adherend provide the bonds. (2) The intimate sticking together of metallic surfaces under compressive stresses by bonds which form as a function of stress, time and temperature. The speed of formation is related to **dislocations**, and may occur virtually instantaneously under high shear stresses. See **cold welding**. (3) Intermolecular forces which hold matter together (US *bond strength*).

adhesive Agent for joining materials by adhesion, usually polymeric material. May be based on thermoplastic resin (eg polystyrene cement) or thermoset (eg epoxy resin). Viscosity is important for gap filling (high, as in epoxies) or surface penetration (low, as in cyanoacrylates). Also *glue, cement, binder*.

adhesive-bonded non-woven fabric A fabric made from a **batt** or web of fibres bonded together with an adhesive.

adhesive bonding Joining of parts with polymeric adhesive. Widely used for assembly of complex composite products (eg helicopter rotor blade). Great potential for routine assembly of car bodies, replacing **spot welding**.

adhesive wear Mechanism of wear due to the welding together and subsequent shearing off of the contact areas between two surfaces sliding over one another.

adiabatic heating Self-heating effect which occurs in extruder or injection moulding barrel from action of rotating screw on polymer melt. Attributed to dissipation of mechanical shear forces as heat. Important in injection moulding of rubbers. See **damping**. Also *shear heating*.

adiabatic process Process which occurs without interchange of heat with surroundings.

adipamide *1,4-Butanedicarboxamide*, $NH_2CO(CH_2)_4CONH_2$. Used in synthetic fibre manufacture.

adipic acid *Butanedicarboxylic acid*, $HOOC.(CH_2)_4.COOH$. Colourless needles; mp 149°C; bp 265°C; formed by the oxidation of cyclo-hexanone, or by the treatment of oleic acid with nitric acid. Used in the manufacture of nylon 6,6 and polyester resin.

Adiprene TN for polyurethane elastomers which combine high abrasion resistance with hardness and resilience.

Admiralty brass See **copper alloys** p. 76.

admixture Property-modifying additive to eg Portland cement.

adsorbate Adsorbed substance. See **adsorption**.

adsorption The taking up of one substance at the surface of another.

adsorption chromatography See **chromatography**.

adsorption surface area The surface area of a powder particle determined by the mass of a specified substance that it can adsorb.

advance metal Copper-base alloy with 45% nickel.

Ae The transformation temperature at equilibrium of the phase changes in iron and steel, subscripts indicating the designated change.

aeolotropic Having physical properties which vary with direction or position. See **anisotropic**.

aerosol (1) A colloidal system, such as a mist or a fog, in which the dispersion medium is a gas. See **colloid**. (2) Pressurized container with built-in spray mechanism used for packaging insecticides, deodorants, paints etc.

afara W African **hardwood** (*Terminalia*) which is pale yellow-brown to straw coloured, with black streaks in the **heartwood**, and has a close, straight-grained texture. Not very durable or decay resistant. Used in joinery, furniture, coffins, light construction, and the black heartwood is used for veneers. Density 480–640 kg m^{-3} (average 550 kg m^{-3}). Also *limba*.

afgalaine An all-wool, plain-weave, dress cloth.

African blackwood E African **hardwood** (*Dalbergia*) yielding a very dense (1200 kg m^{-3}), exceptionally hard, almost black, straight-grained, durable wood. Used for musical instruments, ornaments, truncheons, bearings and pulley blocks.

African mahogany W African **hardwood** (*Khaya*) which seasons well. It is liable to infestation by powder-post beetles. It works easily, finishes to a good surface, and is used for cabinetmaking, veneering, plywood manufacture, high-class joinery and furniture. Density 540–590 kg m^{-3}.

African walnut A tropical W African **hardwood** from the genus *Lovoa*; it is not a true walnut, but finds similar uses in high-class furniture and cabinetmaking, gun stocks, billiard tables and veneers. It has a bronze orange-brown **heartwood**, with an interlocked, sometimes spiral grain and lustrous texture. Density 480–650 kg m^{-3} (average 550 kg m^{-3}).

African whitewood See **obeche**.

afrormosia A very durable **hardwood** from W Africa, highly resistant to termites, and used as a substitute for **teak**. Density 620–780 kg m^{-3} (average 690 kg m^{-3}).

afterglow A gaseous, luminous medium immediately after the cessation of electric current or downstream of an electric discharge. See **persistence**.

agba Tropical W African **hardwood** (*Gossweilerodendron*) whose **heartwood** is pale pinkish-straw to tan coloured, and has a straight to wavy grain and fine texture. Very durable and decay resistant, but **sapwood** attacked by furniture beetle. Used in joinery, furniture, boatbuilding, flooring, coffins, light construction and in veneers. Average density 520 kg m^{-3}.

age hardening The production of structural change spontaneously after some time; normally it is useful in improving mechanical properties in some respect, particularly hardness. See **precipitation hardening** p. 254.

ageing Change in the properties of a material with time. Specifically: (1) a change in the magnetic properties of iron, eg increase of hysteresis loss of sheet-steel laminations; also the process whereby the subpermanent magnetism can be removed in the manufacture of permanent magnets; (2) final stage of **precipitation hardening**, producing an increase in strength and hardness in metal alloys, due to precipitation of second phase particles from supersaturated solid solution over a period of days at room temperature, or several hours at an elevated temperature ('artificial' ageing); (3) the exposure of freshly printed textile fabrics to steam to produce fully developed colours; (4) slow deterioration in polymer products due to oxygen or ozone cracking, increase in crystallinity, relaxation of internal stress etc; (5) deterioration of the properties of ferroelectric materials. See **dielectric and ferroelectric materials** p. 94.

agglomerate Assemblage of particles rigidly joined together, as by partial fusion (**sintering**) or by growing together.

agglutination The coalescing of small suspended particles to form larger masses which are usually precipitated. Cf **coagulation**.

aggregate (1) Assemblage of powder particles which are loosely coherent. (2) Mixture of sand and gravel or crushed rock used in making concrete. Graded aggregate has a graded size distribution so that the particles fit better together, requiring less cement in the mix.

A-glass Designation for a glass fibre of composition (wt %): SiO_2 72, Na_2O 14, CaO 10, MgO 2.5, Al_2O_3 0.6 which is similar to that of the soda-lime-silica glass used for windows and bottles. Its resistance to water, mineral acids and alkalis is much less than that of **C-** and **E-glass** fibres.

air capacitor A capacitor in which the dielectric is nearly all air, for tuning electrical circuits with minimum dielectric loss.

air cooling The cooling of hot bodies by a stream of cold air; distinct from water cooling.

air gap Section of air, usually short, in a magnetic circuit, esp. in a motor or generator, a relay, or a choke. The main flux passes through the gap, with leakage outside depending on dimensions and **permeability** (2).

air-entraining agent A resin **admixture** added to either cement or concrete in order to trap small air bubbles.

air-hardening steel Steel with sufficient carbon and other alloying elements to allow sections over 500 mm (20 in) to harden fully when cooled in air or other gas from above its transformation temperature. Also *self-hardening steel*. See **isothermal transformation diagram** p. 177.

air jet spinning Method of converting staple fibres into yarn: they are spun together by jets of air which strike the fibres tangentially, making them rotate.

air laying Method of forming a **batt** or **web** by collecting fibres from an air stream onto a mesh ready for manufacturing a non-woven fabric.

airline Straight line drawn on the magnetization curve of a motor, or other electrical apparatus, expressing the magnetizing force necessary to maintain the magnetic flux across an **air gap** in the magnetic circuit.

Airy points The optimum points for supporting a beam horizontally to minimize the bending deflection. The distance apart of the points is equal to $l/(n^2-1)$ where l is the length of the beam and n the number of supports.

Ala Symbol for **alanine**.

alabaster A massive form of **gypsum**, often pleasingly blotched and stained. Because of its softness it is easily carved and polished, and is widely used for ornamental purposes. Chemically it is $CaSO_4.2H_2O$. *Oriental alabaster*, onyx marble is a beautifully banded form of stalagmitic *calcite*.

alanine *2-aminopropanoic acid*, $CH_3C(NH_2)COOH$. The L- or S-isomer is a common constituent of proteins. Symbol Ala, short form A.

albert A former standard size of note-paper, 192×102 mm (6×4 in).

Alclad Composite sheets consisting of an alloy of the **Duralumin** type (to give strength) coated with pure aluminium (to give corrosion resistance).

alcohol A general term for compounds formed from hydroxyl groups attached to carbon atoms in place of hydrogen atoms. The general formula is R.OH, where R signifies an aliphatic radical. In particular, ethanol. Hydroxyl groups attached to aromatic

rings give phenols.

Alcomax UK equivalent of **Alnico** permanent magnet alloy.

aldehydes *Alkanals*. A group of compounds containing the CO group attached to both a hydrogen atom and a hydrocarbon radical, viz. R.CHO.

alder A tree (*Alnus*) producing a straight-grained fine-textured **hardwood** noted for its durability under water. It is used for cabinet making, plywood, shoe heels, clogs, bobbins, wooden cogs and small turned items. Density 530 kg m^{-3}.

alignment Process of orientation of eg electric or magnetic dipoles when acted on by an external field. During magnetization, the alignment of domains is changed by the magnetizing field.

aliphatic compounds Methane derivatives of fatty compounds; open-chain or ring carbon compounds not having aromatic properties. Cf **aromatic compounds**.

alite Tricalcium silicate, $CaO.Ca_2SiO_4$. The major constituent (typically 55 wt%) of Portland cement, it hydrates at a medium rate during the setting reaction. See **cement and concrete** p. 52.

alkali A hydroxide which dissolves in water to form an alkaline or basic solution which has pH > 7 and contains hydroxyl ions, OH–.

alkali metals The elements lithium, sodium, potassium, rubidium, caesium and francium, all metals in the first group of the periodic table. See Appendix 1. In most compounds they occur as univalent ions.

alkaline earth metals The elements calcium, strontium, barium and radium, all divalent metals in the second group of the periodic table. See Appendix 1.

alkalinity The extent to which a solution is alkaline. See **pH, acidity**.

alkane General name of hydrocarbons of the methane series, of general formula C_nH_{2n+2}.

Alkathene TN for **polyethylene**.

alkene *Olefin*. General name for unsaturated hydrocarbons of the ethene series, of general formula C_nH_{2n}.

alkyd resins Polyester thermosets derived from glycerol and phthalic anhydride (glyptal resins). Also includes diallyl esters and various polyesters used as resin binders in alkyd moulding materials. Widely used for coatings. See **thermosets** p. 318.

alkyl A general term for monovalent aliphatic hydrocarbon radicals.

alkyne An aliphatic hydrocarbon with a triple bond. The simplest is ethyne or acetylene, HC≡CH.

allo- Prefix from Gk. *allos*, other. In particular it is used to show that a compound is a stereoisomer of a more common compound.

allophanate Type of group occurring in

polyurethanes, formed by reaction of excess isocyanate with a chain urethane group, so giving crosslinks in the resultant material. Compare with **biuret** group.

allotropy The existence of an element in two or more solid, liquid, or gaseous forms, in one phase of matter, called allotropes.

allowed band Range of energy levels permitted to electrons in a molecule or crystal. These may or may not be occupied.

allowed transition Electronic transition between energy levels which is not prohibited by any quantum selection rule.

alloy Mixture of atomic species exhibiting metallic properties and usually prepared by adding other metals or non metals to solvent metal in the liquid state, but may also be formed from sintered powders or by intimate mixing by mechanical means. They may be compounds, eutectic mixtures or solid solutions. If the component metals crystallize in the same form and their atomic radii are within 15% then, provided there is no significant electrochemical difference, complete intersolubility in the solid state may result. However, normally solid solubility is restricted and a number of different solid phases may exist in the system. See **phase diagram** p. 236. Alloys are stronger than their constituent metals and may exhibit a variety of hardening mechanisms. Solid solutions exhibit lower electrical conductivity and reduced ductility compared with the pure components. The term alloy is also used to describe molecular mixtures and solutions in plastics and ceramics.

alloy cast-iron Cast-iron containing alloying elements in addition to carbon and the normal low levels of manganese and silicon, usually some combination of nickel, chromium, copper and molybdenum. These elements may be added to increase the strength of ordinary irons, to facilitate heat treatment, or to obtain martensitic, austenitic or ferritic irons.

alloying Process of making an alloy.

alloy junction One formed by alloying one or more impurity metals with a semiconductor. Small buttons of impurity metal are placed at desired locations on a semiconductor wafer; heating to melting point and rapidly cooling again produces regions of *p-type* or *n-type* conductivity, according to choice of impurity. Also *fused junction*.

alloy reaction limit Concentration in alloy of a specific component, below which corrosion occurs in a given environment.

alloy steel A steel to which elements not present in carbon steel have been added, or in which the content of manganese or silicon is increased above that in carbon steel. See **steels** p. 296.

allyl group The unsaturated monovalent aliphatic group $H_2C{=}CH.CH_2{-}$.

allyl resin One formed by the polymerization of chemical compounds of the allyl group, eg **CR 39**.

Alnico US TN for a high-energy permanent magnet material, an alloy of aluminium, nickel, cobalt, iron and copper.

Aloxite TN designating a proprietary fused alumina and associated abrasive products.

alpaca The fine, strong hair of the alpaca (*Lama pacos*) of South America; the fabric made from such hair. This animal belongs to the camel family and is a close relative of the llama (*L. glama*) and the vicuña (*L. vicugna*).

alpha-beta brass Copper-zinc alloy containing 38–46% (usually 40%) of zinc. It consists of a mixture of the α-constituent (see **alpha brass**) and the β-constituent. See **copper alloys** p. 76.

alpha-brass A copper-zinc alloy containing up to 38% of zinc. Consists constitutionally of a solid solution of zinc in copper. Commercial alpha brasses of several compositions are made. All are used mainly for cold-working. See **copper alloys** p. 76.

alpha-bronze A copper-tin alloy consisting of the alpha solid solution of tin in copper. Commercial forms contain 4 or 5% of tin. This alloy, which differs from gun metal and phosphor bronze in that it can be worked, is used for coinage, springs, turbine blades etc. See **copper alloys** p. 76.

alpha decay Radioactive disintegration resulting in emission of α-particle. Also *alpha disintegration*.

alpha emitter Natural or artificial radioactive isotope which disintegrates through emission of α-rays.

alpha iron One of the polymorphic forms of iron, stable below 1179 K. Has a body-centred cubic lattice, and is magnetic up to 1041 K.

alpha particle Nucleus of helium atom of mass number four, consisting of two neutrons and two protons and so doubly positively charged. Emitted from natural or radioactive isotopes. Often written α-particle.

alpha radiation **Alpha particles** emitted from radioactive isotopes.

alternating copolymer See **copolymer** p. 75.

alternating stress The stress induced in a material by a varying force which acts alternately and periodically in opposite directions.

Alumel TN for an alloy of nickel with up to 5% aluminium, manganese and silicon, used (with **Chromel**) in thermocouples.

alumina Aluminium oxide, Al_2O_3. It is used as an abrasive, and as a structural ceramic.

aluminium alloys

There are two principal classifications based on suitability for specific manufacturing processes, namely casting alloys and wrought alloys, both of which are further subdivided into the categories *heat treatable* and *non-heat treatable*.

Virtually all alloys are initially produced from the liquid state, but when intricate engineering castings are needed as finished shapes with little or no further deformation, the alloy composition is selected to give good casting characteristics (fluidity, accurate mould reproduction, internal soundness etc). For this reason, silicon in amounts between 4 and 13% is almost universally added, the greater the percentage the better the fluidity, but higher values tend to have an embrittling effect. For certain processes a narrow freezing range is desirable (eg *gravity die casting*), so a silicon content near the eutectic (11.4%) is used; other processes (eg *pressure die-casting*) respond better to alloys with a limited freezing range, which tend to contain only 4–6% silicon. See **eutectic**. Other elements are added to impart specific properties such as solid solution strengthening (Mn up to 0.5%), corrosion resistance (Mg 7–10%), grain size control (Ti 0.25%). A useful feature is that the coarseness of the Al-Si eutectic and its associated lack of ductility may be dramatically altered by minute amounts of sodium, a treatment referred to as *modification*.

The yield and tensile strengths of castings which contain only silicon and solid solution strengtheners are limited, but additions can be made which allow precipitation hardening and ageing treatments to give strengths as high as those of wrought materials. Copper (3.5–5%), magnesium (0.3–0.8%), and/or zinc (1–5%) are added for this purpose. See **precipitation hardening**. Nickel imparts elevated temperature strength with low expansion characteristics and 2% is commonly added to piston alloys. Chromium (0.5%), niobium (0.2%) or cobalt (0.5–1%) are also included for specific purposes.

Materials to be hot or cold worked to semifinished products (extrusions, forgings, plate, sheet, wire, etc) must possess ductility; hence casting qualities are of lesser importance in wrought alloys. Silicon is detrimental in excess of the amount necessary to form Mg_2Si for precipitation hardening (up to 1% together with 0.5% magnesium) and the principal alloying ingredients are to promote strength, corrosion resistance or other desirable properties. Otherwise, the same addition elements are made to the wrought alloys as to the casting group, though the quantitites may differ somewhat.

continued on next page

aluminate (1) Salt of aluminic acid, H_3AlO_3, a tautomeric form of aluminium hydroxide, which acts as a weak acid. Ortho-aluminates have the general formula M_3AlO_3 or $M_3Al(OH)_6$, and meta-aluminates, $MAlO_2$ or $MAl(OH)_4$, where M is a monovalent metal. Sodium aluminate, Na_3AlO_3, is used as a coagulant in water purification and softening. (2) Tricalcium aluminate, $2CaO.Ca(Al_2O_2)_2$. A constituent ($\sim$ 12 wt%) of Portland cement, it hydrates at a rapid rate during the setting reaction. See **cement and concrete** p. 52.

alumina trihydrate $Al_2O_3.3H_2O$. Used as a fire-retarding additive in plastics.

aluminium Silver-white metallic element, forming a protective film of oxide. Symbol Al. Obtained from bauxite, it has numerous uses and is the basis of light alloys for use in eg structural work; alloyed with iron and cobalt in many types of permanent magnet. Polished aluminium reflects well beyond the visible spectrum in both directions, and does not corrode in sea water. Foil aluminium is much used for capacitors. The metal can be used as a window in X-ray tubes and as sheathing for reactor fuel rods. Aluminium is produced on a large scale where electric power is cheap when bauxite is electrolysed in fused cryolite. US *aluminum*. See Appendix 1, 2.

aluminium alloys See panel on p. 8.

aluminium anode cell One with an aluminium anode immersed in an electrolyte

---

aluminium alloys (contd)

Several grades of commercially pure (99%+) aluminium fall into the wrought category. These, like the moderate strength solid solution alloys, are only capable of being strengthened by working. As the degree of cold work increases, yield and tensile strength rise but ductility falls. A range of temper designations is used to denote the combination of mechanical properties. In the temper designation 'O' stands for the annealed state with lowest strength and highest ductility; 'H' denotes strain-hardened material and is followed by two digits, the first indicating whether the material is simply strain hardened or has received subsequent stabilizing heat treatment and the second digit is from an eight point scale related to the degree of strain hardening with '8' denoting the greatest strength and least ductility. Thus 'H14' represents non-stabilized material in a half-hard state midway between the softest annealed and the fully work hardened conditions; 'H16' represents three-quarters hard, and so on.

Commercially pure aluminium and solid solution strengthened alloys may readily be joined by welding using appropriate techniques, though this removes the effects of any work hardening indicated in the temper designation. Heat treatable alloys are not usually welded since this results in *over-ageing*; mechanical fasteners or adhesive bonding are therefore preferred.

High-performance aluminium alloys, eg for aircraft applications, are based on wrought materials which are precipitation hardened after working to the desired form. Some of the well-established alloys are still referred to by trade names, though these are now obsolete. Examples are: Duralumin (age hardenable 4% copper together with manganese, magnesium and silicon) and Hiduminium (age hardenable, based on copper, magnesium and silicon). Because the heat treated high strength alloys are susceptible to corrosion, sheet is sometimes roll bonded with pure aluminium on its surfaces, which is given the trade name Alclad.

---

which does not attack aluminium. The cathode may also be of aluminium or some other metal, eg lead. Such cells can be used as rectifiers or as high-capacitance capacitors. See **electrolytic capacitor**.

aluminium antimonide A semiconducting material used for transistors up to a temperature of 500°C.

aluminium brass Brass to which aluminium has been added to increase its resistance to corrosion. Used for condenser tubes in steam plant. Contains 1–6% Al, 24–42% Zn, 55–71% Cu. See **copper alloys** p. 76.

aluminium bronze Copper-aluminium alloys which contain 4–11% aluminium, and may also contain up to 5% each of iron and nickel. These alloys have high tensile strength, are capable of being cast or cold-worked, and are resistant to corrosion. See **copper alloys** p. 76.

aluminium chloride Strong Lewis acid used in cationic polymerization etc.

aluminosilicates Compounds of alumina, silica and bases, with water of hydration in some cases. They include clays, mica, zeolites etc, constituents of glass, **porcelain** and **cement**.

aluminothermic process The reduction

of metallic oxides by the use of finely divided aluminium powder. An intimate mixture of the oxide to be reduced and aluminium powder is placed in a refractory crucible; a mixture of aluminium powder and sodium peroxide is placed over this and the mass fired by means of a fuse or magnesium ribbon. The aluminium is almost instantaneously oxidized, at the same time reducing the metallic oxide to metal. This process, also known as the thermite process, is used esp. for the oxides of metals which are reduced with difficulty (eg titanium, molybdenum). On ignition, the mass may reach a temperature of 3000 K. Magnesium incendiary bombs have thermite as the igniting agent. Also *Thermit process*.

aluminous cement See **high alumina cement**.

aluminum US spelling of **aluminium**.

alums A large number of isomorphous compounds whose general formula is: $R'R'''(SO_4)_2.12H_2O$ or $R'_2SO_4.R'''(SO_4)_3.24H_2O$ where R' represents an atom of a univalent metal or radical – potassium, sodium, ammonium, rubidium, caesium, silver, thallium; and R''' represents an atom of a tervalent metal – aluminium, iron, chromium, manganese, thallium.

amalgam Solution alloy based on mercury metal, commonly with silver or gold, as these are utilized in one method for their extraction, but several other metals also dissolve in mercury. See **alloy**. Specially formulated solid amalgams are used for dental fillings.

amber A natural thermoplastic fossil resin known and highly valued since prehistoric times for its lustre, transparency and yellow colour. Produced by certain species of tree in Tertiary times and occurring in Estonia etc. Rich in succinic acid.

ambi- L. form of Gk *amphi-*.

ambient temperature Term used to denote temperature of surrounding air.

ambipolar Said of any condition or property which applies equally to positive and negative charge carriers (eg positive or negative ions, holes, electrons) in a plasma or semiconductor.

amboyna A **hardwood** (*Pterocarpus*) from the E Indies related to the paduak, whose **heartwood** varies from pale yellow to brick red, with a wavy grain, fine texture and characteristic smell. Very durable and resistant to insects and termites. Used in high-class joinery, furniture and cabinetmaking, instrument cases, and, especially the **burr**, is used for veneers (*amboyna burr*). Average density 600 kg m^{-3}. Also *narra*.

ambroid A semisynthetic **amber** made by compressing and heating small, scrap pieces of amber and used for pipe mouthpieces, beads etc.

American mahogany Hardwood tree (*Swietenia*) from C and S America. **Heartwood** has medium, uniform texture with colour varying from light, reddish-brown to deep, rich red. Durable and used for high-class furniture and cabinetmaking, veneers, patternmaking etc. Density 540–640 kg m^{-3}.

American pitch pine See **pitch pine**.

American plane See **sycamore**.

American red gum *Liquidambar styraciflua*, a **hardwood** tree of silky surface and irregular grain. Its **heartwood** (known as *gum, sweet gum, bilsted* or *satin walnut*) is marketed separately from its **sapwood** (*sap gum* or *hazel pine*). Not very durable; used mainly for furniture, fittings, cooperage and panelling. Varies in colour from pinkish-brown to deep red brown (heartwood) and creamy white (sapwood). Density 560 kg m^{-3}.

American whitewood Hardwood tree (*Liriodendron tulipifera*) with very wide, whitish **sapwood**, yellow- to olive-brown **heartwood**, straight grain and fine texture. Non-durable, so used for interior fittings, joinery, cabinetmaking, etc and in plywood and veneers. Density 510 kg m^{-3}. Also *tulip*

tree, canary whitewood. (NB Not the same as whitewood, which is the UK term for softwood from fir and spruce.)

amide group –CO.NH$_2$– group in amides, when replacing the hydroxyl in a carboxyl group.

amides A group of compounds, eg polyamides, in which the hydroxyl of the carboxyl group of acids has been replaced by the amine group –NH$_2$. It is the group –CO–NH– in polymers.

amines Aminoalkanes. Organic derivatives of ammonia NH$_3$ in which one or more hydrogen atoms are replaced by organic radicals. Also *amino group*.

amino acid The basic chain unit of proteins and polypeptides. There are 20 natural amino acids and all except glycine (Gly) are chiral and comprise a tetravalent asymmetric carbon atom to which are attached four different groups (R–, H–, NH$_2$– and –CO$_2$H). The family of amino acids is created by variation of R, the simplest being glycine (R = H) and alanine (R = CH$_3$). The configuration of the asymmetric carbon atom is L in all natural amino acids.

aminoplastics Those derived from the reaction of urea, thiourea, melamine, or allied compounds (eg cyanamide polymers and diaminotriazines) with aldehydes, particularly formaldehyde (methanal). See **thermosets** p. 318.

ammonium The ion NH$_4^+$, which behaves in many respects like an alkali metal ion.

ammonium dihydroxide phosphate **Piezoelectric crystal** used in microphones and other transducers; it can withstand a temperature higher than can Rochelle salt.

amorphous Descriptive of a material without the periodic, ordered structure of crystalline solids. See **glasses and glassmaking** p. 144.

amorphous carbon See **diamond-like carbon**.

amorphous metal See **metallic glass**.

amorphous polymers Materials having polymer chains which either cannot crystallize due to chain irregularity (eg atactic chains) or have been cooled from the melt so quickly as to inhibit crystallization. Thus polystyrene is amorphous because its chains are **atactic** but polyethylene terephthalate film is amorphous owing to cooling at a high rate. Term must be used with care, since some non-crystalline polymers have a characteristic microstructure. See **rubber toughening** p. 270.

amorphous regions Zones in partly crystalline polymers which have not crystallized due to entanglements etc. If the glass transition temperature, T_g, is below ambient, such regions are elastomeric and help toughen the material. See **crystallization of polymers** p. 82.

amorphous selenium Allotropic form of elemental selenium composed of linear chains of atoms linked together. Brittle, glassy solid at ambient temperature but becomes elastomeric above its glass transition temperature of 70°C. Like plastic sulphur and polymeric tellurium, a member of group six of the periodic table. Used commercially in **xerography**, owing to its photoconductive properties. See Appendix 1.

amorphous semiconductor That prepared in the amorphous state. It tends to have a much lower electrical conductivity than that of its crystalline counterparts, and is typically made from hydrogenated **amorphous silicon** or **chalcogenide glasses**.

amorphous silicon Non-crystalline silicon film made by eg **chemical vapour deposition**, usually in fact containing considerable amounts of chemically bonded hydrogen (hence sometimes αSi:H). Used as a semiconductor in thin film transistors and photocells.

amp Deprecated abbrev. for **ampere**; A is preferred.

ampere SI unit of electric current. Defined as that current which, if maintained in two parallel conductors of infinite length, of negligible cross-section, and placed 1 metre apart in a vacuum, would produce a force between the conductors equal to 2×10^{-7} newtons per metre of length. One of the SI fundamental units.

Ampère's law The relation between the magnetizing field H around a conductor, length l, carrying a current i is

$$\oint H \cdot d l = i.$$

Ampère's rule Rule for the direction of the magnetic field associated with a current. The direction of the field is that of an advancing right hand screw when turning with the current. Alternatively, if the conductor is grasped with the right hand, the thumb pointing in the direction of the current, the fingers will curl around the conductor in the direction of the field.

Ampère's theory of magnetization A theory based on the assumption that the magnetic property of a magnet is due to currents circulating in the atoms of the magnet.

ampere-turn SI unit of magneto-motive force, which drives flux through magnetic circuits, arising from one ampere flowing round one turn of a conductor. Abbrev. At.

ampere-turns per metre SI unit of magnetizing force, magnetic field intensity.

amphi- Prefix from Gk meaning both, on both sides (or ends) or around.

amphoteric Having both acidic and basic properties, eg aluminium oxide, zinc oxide, which form salts with acids and with alkalis.

amu Abbrev. for **atomic mass unit**.

amyl acetate *Pentyl ethanoate*. $CH_3CO.O.C_5H_{11}$. A colourless liquid, of ethereal pear-like odour, bp 138°C, used for fruit essences and as an important solvent for nitrocellulose. Isoamyl acetate is also known as *pear oil*.

amyloid A starch-like **cellulose** compound, produced by treatment of cellulose with concentrated sulphuric acid for a short period.

amylose The sol constituent of starch paste; with linear polymer of glucose units having α-1$\rightarrow$4 glucosidic bonds.

an Prefix from Gk *an*, not.

anabaric crystallization Term for high-pressure (> 3 kbar) crystallization of polymers, esp. polyethylene, when extended chains form.

anaerobic adhesives Adhesives based on **monomers** which will only polymerize (ie *cure*) in the absence of oxygen. They are usually tetrafunctional esters derived from acrylic acid, such as diacrylates and dimethacrylates.

analyser The second filter in a polarimeter; when rotated 90° relative to polarizer it will not allow polarized light to pass.

andiroba Hardwood (*Carapa*) of C and S America, light- to red-brown, and straightish-grained. It is inclined to warp and split during seasoning, but is moderately durable. Used for making furniture, veneers, flooring and roofing and as a structural timber. Mean density 640 kg m^{-3}.

anelasticity (1) Any recoverable deformation which deviates from linear elastic behaviour. (2) Any structural inhomogeneity or discontinuity which would dampen or attenuate an elastic wave propagating in a body.

anelectric Term once used for a body which does not become electrified by friction.

angle iron Mild steel bar rolled to an L-shaped cross-section, much used for light structural work. Also *angle, angle bar, angle steel*.

angle of bite In rolling processes, the maximum angle obtainable between the roll radius where it first contacts the metal and the line joining the centres of the two opposing rolls. Also *angle of nip*.

angle of repose The greatest angle to the horizontal which the inclined surface of a heap of loose material (eg a powder, earth or gravel) can assume and remain stationary.

angle ply laminate Laminated material of wood or fibre-reinforced composites in which the angles between the orientation directions of the laminae are not 90°; commonly used angles are 30°, 45° and 60°.

angora The hair of the angora rabbit or the soft yarn and fabric made from it.

Ångström Unit of wavelength for electromagnetic radiation = 10^{-10}m. Superseded by nanometre (=10^{-9}m).

angular momentum The moment of the linear **momentum** of a particle about an axis. Any rotating body has an angular momentum about its centre of mass, its *spin angular momentum*. The angular momentum of the centre of mass of a body relative to an external axis is its *orbital angular momentum*.

anhydrides Substances, including organic compounds and inorganic oxides, which either combine with water to form acids, or which may be obtained from the latter by the elimination of water.

anhydrite Naturally occurring anhydrous calcium sulphate which readily forms gypsum and from which anhydrite plaster is made by grinding to powder with a suitable accelerator.

anhydrous Containing no water. With crystalline oxides, salts etc it emphasizes that they contain no combined water, eg **water of crystallization**.

animal-sized Paper which has been sized by passing the sheet or web through a bath containing a solution essentially of gelatine and then drying. See **sizing**.

anion Negative ion, ie atom or molecule which has gained one or more electrons in an electrolyte, and is therefore attracted to an anode, the positive electrode. Anions include all non-metallic ions, acid radicals and the hydroxyl ion. In a primary cell, the deposition of anions on an electrode makes it the negative pole. Anions also exist in gaseous discharge. See **cation**.

anionic polymerization Polymerization using anionic catalyst such as butyl lithium. See **chain polymerization** p. 56.

aniso- Prefix from Gk *an*, not; *isos*, equal.

anisotropic Any material whose physical properties depend upon direction relative to some defined axes (eg crystalline axes, fibre orientation, draw direction) in the material. These properties normally include elasticity, thermal and electrical conductivity, permittivity, permeability, refractive index, strength etc. Also said of such processes as **etching** when certain directions are preferred.

anisotropic conductivity The property of having a directionally dependent conductivity (electrical or thermal).

anisotropic etching Describes an etching process which proceeds preferentially in one direction. In semiconductor processing when dry etching is accomplished with energetic ion bombardment, the lateral etch rate may be substantially less than the vertical rate so

that under-cutting is avoided, allowing narrow, steep-sided features to be defined. Cf **isotropic etching**.

anisotropic liquids See **liquid crystals**.

anisotropy Describes a property of a substance when that property depends on direction as revealed by measurement, eg crystals and liquid crystals in which the refractive index is different in different directions, or when magnetic dipoles align with certain crystal axes in magnetic materials. See **anisotropic**.

anneal To heat in a furnace for a period followed by slow cooling in order to bring about softening or relaxation of internal stress Commonly applied to metals or glass processing. See **annealing** p. 13.

annealing See panel on p. 13.

annealing furnace Batch-worked or continuous oven or furnace, often having controllable atmosphere, in which metal, alloy or glass is annealed.

annealing point One of the reference temperatures in **glassmaking** p. 144.

anode Positive electrode in a battery, electrolytic cell or electronic valve or tube. Cf **cathode**.

anode brightening, polishing See **electrolytic polishing**.

anode slime Residual slime left when anode has been electrolytically dissolved. It may contain valuable byproduct metals.

anodic oxidation Oxidation, ie removal of electrons from a substance by placing it in the anodic region of an electrolytic cell. The substance to be oxidized may be either a part of the electrolyte or the anode itself. See **anodizing**.

anodic protection System for passivating steel by making it the anode in a protective circuit. Cf **cathodic protection**.

anodic treatment See **anodizing**.

anodized Description of metal surface protected by chemical or electrolytic action. Commonly refers to aluminium where an oxide layer is produced which acts as a barrier to corrosive agents which would otherwise attack the metal. Thin films may be dyed with bright colours before sealing to produce reflective, highly decorative finishes. Thicker films (engineering finishes) render the surface resistant to abrasion. Anodic films on titanium yield colourful interference films often applied to jewellery and art decorative work.

anodizing Electrolytic process which increases the thickness of the layer of oxide on the surface of metals and alloys when these form the anode of the electrolytic cell. Usually applied to aluminium and its alloys to deposit a hard, non-corroding oxide film. The aluminium is made the anode in a cell

annealing

Essentially a heat treatment process intended to bring about a soft or stress-free state in worked materials. It usually involves heating to a temperature where diffusion or stress relaxation can occur, holding for a period and then cooling slowly so as to minimize thermal gradients which could re-introduce stress by differential thermal contraction. The process will depend on whether only internal stress relief is needed or whether recrystallization is required to restore ductility of work hardened material and permit further cold working.

Maintaining a metal at elevated temperatures reduces **dislocations, vacancies**, frozen-in stress and other metastable conditions. Depending on temperature and time, the stages of *recovery, recrystallization* and *grain growth* can occur. In ferrous alloys, the metal is held at a temperature above the upper critical temperature for a variable time and then cooled at a predetermined rate, depending on the alloy and the particular properties of hardness, machinability etc that are needed. The term is usually qualified, eg *quench annealing, isothermal annealing.*

Recrystallization (as well as recovery) is a thermally activated process which will only occur at a significant rate when the temperature exceeds a critical value, depending upon the alloy composition and prior degree of cold work, known as the *recrystallization temperature.*

Stress relief annealing is carried out at fairly low temperatures ($0.3\ T_m$ for metals) where no microstructural changes take place and mechanical properties are virtually unaltered. Creep and viscous flow (eg in glass) removes differential strains across the body of the object and thus eliminates internal stresses. Applied for example to welded structures and glass artefacts, where residual stress could lead to cracking or failure in service.

Full annealing (or *recrystallization annealing*) is carried out at temperatures in the region of $0.6\ T_m$ in order to bring about formation of a new grain structure in cold worked material (Fig. 1). The increase in dislocation density

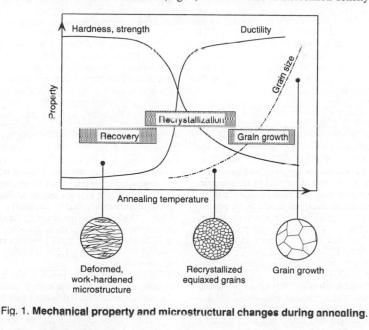

Fig. 1. **Mechanical property and microstructural changes during annealing.**

continued on next page

13

annealing (contd)

during cold working hardens the material and reduces ductility. The first stage in annealing is recovery, where dislocation networks stabilize by climb and cross slip and corresponds to the stress relief stage described above. This is followed by massive annihilation of dislocations and production of nuclei of new grains, which subsequently grow to form an equiaxial structure throughout the section. Hardness falls and ductility rises while a number of microstructural changes take place.

After a new set of grains has formed, continued annealing, especially at temperatures above the minimum for recrystallization, will result in grain growth, caused by the thermodynamic drive to minimize surface energy to volume ratio. Large grains can have undesirable effects in subsequent manufacturing operations (eg a surface blemish known as the *orange peel effect*), so the annealing is usually only carried out for the minimum time necessary to cause the recrystallized grains to replace completely the deformed cold worked structure.

Steels, as well as certain other metals which exhibit allotropic changes, may be annealed at temperatures either above or below the critical transformation range, allowing the new crop of grains to be either austenite or ferrite respectively. However, if recrystallization occurs in the austenite condition, the new grains have to transform to the appropriate ferrite plus pearlite structure as they undergo the eutectoid transformation on cooling. This transformation offers the potential to refine the grain structure of steels without prior cold working, a mechanism denied to alloys which are not allotropic. Whether or not a steel has been previously cold worked, it recrystallizes as it is heated to the austenitic state and the subsequent reverse transformation on cooling produces smaller grains than the original. This leads to a process known as *grain refining*, which takes place regardless of whether or not the original material had been cold worked. It can, for example, be applied to castings.

If the intention is to achieve the softest possible condition in a low to medium carbon steel ($< 0.8\%$), the material will be furnace cooled from the austenitic temperature, giving a *full* or *furnace anneal*. However, if the intention is principally to refine the grain structure and remove any residual working effects, the material may be taken from the furnace and cooled in air, which is known as *normalizing*.

Medium and high carbon steels may be annealed for long periods just below the transformation range, especially if they have been previously hardened, in order to induce the carbides to take up rounded shapes in the ferrite matrix. This produces the softest possible and most ductile condition for such steels and is called a *spheroidizing anneal*.

Low carbon steels ($< 0.15\%$), especially in sheet form, may be treated to remove the effects of cold work by recrystallizing below the ferrite/austenite transformation temperature. This produces equiaxial ferrite grains with the pearlitic carbide strung out in the direction of working. The material is softened and suitable for further cold work and this process offers several economic and technical advantages over the austenitic treatments for such steels. It is known as *sub-critical* or *process annealing*, and is usually carried out at temperatures in the range 900–975 K.

containing chromic (VI) or sulphuric acid. Other metals can be treated, eg titanium. Also *anodic treatment*.

anomalous viscosity A term used to describe liquids which show a decrease in viscosity as their rate of flow (ie velocity gradient or shear strain rate) increases. Also *non-Newtonian fluids* or *pseudoplastic fluids*. Advantage is taken of this behaviour when injection-moulding polymer melts.

anomaly Any departure from the strict characteristics of the type.

anti- Prefix from Gk *anti*, against.

anti-blocking agent Fine powder added to eg low-density polyethylene film after manufacture to prevent the film sticking to itself. Examples include talc or fine silicas.

anti-extrusion ring Nylon or acetal ring fitted to heavy duty rubber seal to prevent extrusion through sealed gap.

anti-ferromagnetism Phenomenon in some magnetically ordered materials in which there is an anti-parallel alignment of spins in two interpenetrating structures so that there is no overall bulk spontaneous magnetization. Anti-ferromagnetics have a positive susceptibility. The anti-parallel alignment is disturbed as the temperature increases until at the **Néel temperature** the material becomes paramagnetic. See **ferromagnetics and ferrimagnetics** p. 128.

anti-friction bearing Used to describe a wide range of bearings such as ball-, roller-, special metallic alloy-, and plastic-based bearings. All designed to reduce friction between moving parts, the choice depending on the duty.

anti-friction metal See **white metal**.

antimony Metallic element, symbol Sb. Used in alloys for cable covers, batteries etc; also as a donor impurity in silicon. See Appendix 1, 2.

antimony alloys Antimony is not used as the basis of important alloys, but it is an essential constituent in type metals, bearing metals (which contain 3–20%), in lead for shrapnel (10%), storage battery plates (4–12%), roofing, gutters and tank linings (6–12%).

anti-oxidants These are substances which delay the oxidation of paints, plastics, rubbers etc. Raw vegetable oils contain natural anti-oxidants which reduce the speed of drying of paints. Deliberately added anti-oxidants, generally phenol derivatives, delay the skinning of paints in the can at the cost of slightly slower drying. Similar substances added to plastics, rubbers, foods and drugs delay degradation by oxidation.

anti-ozonants Materials added to rubbers, esp those with alkene bonds in main chain, to inhibit ozone cracks. Polybutadienes and natural rubber esp. liable to attack. Often based on waxes, which leach to surface.

anti-plasticization Effect produced in a polymer by addition of a specific chemical; the opposite of plasticization, giving a material with higher modulus and lower elongation to break. Beyond a critical concentration, properties revert to those of conventional plasticization.

antique The surface finish originally applied to machine-made papers to imitate handmade printings. The term is now used to describe any rough-surfaced paper which bulks well, eg book or cover paper.

anti-solar glass Glass which absorbs or reflects infrared radiation (heat) from sunshine and reduces glare, but transmits most of the light.

anti-statics Chemicals added to polymers to discourage build-up of static electricity on product surfaces. Include quaternary ammonium salts and polyethylene glycols. Also *anti-stats*.

anti-vibration mounting Rubber spring designed to absorb vibrations from engines etc. Care needed in design and materials selection to match vibration frequency with main damping peak of elastomer.

antung Slub-free, plain-weave fabric made from wild silk.

anvil A block of iron, sometimes steel-faced, on which work is supported during forging.

apex The top or pointed end of anything. Adj. *apical*.

apparent particle density The mass of a particle of powder divided by the volume of the particle, excluding open pores but including closed pores.

apparent (powder) density The mass of a powder divided by the volume occupied by it under specified conditions of packing.

apparent viscosity Term applied to the viscosity of many non-Newtonian fluids (eg **polymers**). Specifically to viscosity calculated using **Poiseuille's law**.

apple European and Asian **hardwood** (*Malus*) whose **heartwood** is pinkish-buff coloured, with a straight-grained, fine texture. Not very durable; used in decorative applications (eg carving and inlay), tool handles, pipes and for veneers. Density 700–720 kg m^{-3}.

aq A symbol for water.

Aquadag TN for a colloidal suspension of graphite in water.

aqua fortis Ancient name for concentrated nitric acid.

aqua regia A mixture consisting of one volume of concentrated nitric acid to three volumes of concentrated hydrochloric acid.

aqueous Consisting largely of water; dissolved in water.

Ar (1) Symbol for **argon**. (2) A general symbol for an aryl, or aromatic, radical. (3) The transformation temperatures of the phase changes during the cooling of iron and steel, subscripts indicating the appropriate change.

Araldite TN for range of epoxy resins used for adhesives, encapsulation of electrical components etc. See **thermosets** p. 318.

aramid fibres Fibres made from linear polymers containing recurring amide groups (–CO–NH–) joined directly to two aromatic rings (aramid is derived from aromatic amide). The fibres have high moduli and are frequently used in composite materials and for rope and textiles. See **high-performance fibres** p. 158.

arbitration bar Test bar, cast from a given furnace charge, to determine whether the main casting is to specification.

Archimedes' principle When a body is wholly or partly immersed in a fluid it experiences an upthrust equal to the weight of fluid it displaces; the upthrust acts vertically through the centre of gravity of the displaced fluid.

arc resistance Ability of an insulator to withstand high-voltage sparking.

arc spraying Method of fusing (and thence depositing) refractory ceramic and metal powders by blowing them through an electric arc or plasma. Used for applying a variety of thin and thick film coatings. Also *plasma spraying*.

argon An element which forms no known compound, one of the noble gases. Symbol Ar. Argon constitutes about 1% by volume of the atmosphere, from which it is obtained by the fractionation of liquid air. It is used in gas-filled electric lamps, radiation counters, fluorescent tubes and as an inert atmosphere for welding, etc. See Appendix 1, 2.

Armco TN for a soft iron with less than 1% impurities.

armour plate Traditionally, specially heavy alloy steel plate hardened on the surface; used for the protection of fighting vehicles and ships. There is also a form of armour plate based on aluminium alloy particularly suitable for fast moving military vehicles (*Chobham armour*).

aromatic compounds Compounds related to benzene. Ring compounds containing conjugated double bonds.

aromatic polymers Those possessing benzene rings either in side groups (eg polystyrene) or in the main backbone chain (eg polycarbonate, aramid fibre).

arrest points Discontinuities in heating and cooling curves, due to absorption of thermal energy during heating or its evolution during cooling, and indicating structural (phase) changes occurring in a material.

Arrhenius's (rate) equation The rate, R, of a thermally activated, physical process is given by:

$$R = R_0 \exp\left(-E_a/kT\right),$$

where R_0 is a constant, E_a is the activation energy, k is **Boltzmann's constant** and T is the absolute temperature.

arris edge Small bevel on the edge of a piece of glass, of width not exceeding 1/16 in (1.5 mm), at an angle of approx. 45° to the surface .

arsenic Symbol As. An element which occurs free and combined in many minerals. An impurity of several commercial metals.

Called grey or γ-arsenic to distinguish it from the other allotropic modifications. Used in alloys and in the manufacture of lead shot. Obtained from the roasting of arsenical ores. It is highly poisonous. Arsenic is an n-type dopant in silicon. See Appendix 1, 2.

arsenical copper Copper containing up to about 0.6% arsenic. This element slightly increases the hardness and strength and raises the recrystallization temperature. See **copper alloys** p. 76.

arsenide Arsenic bonds with most metals to form *arsenides*; eg iron – $FeAs_2$. Arsenides are decomposed by water or dilute acids with the formation of the hydride arsine. See **gallium arsenide**.

artefact, artifact A man-made stone, wood or metal implement.

artificial ageing See **precipitation hardening**.

art paper Paper coated on one or both sides with one or more applications of an aqueous suspension of adhesive and mineral matter, such as china clay, to provide a surface(s) suitable for high-class colour print reproduction.

aryl A term for aromatic monovalent hydrocarbon radicals; eg C_6H_5Cl is an aryl halide.

asbestos Naturally occurring, fine (~ 5 μm dia) mineral fibres which are highly heat-resistant, used in brake linings, thermal insulation, fire-resistant fabrics, asbestos cement etc, though with increasing concern for the health hazards from loose fibres and dust. Derived from chrysotile (serpentine), and the amphiboles actinolite, amosite, anthophyllite and crocidolite (blue asbestos).

ash (1) A **hardwood** tree (*Fraxinus*) with American, European and Japanese varieties, yielding a tough and elastic timber. Generally straight-grained, with a coarse, even texture, it is perishable and liable to insect attack. Typical uses are for ladders, hammer and tool handles, spokes, oars, poles, camp furniture, coffins and artificial limbs. Density 510–830 kg m^{-3} (mean 710 kg m^{-3}). (2) Non-volatile inorganic residue remaining after the ignition or oxidation of an organic material.

aspect ratio (1) Ratio of the length of a fibre or wire to its width or diameter. (2) Ratio of the span to the mean chord of an aerofoil (usually a wing). (3) Ratio of the width to the height of a picture.

asperity Slightly raised parts of a surface which form the actual points of contact between two surfaces at a microscopic level, elastically and plastically flattened to take the load (normal force).

asphalt The name given to various bituminous substances which may be: (1) of natural occurrence in oil-bearing strata from which the volatiles have evaporated; (2) a residue in petroleum distillation; (3) a mixture of asphaltic bitumen and granite chippings, sand or powdered limestone. Asphalt is used extensively for paving, road-making, damp-proof courses, in the manufacture of roofing felt and paints and as the raw material for certain moulded plastics. See **bitumen.**

asphaltenes Such constituents of asphaltic bitumens as are soluble in carbon disulphide but not in petroleum spirit. See **carbenes, malthenes.**

assay value Troy ounces of precious metal per avoirdupois ton of ore.

assembly Construction of product from several or many components. Methods used for attachment include welding, fastening, push-fit, snap-fit, lock-fit, adhesive bonding, ultrasonic welding etc. Many products are now designed for robotic assembly.

A-stage Stage in the cure of a thermosetting resin of the phenol formaldehyde type when is fusible and wholly soluble in alcohols and acetone. See **thermosets** p. 318.

astatine Radioactive element, the heaviest halogen. Symbol At; at. no. 85, r.a.m. 210. Isotopes occur naturally as members of the actinium, uranium or neptunium series, or may be produced by the α-bombardment of bismuth. See Appendix 1, 2.

ASTM Abbrev. for American Society for Testing Materials.

astrakhan A curled-pile woven, warp-knitted, or weft-knitted fabric designed to resemble the fleece of a still-born or very young astrakhan lamb.

asymmetric atom An atom bonded to three or more other atoms in such a way that the arrangement cannot be superimposed on its mirror image. In particular, a carbon atom attached to four different groups. Most chiral molecules can be described in terms of specific asymmetric atoms, eg the alpha carbon atoms in amino acids. See **chirality.**

asymmetry The condition of being irregular and not divisible into equal halves about any plane.

atactic Term used to denote certain linear hydrocarbon polymers in which substituent groups are arranged at random around the main carbon chain. See **stereoregular polymers** p. 299.

athermal solutions Solutions formed without production or absorption of heat on mixing the components.

athermal transformation A solid state reaction eg the **martensitic transformation** of steel, in which thermal activation is not required. The transformation is driven by increasing thermodynamic instability of a metastable phase, which eventually transforms by physical shear of the crystal lattice.

atom The smallest particle of an element

which can take part in a chemical reaction. See **atomic structure, Dalton's atomic theory.**

atomic absorption coefficient For an element, the fractional decrease in intensity of radiation per number of atoms per unit area. Symbol μ_a. Related to the linear absorption coefficient μ by

$$\mu = \frac{1}{V} \sum_i n_i (\mu_a)_i \, ,$$

where the material contains n_i atoms of element i in a volume V.

atomic bond See **bonding** p. 33.

atomic clock A clock whose frequency of operation is controlled by the frequency of an atomic or molecular process. The inversion of the ammonia molecule with a frequency of 2.3786×10^{10} Hz provides the basic oscillations of the *ammonia clock*. The difference in energy between two states of a caesium atom in a magnetic field giving a frequency of 9 192 631 770 Hz is the basis of the *caesium clock* which has an accuracy of better than one in 10^{13}.

atomic heat Product of specific heat capacity and **relative atomic mass** in grams; approx. the same for most solid elements at high temperatures.

atomicity The number of atoms contained in a molecule of an element.

atomic mass unit Abbrev. *amu.* Exactly one twelfth the mass of a neutral atom of the most abundant isotope of carbon, ^{12}C. One amu $= 1.660 \times 10^{-27}$ kg. Before 1960, amu was defined in terms of the mass of the ^{16}O isotope and amu was 1.6599×10^{-27} kg. Also *dalton.* See **atomic weight.**

atomic number The order of an element in the periodic table (Appendix 1), and identified with the number of unit positive charges in the nucleus (independent of the associated neutrons). Equal to the number of external electrons in the neutral state of the atom, it determines its chemistry. Symbol Z.

atomic orbital Strictly a wave function defining the energy of an electron in an atom. Electron energy is quantized, the lowest main shell being K, followed by L, M etc. Atomic orbitals are designated s, p, d and f, sub-shells which can hold 2, 6, 10, and 14 electrons respectively. s orbitals are spherical while the others are lobed and thus directional. It is the p, d and f orbitals that determine **stereochemistry.** See **atomic structure.**

atomic plane A solid is crystalline because its atoms are ordered in intersecting planes (atomic planes) corresponding to the planes of the crystal. See **X-ray crystallography.**

atomic radii Half of the internuclear distance between the nuclei of two identical non-bonded atoms at equilibrium separation.

atomic refraction The contribution made by a **mole** of an element to the **molecular refraction** of a compound.

atomic scattering That of radiation, usually electrons or X-rays, by the individual atoms in the medium through which it passes. The scattering is by the electronic structure of the atom in contrast to nuclear scattering which is by the nucleus.

atomic scattering factor The ratio of the amplitude of coherent scattered X-radiation from an atom to that of a single electron placed at the atomic centre. The atomic scattering factor depends on the electron-density distribution in the atom and is a function of the scattering angle.

atomic spectrum Electronic transitions between the discrete energy states of an atom involve either the emission or absorption of photons. Such emission or absorption spectra are line spectra. The spectrum is characteristic of the atom involved.

atomic structure The chemical behaviour of the various elements arises from the differences in the electron configuration of the atoms in their normal electrically neutral state. Each atom consists of a heavy nucleus with a positive charge produced by a number of protons equal to its atomic number. There is an equal number of electrons outside the nucleus to balance this charge. The nucleus also contains electrically neutral neutrons; protons and neutrons are collectively referred to as nucleons. The **Sommerfeld model**, modified by the wave mechanical concept of orbitals, describes the electron configuration of the atom. Electrons are fermions which must conform to the **Pauli exclusion principle** and no two electrons in the same atom can be in the same quantum state, ie have the same set of four **quantum numbers**. The principle quantum number indicates the shell to which the orbital belongs and varies from 1 (K shell) closest to the nucleus to 7 (Q shell), the most remote. In general, the closer an electron is to the nucleus the greater the coulomb attraction and so the greater the binding energy retaining the electron in the atom. Nuclear binding forces tend to give greatest stability when the neutron number and the proton number are approximately equal. Due to electrostatic repulsion between protons, the heavier nuclei are most stable when more than half their nucleons are neutrons; elements with more than 83 protons are unstable and undergo radioactive disintegration. Those with more than 92 protons are not found naturally on Earth, but can be synthesized in high-energy laboratories. These are the transuranic elements which have short half-lives. Most elements exist with several stable isotopes and the chemical atomic

weight gives the average of a normal mixture of these isotopes.

atomic volume Ratio for an element of the relative atomic mass to the density; this shows a remarkable periodicity with respect to atomic number.

atomic weight See **relative atomic mass**. Mass of atoms of an element in atomic mass units on the unified scale where one amu=1.660×10^{-27} kg. For natural elements with more than one isotope, it is the average for the mixture of isotopes.

atomized powder One produced by the dispersion of molten metal or other material by spraying under conditions such that the material breaks down into powder.

atomizer A nozzle through which a liquid is sprayed under pressure. Its function is to break up the substance into a fine mist which may subsequently solidy to form fine powder particles or be deposited onto a surface and form a coating.

ATR Abbrev. for **Attenuated Total Reflection**.

attenuated total reflection Spectroscopic method of analyzing thin films on reflective substrates, esp. using **infrared radiation**. Abbrev. *ATR*.

at.wt. Abbrev. for **atomic weight**, now **relative atomic mass**.

aufbau Filling of successive electron shells around an atom, electron by electron. Each electron occupies the lowest energy level or atomic orbital available, so creating an electron structure for all the elements. The principle explains the structure of the periodic table.

Auger effect An atom ionized by the ejection of an inner electron can lose energy either by the emission of an X-ray photon as an outer electron makes a transition to the vacancy in the inner shell *or* by the ejection of an outer electron, the *Auger effect*. The energies of the Auger electrons emitted are characteristic of the atomic energy levels. The energy spectrum of Auger electrons is sufficiently characteristic of the atoms in a surface that surface composition and character can usefully be probed in this way.

ausforming Working an alloy steel in the metastable **austenite** condition. The material is first heated to a temperature where the austenite is stable, ie above the Ac₃ temperature, and is then cooled rapidly to the region of 550°C and worked to shape before any transformation to **pearlite** or **bainite** takes place. See **isothermal transformation diagram**, p. 177. It transforms to **martensite** on cooling at ambient temperature and is then tempered. Strength and toughness are enhanced compared with the same material worked conventionally in the austenite region and quenched and tempered as separate operations. See **steels** p. 298.

austempering Heating a steel to transform it to **austenite**, followed by cooling rapidly to a temperature above the martensitic change point, but below the critical range, so that the austenite isothermally transforms to **bainite**, which has properties resembling a quenched and tempered steel of the same composition. See **isothermal transformation diagram** p. 177.

austenite The higher density, high temperature, FCC, γ form of iron and of solid solutions based on it. In pure iron it is stable between 1183 K and 1663 K

austenite bay The shape of the region around 550°C in an **isothermal transformation diagram** (p. 177) which defines the zone where **austenite** is metastable and remains in that condition pending transformation to **pearlite** or **bainite**.

austenitic steels Steels containing sufficient amounts of nickel, nickel and chromium, or manganese to retain austenite at atmospheric temperature; eg *austenitic stainless steel, Hadfield's manganese steel*. See **steels** p. 296.

Australian blackwood Hardwood tree (*Acacia*), whose wood is golden to reddish-brown, with a straight to wavy grain and a fine and even texture. Typical uses include furniture-making, interior fittings, high-class joinery, gun stocks, cooperage and in veneers. Average density 665 kg m⁻³.

autoadhesion Bonding together of identical surfaces, as with contact adhesives.

autoclave A vessel, constructed of thick walled steel (usually alloy steel or frequently nickel alloys), for carrying out chemical reactions under pressure and at elevated temperatures.

automation Industrial closed-loop control system in which manual operation of controls is replaced by automatic machinery.

automotive gas oil Abbrev. *AGO*. A US term for gas oil used mainly as diesel fuel; same as the UK term *DERV*.

autoxidation (1) The slow oxidation of certain substances on exposure to air. (2) Oxidation which is induced by the presence of a second substance, which is itself undergoing oxidation.

autoxidator An alkene-oxygen compound acting as a carrier or intermediate agent during oxidation, in particular during autoxidation.

avalanche diode A semiconductor breakdown diode, usually silicon, in which avalanche breakdown occurs across the entire p–n junction, giving a voltage drop which is

constant and independent of current. Avalanche diodes break down much more sharply than **Zener diodes**. Used in high-speed switching circuits and microwave oscillators.

avalanche effect Cumulative multiplication of carriers in a semiconductor because of avalanche breakdown. This occurs when the electric field across the barrier region is strong enough to allow production and cumulative multiplication of carriers by ionization.

avalanche transistor One depending on avalanche breakdown to produce hole-electron pairs. It can give very high gain in the common-emitter mode or very rapid switching.

avodiré Tropical W African **hardwood** (*Turraeanthus*) whose **heartwood** is golden yellow with a straight to wavy grain and fine texture. Not very durable or resistant to termites. Used in high-class joinery, plywood and for veneers. Average density 550 kg m⁻³.

Avogadro constant See **Avogadro's number**.

Avogadro's number The number of atoms in 12 g of the pure isotope ¹²C, ie the reciprocal of the **atomic mass unit** in grams. It is also by definition the number of molecules (or atoms, ions, electrons) in a **mole** of any substance with the value $6.022\,52{\times}10^{23}$ mol⁻¹. Symbol N_A or L. Also *Avogadro constant*.

Avogadro's law Equal volumes of different gases at the same temperature and pressure contain the same number of molecules.

Avrami equation Empirical equation for describing crystallization of polymers. Relates degree of crystallinity (ξ) to time, t in equation of form: $1 - \xi = \exp(- kt)$, where k is a constant.

axes Plural of *axis*.

azeotropic mixtures Liquid compounds whose boiling point, and hence composition, does not change as vapour is generated and removed on boiling. The boiling point of the azeotropic mixture may be lower (eg water-ethanol) or higher (eg water-hydrochloric acid) than those of its components. Also *constant-boiling mixtures*.

azodicarbonamide Blowing agent used in structural foam moulding to create foam core. Decomposes at about 190°C to give CO, CO_2 and N_2 gases. Abbrev. *ADC*.

azo group The group —N=N—, generally combined with two aromatic radicals. The azo group is a chromophore, and a whole class of dyestuffs is characterized by the presence of this group.

B

b- A symbol for: (1) substitution on the carbon atom of a chain next but one to the functional group; (2) substitution on a carbon atom next but one to an atom common to two condensed aromatic nuclei; (3) substitution on the carbon atom next but one to the hetero-atom in a heterocyclic compound; (4) a stereo-isomer of a sugar.

β- For β-brass, function, particles, waves etc, see under beta. Symbol for: phase constant, ratio of velocity to velocity of light.

B Symbol for **boron**.

B Symbol for: (1) susceptance in an a.c. circuit (unit = siemens; measured by the negative of the reactive component of the admittance); (2) magnetic flux density in a magnetic circuit (unit = tesla = Wb m^{-2} = Vs m^{-2}).

BA Abbrev. for **British Association screw-thread**.

Babbitt's metal A bearing alloy originally patented by Isaac Babbitt, composed of 50 parts tin, 5 antimony and 1 copper. Addition of lead greatly extends range of service. Composition varies widely, with tin 5–90%; copper 1.5–6%; antimony 7–10%; lead 5–48.5%.

back annealing Controlling the softening of a fully work hardened metal so as to produce the desired degree of temper by partial recrystallization. See **annealing** p. 13, **temper**.

backhand welding That in which the torch or electrode hand faces the direction of travel, thus post-heating the existing weld. Cf **forehand welding**.

backlight The light source (often a cold cathode discharge in a flat fluorescent envelope) used in some light-modulating flat panel displays such as those based on **liquid crystals**.

back rake In a lathe tool, the inclination of the top surface or face to a plane parallel to the base of the tool.

bag moulding Use of a flexible membrane (the 'bag') to exert pressure, usually about one atmosphere, on a thermosetting composite **laminate** or sandwich component while it is curing at ambient temperature in an open mould. Pressure can be generated either by evacuating the inside of the bag (*vacuum bag moulding*) or by pressurizing its outer surface (*pressure bag moulding*).

bainite A microstructural product formed in steels when cooled from the austenite state at rates or transformation temperatures intermediate between those which form **pearlite** and **martensite**, ie between about 800 and 500 K. It is an acicular structure of supersaturated ferrite containing particles of carbide, the dispersions of the latter depending on the formation temperature. Its hardness is intermediate between that of pearlite and martensite and exhibits mechanical properties similar to those of tempered martensite in a steel of the same carbon content. See **isothermal transformation diagram** p. 177.

baize A light-weight woollen felt used to cover tables (eg for billiards) and noticeboards.

Bakelite TN for phenol-formaldehyde resin (named after LH Baekeland).

balanced equation The equation for a chemical reaction in which the correct relative numbers of moles of each reactant and product are shown.

balanced laminate A symmetrical laminated material in which the sequence of laminae above the centre plane is the mirror image of that below it.

balanced weave A weave in which the length of free yarn between the intersections is the same in the warp and weft directions and on both sides of the fabric.

balata The coagulated **latex** of the bullet tree of S America, tapped in the same way as natural rubber. It consists mainly of trans–1,4–polyisoprene together with natural resins. After removal of the resin, the material can be shaped and vulcanized. It was used for high-quality golf ball covers, but has now been largely replaced by synthetic ionomer resins. See **gutta percha**.

ball-bearing A shaft bearing consisting of a number of hardened steel balls which roll in spherical grooves (ball tracks) formed in an inner race fitted to the shaft and in an outer race carried in a housing. Balls are spaced and held by a light metal or plastic cage. Also the balls themselves.

balling (1) A process that occurs in the cementite constituent of steels on prolonged annealing at 650–700°C. (2) The operation of forming balls in a **puddling furnace**.

ballotini Small, solid glass spheres or beads used as a filler for plastics and to increase reflectivity in paints and printing inks.

ball-pein hammer A fitter's hammer, the head of which has a flat face at one end, and a smaller hemispherical face or pein at the other; used chiefly in riveting.

ball race (1) The inner or outer steel ring forming one of the ball-tracks of a **ball-bearing**. (2) Commonly, the complete ball-bearing.

ball sizing Forcing a suitable ball through a

hole to finish size it, usually part of a broach with a series of spherical lands of increasing size arranged along it.

balsa wood The wood of *Ochroma lagopus* (W Indian corkwood); it is highly porous, and has the lowest density ($100–250$ kg m^{-3}, average 160 kg m^{-3}) of any **hardwood**; used for heat, sound and vibration isolation, for its buoyancy, and in model making.

Baltic redwood See **Scots pine**.

bamboo Genus of fast-growing, giant grasses (*Bambusa*) common in tropical countries, whose hollow stems become 'woody'. In this form they are used for scaffolding poles, water pipes, blowpipes, cane furniture, parasols etc.

banak See **virola**.

Banbury mixer Type of machine used for compounding rubber with vulcanizing ingredients and carbon black.

B and BB Used to be brand-marks signifying *Best* and *Best Best*, placed on wrought iron to indicate the maker's opinion of its quality. Now obsolete as wrought iron is no longer manufactured.

band edge energy The energy of the edge of the **conduction band** or **valence band** in a solid, measured with respect to some convenient reference or else used as the reference level for other energy states. See **band theory of solids**.

band gap The range of energies which correspond with those values which are forbidden for delocalized states, according to the **band theory of solids**. Localized states such as those associated with ionized dopants, impurity atoms or crystal imperfections exist in the gap. The generation of pairs of electrons and holes requires quanta of at least the energy of the band gap. Direct recombination likewise furnishes quanta with energies at least equal to the band gap. See **optoelectronics** p. 226.

banding A structural feature of wrought metallic materials revealed by etching, resulting from microstructural segregates and constitutional differences within the grain structure becoming drawn out in the direction of working.

bandsaw A narrow endless strip of saw-blading running over and driven by pulleys, as a belt; the strip passes a work table placed normal to the straight part of the blade. The work piece is forced against the blade and intricate shapes can be cut. Also used for cutting animal carcases in butchery.

band spectrum Molecular optical spectrum consisting of numerous very closely spaced lines which are spread through a limited band of frequencies.

band theory of solids For atoms brought together to form a crystalline solid, their out-

ermost electrons are influenced by a *periodic* potential function, so that their possible energies form *bands* of allowed values separated by bands of forbidden values (in contrast to the discrete energy states of an isolated atom). These electrons are not localized or associated with any particular atom in the solid. This band structure is of fundamental importance in explaining the properties of metals, semiconductors and insulators. See **energy band, conduction band, valence band**.

bank paper A thin writing paper of less than 50 g m^{-2}, intended for typewriting or correspondence purposes.

bar Material of uniform cross-section, which may be cast, rolled or extruded.

bar-and-yoke Method of magnetic testing in which the sample is in the form of a bar, clamped into a yoke of relatively large cross-section, which forms a low-reluctance return path for the flux.

barathea Woven fabric used for coats and suits and made from silk, worsted or man-made fibres. Characteristic surface appearance arising from the twill or broken-rib weave used in its manufacture.

Barba's law Concerned with the plastic deformation of metal test pieces when strained to fracture in a tensile test. It states that test pieces of identical size deform in a similar manner.

bare A term signifying slightly smaller than the specified dimension. Cf **full**.

barium A heavy element in the second group of the periodic system, an alkaline earth metal. Symbol Ba. See Appendix 1, 2. In most of its compounds it occurs as Ba^{2+}.

barium ferrite See **ferrite**.

barium oxide BaO. When freshly obtained from the calcined carbonate it is even more reactive with water than calcium oxide and forms barium hydroxide (alkaline). Also *baryta*.

barium sulphate BaSO$_4$. Formed as a heavy white precipitate when sulphuric acid is added to a solution of a barium salt. Very nearly insoluble in water. Although of little pigmentary value, it is much used in paint manufacture and in the preparation of lake pigments. Used in barium meals. Also *barytes*.

barium titanate BaTiO$_3$. A crystalline ceramic with outstanding dielectric, piezoelectric and ferroelectric properties. Used in capacitors and as a piezoelectric transducer. Has a higher **Curie point** than Rochelle salt. See **dielectrics and ferroelectrics** p. 92, **dielectric and ferroelectric materials** p. 94.

bark Outer layer of wood which surrounds the **cambium**.

Barkhausen effect The phenomenon of

discontinuous changes in the magnetization of a magnetic material while the magnetizing field is smoothly varied. It is the consequence of sudden changes in the domain structure as domain walls overcome various pinning defects and to a lesser extent as domain orientations discontinuosly rotate away from preferred crystal axes. Barkhausen (in 1919) detected voltage pulses induced in coils surrounding a magnetic sample as it was magnetized. Analogous ultrasonic emissions are also associated with the magnetization of magnetostrictive materials. The character of Barkhausen emissions is strongly dependent on microstructure and stress.

bar magnet A straight bar-shaped permanent magnet, with a **pole** at each end.

bar mill A rolling mill with grooved rolls, for producing round, square or other forms of bar iron of small section.

barn Unit of effective cross-sectional area of nucleus equal to 10^{-28} m^2. So called, because it was pointed out that although one barn is a very small unit of area, to an elementary particle the size of an atom which could capture it is 'as big as a barn door'.

Barnett effect Magnetization of a ferromagnetic material by rapid rotation of the specimen. Used to measure magnetic susceptibility. See **Einstein–de Haas effect**.

barrel (1) A hollow, usually cylindrical, machine part; often revolving, sometimes with wall apertures. (2) The main cylinder in which molten polymer is prepared for extrusion or injection into moulds. See **injection moulding** p. 169.

barrel etcher A device usually used to oxidize and thereby strip away hardened photoresist materials during semiconductor processing. In it a batch of wafers is exposed to a low-pressure oxygen plasma. See **semiconductor device processing** p. 280.

barrel hopper A machine for unscrambling, orienting and feeding small components during a manufacturing process, in which a revolving barrel tumbles the components on to a sloping, vibrating feeding-blade.

barrel plating Electroplating of many small items by placing them in a perforated barrel revolving in a vat filled with an appropriate plating solution. The barrel is made the **cathode** in the cell and the articles tumble against each other during rotation, continually touching at different places, and so become uniformly coated with the electrodeposit.

barrel temperatures Temperatures at which an extrusion or injection moulding barrel is kept, usually rising to a peak at the nozzle. The range is determined by the poly-

mer type and its melt viscosity. See **injection moulding** p. 169.

barrier layer (1) In general a layer so placed as to inhibit interdiffusion of heat, matter etc. (2) In semiconductor junctions, see **depletion layer**. (3) In an **optical-fibre** cable, an intermediate layer of glass between the low refractive index core and the high refractive index cladding.

barrier-layer capacitance Same as *depletion-layer capacitance*. See **depletion layer**.

baryta paper Paper coated on one side with an emulsion of barium sulphate and gelatine. Used in moving-pointer recording apparatus and for photographic printing papers.

basal planes The name applied to the faces representing the terminating **pinacoid** in all the crystal systems exclusive of the cubic system.

base (1) A substance which tends to donate an electron pair or co-ordinate an electron. In particular, a substance which dissolves in water with the formation of hydroxyl ions and reacts with acids to form salts. (2) The region between the emitter and the collector of a **bipolar** junction **transistor**, into which minority carriers are injected. It is essentially the control electrode of the transistor.

base bullion Ingot base metal containing sufficient silver or gold to repay recovery, eg *argentiferous lead*.

base exchange Chemical method used in soil mechanics to strengthen clays by replacing their H ions with Na ions.

base metal The common metals, towards the electronegative end of the **electrochemical series**, remote from the **noble metals**. They have a relatively negative electrode potential (on the IUPAC system).

base unit The International System of Units (SI) is a coherent system based on seven base units. All derived units are obtained from the base units by multiplication without introducing numerical factors, and approved prefixes are used in the construction of submultiples and multiples. There is only one base or derived unit for each physical quantity. The base units are **metre**, **kilogram**, **second**, **ampere**, **kelvin**, **candela** and **mole**. See Appendix 3.

base resistance Total resistance to base current in a transistor, including *spreading effect*.

basi- Prefix from Gk *basis*, base.

basicity The number of hydrogen ions of an acid which can be neutralized by a base.

basic lead carbonate Approximate composition $2PbCO_3.Pb(OH)_2$. See **white lead**.

basic process A steel-making or melting process in which the furnace is lined with basic refractory, a slag rich in lime being formed, enabling phosphorus to be removed

basic slag Furnace slag rich in phosphorus (as calcium phosphate) which, with silicate and lime, is produced in steel making, Ground and sold as an agricultural fertilizer.

basic steel Steel which has reacted with a basic lining or additive to produce a phosphorus-rich slag and a low-phosphorus steel.

basis weight US method for identifying various papers The basis weight is the weight in pounds of a ream (500 sheets) of a particular paper in the basic size for the grade. The **grammage**, expressed as $g \, m^{-2}$ is now the preferred system.

basket-weave structure See **Widmann-stätten structure**.

basswood A N American tree (*Tilia*) that may grow to over 30 m giving a **hardwood** with straight-grained fine and uniform texture, creamy white to lightish brown in colour. The wood shrinks considerably during seasoning, but finally stabilizes and is suitable for such purposes as pattern making. Density $420 \, kg \, m^{-3}$.

bastard A general term for anything abnormal in shape, size, appearance etc. Paper or board not of a standard size.

bast fibre Cellulose fibre obtained from the stems of various plants often by a rotting (**retting**) stage followed by **beating** (scutching). Examples are flax, hemp and jute.

batch The mixture of raw materials from which glass is produced in the melting furnace. A proportion of **cullet** is either added to the mixture, or placed in the furnace previous to the batch. Also *charge*.

batch furnace A furnace in which the charge is placed and heated to the requisite temperature. The furnace may be maintained at the operating temperature, or heated and cooled with the charge. Distinguished from continuous furnace.

batch process Any process or manufacture in which operations are completely carried out on specific quantities or a limited number of articles, as contrasted to continuous or mass-production. In semiconductor manufacture, one in which several wafers are treated simultaneously as distinct from stages in which wafers are processed singly.

batik dyeing, printing The fabric is treated with wax to form a pattern that is left unaffected by a dye. The wax may then be removed and a different dye applied to give interesting colour effects.

batiste A soft, fine plain-woven fabric often of flax or cotton.

batt Loosely coherent sheet of fibres used for the manufacture of nonwoven fabrics. Also *web*.

battery General term for a number of objects co-operating together, eg a number of accumulator cells, dry cells, capacitors, radars, boilers etc.

baulk A piece of timber square-sawn from the log to a size greater than $6 \times 6 \, in^2$ ($150 \times 150 \, mm^2$).

bauxite A residual rock composed almost entirely of aluminium hydroxides formed by weathering in tropical regions. The most important ore of aluminium.

Bayer process A process for the purification of bauxite, as the first stage in the production of aluminium. Bauxite is digested with a sodium hydroxide solution which dissolves the alumina and precipitates oxides of iron, silicon, titanium etc. The solution is filtered and the aluminium precipitated as the hydroxide.

baywood See **Honduras mahogany**.

BBB polymers Abbrev. for *PolyBisBenzimidazoBenzophenanthrolines polymers*. See **polybenzimidazoles**.

BBL polymers Ladder polymers with a chain structure very similar to **polybenzimidazoles**, with stability to over 600°C.

BCC Abbrev. for **Body-Centred Cubic crystal**.

B-class insulation A class of insulating material which will withstand temperatures up to 130°C.

BCT Abbrev. for **Body-Centred Tetragonal crystal**.

beach marks Fracture surface markings associated with fatigue crack propagation. See **fatigue** p. 123, **fractography**.

bead-coil See **tyre technology** p. 325.

beam (1) A bar or member which is loaded transversely, predominantly in bending. (2) Rolled or extruded sections of certain profiles, eg I-beam. (3) A directed flow of particles or radiation.

beam lead An integrated circuit bonding option for high-frequency applications in which material is etched clear of part of the metallization layer to provide a short beam of metal (usually gold). The chip is then inverted and the beam is bonded direct to conducting tracks.

bearing metals Metals, chiefly alloys, used for that part of a bearing which is in contact with the journal; eg bronze or white metal, used on account of their low coefficient of friction when used with a steel shaft.

bearings Supports provided to locate a revolving or reciprocating shaft.

bearing surface That portion of a bearing in direct contact with the journal; the surface of the journal. See **brasses**.

beater Machine for **beating**.

beating (1) Process for partially breaking down the cell-wall structure of cellulose fibres in water before forming paper sheet (see **paper and papermaking** p. 231). (2) Process for removing heavy impurities from matted, raw natural textile fibres in the opening and **scutching** process. (3) The spare threads available during the weaving of wool to replace missing warp threads in the mending process.

beaver cloth Heavy woollen woven overcoating simulating the lustrous nap of the skin of the beaver by milling and raising the fibres, cutting them level and laying them in the same direction.

Bedford cord Cloth with rounded cords separated by fine sunken lines and running in the warp direction. Made from wool for riding breeches and worsted yarns for suiting materials.

beech A tree (*Fagus*) yielding a **hardwood** with straight grain and uniform texture. Its colour ranges from whitish to a light reddish-brown. It bends well and is excellent for turning. Used for furniture, pulley-blocks, tool handles, athletic goods, gymnasium equipment and cabinet making. Average density 720 kg m^{-3}.

Beer's law The degree of absorption of light varies exponentially with the thickness of the layer of absorbing medium, its molar concentration and **extinction coefficient**.

beetle A machine consisting of a row of wooden or metal hammers, which fall on a roll of damp cloth as it revolves. The operation closes the spaces between the warp and the weft yarns, and imparts a soft glossy finish to cotton and linen.

belite Dicalcium silicate, Ca_2SiO_4. A constituent (~ 20 wt%) of Portland cement, it hydrates at a slow rate during the setting reaction. See **cement and concrete** p. 52.

bell metal High tin bronze, containing up to 30% tin and some zinc and lead. Used in casting bells. See **copper alloys** p. 74.

bellows A flexible, corrugated tubular machine element used for pumping, for transmitting motion, as an expansion joint etc.

bell-type furnace A portable inverted furnace or heated cover operated in conjunction with a series of bases upon which the work to be heated can be loaded and then left to cool after heat treatment. Used chiefly for bright-annealing of non-ferrous metals and bright-hardening of steels.

belt A strip of leather, cotton, plastic, reinforced rubber etc, generally of rectangular cross-section, used for lifting slings and strengthening bands. In endless form used as driving-, conveyor-, abrasive- and other belts.

belt drive The transmission of power from one shaft to another by means of an endless belt running over pulleys having correspondingly shaped rims.

bend (1) To form into a curved or angular shape. (2) A curved length of tubing or conduit used to connect the ends of two adjacent straight lengths which are at an angle to one another.

bending moment The bending moment at any imaginary transverse section of a beam is equal to the algebraic sum of the moments of all the forces to either side of the section.

bending moment diagram One representing the variation of bending moment along a beam. It is a graph of bending moment (*y*-axis) against distance along the beam axis (*x*-axis).

bending rolls Usually three rolls with axes arranged in a triangle so that adjusting one relative to the others forms a curve on a strip or sheet of metal passed between them.

bending strength Also *flexural strength*. See **strength measures** p. 301.

bending test (1) A test made on a beam to determine its deflection and strength under bending load. The most usual forms are symmetrical three-point and symmetrical four-point bending, the advantage of the latter being that a constant bending moment is imposed between the two central loading points. Also *flexural test*. (2) A forge test in which flat bars etc are bent through 180° as a test of ductility.

bentonite A valuable clay, similar in its properties to fuller's earth, formed by the decomposition of volcanic glass, under water. Consists largely of montmorillonite. Used as a bond for sand, asbestos etc; also in the paper, soap and pharmaceutical industries. Thixotropic properties exploited for altering the viscosity of oil drilling muds.

benzene C_6H_6, mp 5°C, bp 80°C, rel.d. 0.879; a colourless liquid, soluble in alcohol, ether, acetone, insoluble in water. Produced from crude oil. A solvent for fats, resins etc. Very flammable. Benzene is the simplest member of the aromatic series of hydrocarbons. Carcinogenic.

benzoyl peroxide $C_6H_5.CO.OO.CO.C_6H_5$. Catalyst for free radical reactions, eg polymerization. Mp 108°C. Prepared by the action of sodium peroxide on benzoyl chloride.

benzyl The aromatic group, $C_6H_5CH_2-$.

berber A carpet square hand-woven by N Africans from hand-spun yarns from the natural coloured wool of local sheep. Commonly misused to describe machine-made carpets considered to have a similar appearance.

beryllides Compounds of other metals with beryllium.

beryllium Steely uncorrodible white metallic element Symbol Be. Main use is for windows in X-ray tubes and as an alloy for hardening copper. Used as a powder for fluorescent tubes until found poisonous. The metal can be evaporated on to glass, forming a mirror for ultraviolet light. Highly toxic. See Appendix 1, 2.

beryllium bronze, copper　A copper-base alloy containing 2.25% of beryllium. Develops high hardness (ie 300–400 Brinell) after quenching from 800°C followed by heating to 300°C. See **copper alloys** p. 76, **precipitation hardening** p. 254.

Bessemer converter　Large barrel-shaped tilting furnace, charged while fairly vertical with molten metal, and 'blown' by air introduced below through *tuyères*. Discharged by tilting. Now obsolete but replaced by variety of similar shaped but smaller vessels operating in slightly different ways and using oxygen in place of air.

Bessemer process　Removal of impurities from molten metal or matte by blowing air through molten charge in Bessemer converter. Used to remove carbon and phosphorus from steel, sulphur and iron from copper matte.

Best and Best Best　See **B and BB**.

best available technology　Abbrev. *BAT*. A US term for the process giving the maximum abatement of pollution without regard to cost or proven necessity. Also *best avail able control technology (BACT)*.

best practical environmental option　The concept recognizing that treatment of pollutants in one medium of the environment (air, land, water) may simply transfer them into another. Thus removing sulphur dioxide from flue gases may cause the calcium sulphate produced to have polluting effects at its disposal sites. Also the *cross-media approach*, it is a co-ordinated approach to pollution pathways, media and disposal routes. Abbrev. *BPEO*.

best practical means　Term with statutory force since 1863, and the basis for control of atmospheric pollution in the UK. Defined as the best practicable means with regard to local conditions, financial implications and current technical knowledge, and includes the provision, maintenance and correct use of plant. Abbrev. *BPM*.

best technical means available　A term from the European Commission which requires consideration of the economic availability of the means of pollution abatement. It approaches the UK term **best practical means**. Abbrev. *BTMA*.

beta brass　Copper-zinc alloys, containing 46–49% of zinc, which consist (at room temperature) of the intermediate constituent (or intermetallic compound) known as beta phase. See **copper alloys** p. 76.

beta decay　Radioactive disintegration with the emission of an electron or positron accompanied by an uncharged antineutrino or neutrino. The mass number of the nucleus remains unchanged but the atomic number is increased by one or decreased by one depending on whether an electron or positron is emitted.

beta-iron　Iron in the temperature range 750–860°C, in which a change from the magnetic (alpha) state to the paramagnetic occurs at about 760°C. With carbon in solution the transition is lowered toward 720°C, and, when cooling, **recalescence** is more marked.

beta particle　An electron or positron emitted in beta decay from a radioactive isotope. Also β-*particle*.

beta radiation　Beta particles emitted from a radioactive source.

Betts process　An electrolytic process for refining lead after drossing. The electrolyte is a solution of lead silica fluoride and hydrofluorsilicic acid, and both contain some gelatine. Impurities are all more noble than lead and remain on the anode. Gold and silver are recovered from the anode sponge.

B(H) curve　Also *B-H* and *B/H curve*. See **magnetization curve**, **characterizing magnetic materials** p. 57.

B(H) loop　Also *B-H* and *B/H loop*. See **characterizing magnetic materials** p. 57.

BHN　Abbrev. for **Brinell Hardness Number**, obtained in the **Brinell hardness test**. Preferred term is now H_B after the hardness number. Obtained by forcing a round steel ball into the surface of the object to be tested under a known load and subsequently measuring the diameter of the indentation so produced. See **hardness testing** p. 153.

bias　The application of a potential difference across, or electric currents through, an electronic device to set an operating condition upon which signals are superimposed. See **true bias**.

biaxial　Said of a crystal having two optical axes. Minerals crystallizing in the orthorhombic, monoclinic and triclinic systems are biaxial. Cf **copper alloys** p. 76, **uniaxial**.

biaxial orientation　State of polymer orientation where molecules are oriented in two orthogonal directions, esp. in blow moulded products such as polyethylene terephthalate bottles. The orientation is in the plane of the wall and helps toughen it.

Bible paper　Heavily loaded, strong, thin, printing paper, generally of 20–40 g m^{-2}.

bicomponent fibre　A synthetic fibre made from two fibre-forming polymers which may be arranged to lie side-by-side or as a sheath surrounding a core. By suitable heat treatment a crimped fibre is produced because of the different shrinkage properties of the two polymers. Advantage may also be taken of the sheath having a lower softening temperature than the core to produce a non-woven fabric.

bifurcated rivet A rivet with a split shank, used for holding together sheets of light material; it is closed by opening and tapping down the two halves of the shank.

BIIR Abbrev. for *Brominated Isoprene Isobutene Rubber*. See **bromobutyl rubber**.

billet Semifinished solid metallic product which has been hot-worked by extrusion, forging and rolling. Smaller than a **bloom**.

billet mills The rolling-mills used in reducing steel ingots to billets. Also *billet rolls*.

billiard cloth Woollen cloth manufactured from finest quality wool, with a closely cropped **dress-face finish** to render it perfectly smooth and damp resistant.

billion In the US and now more generally, a thousand million, or 10^9. Previously elsewhere, a million million or 10^{12}.

bimetallic strip Bonded strip composed of two metals with differing **coefficients of thermal expansion**; the strip deflects when one side of the strip expands more than the other Used in thermal switches etc.

binary Consisting of two components.

binary system and diagram Alloys formed by two metals constitute a binary alloy system, which is represented by the binary constitutional diagram for the system. In general, any two-component system. See **phase diagram** p. 236.

binder Carbon products, organic brake linings, sintered metals, tar macadam etc, employ binder components in the mix to impart cohesion to the body to be formed. The binder may have cold setting properties, or subsequently be heat-treated to give it permanent properties as part of the body or to remove it by volatilization.

binding energy That required to remove a particle from a system, eg outermost electron from an atom, when it is the **ionization potential**.

binding energy of a nucleus All nuclei have *rest* masses less than the total rest mass of their constituent protons and neutrons. The mass difference m is the *mass decrement* or *mass defect*. This arises because all nucleons bound to the nuclei must have negative energy (potential well). So if a system of free nucleons is combined to form a nucleus the total energy of the system must decrease by an amount B, the binding energy of the nucleus. The decrease B is accompanied by a decrease in the mass, the mass decrement, $m = B/c^2$ (the mass-energy equation) where c is the velocity of light. The binding energy per nucleon is B/A where A is the atomic mass number.

Bingham solid Material which shows lit-

tle tendency to flow until a critical stress is reached (eg toothpaste or modelling clay). Such materials may be *Newtonian, dilatant* or *pseudoplastic*.

biocomposites (1) **Composite materials** which occur in and are made by living organisms, such as bone, leather. (2) Composite materials which replace the function of living tissues or organs in mass, such as carbon fibre/epoxy artificial limbs.

biodegradation Breaking down of materials by bacteria, fungi and other organisms.

bioengineering (1) Application of scientific study of human body to improve, aid or assist impaired limbs or organs. (2) Provision of artificial means with the use of synthetic materials, electronic devices etc, to assist defective body functions or parts, such as hip joint implants.

biological polymers See panel on p. 27.

biomaterials Solid materials which occur in and are made by living organisms, such as **chitin, fibroin** or bone. (2) Any materials which replace the function of living tissues or organs in man, such as the alloy Vitallium, or synthetic polymers such as Dacron polyester. See **implant, prosthesis**.

biomechanics Study of motion and energetics of living organisms, esp. human motion; a method used in ergonomics.

Biopol TN for biodegradable polyester synthesized by bacteria under industrial control. The homopolymer is *polyhydroxybutyrate* (PHB), which is highly crystalline and brittle. Copolymerization with *polyhydroxyvalerate* (PHV) reduces the degree of crystallinity and toughens the material. Repeat units are

$$PHB \quad -CO.CH_2.\underset{\underset{CH_3}{|}}{CH}-O-$$

$$PHV \quad -CO.CH_2.\underset{\underset{C_3H_5}{|}}{CH}-O-$$

The material is intended to compete with synthetic thermoplastics like polyethylene etc in packaging as well as structural applications.

biopolymers Naturally occurring long chain molecules eg polysaccharides, proteins, DNA. See **biological polymers** p. 27.

Biot–Fourier equation The equation representing the non-steady conduction of heat through a solid. In the one-dimensional case,

$$\frac{\partial \varphi}{\partial t} = a \frac{\partial^2 \varphi}{\partial x^2},$$

where temperature of a section at right angles to the flow is φ, distance in flow direction is x, time is t, and thermal diffusivity ($k/\rho s$, where k is thermal conductivity, ρ is density and s is specific heat capacity) is a.

biological polymers

Materials of biological origin (wood, bone, cotton, natural rubber, etc) have long been used for engineering purposes, ranging from buildings to rope for ships' cables. Many of them still find wide application although now supplemented by an increasing range of man-made and synthetic materials. Some are natural biocomposites, a combination of hard and stiff materials embedded in a softer, ductile matrix. Thus bone is an intimate mixture of inorganic hydroxyapatite and the protein collagen, the mechanical properties being greater than either component alone. Others are relatively pure: cotton fibre is almost pure β-cellulose, comprising very high molecular mass glucose units linked together by oxygen atoms.

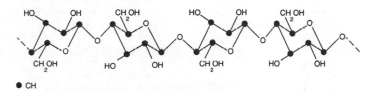

● CH

Fig. 1. **The β-cellulose chain of cotton fibres.**

The chains are highly crystalline and further stabilized by intra- and interchain hydrogen bonds giving a ladder polymer structure:

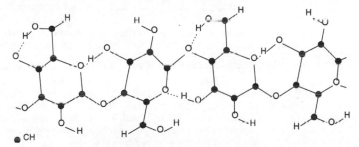

● CH

Fig. 2. **Hydrogen bonding between cellulose chains.**

The chains are aligned along the fibre axis, giving a high tensile modulus when strained. In the cotton *boll*, the cellulose fibres act not as structural reinforcement (as in wood or plant stems) but as a lightweight 'wing' for the seed pod, helping it to be blown about easily. The fibres are short (ca 2.5 cm) so must be spun into yarn before being woven into cloth. Both spinning and weaving increase the flexibility of the product, so modifying a very stiff fibre for practical use.

continued on next page

Biot laws The rotation produced by optically active media is proportional to the length of path, to the concentration (for solutions) and to the inverse square of the wavelength of the light.

Biot modulus The heat transfer to a wall by a flowing medium, giving the ratio of heat transfer by convection to that by conduction. Defined as $\alpha\theta/\lambda$ where heat transfer coefficient is α, thermal conductivity of medium is λ, characteristic length of apparatus is θ.

bipolar transistor A transistor that uses both positive and negative charge carriers. Both p-n-p and n-p-n types of bipolar transistor can be manufactured, as discrete devices or in **integrated circuits**.

biological polymers (contd)

By contrast, silk is the natural product of spiders, silkworms etc and has the specific function of structural support for webs and as a cocoon wrapping. This strong, continuous fibre is an almost regular alternating copolymer of two amino acids, glycine and alanine which crystallizes into a pleated sheet structure with main chains aligned along the fibre axes. The structure is supported laterally by hydrogen bonds very like that found in nylon 6,6.

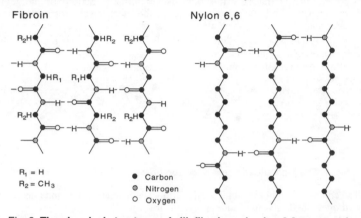

Fibroin   Nylon 6,6

$R_1 = H$
$R_2 = CH_3$

● Carbon
◉ Nitrogen
○ Oxygen

Fig. 3. **The chemical structures of silk fibroin and nylon 6,6 compared**.

Stiffness is not always a desirable property of a fibre: insulation is a desirable additional property to the flexibility and toughness of body hair, for example. Keratin, the basic protein constituent of wool and hair, possesses a much more complex amino acid structure than silk and is less crystalline. The crystalline parts (some 40% compared to 50% in silk) are formed from α-helices loosely packed together. They can easily be unwound when stressed, so wool is much more extensible than silk. Like all natural materials, the full tertiary structure is cellular in origin, with a core surrounded by a cuticle. The cuticle consists of scales oriented in one direction, the basic origin of felting since they act as a ratchet against the scales of neighbouring wool fibres (Fig. 4).

Fig. 4.   **Macrostructure of wool**.

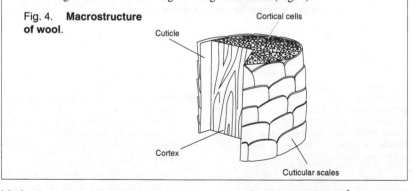

Cortical cells
Cuticle
Cortex
Cuticular scales

birch Common **hardwood** European tree (*Betula*) yielding a lightish to very light brown wood which polishes satisfactorily; due to its lack of colour and figuring is useful for imitating superior wood. It also turns well. Density 600–700 kg m^{-3} (average 660 kg m^{-3}).

bird's beak In microelectronic fabrication, descriptive of the shape of that part of a silicon dioxide layer grown on a silicon wafer near the edge of a region which is protected from oxidation by a diffusion barrier.

birefringence The same as **double refraction** in birefringent materials whose refractive index varies with direction. A measure of birefringence is the difference Δn, between the greatest and least value of refractive index. Arises from anisotropy in the material, eg crystalline anisotropy, molecular orientation, frozen-in or imposed strains. The effect is used in mineralogy, in stress analysis and to determine degree of orientation in plastics and in textile fibres.

Birmingham gauge, Birmingham wire gauge. Systems of designating the diameters of rods and wires by numbers. Obsolescent, being replaced by preferred metric dimensions. Abbrev. *BG, BWG*.

Birox resistor One made from a thick film of bismuth ruthenate fired with a glass; noted for stability.

bismuth Element, symbol Bi. Used as a component of fusible alloys with lead. See Appendix 1, 2.

bisphenol A *1,2-bis(4-hydroxyphenyl)-propane.* Manufacturing intermediate of polycarbonate and epoxy resins. Also *phenol A*.

Bitter pattern A pattern showing boundaries of magnetic domains on the surface of a magnetic material, formed by applying a colloidal suspension of a magnetic powder. The particles accumulate where the domain boundaries intersect the surface.

bitumen Tarry, non-mineralized substances of coal, lignite etc, and their distillation residues, used in paints, varnishes, roofing felts, damp-proofing, etc. See **asphalt**.

bitumen of Judea Natural asphalt from Middle East, with **photoresist** properties. Used by Niepce in 1826 to create first photograph. Also *oil of Judea*.

bituminous plastics Compression moulding materials based on natural asphalts or man-made pitches reinforced with asbestos or cotton fibre, ground wood or cork, talc, slate dust or china clay. Formerly used for products of simple shape eg toilet cisterns, battery cases.

biuret Type of group formed in polyurethanes by reaction of excess isocyanate with urea groups, so giving crosslinks.

black Of parts of castings and forgings not finished by machining; it refers to the dark coating of iron-oxide retained by the surface.

blackbean Hardwood tree (*Castanospermum*) from Australia, with a hard, chocolate-coloured wood, liable to warping and collapse, but naturally resistant to the attack of wood-rotting fungi and to termites. Used for internal fittings and high-class furniture. Average density 700 kg m^{-3}.

black body A body which completely absorbs any heat or light radiation falling upon it. A black body maintained at a steady temperature is a full radiator at that temperature, since any black body remains in equilibrium with the radiation reaching and leaving it.

black-body radiation The radiation that would be radiated from an ideal black body. The energy distribution is dependent only on the temperature and is described by **Planck's radiation law**.

black box A generalized colloquial term for a self contained unit of electronic circuitry; not necessarily black.

blackbutt Hardwood tree (*Eucalyptus*) from Australia, whose wood is hard, of interlocked grain and uniform texture. Colour light brown, with pinkish markings. The wood both looks and feels greasy and yields an abundant supply of oil. Used for heavy-duty flooring, wood blocks, cabinet work and panelling. Average density 880 kg m^{-3}.

black copper Impure metal, carrying some iron, lead and sulphur. Produced from copper ores by blast furnace reduction.

blackheart (1) An abnormal black or dark brown coloration that may occur in the **heartwood** of certain timbers. (2) A form of malleable iron, in which the core contains rosettes of graphite which appear as a dark area on fracture surfaces.

Blackman theory of specific heats of solids A theory based on the dynamics of a crystal lattice of particles developed by Born and von Kármán. It is more exact, but much more complicated than the **Debye theory of specific heats of solids** which treats the crystal as a continuous isotropic medium.

Black Orlon TN for precursor to carbon fibre. See **high-perfomance fibres** p. 158.

black red heat Temperature at which hot metal is just seen to glow in subdued daylight, about 540°C.

blackwood See **African blackwood, Australian blackwood**.

blank A piece of metal, shaped roughly to the required size, on which finishing processes are carried out.

blanket A thick woven, knitted, or nonwoven fabric giving good thermal insulation. Traditionally blankets were made from wool fabrics that were milled and raised but cotton materials of an open construction are also in common use.

blast Air under pressure, blown into a furnace.

blast-furnace Vertical shaft furnace into the top of which ore mineral or scrap metal, fuel and slag-forming rock (**flux**) is charged. Air, sometimes oxygen-enriched and preheated, is blown through from below and products are separately tapped (slag higher and metal lower). Used to smelt iron ore, copper, lead, zinc and other minerals.

blast main The main blast air-pipes supplying air to a furnace.

blazer cloth An all-wool milled and raised fabric used in the manufacture of jackets.

bleaching (1) Part of the purification process for paper pulp to attain the desired whiteness. Either oxidation (eg using free and/or combined chlorine) or reduction methods may be used. (2) A series of wet processes for removing residual impurities, colour and fatty or waxy substances from fibres, yarns, or fabrics. This improves the whiteness and promotes brighter colours after dyeing or printing. Hydrogen peroxide is often used for this purpose.

bleeding In fibres, yarns or fabrics of two or more colours, the running of the darker colours, and consequent staining of the lighter colours, during finishing, washing or solvent cleaning.

blend (1) An intimate mixture of different qualities or kinds of natural or man-made staple fibres. (2) Physical mixture of thermoplastic polymers.

blending Process where mixing of polymers is achieved by adding fillers etc to polymer solution. See **Cowles dissolver**. Term also applied to mixing of dry powders. See **Henschel mixer**.

blind rivet A type of rivet which can be clinched as well as placed by access to one side only of a structure. Usually based on a tubular or semitubular rivet design, eg **explosive rivet**.

blister A raised area on the surface of solid metal produced by the emanation of gas from within the metal while it is hot and plastic.

blister bar Now obsolete. Wrought-iron bars, impregnated with carbon by heating in charcoal. Used in making *crucible steel*.

blister copper An intermediate product in the manufacture of copper. It is produced in a converter, contains 98.5–99.5% of copper, and is subsequently refined to give commercial varieties, eg tough pitch, deoxidized copper.

blister pack Transparent, thin sheet of plastic thermoformed to cover product for display purposes. Also *bubble pack*.

blister steel Now obsolete. Wrought-iron bars impregnated with carbon by heating in charcoal. Before 1740 this was the only steel available.

Bloch function The electrons in a crystalline solid move under the influence of a periodic potential function. The solutions of the **Schrödinger equation** must therefore include a factor which has the same periodicity as the potential; these are called *Bloch func-*

tions. See **band theory of solids**.

Bloch wall The wall of a **magnetic domain**.

blockboard Board composed of **softwood** strips bonded together with polyvinyl alcohol and sandwiched between two outer layers of veneer or **hardboard**.

block copolymer See **copolymer** p. 75.

blocking Tendency of polymer film to adhere to itself, a problem in manufacture. Inhibited with surface coating of an anti-blocking agent.

blood red heat Dark red glow from heated metal, in temperature range 550–630°C.

bloom (1) Surface film on glass; either (a) the thin dielectric layers vacuum deposited on a lens to alter its reflectance properties, hence *blooming* (see **coated lens**), (b) the film of sulphites and sulphates formed during the annealing process or (c) the film caused by weathering. (2) Semifinished piece of metal, rectangular in cross-section and for steel not more than twice as long as it is thick. Cf **billet**.

blooming mills The rolling mills used in reducing steel ingots to blooms. Called cogging mills in UK, and not always distinguished from billet (slab) mills.

blotting paper Weak, free beaten, unsized paper intended for the absorption of aqueous inks from the surface of documents.

blow In a **Bessemer converter**, passage of air through the molten charge.

blow-and-blow machines Machines in which glass is shaped in two stages, but each time by blowing, as opposed, for example, to pressing or sucking. Cf **blow moulding**.

blowholes Gas-filled cavities in solid metals. They are usually formed by the trapping of bubbles of gas evolved during solidification (see **gas evolution**), but may also be caused by steam generated at the mould surface, air entrapped by the incoming metal, or gas given off by inflammable mould dressings.

blowing agent Speciality chemical (eg **azodicarbonamide**) added to polymers which decomposes at a specific elevated temperature during moulding to produce gas. The foam creates a lightweight core which improves the stiffness- and strength-to-weight ratio of the product. Sodium bicarbonate is a simple blowing agent for cellular rubbers and bread.

blowing engine The combined steam- or gas-engine and large reciprocating air-blower for supplying air to a **blast-furnace**.

blowing-iron See **blowpipe**.

blowing-out The operation of stopping down a blast-furnace.

blow moulding Two-stage route for making hollow products, eg bottles, surfboards (when filled with foam). Extruded **parison**

is dropped into the split, female-only, mould and blown to shape by air pressure.

blown film Polymer process where molten low-density polyethylene extrudate is blown by air pressure to make thin film for packaging etc. See **extrusion** p. 121.

blow pin Device through which air is blown in final stage of **blow moulding**. Usually ascends into base of descending **parison**.

blowpipe A metal tube, some 2 m in length, with a bore of 2–4 mm and a thickened nose which is dipped into molten glass and withdrawn from the furnace. The glass is subsequently manipulated on the end of the blowpipe and blown out to shape. Also *blowing-iron*.

blue brittleness Embrittlement of medium and high carbon steels during tempering in the range between 205 and 315°C, so named because the surface of the steel becomes coated with blue coloured oxidation film.

blue gum See **Red River gum**.

blueing The production of a blue oxide film on polished steel by heating in contact with saltpetre or wood ash; either to form a protective coating, or incidental to annealing.

blueing salts Caustic solution of sodium nitrate, used hot to produce a blue oxide film on surface of steel.

blue metal Condensed metallic fume resulting from distillation of zinc from its ore concentrates. Blue tint is due to slight surface-oxidation of the fine particles.

blueprint paper A paper coated with a solution of potassium ferricyanide and ammonium ferric citrate bound together with gelatine or gum arabic, used for copying engineering drawings.

blue stain A form of sapstain producing a bluish discoloration; caused by the growth of fungi which, however, does not greatly affect the strength of the wood.

Blu-tack TN for filled polymer dough with tack sufficient to grip vertical surfaces.

BMC US abbrev. for *Bulk Moulding Compound*, the same as **dough moulding compound** (DMC).

board (1) Timber cut to a thickness of less than 2 in (50 mm), and to any width from 4 in (100 mm) upwards. (2) Stiff, thick paper, generally of 220 or 250 g m^{-2} or more.

board foot Volume measure for timber, 1 bd ft = 144 in^3 or 2.36×10^{-3} m^3.

boarding Heat treatment of knitted garments, esp. nylon stockings, on a former in order to give the desired shape and size.

board measure Area measure for wooden boards. See **foot super**.

bobbin A light spool on which **slubbings**, **roving** or yarn is wound ready for the next process. A weft bobbin or pirn is loaded with yarn suitable for use as the weft of woven

fabrics. Bobbin lace is a hand-made lace produced from threads fed from small bobbins.

body-centred cubic A crystal lattice with a cubic unit cell, the centre of which is identical in environment and orientation to its vertices. Specifically a common structure of metals, in which the unit cell contains two atoms, based on this lattice. See **close-packing** p. 62, Appendix 2. Abbrev. *BCC*.

body-centred tetragonal structure A distorted form of **body-centred cubic** crystal formed esp. in **martensite** in steel, the amount of distortion depending on the carbon content of the steel. Abbrev. *BCT*. The BCT lattice does not have the five independent slip systems (see **von Mises criterion**) necessary for ductility, thus it contributes to the hardness of matensite by impeding dislocation movement.

Bohr atom Concept of the atom, with electrons moving in a limited number of circular orbits about the nucleus. These are stationary states. Emission or absorption of electromagnetic radiation only results when there is a transition from one orbit (state) to another.

Bohr magneton Unit of magnetic moment, for electron, defined by

$$\mu_B = e h / 4\pi m_e ,$$

where e = charge, h = Planck's constant, and m_e = rest mass, so that

$$\mu_B = 9.27 \times 10^{-24} \text{ J T}^{-1} .$$

The nuclear Bohr magneton is defined by

$$\mu_N = \frac{eh}{4\pi M} = \frac{\mu_B}{1836} = 5.05 \times 10^{-27} \text{ J T}^{-1},$$

M being the rest mass of the proton.

Bohr model A combination of the Rutherford model of the atom with the quantum theory. The Bohr model is based on the following four postulates. (1) An electron in an atom moves in a circular orbit about the nucleus under the influence of the electrostatic attraction between the electron and the nucleus. (2) An electron can only move in an orbit for which its orbital angular momentum is an integral multiple of $h/2\pi$, where h is Planck's constant. (3) An electron moving in such an orbit does not radiate electromagnetic energy and so its total energy E remains constant. (4) Electromagnetic radiation is emitted if an electron makes a transition from an orbit of energy E_i to one of lower energy E_f, and the frequency of the emitted radiation is $v = (E_i - E_f)/h$.

Bohr radius According to the Bohr model of the hydrogen atom, the electron when in its lowest energy state, moves round the nucleus in a circular orbit of radius

$$a_0 = \frac{4\pi\,\varepsilon_0\,\hbar^2}{m_e\,e^2} = 5.292 \times 10^{-11}\ \mathrm{m}$$

where $\hbar$ is Planck's constant divided by 2π, m_e is the mass of the electron, e is the electronic charge and ε_0 the permittivity of free space. The Bohr radius is a fundamental distance in atomic phenomena.

Bohr–Sommerfeld atom Atom obeying modifications of Bohr's postulates suggested by Sommerfeld and allowing for possibility of elliptical electron orbits.

Bohr theory See **Bohr model**.

boiler Describes a wide range of pressure vessels in which water or other fluid is heated and then discharged, eg either as hot water for heating or as high pressure steam for power generation.

boiler plate Mild steel plate, generally produced by the open-hearth process; used mainly for the shells and drums of steam-boilers.

boiler scale A hard coating, chiefly calcium sulphate, deposited on the surfaces of plates and tubes in contact with the water in a steam-boiler. If excessive, it leads to overheating of the metal and ultimate failure.

boiler tubes Steel tubes forming part of the heating surface in a boiler. In water-tube boilers the hot gases surround the tube; in fire-tube boilers the hot gases pass through the tube.

boiling The very rapid conversion of a liquid into vapour by the violent evolution of bubbles. It occurs when the temperature reaches such a value that the saturated vapour pressure of the liquid equals the pressure of the atmosphere.

boiling point The temperature at which a liquid boils when exposed to the atmosphere. Since, at the boiling point, the saturated vapour pressure of a liquid equals the pressure of the atmosphere, the boiling point varies with pressure; it is usual, therefore, to state its value at the standard atmospheric pressure of $101.325\ \mathrm{kN\ m^{-2}}$. Abbrev. *bp*.

boll The seed case of the cotton plant that opens as ripening proceeds. It contains the cotton seeds and the attached fine fibres or lint.

bolometer A device for measuring microwave or infra-red energy, consisting eg of a temperature dependent resistance used in a bridge circuit which gives an indication when power heats the resistor. Used for power measurement, standing wave detectors and infrared search and guidance systems.

bolster (1) A steel block which supports the lower part of the die in a pressing or punching machine or a moulding tool. (2) The

rocking steel frame by which the bogie (US truck) supports the weight of a locomotive or other rolling stock.

bolt A cylindrical, partly screwed bar provided with a head. With a nut, a common means of fastening two parts together.

Boltzmann principle See **Boltzmann's distribution**.

Boltzmann's constant Fundamental physical constant, given by $k = R/N_A = 1.380\ 5 \times 10^{-23}\ \mathrm{J\ K^{-1}}$, where R = ideal gas constant, N_A = Avogadro's number (Avogadro constant).

Boltzmann's distribution Statistical distribution of large numbers of small particles when subjected to thermal agitation and acted upon by electric, magnetic or gravitational fields. In statistical equilibrium, the number of particles n per unit volume in any region is given by $n = n_0 \exp(-E/kT)$, where k = Boltzmann's constant, T = absolute temperature, E = potential energy of a particle in given region relative to where $n = n_0$. Also *Boltzmann principle*.

Boltzmann's superposition principle In a linear viscoelastic material, the accumulated viscoelastic creep strain resulting from a series of stress increments is the superposed sum of the creep responses to the individual increments.

bond Link between atoms, considered to be electrical and arising from electrons as distributed around the nucleus of atoms so bonded. See **chemical bond**.

bond angle The angle between the lines connecting the nucleus of one atom to the nuclei of two other atoms bonded to it, eg in water the H–O–H angle is about $105°$.

bond distance See **bond length**.

bonded fabrics A material made by fabricating fibres into sheet form with the aid of a binder. Used for polishing cloths, curtains, filter cloths etc.

bonded-fibre fabric A non-woven fabric made from a mass of fibres held together by adhesive or by processes such as needling or stitching. See **adhesive-bonded non-woven fabric**.

bond energy The energy in joules released on the formation of a chemical bond between atoms, and absorbed on its breaking.

bonding See panel on p. 33.

bond length (1) Distance between bonded atoms in a molecule. Specifically used for distances between atoms in a covalent compound. Typical lengths are O–H 96 pm, C–H 107 pm, C–C 154 pm, C=C 133 pm, C≡C 120 pm. (2) In civil engineering the minimum length of reinforcing bar required to be embedded in concrete to ensure that the bond is sufficient for anchorage purposes.

bond paper A paper similar to bank paper

bonding

Interactions between individual atoms and molecules are classified according to the way in which the outermost orbital electrons behave. Most of the elements in the periodic table (Appendix 1) are metals: the outer electrons are very loosely held by the nuclei and so form a mobile 'sea of electrons' between the closely packed atoms. This is essentially why metals are good electrical conductors. The metallic bond is relatively weak, which is the reason why pure metals are ductile and malleable. Alloying metals with other elements inhibits dislocation motion, so improving mechanical properties like tensile modulus and yield strength.

In energetic terms, the strongest bond is the covalent chemical bond formed by electron-sharing between neighbouring atoms. It is exemplified by bonding in elemental carbon, particularly diamond and graphite. In diamond, each carbon atom forms four covalent bonds (so-called sp^3 hybrids) by sharing its four outer electrons with four neighbouring atoms in a tetrahedral configuration (Fig. 1).

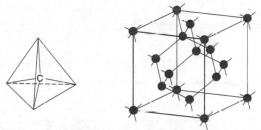

Fig. 1. **Carbon-carbon bonds in the tetrahedral structure of diamond.**

When closely packed, the tetrahedral atoms form the highly symmetrical diamond structure. The carbon-carbon bonds are the strongest known with a bond energy of about 330 kJ per mole of bonds, helping to explain the hardness of the mineral (see **hardness measurements**). Perhaps the weakest covalent bond is the hydrogen bond formed between hydrogen atoms which are strongly bonded to oxygen or nitrogen (Fig. 2).

Fig. 2. **Hydrogen bonding in water**.

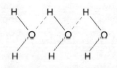

It helps explain the anomalous properties of water (high T_m and T_B, ice less dense than water etc) where groups of three molecules are loosely held by the bond. It is also important in biological molecules such as DNA, proteins and cellulose. See **biomaterials**.

Ionic bonding, such as that which exists in refractories (eg MgO), ceramics (eg silicates), salts (eg sodium fluoride) and glasses is formed by transfer of electrons between atoms of different electronegativity. See Appendix 2. Highly electronegative non-metals like fluorine gain in electron(s) to form anions (eg F^-), with a complete set of paired electrons in its outer orbitals. Metals easily lose electrons to form cations (eg Na^+) and will supply exactly the number needed in sodium fluoride (NaF) where the numbers of cations and anions are equal. The anions are larger than the cations (Appendix 2) so they close pack to form a crystal lattice, and the cations fit into the octahedral or tetrahedral interstitial sites (see **crystal structure**). Unlike covalent bonds,

continued on next page

bonding (contd)

ionic bonds which hold the solid together are non-directional, and electrostatic in nature. They can be very strong, up to 280 kJ mole^{-1} in MgO for example, explaining their very high melting points which are exploited in high-temperature-resistant refractories.

Bonding types can be described more exactly by the interatomic potential energy curve (Fig. 3), a graph of interatomic distance (r) versus potential energy :

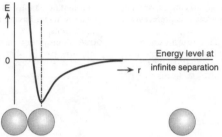

Fig. 3. **Interatomic potential energy curve**.

The equilibrium distance between atoms in solids is a result of two opposing forces: repulsion when close and attraction when separated. The depth of the potential energy well is a measure of the bond strength, so that a very weak bond like the van der Waals bond possesses a shallow well at greater equilibrium distances than a covalent bond.

but of 50 g m^{-2} or more.

bond strength Strictly, bond dissociation energy or the energy needed to separate a pair of bonded atoms from their equilibrium position to infinity. Carbon-carbon bonds are the strongest, with bond energies of about 350 kJ mole^{-1} of bonds for aliphatic compounds and about 410 kJ mole^{-1} for aromatics. US for intermolecular forces which hold matter together. Cf **adhesion**. See **bonding** p. 33.

bone A structural **biocomposite** of ca 70% wt inorganic calcium salts embedded in collagen fibres. Most of the inorganic phase consists of hydroxyapatite (calcium phosphate) but large amounts of carbonate, citrate and fluoride amines are also present. Long bones like the femur or thigh bone are composed of a harder, compact composite outer layer surrounding a spongy interior (*cancellous bone*) which improves the stiffness- and strength-to-weight indices for the material. Like most biocomposites, bone exhibits viscoelastic properties and is therefore sensitive to rate of loading etc.

bone china Form of 'soft' (ie lower firing temperature) **porcelain** produced in the UK. Based on approximately equal amounts of bone ash (mainly calcium phosphate), China

clay (**kaolin**) and Cornish stone (a potash feldspar partially converted into *kaolinite*, a hydrated aluminium silicate), yielding a white, translucent, low-porosity material after firing to about 1250°C.

bone-dry paper Paper dried completely to contain no moisture. Also *oven-dry paper*.

boot See **potette**.

borax bead Borax, when heated, fuses to a clear glass. Fused borax dissolves some metal oxides giving glasses with a characteristic colour. The use of borax bead in chemical analysis is based on this fact.

borazon See **boron nitride**.

bore The circular hole along the axis of a pipe or machine part; the internal wall of a cylinder; the diameter of such a hole.

boric acid H_3BO_3. A tribasic acid. On heating it loses water and forms metaboric acid, $H_2B_4O_4$, and on further heating it forms tetraboric acid, or the so-called pyroboric acid, $H_2B_4O_7$. On heating at a still higher temperature it forms anhydrous boron (III) oxide, or boric oxide. It occurs as tabular triclinic crystals deposited in the neighbourhood of fumaroles, and known also in solution in the hot lagoons of Tuscany and elsewhere. Also *boracic acid, sassolite*.

boric oxide *Boron (III) oxide*, B_2O_3. An 'intermediate oxide' like aluminium oxide, having feeble acidic and basic properties. As a weak acid it forms a series of borates. See **boric acid**.

boride Any of a class of substances, some of which are extremely hard and heat resistant, made by combining boron chemically with a metal.

boring The process of machining a cylindrical hole, performed in a lathe, boring machine or boring mill; for large holes, or when great accuracy is required, it is preferable to using a drill.

Born–Haber cycle The thermochemical calculation of crystal lattice energy by application of **Hess's law** using **standard heat of formation, electron affinity, ionization potential** etc.

Born–Oppenheimer approximation Used in considering the electronic behaviour of molecules. The problems of the electronic and nuclear motion are treated separately.

boron Amorphous yellowish-brown element. Symbol B. Can be formed into a conducting metal. In silicon, boron is a p-type dopant. See Appendix 1, 2.

boron carbide B_4C. Obtained from B_2O_3 and coke at about 2500°C. Very hard material, and for this reason used as an abrasive in cutting tools where extreme hardness is required. Extremely resistant to chemical reagents at ordinary temperatures.

boron nitride BN, a compound isoelectronic with elemental carbon, and having two polymorphs, one similar to graphite and the other (called *borazon*) similar to diamond.

borosilicate glass Family of glasses based on silica and borax, B_2O_3 which have a higher resistance to thermal shock (due to their lower **thermal expansion coefficient** and are more chemically resistant than soda-lime-silica glasses. Used for domestic ovenware and laboratory glassware, typical composition (wt %): SiO_2 80.8, B_2O_3 12.0, Na_2O 4.2, Al_2O_3 2.2, K_2O 0.6, CaO 0.3, MgO 0.3. Different compositions are used for glass-to-metal seals.

Bose–Einstein distribution law A *distribution law* of *statistical mechanics* which is applicable to a system of particles with symmetric wave functions unchanged when two particles are interchanged, this being the characteristic of most neutral gas molecules. It can be stated as

$$\bar{n}_i = \frac{g_i}{\frac{1}{A}\exp\left(\frac{E_i}{kT} - 1\right)},$$

where $\bar{n}_i$ = average number of molecules with energy E_i, g_i = degeneracy factor, A = constant. See **boson**.

bosh (1) A water tank for cooling metal and glassmaking tools. (2) The tapering portion of a blast-furnace, between the largest diameter (at the bottom of the stack) and the smaller diameter (at the top of the hearth).

boson A particle which obeys Bose–Einstein statistics but not the **Pauli exclusion principle**. Bosons have a total spin angular momentum of $n\,\hbar$ where n is an integer and $\hbar$ is the Dirac constant (Planck's constant divided by 2π). Photons, α-particles and all nuclei having an even mass number are bosons.

boss A projection, usually cylindrical, on a machine part in which a shaft or pin is to be supported; eg the thickened part at the end of a lever, provided to give a longer bearing to the pin.

bottle glass Soda-lime-silica glass used for the manufacture of common bottles with a typical composition (wt %) of: SiO_2 74.0, Na_2O 16.4, CaO 9.0, Al_2O_3 0.6.

bottle-making machines Operated in various ways, the glass bottle is formed in two stages, ie the **parison** and the finished bottle. Wide-mouth ware may be formed by pressing the parison and then blowing, narrow-mouth by blowing or sucking and blowing. In the last method, the glass is gathered by suction into the parison mould, in the other two it is dropped by a mechanical-feeding device, hence the terms suction-fed and feeder-fed machines.

bouclé Fabric made from fancy yarns and having a rough, textured surface, mainly used for women's garments. Yarn made from a core thread with an outer yarn wrapped round it to give a knobbly appearance.

boundary films Films of one constituent of an alloy surrounding the crystals of another.

boundary lubrication A state of partial lubrication which may exist between two surfaces in the absence of a fluid oil film, due to the existence of adsorbed mono-molecular layers of lubricant on the surfaces.

bound charge That induced static charge which is 'bound' by the presence of the charge of opposite polarity which induces it. Also, in a dielectric, the charge arising from polarization. Also *surface charge*. Cf **free charge**.

bound state Quantum mechanical state of a system in which the energy is *discrete* and the wave function is localized, eg that of an electron in an atom, where transitions between the bound states give rise to atomic spectral lines.

box annealing Heating to soften work hardened material by placing the work in a sealed box inside the furnace in order to exclude air.

box cloth A woven woollen fabric, milled

and finished with a smooth surface like felt. **Billiard cloth** is an example.

boxwood The pale-yellow, close grained, hard and tough wood of the European box tree (*Buxus sempervirens*) and related **hardwood** species, used for drawing scales, tool handles, blocks for wood-cuts etc; it requires several years of seasoning.

BR Abbrev. for **Butadiene Rubber**.

Brabender mixer Laboratory-scale mixer for plastics and rubbers comprising an internal chamber fitted with contra-rotating rotors, very similar in function to the industrial scale **Banbury mixer**.

brachy- Prefix from Gk *brachus*, short.

brady- Prefix from Gk *bradys*, slow.

Bragg angle The angle the incident and diffracted X-rays make with a crystal plane when the **Bragg equation** is satisfied for maximum diffracted intensity.

Bragg diffraction Diffraction of X-rays according to the Bragg equation.

Bragg equation If X-rays of wavelength λ are incident on a crystal, diffracted beams of maximum intensity occur in only those directions in which constructive interference takes place between the X-rays scattered by successive layers of atomic planes. If d is the interplanar spacing, the Bragg equation, $n\lambda = 2d \sin \theta$, gives the condition for these diffracted beams; θ is the angle between the incident and diffracted beams and the planes, and n is an integer. Also applied to electron, neutron and proton diffraction.

Bragg's law See **Bragg equation**.

braids (1) A wide range of narrow fabrics woven on smallware looms and used as a trimming for dress material, upholstery or coach and car interiors. (2) The product obtained by **braiding**.

braiding The process of plaiting in which three or more threads are interlaced to give a flat or tubular fabric.

brake drum A steel or cast-iron drum attached to a wheel or shaft so that its motion may be retarded by the application of an external band or internal brake shoes.

branched polymer Polymer molecules possessing side chains made by a branching reaction. See **chain polymerization** p. 56.

brass Primarily, name applied to an alloy of copper and zinc, but other elements such as aluminium, iron, manganese, nickel, tin and lead are frequently added. There are numerous varieties. See **copper alloys** p. 76.

Brazilian mahogany See (1) **American mahogany**, (2) **vinhatico**.

Brazilian rosewood Highly prized, decorative cabinet-wood from the Brazilian **hardwood** (*Dalbergia nigra*). **Heartwood** is black-streaked violet to chocolate with straight to wavy grain and coarse texture;

mean density 850 kg m^{-3}. Very durable and used in top quality furniture, cabinetmaking, panelling, veneers, etc. Also *jacaranda*, *rio rosewood*.

Brazilian tulipwood Fragrantly scented, tropical S American **hardwood** (*Dalbergia frutescens*) with yellowish-pink **heartwood** striped with shades from violet to light pink, irregular, straightish grain and fine texture; density about 960 kg m^{-3}. Used for all manner of fancy and decorative goods, and in decorative veneers. Also *jacaranda rosa*.

brazing The process of joining two pieces of metal by fusing a layer of brass, **spelter** or **brazing solder** between the adjoining surfaces.

brazing solders Alloys used for brazing. They include copper-zinc (50–55% copper), copper-zinc-silver (16–52% copper, 4–38% zinc and 10–80% silver) and nickel-silver alloys.

breakdown diode See **Zener diode**.

breakdown voltage The potential difference at which a marked increase in the current through an insulator or a semiconductor occurs.

breaker See **tyre technology** p. 325.

breaker fabric In conveyor belts and vehicle tyres, a layer of fabric placed between the main fabric of the belt or tyre and the outer rubber or plastic surface.

breaker plate Device fitted in front of the extruder screw to aid mixing. See **extrusion** p. 121.

breaking elongation, extension The maximum elongation of fibre, yarn, or fabric just before breaking. Cf **elongation**.

breaking length The length beyond which a rope or a strip of paper of uniform width would break under its own weight if suspended from one end. Usually expressed in metres.

breaking stress The stress necessary to break a material, either in tension or compression. See **strength measures**.

breeze block Lightweight building brick made from breeze, ie furnace ashes or clinker, bound with Portland cement.

bremsstrahlung Electromagnetic radiation emitted when a charged particle slows down. Thus when electrons collide with a target and suffer large decelerations, the X-radiation emitted constitutes the continuous *X-ray spectrum*. From Ger. 'braking radiation'.

brevi- Prefix from L. *brevis*, short.

bridge bearing Bearing on which the decks of bridges rest to accommodate esp. horizontal thermal expansion and contraction. Usually a steel plate on steel rollers or a laminated structure of steel plates sandwiching blocks of rubber. See **expansion rollers**, **laminated bearing**.

bridges and materials See panel on p. 38.

Bridgman process A method of growing a large single crystal in a crucible. A seed crystal is dipped into a melt which is being slowly withdrawn from a furnace. Cf **Czochralski process, float zone process.**

bright annealing The heating and slow cooling of steel or other alloys in a carefully controlled atmosphere, so that oxidation of the surface is reduced to a minimum and the metal surface retains its bright appearance. See **box annealing.**

brightening agent A compound that on addition to a white or coloured textile material increases its brightness by converting some of the incident ultraviolet radiation into visible light. Also *fluorescent brightener, optical brightener.* In metal electroplating, an agent which produces a bright reflective deposit. Cf **fluorescent whitening agents.**

Brillouin formula A quantum mechanical analogue in paramagnetism of the Langevin equation in classical theory of magnetism.

Brillouin scattering The scattering of light by the acoustic modes of vibration in a crystal, ie *photon-phonon* scattering.

Brillouin zone Polyhedron in k-space, k being the position wave vector of the groups or bands of electron energy states in the band theory of solids. Often constructed by consideration of crystal lattices and their symmetries.

Brinell hardness test A method of measuring the hardness of a material by measuring the area of indentation produced by a hard steel ball under standard conditions of loading. Expressed as Brinell Hardness Number (BHN or H_B) which is the quotient of the load on the ball in kilogram force divided by the area of indentation in mm^2. See **hardness measurements** p. 153.

Bristol board A fine quality cardboard made by pasting several sheets together, the middle sheets usually being of an inferior grade.

Britannia metal Alloy series of tin (80–90%) with antimony, copper, lead or zinc or a mixture of these.

British Association (BA) screw thread A system of symmetrical vee threads of 47° included angle with rounded roots and crests. It is designated by numbers from 0 to 25, ranging from 6.00 mm to 0.250 mm in diameter and from 10 mm to 0.070 mm pitch. Used in instrument work, but now being superseded by standard metric sizes. Even numbers are preferred sizes.

British Columbian pine See **Douglas fir.**

British Standards Institution UK national organization for the preparation and issue of standard specifications. Abbrev. *BSI.*

British Standard specification A specification of efficiency, grade, size etc drawn up by the British Standards Institution, referenced so that the material required can be briefly described in a bill or schedule of quantities. The definitions are legally acceptable. Abbrev. *BSS.*

British Standard Whitworth (BSW) thread The pre-metric UK screw thread, still widely used in the US, having a profile angle of 55° and a radius at root and crest of $0.1373 \times$ pitch; 1/6th of the thread cut off. The pitch is standardized with respect to the diameter of the bar on which it is cut.

brittle fracture Fracture which occurs with no discernable plastic deformation, ie in the elastic region of the stress-strain curve. Caused by propagation of a crack as distinct from yielding. In metals it may be either intergranular or by cleavage along certain crystal planes. See **strength measures** p. 301.

brittleheart Defect in wood, esp. in low-density, tropical hardwoods, in which circumferential shrinkage stresses in outer layers become large enough to exceed compressive strength of core wood; resulting yield produces shear lines in the timber.

brittleness The tendency to brittle fracture, ie without significant plastic deformation. Loosely used as the opposite of **toughness**, but more precisely means having low values of toughness or **fracture toughness**. See **strength measures** p. 301.

brittle temperature Point at which a material changes in fracture behaviour, from ductile to brittle. For polymers, it is often a little below T_g, the glass transition temperature. Sometimes denoted T_B. Sensitive to sample geometry (eg stress concentrations) and rate effects, such as occur in impact tests. Also *ductile brittle transition temperature.*

broach A metal-cutting tool for machining holes, often non-circular; it consists of a tapered shaft carrying transverse cutting edges, which is driven or pulled through the roughly finished hole.

broadcloth (1) A suiting cloth at least 1.35 m wide. (2) A woollen cloth, woven from fine **merino** yarns in a twill weave, heavily milled, and finished with a **dress-face**. (3) In US a lightweight poplin shirting fabric.

broadsides See **paper sizes.**

brocade Dress or furnishing fabrics produced by **Jacquard** or dobby weaving The design is developed by floating the warp and/or weft threads in irregular order on a simple ground fabric, such as satin.

broderie anglaise Machine-embroidered, lightweight woven cloth which includes holes in the pattern.

broke Wet or dry paper removed during the

bridges and materials

The largest bridges are spectacular structures involving huge quantities of material. The Humber Bridge (Yorkshire, UK) has in 1993 the world's longest span of 1410 m (see Fig. 1) and needed 500 000 tonnes of concrete for its towers and anchorages and 30 000 tonnes of steel for its deck and cables.

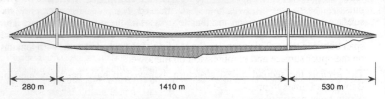

Fig. 1. **A schematic drawing of the suspension bridge over the River Humber**.

Apart from the very biggest, bridges are such a common feature of our landscape and townscape that they tend to get taken for granted, but there are 200 000 of them in the UK alone and more are added every year. Only about 1% have spans greater than 30 m. Their structural forms are usually easily visible and have evolved in parallel with developments in materials and in the understanding of how to use them. This is well illustrated by the progression of longest spans which reflect the limits of what was feasible in their day (see Fig. 2). The main limit was and still is the highest value of **tensile strength**, or more precisely **specific tensile strength**, offered in materials that were relatively cheap, available in tonnage quantities, could be readily handled and were sufficiently durable to withstand the environment for many years (the current British Standard specifies 120 years).

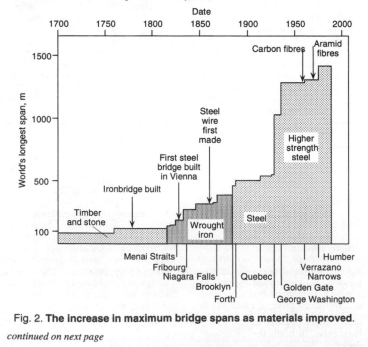

Fig. 2. **The increase in maximum bridge spans as materials improved**.

continued on next page

38

Classification of bridge structures

Bridge structures can be classified into five groups (Fig. 3), each of which imposes loads on their materials in different ways. The earliest of these was the *beam*, initially just a tree trunk or slab of rock laid across a stream or gully. The loading (*dead* due to the weight of the structure plus *live* due to all other forces) puts the beam into bending between the supports, inducing compressive stresses on the upper surface and tensile stresses on the lower at mid span. The *cantilever* bridge comprises two centrally supported beams carrying a central span between them. This reverses the loading of the simple beam, with tension on the upper surface and compression on the lower.

The *arch* is the second oldest structure, believed to have been invented in Babylonian times to provide a structure which could be built with clay bricks, which have negligible tensile strength. Thus the main load-bearing elements around the ring of the arch had to remain in compression. *Suspension* bridges (eg Fig. 1) are effectively inverted arches with the signs of the loads reversed so that their cables are in tension. Although light suspension bridges with ropes of natural fibres were long a feature of many cultures, a significant load-bearing capacity had to await the advent of high strength materials – initially wrought iron, then steel wires and latterly high-strength steel wire. Future developments will certainly include the synthetic, low-density, high-strength fibres such as **aramid**, **carbon** and **gel-spun polyethylene**. See **high-performance fibres** p. 158.

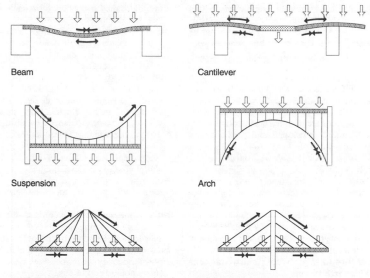

Beam   Cantilever

Suspension   Arch

Cable stayed, fan   Cable stayed, harp

Fig. 3. **Loads and induced stresses in various bridge types**. The load is shown by open arrows and the induced stresses by filled arrows.

Finally there is the *cable-stayed* or *bridle-chord* bridge which is becoming increasingly popular for medium spans. These can be of the fan type or the harp type. In both, the tension in the cables induces a compression in the decking which is particularly useful during construction.

paper making or finishing processes and re-used within the mill. Cf **cullet**.

broken ends Warp threads which have broken during weaving.

broken picks Defects in weaving due to breaking of the weft.

bromine A non-metallic element in the seventh group of the periodic table, one of the halogens. See Appendix 1, 2. Symbol Br, a dark red liquid, giving off a poisonous vapour, Br_2, with an irritating smell.

bromobutyl rubber Type of butyl rubber used in tyre industry for linings etc, made by treating isoprene-isobutene rubber with bromine to enhance reactivity during vulcanization.

bronze Primarily an alloy of copper and tin, but the name is now applied to other alloys not containing tin, eg aluminium bronze, manganese bronze and beryllium bronze. See **copper alloys** p. 76.

Brownian motion, movement Small movements of light suspended bodies such as galvanometer coils, or a colloid in a solution, due to statistical fluctuations in the bombardment by surrounding molecules of the dispersion medium. See **colloidal state**.

brushed fabric A fabric, usually woven, that has been brushed or plucked so that some of the fibres stand out from the constituent yarns. The process is also known as raising and is carried out by machines with rollers covered with wire hooks, emery paper, or teazles.

brush plating Method of electroplating in which the anode carries a pad or brush containing concentrated electrolyte or gel which is worked over the surface to be plated. Similar methods are used for brush polishing.

BS Standard developed and published by **British Standards Institution**, eg BS 5750.

BSI Abbrev. for **British Standards Institution**.

B-stage Transition stage through which a thermosetting synthetic resin of the phenol formaldehyde type passes during the curing process, characterized by softening to rubber-like consistency when heated, and insolubility in ethanol or acetone (propanone). See **thermosets** p. 318.

BSW thread Abbrev. for **British Standard Whitworth thread**.

bubble film Duplex polymer film with regular array of bubbles thermoformed into one side, used for crush-proof packaging.

bubble stabilization Tendency of polymer bubble blown after extrusion in manufacture of film, to maintain a constant shape without breaking or collapsing. Depends on tension-stiffening behaviour of molten polymer. See **Troutonian fluids**.

bubble store See **magnetic bubble memory**.

buckle (1) To twist or bend out of shape; said usually of plates or of the deformation of a structural member under compressive load. (2) A metal strap. (3) In foundrywork, a swelling on the surface of a sand mould due to steam generated below the surface.

buckling Mode of deformation in which an elastic instability occurs in a plate or a structural member under compressive load, resulting in a twisting or bending out of shape. Usually leads to plastic deformation and eventual collapse. See **Euler buckling limit**.

buckminsterfullerene Allotrope of carbon, having 60 carbon atoms linked together in hexagons and pentagon rings to form a closed, near spherical and very stable structure. Soluble in benzene etc forming deep-red solution. Named after eponymous architect of geodesic domes. Also *fullerene*.

buckram A strong linen or cotton fabric stiffened by starch, gum, or latex and used for linings, hats, and in bookbinding.

buff A revolving disk composed of layers of cloth charged with abrasive powder; used for polishing metals.

buffed crumb Flakes of rubber produced by abrading treads of worn tyres for retreading; of limited use as recycled material for new tyre compounds.

build-up sequence The order in which the welding beads are applied along a chamfered seam joining thick metals in order to give maximum strength.

built-in voltage The potential difference which arises across an unbiased **depletion layer** in a semiconductor device. The drift of charges in the electric field associated with it, is exactly balanced by diffusion in the local charge concentration gradients. In effect the contact potential between regions of extrinsic semiconductor where there are abrupt changes in the doping.

bulb bar A rolled, or extruded, bar of strip form in which the section is thickened along one edge.

bulk A measure of the reciprocal of the density of paper, being the ratio of thickness to substance. A loose synonym for thickness.

bulk density The value of the apparent powder density when measured under stated freely poured conditions.

bulked yarn See **textured yarn**.

bulk factor Ratio of the density of sound moulding to the powder density of the moulding material from which it was made.

bulkiness A term used to describe the properties of a bed of powder. It is defined as the reciprocal of the apparent density of the powder under the stated conditions.

bulk modulus One of the four basic elastic constants for elastically isotropic materials, it is defined as the ratio of the applied, uniform triaxial stress (eg hydrostatic stress) to volumetric strain in a body. Symbol K. Related to **Young's modulus**, E, **Poisson's ratio**, v and **shear modulus**, G, by:

$$K = \frac{E}{3(1-2v)} = \frac{EG}{3(3G-E)}.$$

bulletwood Timber from a tree of the genus, *Mimusops*, which also yields **balata**, a soft rubber-like material used in golf balls and for impregnating belts. Renowned for its strength and durability, it is used for structural work, boat building, furniture and cabinet making, tool handles, wheel spokes and railway sleepers.

bullion (1) Gold or silver in bulk, ie as produced at the refineries, not in the form of coin. (2) The gold-silver alloy produced before the metals are separated.

bullion content In parcel of metal or minerals being sold, where the main value is that of the base metal which forms the bulk of the parcel, the contained gold or other precious metal of minor value included in the sale.

bullion point The centre piece of a sheet of glass made by the old method of spinning a hot glass vessel in a furnace until it opened out under centrifugal action to a circular sheet. The centre piece bears the mark of attachment to the rod used to spin the sheet. The method is obsolete now, but is revived for 'antique' effects. Also *bull's eye*.

Buna Former name for family of polybutadienes.

buoyancy The apparent loss in weight of a body when wholly or partly immersed in a fluid; due to the upthrust exerted by the fluid. See **Archimedes' principle**.

Burgers vector A crystal vector which denotes the amount and direction of atomic displacement which will occur within a crystal when a dislocation moves. See **dislocation**.

buried layer A high-conductivity layer diffused into active regions of a semiconductor wafer before growth of the **epitaxial** layer in which devices are defined. It is used to decrease the collector resistance of certain bipolar junction transistors.

burl See **burr**.

burlap A coarse jute, hemp or flax fabric.

burning (1) The heating of an alloy to too high a temperature, causing local fusion or excessive penetration of oxide, and rendering the alloy weak and brittle. (2) Changing the colour of certain precious stones by exposing them to heat.

burn mark Moulding defect found on polymer surfaces caused by adiabatic compression of gas trapped in mould cavity by advancing melt front.

burnt metal Metal which has become internally oxidized by overheating, and so is rendered useless for engineering purposes.

burr (1) The rough edge or ridge on a material resulting from various operations like punching and cutting. (2) A rotary tool with cutting teeth like a file. (3) A knob or knot in a tree which, when sliced, produces strong contrasts in the form and colour of the markings which are prized for their decorative effect in edge veneers. Also *bur, burl*.

burst test Physical test method for characterizing paper by the limiting pressure (applied normally to the paper surface by means of a rubber diaphragm) that a test piece will withstand when fixed horizontally between two clamps, under prescribed conditions of test.

bush (1) A cylindrical sleeve, usually inserted in a machine part to form a bearing surface for a pin or shaft. (2) A hardened cylindrical insert in a drilling jig to position a drill or reamer accurately.

bushing A small electric melting unit, usually made of platinum, with numerous holes (usually in multiples of 204) in the base, used for the manufacture of glass fibres.

butadiene *Butan 1,2 : 3,4 diene*. $CH_2=CH.CH=CH_2$, a di-alkene with conjugate linking. An isoprene homologue, important in the manufacture of synthetic rubbers etc. See **polymers** p. 247.

butadiene polymers See **polymers** p. 247.

butadiene rubber *Polybutadiene*. Abbrev. *BR*. See **polymers** p. 247, **elastomers** p. 106.

butane C_4H_{10}, an alkane hydrocarbon, bp 1°C, rel.d. at 0°C 0.600, contained in natural petroleum, obtained from casing head gases in petroleum distillation. Used commercially in compressed form, and supplied in steel cylinders for domestic and industrial purposes, eg Calor gas.

butenes C_4H_8, alkene hydrocarbons, the next higher homologues to propylene. Three isomers are possible and known, normally gaseous, bp between 6°C and –3°C. Monomers for various polymers. Also *butylenes*.

butterfly Term used in metal extrusion where an open 'U' shape is first made and the sides then folded closer to make a vertically sided 'U'. This enables the die to be much stronger because the narrow section to form the inside of the 'U' can have a wider base

butterfly curve The strain vs applied field curve of ferroelectrics. See **dielectric and ferroelectric materials** p. 94

butternut A N American **hardwood** of the walnut (*Juglans*) family, of economic importance for its fruits as well as its wood.

It is brownish in colour with normally straight grain but a somewhat coarse texture. It is mostly used for cabinet making, carving, panelling and veneers. Mean density 450 kg m^{-3}.

butt joint A joint between two plates whose edges abut. Rivetted or bolted joints may be covered by a narrow strip or *strap*. Welded joints, except between thin sheets, require bevelling so that their edges form a V-shape for the filler metal.

buttonwood See **sycamore**.

butt-welded tube Tube made by drawing mild steel strip through a bell-shaped die so that the strip is coiled into a tube, the edges then being pressed together and welded.

butt-welding The joining of two plates or surfaces by placing them together, edge to edge, and welding along the seam thus formed. See **welding**.

butyl group The aliphatic group C_4H_9—, with four isomeric forms: primary, secondary, iso- and tertiary.

butyl lithium Organo-metallic compound, C_3H_7Li, used in anionic polymerization.

butyl rubber Copolymer made from isobutylene and a small amount of isoprene (~1%) to aid **vulcanization**. Cationically polymerized at low temperature. Outstanding properties include very low gas permeability (so used for tyre linings) and very broad **damping** peak (so useful for anti-vibration products). Very low **rebound resilience** at ambient temperature.

BWG Abbrev. for **Birmingham Wire Gauge**.

Bz A symbol for the benzoyl (benzenecarboxyl) radical $C_6H_5.CO$.

C

c A symbol for: (1) **concentration**; (2) the velocity of electromagnetic radiation *in vacuo*. Its value, according to the most accurate recent measurements, is 2.99792456×10^8 m s^{-1}.

c- Abbrev. for: (1) *cyclo-*, ie containing an alicyclic ring; (2) *cis-*, ie containing the two groups on the same side of the plane of the double bond or ring.

C The symbol for: (1) carbon; (2) coulomb; (3) When used after a number of degrees thus: 45°C, the symbol indicates a temperature on the Celsius or Centigrade scale..

C A symbol for: (1) **concentration**; (2) (with subscript) **molar heat capacity**: C_p, at constant pressure: C_v, at constant volume.

C- Containing the radical attached to a carbon atom.

Ca The symbol for **calcium**.

CAB Abbrev. for **Cellulose Acetate Butyrate**.

cable A general term for rope or chain used for engineering purposes. Specifically, a ship's anchor cable.

cable stitch In knitting, the rope-like appearance obtained by passing groups of adjacent wales under and over one another. See **fibre assemblies** p. 130.

cabling Twisting together two or more doubled or folded yarns. The result in most cases is a balanced cord of four, six or more yarns. Tyre cord is one example.

CAD Abbrev. for **Computer Aided Design**.

cadmium White metallic element, symbol Cd. See Appendix 1, 2. Cadmium plating is widely used as a corrosion protective for steel and its alloys. Films of cadmium are photosensitive in the ultraviolet between 250 and 295 nm, with a peak at 260 nm.

cadmium copper A variety of copper containing 0.7–1.0% of cadmium. Used for trolley, telephone and telegraph wires because it gives high strength in cold-drawn condition combined with good conductivity. See **copper alloys** p. 76.

cadmium photocell A photoconductive cell using cadmium disulphide or cadmium selenide as the photosensitive semiconductor. Sensitive to longer wavelengths and infrared. It has a rapid response to changes in light intensity.

cadmium-sulphide detector Radiation detector equivalent to a solid-state ionization chamber, but with amplifier effect (due to hole trapping).

CAE Abbrev. for **Computer Aided Engineering**.

caesium Metallic element, symbol Cs. As a photosensitor it has a peak response at 800 nm in the infrared, both thermal- and photoemission being high. Caesium, when alloyed with antimony, gallium, indium and thorium, is generally photosensitive. See Appendix 1, 2.

cage A regular arrangement of anions in a crystal structure within which smaller cations may be held. Four anions form a tetrahedral cage, six an octahedral one, eight a cubic one and 12 a cubeoctahedral one.

cake Rectangular casting of copper or its alloys before rolling into sheet or strip.

calcareous Containing compounds of calcium, particularly minerals.

calcination The subjection of a material to prolonged heating at fairly high temperatures.

calcined powder One produced or modified by heating to a high temperature.

calcium Metallic element, symbol Ca. Occurs in nature in the form of several compounds, the carbonate predominating. Produced by electrolysis of fused calcium chloride. Used as a reducing metal and as a getter in low-noise valves. See Appendix 1, 2.

calcium carbonate $CaCO_3$. Very abundant in nature as chalk, limestone and calcite. Almost insoluble in water, unless the water contains dissolved carbon dioxide, when solution results in the form of calcium hydrogen carbonate, causing the temporary hardness of water.

calcium fluoride CaF_2. In the form of fluorspar it is used for the manufacture of hydrofluoric acid. It is also an important constituent of opal glass. Exhibits high **thermoluminescence**.

calcium tungstate screen A fluorescent screen used in a **cathode ray oscilloscope**; it gives a blue and ultraviolet **luminescence**.

calender Rolling machine with horizontal-axis rolls of metal or fibrous composition, stacked vertically and carried in side frames. Material is fed through the gaps between the rolls, known as nips, to impart the required degree of finish or to control its thickness and compression. Used in paper, textile, plastics and rubber processing.

calendered paper Paper that has been calendered. If the calenders were part of the paper making machine the resultant effect is known as machine-finished (*mf*). A higher degree of gloss can be obtained by means of a supercalender, separate from the paper machine and containing some rolls of fibrous composition. This is known as super-calendered (*sc*) finish.

calico A plain cotton cloth heavier than **muslin**.

calomel *Mercury* (I) *chloride*, Hg_2Cl_2; found naturally in whitish or greyish masses, associated with cinnabar. Used in physical chemistry as a reference electrode (ie a half-cell comprising a mercury electrode in a solution of potassium chloride saturated with calomel).

calorescence The absorption of radiation of a certain wavelength by a body, and its re-emission as radiation of shorter wavelength. The effect is familiar in the emission of visible rays by a body which has been heated to redness by focusing infrared radiation on to it.

calorie The unit of quantity of heat in the CGS system. The 15°C calorie is the quantity of heat required to raise the temperature of 1 g of pure water by 1°C at 15°C; this equals 4.1855 J. By agreement, the International Table calorie (cal_{IT}) equals 4.186 J exactly, the thermochemical calorie equals 4.184 J exactly. There are other designations, eg gram calorie, mean calorie, and large or kilocalorie (= 1000 cal, used particularly in nutritional work). The calorie has now been largely replaced by the SI unit of the joule (J).

calorimeter The vessel containing the liquid used in calorimetry. The name is also applied to the complete apparatus used in measuring thermal quantities.

calorimetry The measurement of thermal constants, such as specific heat, latent heat, or calorific value. Such measurements usually necessitate the determination of a quantity of heat, by observing the rise of temperature it produces in a known quantity of water or other liquid.

calorizing A process of rendering the surface of steel or iron resistant to oxidation by spraying the surface with aluminium and heating to a temperature of 800–1000°C.

calx Burnt or quick-lime.

CAM Abbreviation for **Computer Aided Machining**, **Computer Aided Manufacture**.

cambium Zone of living cells in wood lying between the bark and sapwood; it is here that growth of the tree through cell division occurs. See **structure of wood** p. 305.

cambric A closely woven fine linen cloth, used chiefly for handkerchiefs; the name is also applied to a plain weave fine quality cotton cloth.

camel hair A silky fibre from the haunch and underpart of the camel or dromedary; used for dress fabrics, warm coverings, artists' paint brushes etc.

camphor Natural or synthetic ketone of formula $C_{10}H_{16}O$ and melting point 175°C;

used as solid plasticizer for cellulose nitrate in celluloid.

CAMPUS TN for database of properties of polymers supplied by a range of European manufacturers.

Canada balsam Balsam of fir, or Canada turpentine. A yellowish liquid, of pine-like odour, soluble in ethoxyethane, trichloromethane, benzene; obtained from *Abies balsamica*. Used for lacquers and varnishes, and as an adhesive for lenses, instruments etc, its refractive index being approximately the same as that of most optical glasses.

Canadian spruce Name for the wood of several trees of the *Picea* genus, the most important commercial timber in Canada. Typical uses include paper, sounding boards for musical instruments, food containers etc. See **sitka spruce, whitewood.**

candlewick A coarse folded yarn made from cotton. As well as being used for wicks in candles the yarn is used for the manufacture of fabrics suitable for bedspreads.

canvas A heavy closely woven fabric made from cotton, flax, hemp or jute for uses where strength and firmness are required eg interlinings, sails, tents.

caoutchouc French name for natural rubber, derived from Carib indian term for coagulated latex tapped form various species of plants and trees.

capacitance The electrical capacitance of an isolated conductor is defined as the ratio of the total charge on it to its potential. $C = Q/V$. See **farad.**

capacitor An electric component having **capacitance**; formed by conductors (usually thin and extended) separated by a dielectric, which may be vacuum, paper (waxed or oiled), mica, glass, plastic foil, fused ceramic, or air etc. Maximum potential difference which can be applied depends on the electrical breakdown of dielectric used. Modern construction uses sheets of metal foil and insulating material wound into a compact assembly. Air capacitors, of adjustable parallel vanes, are used for tuning high-frequency oscillators. Formerly *condenser*.

Cape olive, walnut See **stinkwood.**

caprolactam Cyclic amide which polymerizes to nylon 6. See **step polymerization** p. 298.

CAPS Acronym for *Computer Aided Polymer Selection* methods using large databases of polymers from different manufacturers.

carat Also *karat*. (1) A standard mass for precious stones. The *metric carat*, standardized in 1932, equals 200 mg (3.086 grain). (2) The standard of fineness for gold. The standard for pure gold is 24 carats; 22 carat gold has two parts of alloy; 18 carat gold six parts of alloy.

carbamic acid $NH_2.CO.OH$; is not known to occur free, being known only in the form of derivatives, eg the ammonium salt, NH_2COONH_4. The esters are known as urethanes.

carbanion A short-lived, negatively charged intermediate formed by the removal of a proton from a C–H bond, eg in **butyl lithium**.

carbenes (1) Reactive uncharged intermediates of formula CXY, where X and Y are organic radicals or halogen atoms. (2) Such constituents of asphaltic material as are soluble in carbon disulphide but not in carbon tetrachloride. See **asphaltenes, malthenes.**

carbides Binary compounds of metals with carbon. Carbides of group IV to VI metals (eg silicon, iron, tungsten) are exceptionally hard and refractory. In groups I and II, calcium carbide (ethynide) is the most useful. See **cemented carbides, cementite.**

carbide tools Cutting and forming tools used for hard materials or at high temperatures. They are made of carbides of tungsten, tantalum and other metals held in a matrix of cobalt, nickel etc, and are very hard with good compressive strength.

carbodi-imide resins Thermoset foams formed by self-polymerization reaction of di- isocyanates, Crosslinking occurs by reaction between free isocyanate groups and di-imide (–N=C=N–) group.

car body materials See panel on p. 46.

carbohydrates A group of compounds represented by the general formula $C_x(H_2O)_y$. Substances found in plants and animals, eg sugars, starch, **cellulose**. The carbohydrates also comprise other compounds of a different general formula but closely related to the above substances, eg rhamnose, $C_6H_{12}O_5$. Carbohydrates are divided into monosaccharides, oligosaccharides, polysaccharides. The carbohydrate element in diet supplies energy, provided by the oxidation of the constituent elements.

carbon Element, symbol C. See Appendix 1, 2. Its allotropic modifications are **diamond** and **graphite**. In diamond, the atom is tetravalent, the bonds being directed towards the vertices of a regular tetrahedron. Widely used in brushes for electric generators and motors, and alloyed with iron for steel. Colloidal carbon or graphite is used to coat cathode-ray tubes and electrodes in valves, to inhibit photoelectrons and secondary electrons. Highpurity carbon, crystallized to graphite in a coke furnace for many days, is used in many types of nuclear reactors, particularly for moderation of neutrons. See **carbon compounds, carbon fibre.**

carbon black Finely divided carbon produced by burning hydrocarbons (eg methane) in conditions in which combustion is incomplete. Widely used in the rubber, paint, plastic, ink and other industries. It forms a very fine pigment containing up to 95% carbon, giving a very intense black; prepared by burning natural gas or oil and letting the flame impinge on a cool surface. Various grades are identified by particle size. Thus HAF is High Abrasive Fine and is a most important reinforcing filler for rubbers. Also *channel black, gas black*. See **tyre technology** p. 325.

carbon compounds Compounds containing one or more carbon atoms in the molecule. They comprise all organic compounds and include also compounds, eg carbides, carbonates, carbon dioxide etc, which are usually dealt with in inorganic chemistry. Carbon compounds are the basis of all living matter.

carbon dioxide CO_2. A colourless gas; density at s.t.p. 1.976 kg m^{-3}, about 1.5 times that of air. It plays an essential part in metabolism, being exhaled by animals and absorbed by plants. High-pressure carbon dioxide has found a considerable use as a coolant in nuclear reactors.

carbon-dioxide welding Metal arc welding using CO_2 as the shielding gas.

carbon fibre A high modulus, highly oriented fibre of about 8 μm in diameter, consisting almost exclusively of carbon atoms. It is made as continuous filament by the pyrolysis in an inert atmosphere of organic fibres such as cellulose but usually of polyacrylonitrile. It is used as a reinforcing material with epoxy or polyester resins to form composites which have a higher strength/weight ratio than metals. Boron, **aramid** and glass are alternative fibre materials. See **high-performance fibres** p. 158.

carbon-fibre reinforced plastic Class of **composite materials** comprising a polymeric **matrix** (frequently an **epoxy** resin) reinforced with carbon fibre. Increasingly used as a structural material for its high specific stiffness and **specific strength** in applications varying from aerospace to sports equipment. Abbrev. *CFRP*.

carbonization (1) The destructive distillation of substances out of contact with air, accompanied by the formation of carbon, in addition to liquid and gaseous products. Coal yields coke, while wood, sugar etc yield charcoal. (2) The steeping of wool in a dilute solution of sulphuric acid, or its treatment by hydrochloric acid gas (dry process). This converts any cellulosic impurities into carbon dust and thereby facilitates their removal. (3) See **cementation.**

carbonized filament Thoriated tungsten filament coated with tungsten carbide to re-

car body materials

Until the 1930s the structure of motor cars had developed little from their horse-drawn precursors. The main load-bearing element (including resistance to **bending** and **torsion**) was the *chassis*, a frame with two steel (or sometimes wood) members running the length of the car which were joined to each other by cross members, and to which the body and all other parts of the vehicle were attached (see Fig. 1a).

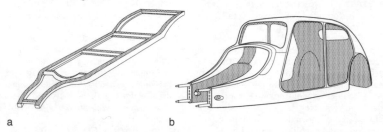

a　　　　　　　　　　　　　　　　b

Fig. 1. (a) **A schematic drawing of a steel car chassis**. (b) **The Citroën monocoque body shell of 1934**.

The early body panels were of wood or fabric fastened to wooden frames. Later these were replaced by **mild steel** panels (or aluminium on more expensive models), and then an increasing proportion of the panels were welded together, reducing the need for a wooden frame and producing a stiffer structure. These developments culminated in 1934 with the introduction of the first chassisless body, the Citroën Light 15 (see Fig. 1b) in which all the stiffness resided in the **monocoque** body shell. Two important prerequisites were the development of **rolling** technology to produce thin sheet steel of sufficient width and the introduction of **resistance spot welding** for joining the thin sheets. This type of body not only has production advantages with less hand work and easier automation but also structural ones. It has become the dominant structure for mass-produced cars ever since.

With the increasing pressure for longer-lived, more fuel-efficient cars, attention has focused on alternative materials to mild steel. One approach has led to the introduction of **high-strength low-alloy** (HSLA) or **micro-alloyed steels** for critical body panels (eg as in Fig. 2), which have up to 80% higher **yield stress** and 60% higher **tensile strength** than mild steel, thus requiring less material for the same performance. Aluminium and its alloys are another

continued on next page

duce loss of thorium from the surface.

carbon monoxide CO. Formed when carbon is heated in a limited supply of air. Poisonous. Its properties as a reducing agent render it valuable in industrial processes.

carbon paper Paper coated with waxes containing dyes or carbon black, used for making duplicate copies in typewriting etc.

carbon resistor Negative temperature coefficient, non-inductive resistor formed of powdered carbon with ceramic binding material. Used for low-temperature measurements because of the large increase in resistance as temperature decreases.

carbon steel A steel whose properties are determined principally by the amount of carbon present and contains no other deliberate alloying ingredient except those necessary to ensure deoxidation and physical quality. Also *plain carbon steel*. See **steels** p. 296.

carbonyl powders Metal powders produced by reacting carbon monoxide with the metal to form the gaseous carbonyl. This is then decomposed by heat to yield powder of high purity.

Carborundum　　　TN for silicon carbide

car body materials (contd)

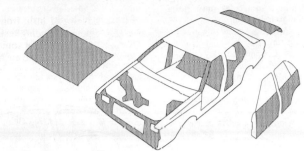

Fig. 2. **The use of high-strength low alloy steel in critical body panels**.

possible alternative, with significant weight saving and better corrosion resistance, but more expensive and more difficult to weld than steel. Specialist car makers have long used **polyester/glass-fibre composites** for their bodies, panels of which are mounted on a tubular steel space frame. Practically all Formula 1 racing cars have **carbon-fibre/epoxy** panels and aluminium **honeycomb** monocoque body shells. These are not, however, really suited to mass production.

Instead, volume car manufacturers are increasingly using plastics and plastic composites in their car exteriors. Materials include sheet moulding compound (**SMC**), the long fibre (**ZMC**), reaction injection moulded (**RIM**) and reinforced reaction injection moulded (**RRIM**) polyurethane for bumpers and body panels, polycarbonate blends and elastomer-modified nylon for bumpers. These trends are certain to continue.

abrasives.

carborundum wheel See **grinding wheel**.

carboxyl group The acid group –COOH.

carboxylic acid R–$(COOH)_n$. An organic compound having one or more carboxyl groups.

carboy Large, narrow-necked glass container, usually of balloon shape, having a capacity of 20 l or more.

carburizing A method of **case hardening** low carbon steel in which the metal component is heated above its ferrite-austenite transition in a suitable carbonaceous atmosphere. Carbon diffuses into the surface and establishes a concentration gradient. The steel can subsequently be hardened by quenching either directly or after re-heating to refine the grain structure. It is usually lightly tempered afterwards, producing a hard case over a tough core.

carcass, tyre carcass Body of tyre, comprising bead coil, inner lining, side wall, breaker and plies often without tyre tread. Term applied to used tyres from which worn tread has been buffed ready for retreading or to green tyre prior to shaping and **vulcaniza-**

tion. See **tyre technology** p. 325.

card cloth Strong material (eg a fabric-rubber laminate) fitted with masses of strong, flexible wire teeth, pins or spikes. Often in the form of flats or rotating cylinders, and used for **carding**.

carded yarns Yarns spun from **slivers** directly after **carding**.

carding The process of passing fibres through a machine called a *card* which disentangles them and makes them lie fairly straight, to form a light fluffy web or **sliver**.

carrier A real or imaginary particle responsible for the transport of electric charge in a material. In semiconductors electrons in the conduction band and positive holes in the valence band are the carriers; in oxide ceramics, electrons hopping between ions, diffusing oxygen ions and mobile cations can also transport charge.

carrier mobility The mean drift velocity of the charge carriers in a material per unit electric field.

cartilage A form of connective tissue in which the cells are embedded in a firm matrix of chondroitin or chondrin. Two functions: maintenance of shape (ear, nose, etc) and bearing surfaces in joints. Replaced by sintered **ultra-high molecular mass polyethylene** in hip joint implant.

carton board Any paper board intended for conversion into cartons.

cartridge brass Copper-zinc alloy containing approximately 30% zinc. Possesses high ductility; capable of being heavily cold-worked. Widely used for cold pressings, cartridges, tubes etc. See **copper alloys** p. 76.

cartridge paper Originally a tough paper intended for winding the tubes of shotgun cartridges (*ammunition cartridge*). Now also a paper made for drawing purposes (*drawing cartridge*) or for lithographic printing (*offset cartridge* or *matt-coated cartridge*).

case That part near the surface of a ferrous alloy which has been so altered as to allow case hardening.

case-hardening (1) The production of a hard surface layer in steel either (a) by heating in a carbonaceous medium to increase the carbon content, followed by quenching and lightly tempering or (b) by rapidly heating the surface of a medium/high carbon steel to above the ferrite/austenite transformation temperature and then quenching and tempering, as in flame and induction hardening. (2) In seasoning of wood, a condition in which the surface of timber becomes set in an expanded condition and remains under compression while the interior is in tension.

casein Protein present in micellar form in skimmed milk and precipitated from it by acidification. It contains inorganic phosphate salts which aid processing into shaped products. Plasticized with water and cross-linked with formaldelyde, it was formerly used widely for decorative products, but now mainly confined to buttons.

casement A plain-woven cotton (or man-made) fabric used for curtains.

cashmere Fine, down-like fibres obtained from the undercoat of the Asiatic goat, or similar material obtained from goats in Australasia and Scotland. Frequently blended with wool and used for the manufacture of cardigans and sweaters.

casting copper Metal of lower purity than best selected copper. Generally contains about 99.4% of copper.

casting resins Term applied to liquid resins which can be introduced into a moulded shape and polymerized catalytically in situ. Usually polymethyl methacrylate, polyester, epoxy and polyurethane resins.

casting wheel Large wheel on which ingot moulds are arranged peripherally and filled from stream of molten metal issuing from furnace or pouring ladle.

cast iron Any iron-carbon alloy in which the carbon content exceeds the solubility of carbon in austenite at the eutectic temperature. Widely used in engineering on account of their high fluidity and excellent casting characteristics. Carbon content usually in the range 2–4.3%. Some kinds are brittle and others difficult to machine, but see **grey iron, ductile cast iron.**

cast steel Shapes that have been formed directly from liquid by casting into a mould. Formerly applied to wrought objects produced by working steel made by the crucible process to distinguish from that made by cementation of wrought-iron, but both of these methods are long obsolete.

catalysis The acceleration or retardation of a chemical reaction by a substance which itself undergoes no permanent chemical change, or which can be recovered when the chemical reaction is completed. It lowers the **activation energy**.

catalyst A substance which catalyses a reaction. See **catalysis**.

catalytic converter A device fitted into the exhaust system of petrol-engined vehicles to reduce the emissions of carbon monoxide (CO), nitrogen oxides (NO_X) and hydrocarbons. A *three-way* catalyst of platinum, palladium and rhodium on a ceramic lattice oxidizes CO to carbon dioxide (CO_2), hydrocarbons to CO_2 and water, and reduces NO_X to nitrogen. An *oxidation catalyst* oxidizes CO and hydrocarbons but does not affect NO_X: this has to be reduced by engine design using *lean-burn* principles.

catalytic poison A substance which inhibits the activity of a catalyst. Also *anticatalyst*.

catenation See **chain**.

cathode (1) In an electronic tube or valve, an electrode through which a primary stream of electrons enters the inter-electrode space. During conduction, the cathode is negative with respect to the anode. Such a cathode may be *cold*, electron emission being due to electric fields, photo-emission, or impact by other particles, or *thermionic*, where the cathode is heated by some means. (2) In a semiconductor diode, the electrode to which the forward current flows. (3) In a thyristor, the electrode by which current leaves the thyristor when it is in the ON state. (4) In a light-emitting diode, the electrode to which forward current flows within the device. (5) In electrolytic applications, the electrode at which positive ions are discharged, or negative ions formed.

cathode copper The product of electrolytic refining, after which the cathodes are melted, oxidized, poled, and cast into wirebars, cakes, billets etc.

cathode poisoning Reduction of thermionic emission from a cathode as a result of minute traces of adsorbed impurities.

cathode ray A stream of negatively charged particles (electrons) emitted normally from the surface of a cathode in a vacuum or low-pressure gas. The velocity of the electrons is proportional to the square root of the accelerating potential, being 6×10^5 m s^{-1} for one volt. They can be deflected and formed into beams by the application of electric or magnetic fields, or a combination of both, and are widely used in oscilloscopes and TV (in cathode ray tubes), electron microscopes and electron-beam welding, and electron-beam tubes for high-frequency amplifiers and oscillators.

cathode ray oscilloscope Device for displaying electronic signals by modulating a beam of electrons before it impinges on a **fluorescent screen**. Abbrev. *CRO*.

cathodic protection Protection of a metal structure against corrosion by either coupling it with a more electronegative metal (sacrificial anode) or by impressing an e.m.f. (voltage) which makes the structure the cathode in an electrolytic corrosion cell.

cathodoluminescence The emission of light, with a possible afterglow, from a material when irradiated by an electron beam, such as occurs in the phosphor of a cathode ray tube.

cation Ion in an electrolyte which carries a positive charge and which migrates towards the cathode under the influence of a potential gradient in electrolysis. It is the deposition of the cation in a primary cell which determines the *positive terminal*.

cationic polymerization Polymerization using cationic catalyst such as aluminium chloride. Commercial polymers include butyl rubber. See **chain polymerization** p. 56.

cat's eye (1) A crescent-shaped blister in glass. (2) A reflecting stud containing glass lenses set in rubber and used for road-marking.

caustic embrittlement The intergranular corrosion of steel in hot alkaline solutions, eg in boilers.

caustic lime The residue of calcium oxide, obtained from freshly calcined calcium carbonate; it reacts with water, evolving much heat and producing slaked lime (calcium hydroxide, hydrate of lime, or hydrated lime). Also *quicklime*. See **lime**.

caustic potash Potassium hydroxide. The name potash is derived from 'ash' (meaning the ash from wood) and 'pot' from the pots in which the aqueous extract of the ash was formerly evaporated. Highly alkaline.

caustic soda See **sodium hydroxide**.

cavalry twill Firm, warp-faced cloth, often wool, with steep double-twill lines, made in different weights for trousers, breeches, rainwear and dresses.

cavitation Generation of cavities (eg bubbles) in liquids by rapid pressure changes such as those induced by ultrasound. When cavity bubbles implode, they produce shock waves in the liquid. Components can be damaged by cavitation if it is induced by turbulent flow.

cavity Hollow space within moulding tool which will form final product.

Cavity Transfer Mixing TN for polymer mixing process using specially designed head fitted to extruder output. The head comprises a cylinder and loose-fitting rotating core (attached to the screw). Each mating surface possesses a number of hemi-spherical hollows, by means of which the melt is repeatedly turned and thoroughly mixed with pigments etc. See **extrusion**.

cc An abbrev. for *cubic centimetre*, the unit of volume in the CGS metric system. Also cm^3.

CCD See **charge-coupled device**.

CCD array An array of many thousands of photodiodes, whose response to an image focused on the surface of the array can be converted into a video signal by employing CCD electronic circuits. An alternative to vacuum tubes in television cameras.

C-class insulation A class of insulating material which will withstand a temperature of over 180°C.

CCP Abbrev. for **Cubic Close Packing**.

CED Abbrev. for **Cohesive Energy Density**.

cedar Also *deodar*. **Softwood** tree (*Cedrus*), of European (*C. atlantica*) and Asian (*C. deodara*) origin, whose wood is light brown, straight-grained, with a medium fine and uniform texture, and a distinctive 'cedary' smell. It is used for pattern-making, boat building and cheap furniture, and also makes good structural timber. It is not resistant to termite attack although it is resistant to powder-post beetle infestation. Mean density 580 kg m^{-3}.

ceiling temperature Critical temperature above which polymerization cannot occur. Of the order of hundreds of °C for most common polymers, it is the result of the high and negative entropy of polymerization competing against the high, negative heat of polymerization. Also *floor temperature*.

Celite TN for a form of diatomite used as an insulating material and filter aid.

cell (1) Chemical generator of emf. (2) Small item forming part of experimental assembly, eg Kerr cell, dielectric test cell etc.

cell constant The conversion factor relating the electrical **conductance** of a conductivity cell to the conductivity of the liquid in it.

Cellidor TN for **cellulose acetate-butyrate** (Euro).

Cellophane TN for thin, transparent packaging film (eg on cigarette packets) made from regenerated cellulose by modified viscose process.

Cellosolve *Hydroxy-ether, 2-ethoxy-ethan-1-ol*, $C_2H_5O.CH_2.CH_2OH$. A colourless liquid used as a solvent in the plastics industry. It is miscible with water, ethanol and ethoxyethane, and boils at 135.3°C.

cellular fabric The name applied to a woven or knitted fabric featuring an open, or cell-like, structure used mainly for shirts, blouses and underwear .

cellular glass Glass foam often made with H_2S as a blowing agent, hence its dark colour and strong sulphide smell; a very moisture-resistant, low-temperature insulation material.

cellular silica Inorganic silicates containing numerous air cells and low permeability to water vapour.

cellular solid Generic term for materials comprising an assembly of cells with solid edges or faces, packed together to fill space. Occur naturally as eg wood (see **structure of wood** p. 305), cancellous bone, **cork**, coral and sponge. Also manufactured as eg **foam** and **honeycomb** structures.

cellular structure See **expanded plastics**, **foam**. Cf **network structure**.

celluloid Man-made thermoplastic consisting of nitrocellulose polymer plasticized with camphor. Formerly widely used for moulded products but now restricted to ping-pong balls and spectacle frames due to competition from lower cost and less flammable synthetic plastics.

cellulose Linear homopolymer of glucose. The most abundant organic polymer on Earth. The glucose repeat units are linked in the β configuration (see **biological polymers** p. 27), by contrast with the α configuration in starch. The β configuration allows the chains to crystallize in a linear conformation, so natural cellulose as found in wood and natural fibres such as flax and cotton is highly crystalline (90–95%). The crystallites are organized in fibrils which are aligned along the fibre axis, producing a high tensile modulus of ca 100 GPa when stressed. The molecular mass probably exceeds one million but natural cellulose is difficult to extract without chain cleavage, whether by natural biodegradation or chemical treatment. Cotton fibre is almost pure cellulose but wood is a natural biocomposite, with 50–60% cellulose content embedded in lignin. Cellulose is the most important constituent of paper and cardboard, as well as providing man made fibres (principally rayon) and cellulosic thermoplastics (nitrocellulose, celluloid, celluose acetate).

cellulose acetate Man-made derivative of **cellulose** made by treating it with acetic anhydride/methylene chloride/sulphuric acid mixture. Up to three hydroxyl groups in the repeat unit are esterified (eg triacetate), and the product is soluble in acetone or chloroform owing to the breakdown of the crystal structure, lowering of molecular mass and acetylation. Injection moulded products (eg combs) now less popular than extrudates (eg photographic film).

cellulose acetate-butyrate Random copolymer of cellulose acetate and butyrate. TNs Tenite, Cellidor.

cellulose esters Generic term for the man-made derivatives of **cellulose** where hydroxyl groups on the repeat unit are esterified eg cellulose acetate. Other esters include acetate butyrate (CAB), acetatepropianate and cellulose propionate.

cellulose ethers Generic term for man-made derivations of **cellulose** where hydroxyl groups (–OH) on the molecule are replaced by ether groups (–OR) such as ethyl cellulose (R = –C_2H_5) used for protective films. Also methyl cellulose (R = –OCH_3), hydroxyethyl cellulose (R = –O_2CH_4OH), hydroxyl propyl cellulose (R = –OC_3H_6OH), which are water-soluble polymers used as thickening agents, paper sizes etc.

cellulose lacquers Lacquers prepared by dissolving nitrocellulose or acetylcellulose in a mixture of suitable solvents, with the admixture of resins, plasticizers and pigments.

cellulose nitrate Nitro esters of cellulose formed by a mixture of nitric and sulphuric acid acting on cotton fibre. Nitrocellulose with up to four nitro groups per repeat unit is not explosive and is plasticized with camphor to produce thermoplastic products (celluloid). Can also be dissolved in ester solvents for films and coatings eg of playing cards, staples etc. Much reduced usage owing to competition from synthetic plastics, cellulose acetate etc.

cellulose xanthate Sodium salt of cellulose-dithiocarbonic acid produced by treating cellulose with caustic soda/carbon disulphide solution. Hydroxyl groups (–OH) on the glucose repeat units are so converted.

cellulosics General name for useful man-made materials derived from natural **cellulose**.

cement (1) Generic term for any binding agent, adhesive or glue. (2) Binding agent for concrete, often Portland cement, which hardens as it reacts slowly with water. See panel on p. 52. (3) In mammalian teeth, a layer resembling bone covering the dentine beyond the enamel.

cement and concrete See panel on p. 52.

cementation Any process in which the surface of a metal is impregnated at high temperature by another substance. Also *carbonization, carburization*. See **case-hardening**.

cement copper Impure copper, obtained when the metal is precipitated by means of iron from dilute solutions arising from leaching. Also *cementation copper*.

cemented carbides Sintered carbides. See **carbide tools**.

cementite The iron carbide (Fe_3C) constituent of steel and cast-iron (particularly white cast iron) containing 6.67 wt% carbon. It has a complex hexagonal crystal structure with carbon in interstitial sites. Very hard (harder than **martensite**) and brittle.

centi- Prefix meaning one-hundredth.

centrifuge Rotating machine which uses centrifugal force to separate molecules from solution, particles and solids from liquids, and immiscible liquids from each other. Depends on differences in the relative densities of the substances to be separated. Used widely in science where accelerations up to $500\,000\,g$ may be obtained in an ultracentrifuge. In industry they are used in sugar and cream production, separating water from fuel and swarf from cutting oil etc.

ceramic capacitor One using a high permittivity dielectric such as barium titanate to provide a high capacitance/unit volume.

ceramic insulator An insulator made of ceramic material, eg **porcelain**; generally used for outdoor installations.

ceramics processing See panel on p. 54.

ceramics The art and science of non-organic non-metallic materials. The term covers the purification of raw materials, the study and production of the chemical compounds concerned, their formation into components, the study of structure, constitution and properties. See **alumina, carbides.**

ceramic transducer One based on the electrical properties of ceramics such as **piezoelectricity**.

cerammed glass See **glass-ceramic**.

cerium A steel-grey metallic element, one of the *rare earth* metals. Symbol Ce. Electrical resistivity $0.78\,\mu\Omega$ m. When alloyed with iron and several rare elements, it is used as the sparking component in automatic lighters and other ignition devices. It is also a constituent (0.15%) in the aluminium base alloy ceralumin and is photosensitive in the ultraviolet region. Used on tracer bullets, and formerly for flashlight powders and gas mantles. See Appendix 1, 2.

cermet Ceramic articles bonded with metal. Composite materials combining the hardness and high temperature characteristics of ceramics with the mechanical properties of metal, eg **cemented carbides**.

Ceylon satinwood A lustrous, golden yellow, **hardwood** timber with a narrowly interlocked grain, and a fine and uniform texture. From *Chloroxylon swietenia*, a native of India and Ceylon. It is naturally durable when used in exposed positions, but is regarded as a cabinet wood and source of veneers. Mean density 980 kg m^{-3}. Also *East Indian satinwood*.

CFRP Abbrev for **Carbon-Fibre Reinforced Plastic**.

C-glass Designation for a chemically-resistant grade of glass fibre of composition (wt %): SiO_2 65, CaO 14, Na_2O 8, B_2O_3 6, Al_2O_3 4, MgO 3, Fe_2O_3 0.3.

chafer Triangular strip of rubber between bead-coil and side-wall of tyre. See **tyre technology** p. 325.

chain (1) A series of linked atoms, generally in an organic molecule (*catenation*). Chains may consist of one kind of atom only (eg carbon chains), or of several kinds of atoms (eg carbon nitrogen chains). There are open-chain and closed-chain compounds, the latter being called ring or cyclic compounds. (2) A series of interconnected links forming a flexible cable, used for sustaining a tensile load.

chain branching Termination reaction which leads to formation of branched polymer (eg low-density polyethylene). See **chain polymerization** p. 56.

chain defects Any kind of deviation from the normal molecular structure of the backbone of a polymer chain, as expressed by the repeat unit. They include chain ends, head-to-head units, oxidized units (eg carbonyl groups), tertiary carbon atoms and itinerant copolymer repeat units.

chain end Possible weak point in polymer chain to act as a point of initiation of degradation or depolymerization. Some polymers may need capping with a protective end group to inhibit such instability.

chain extending agent Chemical compound which can react with active chain ends to give much larger polymer molecule.

chain flexibility Ease or difficulty of movement of a polymer chain, critically dependant on structure (or configuration) of repeat unit and temperature. Polyethylene and silicone polymers for example have very flexible single chains, so have very low T_gs and are good low-temperature materials if non-crystalline. Large side-groups or aromatic rings in the backbone chain increase rigidity, so increasing T_g and T_m (if chain is regular).

chain-folded crystal Conformation of

cement and concrete

Concrete, whether reinforced or not, has become the most widely used civil engineering material. It is made of cement and aggregate, which may have a graded size distribution (Fig. 1), mixed with water. The mixing-in of water initiates the setting reaction between it and the cement, binding the whole mass together to form concrete. During the first few hours of setting the material remains pourable and workable. This is its great advantage as an engineering material: as a fluid, it can readily be moulded into complex shapes *in situ*, yet subsequently it acquires properties similar to those of natural stone. The most frequently used cement, especially in the UK, is *Portland cement*, although different materials are used in other parts of the world.

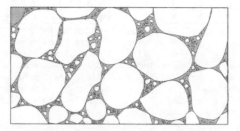

Fig. 1. **Concrete made with graded aggregate.**

Portland cement is made from a mixture of about 75 wt % limestone, $CaCO_3$, and 25 wt % clay, principally aluminosilicate, but with a significant iron oxide and alkali oxide content. These are ground together and fired with coal in air at up to 1500°C to produce a clinker, which, in turn, is mixed with 3–5 wt % of gypsum and ground again to give cement powder, with a mean particle size of under 10 μm. The particle size is important – the smaller it is, the faster is the setting. The powder is a complex mixture of multicomponent, mineral solid solutions, the principal constituents of which are (approx. wt %): *alite* (tricalcium silicate, $CaO.Ca_2SiO_4$) 55, *belite* (dicalcium silicate, Ca_2SiO_4) 20, *aluminate* (tricalcium aluminate, $2CaO.Ca(AlO_2)_2$) 12, *ferrite* (tetracalcium aluminoferrite, $4CaO.Al_2O_3.Fe_2O_3$) 8, *gypsum* (hydrated calcium sulphate, $CaSO_4.2H_2O$) 3.5 and *oxides* (K_2O, Na_2O, CaO) 1.5.

The sequence and variety of hydration reactions during setting and hardening are also complex, and are still being elucidated. The polymerization of silica, to form an interlocking network of foils and fibres of silica gel around each cement particle, plays an important role, as do the formation and interlocking of needle-like crystals of *ettringite*, a heavily hydrated calcium aluminosulphate with plate-like crystals of *portlandite*, $Ca(OH)_2$. The initial setting reaction is very rapid, with a high exotherm , 200 W kg^{-1}, within 30 s. This falls to about 1 W kg^{-1} after an hour, and the cement continues hardening at an increasingly slow rate with changes still detectable after 30 years! Some 40 wt % of water is required for complete hydration.

The tensile strength of hardened cement (about 3–5 MN m^{-2}) is only about 10% of its compressive strength, largely due to porosity, the pores of up to 1 mm in size acting as *stress concentrators*. Its Young's modulus is about 30 GN m^{-2}, but it shows partially viscoelastic behaviour. Additives, known as *admixtures*, are used to control such properties as viscosity, setting time, freezing temperature, pore size and permeability to water. *Macro-defect-free* (MDF) cement is made by adding about 5% of polyvinyl alcohol, mixing like a dough and compressing. The material has a maximum pore size of 15 μm and a tensile strength of about 60 MN m^{-2}.

continued on next page

cement and concrete (contd)

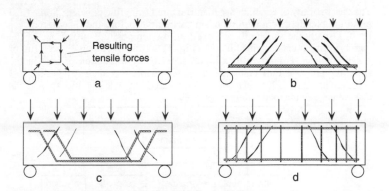

Fig. 2. **Reinforcement in concrete beams**. Shear and bending stresses interact to give tensile forces which act diagonally (a) and result in diagonal cracks (b). Two different arrangements of reinforcing bars to counteract the tensile forces are shown (c and d).

Concrete is made from about 80 vol % of (graded) aggregate, containing three parts sand to two parts gravel, and 20 vol % cement. In reinforced concrete, the concrete is poured around an assemblage of heavily cold-worked, mild steel reinforcing bars, held together with links made from a soft mild steel. The steel bars are positioned to resist the anticipated tensile stresses in the structure, some of which can be the resultant of shear and bending (Fig. 2). Beams are designed so that the steel carries all the tension, the concrete being allowed to crack in tension, but required to resist loads in compression. *Prestressed concrete beams* are arranged so that all the concrete is under compression. This is applied by high-tensile steel wires which are either held in tension as the concrete is cast around them and then unloaded after the concrete has set (*pretensioned*), or the wires are fed into ducts cast in the concrete and tensioned after the concrete has set (*post-tensioned*). For a given bending stiffness, prestressed concrete only requires about half the mass of concrete and a third the mass of steel compared to a reinforced concrete beam.

polymer chains in lamellar crystals. Chains are oriented perpendicular to main flat surface, where they fold over to maintain continuity. Perfect single crystals can be made by cooling dilute polymer solutions, but are not typical of bulk crystallized polymer, where surface chains may be entangled and connect with adjacent single lamellae (as in the switchboard model). See **crystallization of polymers** p. 82.

chain lines The more widely spaced continuous lines in the watermark of a **laid paper**.

chain polymerization See panel on p. 56. Also *addition polymerization*.

chain reaction (1) Chemical or atomic process in which the products of the reaction assist in promoting the process itself such as ordinary fire or combustion, or atomic fis-

sion. (2) Reaction producing polymers. See **chain polymerization** p. 56.

chain transfer Process occurring during free radical polymerization where active chain end terminates by abstracting hydrogen atom from dead chain or solvent or initiator. A new active chain end is thus created, and may lead to graft or branched polymer. See **chain polymerization** p. 56.

chair (1) The 'chair' with long arms on which the glass-maker rolls the blowpipe while fashioning the ware. (2) The group of glassmakers who work together in the process of hand fabrication.

chalcogenide Compound containing an element, or elements from Group VI(B) of the periodic table, ie sulphur, selenium or tellurium.

chalcogenide glass Generic term for ma-

ceramics processing

The vast majority of ceramics are manufactured from fine powders. Some final grinding, blending or mixing is often carried out before the shaping process itself. Usually the shaping is achieved at low temperatures using one of the variety of methods described below (Fig. 1) and then the porous preform, or *green compact*, is brought to a higher density by the action of heat in the sintering process. Full compaction to the theoretical density is often not achieved and the product then contains **porosity**, which is usually expressed as a percentage of the total volume.

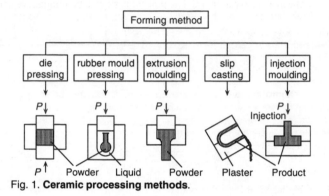

Fig. 1. **Ceramic processing methods**.

Die pressing is by far the most common fabrication process for small ceramic components. High production rates of up to 5000 pieces per minute can be achieved. Sizes can range from about 0.1 to 100 mm with reasonably good dimensional tolerance on the final product ($\pm 1\%$). The main limitations are to the complexity of shape which can be achieved and to the uniformity of the product's density.

Isostatic pressing achieves a more uniform density by using a flexible membrane to transmit hydrostatic pressure from a compressed fluid to the piece, thereby ensuring uniform compaction throughout the body. This process is best used for parts with cylindrical symmetry, such as spark plug insulators or tubes and rates of production of around 1500 pieces per hour are possible. Quite often some final machining will be required because of the imprecise control of dimensions that arises from using a compressible membrane.

Extrusion through a suitable die is an excellent method of producing components with a constant cross-section, provided a batch material with adequate plasticity is available. For clay-based ceramics the plasticity is achieved simply by controlling the amount of water present. For powders, suitable organic plasticizers are used to provide the correct consistency.

Injection moulding is a technique borrowed from the plastics industry. The batch used contains up to 60–70% by volume of fine ceramic powder in a polymeric matrix. The mix is heated to a plastic state and is then injected into a mould under pressure where it cools and sets prior to extraction. The next, delicate step is to eliminate the polymeric material without damaging or disrupting the preform. Long controlled burn-out treatments of, say, 36–48 hours are often required. The method is very good for mass production of complicated shapes, provided the maximum wall thickness of the component is not

continued on next page

too great and the initial high capital cost of the injection machine and dies can be justified.

In *slip casting*, the 'slip' is a colloidal suspension of ceramic powder in a liquid which is usually, but not always, water. The slip is poured into a mould which is microporous and slowly draws out the liquid from the slip by capillary action. The result is that a layer of fairly solid material is built up against the mould wall over a period of a few hours. The excess slip can then be poured out to leave a hollow component of uniform wall thickness. The moulds are usually *plaster of Paris* (hydrated calcium sulphate or gypsum) and are relatively cheap and easy to make. The process is slow, labour intensive and lacks precision, but is nevertheless used for some engineering components.

All the above processes are followed by drying, and then a sintering process at an appropriate temperature, often around 0.7 T_m, to bring about densification. The temperature required is high because sintering occurs by solid state diffusion, either along the external surfaces or through the crystalline interior. In the latter case, crystal defects such as grain boundaries are preferred paths. Sometimes impurities in the batch or deliberate additions (*fluxes*) can speed up the sintering processes or allow lower temperatures to be used.

Liquid phase sintering can occur at lower temperatures than solid phase sintering when a liquid phase is present to the extent of 2–20% by volume. The formation of this phase depends on having appropriate additives in the powder mix. In this mode of sintering, dissolution of atoms from the powder into the liquid occurs followed by reprecipitation.

In *reaction sintering* the densification is aided by a chemical reaction, usually between a solid powder, and a liquid or gas. The commonest examples are the production of silicon nitride by firing a silicon powder compact in nitrogen gas and the production of silicon carbide by reacting graphite with liquid silicon in the presence of some silicon carbide particles.

Sometimes both shaping and sintering are done simultaneously by die pressing at high temperatures, known as *hot pressing*. It is a slow and expensive process capable of producing only simple shapes of limited dimensions. A related process is known as *hot isostatic pressing* (HIP), in which a batch of the order of 10–100 ceramic components can be densified simultaneously in a high-temperature, high-pressure gas chamber (Fig. 2). The gas, usually nitrogen or argon, is the medium by which pressure is transmitted to the preshaped, partially densified components. The economics of this process are much better than those of hot die-pressing, but nevertheless it is used only for products of high value, such as ceramic cutting tools or turbocharger rotors.

Sol-gel processing is being developed for ceramics as well as inorganic glasses. For other procedures see **electroceramic processing** p. 109.

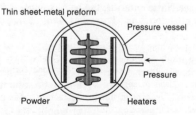

Fig. 2. **Hot isostatic pressing.**

chain polymerization

There are two basic ways of making high-molecular mass polymer chains: *chain-growth* and *step-growth*. Each mechanism gives rise to quite different types of polymer with characteristic properties. Chain-growth polymers include polyethylene, vinyl polymers and polybutadiene; they are formed by activating a small number of monomer molecules using catalysts such as benzoyl peroxide. When heated, the peroxide splits to give two free radicals which react with the monomer to create another free radical (*initiation*). This activated monomer molecule then reacts very rapidly in a chain reaction with further monomer units to make the polymer chain (*propagation*). Chain termination occurs by two free radical chain ends reacting together (*recombination* and/or *disproportionation*) or by a free radical reacting with dead chain to form one chain and a branching chain (*chain branching*). High molecular mass polymer is formed very quickly, and specific chemicals (*terminators*) are used commercially to control such polymerizations. This is shown schematically in the figure (T = time).

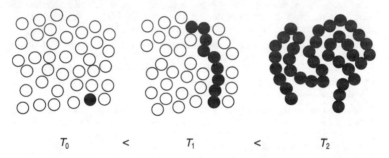

$$T_0 \quad < \quad T_1 \quad < \quad T_2$$

Extra control of the very rapid propagation step is achieved by conducting the reaction as an emulsion or as a suspension polymerization in a water matrix. This also aids the removal of the large amounts of heat produced, since all polymerization reactions are exothermic.

Control of polymer chain configuration with free radical ends is not easy, and it tends to produce atactic polymers (eg vinyls), branched polymers (eg low-density polyethylene) or random copolymers (with two or more monomers). The proportion of the different repeat units in a copolymer is determined by the **copolymer equation**; the distribution of molecular mass is usually very broad. Altering the ionic charge of the catalyst enables a greater control over the polymer structure. Anionic polymerization allows block copolymers to be produced easily since the negatively charged chain end cannot branch and will only react with specific compounds. Monodisperse polymers can also be easily produced. Control of tacticity is achieved using co-ordination polymerization with a solid **Ziegler-Natta catalyst**; the best example is isotactic polypropylene. Chain polymerization is also known as *addition polymerization*.

terials made by melt-quenching or vapour deposition of chalcogenide compounds. Generally semiconductors, with high transparency in the infrared. Used in **xerography**, and as solid electrolytes esp. in thin film batteries. Also *chalconide glass*.

chambray Plain weave cotton cloth with dyed warp and undyed weft producing a speckled effect; used for dresses.

channel A standard form of rolled-steel section, consisting of three sides at right-angles, in channel form. See **rolled-steel sections**.

characterizing magnetic materials See panel on p. 57.

charcoal The residue from the destructive distillation of wood or animal matter with exclusion of air; contains carbon and inorganic matter.

characterizing magnetic materials

A magnetic material is characterized in terms of the degree to which it is magnetized by an external magnetizing field. A magnetizing field (H) is associated with electric currents and in free space leads to a magnetic flux density (B). The relationship is linear with the constant μ_0 known as the *permeability* of free space, which has the value $4\pi \times 10^{-7}$ H m^{-1}:

$$dB = \mu_0 \, dH.$$

Strictly B and H are vector quantities. When materials are present the consequence of their becoming magnetized by the field, H, can be incorporated in a relative permeability, μ_r, so that

$$B = \mu_r \mu_0 H.$$

Paramagnetic materials have μ_r slightly greater than unity and diamagnetic substances have μ_r slightly less than unity (see Fig. 1a).

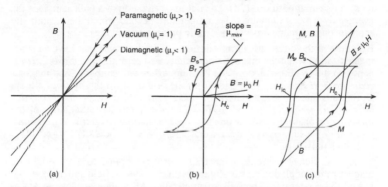

Fig. 1. *B–H* curves for (a) **non-magnets**, (b) **soft magnets** and (c) **hard magnets**.

The magnetic susceptibility, χ, reflects the degree to which a material has become magnetized. The magnetization, M, which results from the application of the magnetizing field, H, is given by

$$M = \chi H.$$

Thus, the magnetic flux density when a material is present can also be written

$$B = \mu_0(H + M) = \mu_0(1 + \chi)H \, ,$$

so that the relative permeability and the susceptibility are simply related by

$$\mu_r = (1 + \chi) \, .$$

The common magnetic materials (see **ferromagnetics and ferrimagnetics** p. 128) do not have unique values of relative permeability and susceptibility. These materials exhibit hysteresis, that is the magnetization, M, depends not only on the magnetizing field, H, but also on the previous state of magnetization: susceptibility and permeability are not constants. These materials are therefore characterized by the magnetization curve $M(H)$, although it is often presented in terms of the associated flux density curve $B(H)$. The relationship is as follows:

$$B(H) = \mu_0 [H + M(H)] \, .$$

The curves $B(H)$ and $M(H)$ are customarily referred to as *hysteresis loops* (see

continued on next page

Fig. 1). The distinction between $B(H)$ and $M(H)$ is more important for hard magnetic materials. Soft magnetic materials are so readily magnetized that in normal use $H \ll M$.

In magnetic materials, when there is no magnetizing field present ($H= 0$), the magnetization, M, arises from the alignment of internal atomic magnetic moments. If the number density of atomic magnet moments is n per unit volume and their strength is m, then the maximum achievable magnetization is given by

$$M_0 = n\,m\,.$$

The magnetization of a bulk sample may fall below this value as a consequence of the thermal disruptions of the perfect alignment of magnetic moments together with the effects which lead to the arrangement into locally aligned *domains*, preferentially oriented by microstructural features (see **ferromagnetics and ferrimagnetics** p. 128). The net magnetization of a bulk sample is the average over many domains. The temperature beyond which there can be no spontaneous magnetic order is the **Curie point** (temperature).

By virtue of the arrangement into domains, a bulk sample can have zero net magnetization when the magnetizing field is also zero; this point ($M=0$, $H=0$) is not on the hysteresis loop. When an unmagnetized sample is first magnetized, the B–H curve traces a path from the origin out to the hysteresis loop. During this initial phase a material is characterized also by its *initial (relative differential) permeability* ($dB/d\mu_0 H$), which measures the initial slope of the magnetizing phase. Elsewhere on the loop, the *maximum (relative differential) permeability* is also used as a chacteristic quantity, particularly for soft magnetic materials.

When a sufficiently strong magnetizing field is applied to a material, the magnetization of all domains is aligned with the external field and the material is said to be saturated. The bulk magnetization M_s is then as close to M_0 as thermal effects will allow.

The value of the magnetization of a bulk sample when a magnetizing field is restored to zero after having achieved saturation is termed the remanent magnetization; the corresponding value of the flux density is called the **remanence**, B_r. The value of the magnetizing field necessary to further reduce the bulk magnetization to zero is called the *intrinsic coercivity*, H_{ic}, while that necessary only to reduce the net flux density to zero is called simply the *coercivity* or the coercive force, H_c.

The area of a B–H hysteresis loop, or *hysteresis loss*, W_H, represents the energy per unit volume dissipated as heat in rearranging the domain structure as the material is taken round one cycle from saturation in one direction to saturation in the opposite direction and back again. Transformer cores are generally required to be constructed from materials with low values of W_H, that is with narrow loops. Suitable materials are often described in terms of an apparent relative permeability (actually the maximum differenial value), μ_r, the hysteresis loss, W_H, the saturation flux density, B_s, corresponding with the saturation magnetization and the bulk electrical resistivity. This last quantity is important in connection with eddy currents induced by changing magnetic flux.

The hysteresis characteristics of hard magnetic materials are often summarized by reference to the remanent flux density, B_r, the coercivity, H_c, and the so-called *energy product* or BH_{max}. The energy product reflects a material's ability to remain magnetized in the presence of a demagnetizing field. It is the maximum value of the product BH_{max} associated with the upper left quadrant of a hysteresis loop. For magnetic recording media, the resilience to demagnetization is measured with reference to the squareness of the M–H loop.

charcoal iron Pig-iron made in a blast-furnace using charcoal instead of coke. Sometimes wrought-iron is also made from this. Now obsolete.

charge (1) Electrical energy stored in chemical form in secondary cell. (2) Quantity of unbalanced electricity in a body, ie the excess or deficiency of electrons, giving the body negative or positive electrification, respectively. See **negative, positive**. The charge of an ion is one or more times that of an electron, of either sign. (3) Synonym for **batch**. (4) The quantity of glass required to fill a pot, the fireclay vessel containing the glass melt in a pot furnace.

charge-coupled device A semiconductor device which relies on the short-term storage of minority carriers in spatially defined depletion zones on its surface. The charges thus stored can be moved about by the application of control voltages via metallic conductors to the storage points, in the manner of a shift register. Abbrev. *CCD*.

charged-pair complex Association of anion and cation, usually in dilute solution.

Charpy test A flexed beam, notched-specimen, impact test in which both ends of a notched specimen are supported and a striker carried on a pendulum impacts the specimen centrally on the face opposite the notch; the energy absorbed in fracture is then calculated from the height to which the pendulum rises as it continues its swing. See **impact tests** p. 165.

cheese Densely wound cylindrical package of yarn made on various winding machines for warping and other purposes.

cheese cloth A lightweight, open cotton fabric used in cheese manufacture but sometimes adapted for shirts and blouses.

chemical bond The electric forces linking atoms in molecules or non-molecular solid phases. Three basic types of bond are usually distinguished: (1) ionic or electrostatic bonding, in which valence electrons are lost or gained, and atoms which are oppositely charged are held together by coulombic forces; (2) covalent bonding, in which valence electrons are associated with two nuclei, the resulting bond being described as polar if the atoms are of differing electronegativity; (3) metallic bonding, in which valence electrons are shared over many nuclei, and electronic conduction occurs. See **bonding** p. 33.

chemical compound A substance composed of two or more elements in definite proportions by weight, which are independent of its mode of preparation. Thus the ratio of oxygen to carbon in pure carbon monoxide is the same whether the gas is obtained by the oxidation of carbon, or by

the reduction of carbon dioxide.

chemical constitution See **constitution**.

chemical elements See table of chemical elements in Appendix 2.

chemical engineering Design, construction and operation of plant and works in which matter undergoes changes of state and composition.

chemical equation A quantitative symbolic representation of the changes occurring in a chemical reaction, based on the requirement that matter is neither added nor removed during the reaction.

chemical kinetics The study of the rates of chemical reactions. See **order of reaction**.

chemical lead Lead of purity exceeding 99.9%; suitable for the lining of vessels used to hold sulphuric acid and other chemicals.

chemical pulp Pulp in which the fibres have been resolved by chemical, as distinct from mechanical means, with the removal of the greater part of the lignin and other non-cellulose material. Cf **mechanical pulp**.

chemical reaction A process in which at least one substance is changed into another.

chemical symbol A single capital letter, or a combination of a capital letter and a small one, which is used to represent either an atom or a mole of a chemical element; eg the symbol for sodium is Na, for sulphur is S etc. See list of naturally occurring elements and their properties in Appendix 2.

chemical vapour deposition The deposition of solid material, usually as a thin film, from precursors in the gas phase. Abbrev. *CVD*. Cf **physical vapour deposition**.

chemiluminescence A process in which visible light is produced by a chemical reaction.

chemisorption Irreversible adsorption in which the absorbed surface is held on the substance by chemical forces.

chemistry The study of the composition of substances and of the changes of composition which they undergo. The main branches of the subject are often considered to be inorganic, organic and physical chemistry.

chenille Type of yarn with fibres projecting all round a central core of threads, and cloth woven with such yarn in the weft.

cherry Hardwood tree (*Prunus*), economically not regarded as a timber tree, though the wood has certain minor uses. It carves and turns well, and is used for picture frames, inlaying, tobacco pipes, walking-sticks and veneers. Density about 610 kg m^{-3}.

chestnut A light- to dark-brown wood resembling oak; much used for fencing, posts and rails.

chicle The natural **latex** of *Sapota achras* composed of a mixture of cis and trans polyisoprene with natural resins. De-resinated chicle is one base for chewing gum. See **jelutong**.

chiffon Very fine, soft woven dress material of silk or man-made fibres; the word is also used in other cloth descriptions to indicate lightness, eg chiffon velvet, chiffon taffeta.

chill crystals Small crystals formed by the rapid freezing of molten metal when it comes into contact with the surface of a cold metal mould.

chilled iron Cast-iron cast in moulds constructed wholly or partly of metal, so that the surface of the casting is white and hard while the interior is grey.

china clay A clay consisting mainly of kaolinite, one of the most important raw materials of the ceramic industry. China clay is obtained from kaolinized granite, for example in SW England, and is separated from the other constituents of the granite (quartz and mica) by washing out with high-pressure jets of water. Also *kaolin, porcelain clay.*

chintz Plain-woven cotton fabric that has been printed and then glazed by calendering. It may also be stiffened by the addition of starch.

chip (1) Popular name for an **integrated circuit**. The term derives from the method of manufacture, as each chip is made as part of a *wafer*, a flat sheet of silicon, impregnated with impurities in a pattern to form an array of transistors and resistors with the electrical interconnections made by depositing thin layers of gold or aluminium. Many copies of the integrated circuit are formed simultaneously and then have to be broken apart. See **semiconductor device processing** p. 280. (2) See **swarf**.

chipboard (1) A wood-based **composite material** made by compression moulding chips of **urea-formaldehyde** resin coated, waste timber into sheets. (2) A board, usually made from waste paper, used in box making. Also *particle board.*

chirality Absence of symmetry involving reflection or inversion. A molecule is chiral if no stable configuration of it can be superimposed on its mirror image. See **stereoregular polymers** p. 299.

chitin A natural nitrogenous polysaccharide with the formula $(C_8H_{13}N_5)_n$ occurring as external skeletal material in most invertebrates, such as insects and arthropods (eg lobster). Usually occurs in **biocomposite** form, reinforced by inorganic minerals.

chlorinated polyvinyl chloride. Polymer used in solvent cements. Abbrev. *CPVC.* See **solvent welding**.

chlorine Element, symbol Cl. The second halogen, molecular chlorine (Cl_2) is a greenish yellow diatomic gas, with an irritating smell and a destructive effect on the respiratory tract. Chlorine is used in the organic chemicals industry to produce tetrachloromethane, trichloromethane, polyvinyl chloride etc. See Appendix 1, 2.

chlorobutadiene See **chloroprene**.

chlorobutyl rubber Type of butyl rubber used in tyre linings, with enhanced reactivity during **vulcanization**.

chlorofibre Fibres made from copolymers with vinyl chloride or vinylidene chloride predominating. Other constituent monomers include acrylonitrile. The material has markedly hydrophobic properties.

chloroprene *2-Chlorobutan 1,2 : 3,4-diene.* Starting point for the manufacture of **chloroprene rubber**.

chloroprene rubber Chemically resistant elastomer made by polymerizing chloroprene monomer. TN Neoprene.

chondroitin Natural linear polysaccharide fluid which binds with proteins to form proteoglycans. Found in the cornea of the eye as well as structural tissue like cartilage. Derived biomaterial, chondroitin sulphate. See **hyaluronic acid**. Synonym *chondrin.*

CHR Abbrev. for **epiChlorHydrin Rubber**.

chroma In the Munsell colour system a term to indicate degree of saturation; zero represents neutral grey, and, depending on the hue, the numbers 10 to 18 represent complete saturation.

chromadizing Treating aluminium or its alloys with chromic acid to improve paint adhesion.

chromate treatment Treating a metal with a hexavalent chromium compound to produce a conversion coating, so altering its surface properties.

chromatography Method of separating (often complex) mixtures. Adsorption chromatography depends on using solid adsorbents which have specific affinities for the adsolved substances. The mixture is introduced on to a column of the adsorbent, eg alumina, and the components eluted with a solvent or series of solvents and detected by physical or chemical methods. Partition chromatography applies the principle of countercurrent distribution to columns and involves the use of two immiscible solvent systems: one solvent system, the stationary phase, is supported on a suitable medium in a column and the mixture introduced in this system at the top of the column; the components are eluted by the other system, the mobile phase. See **gas-, liquid-, paper-, gel permeation-chromatography**.

chrom-, chromo- Prefixes from Gk *chroma, chromatos,* colour. Also *chromato-.*

chrome brick Brick incorporating chromite, used as a refractory lining in steelmaking furnaces.

Chromel Tradename for a nickel-chromium alloy (90/10) used in heating elements and (with **Alumel**) in thermocouples. Alloy used in heating elements, based on nickel with about 10% chromium.

chromic (VI) acid H_2CrO_4. An aqueous solution of chromic oxide. Often applied as a solution of sodium dichromate in sulphuric acid, used for cleaning glassware and etching plastics etc.

chromium Metallic element, symbol Cr. Electrical resistivity at 20°C is 0.131 $\mu\Omega$ m. Obtained from chromite. Alloyed with nickel in heat-resisting alloys and with iron or iron and nickel in stainless and heat-resisting steels. Also used as a corrosion-resistant plating. See Appendix 1, 2.

chromo paper Paper which is more heavily coated than art paper; used for chromolithography.

CHU, Chu Abbrev. for **Centigrade Heat Unit**, the same as the pound-calorie.

CIIR Abbrev. for *Chlorinated Isoprene Isobutene Rubber*. See **chlorobutyl rubber**.

cinder pig Pig iron made from a charge containing a considerable proportion of slag from puddling or reheating furnaces (now obsolete).

Cinpres TN for **hollow moulding** process.

circuit See **printed, hybrid and integrated circuits** p. 257.

circularly polarized Term applied to particular type of polarized **electromagnetic radiation**, esp. visible light, where the plane of vibration is effectively helical. Produced by circularly polarizing filters, esp. **Polaroid**, and used for photoelastic analysis of isochromatics. Not to be confused with plane polarized light.

circular saw A steel disk carrying teeth on its periphery, used for sawing wood, metal or other materials; usually power driven.

ciré Highly lustrous fabric produced by waxing and mechanical polishing, enhanced if satin weave is used.

cire perdue See **investment casting**.

cis- A prefix indicating that geometrical isomer in which the two radicals are adjacent in a metal complex or on the same side of a double bond or alicyclic ring.

clad metal One with two or three layers bonded together to form a composite with eg a corrosion resistant layer formed over a stronger core by co-rolling, heavy plating, chemical deposition etc. See **laminate**.

clamp force See **locking force**.

class-A, -B, -C etc insulating materials Classification of thermal insulation materials according to the temperature which they may be expected to withstand.

clathrate Form of compound in which one component is combined with another by the enclosure of one kind of molecule by the structure of another; eg rare gas in 1,4-dihydroxybenzene.

Clausius–Clapeyron equation This shows the influence of pressure on the temperature at which a change of state occurs:

$$\frac{dp}{dT} = \frac{\lambda}{T\,\Delta V},$$

where λ is the heat absorbed (latent heat), T is the absolute temperature, ΔV is the change in volume.

Clausius–Mosotti equation One relating electrical polarizability to permittivity, principally for fluid dielectrics.

cleavage (1) The splitting of a crystal along certain planes parallel to certain actual or possible crystal faces, when subjected to tension. (2) The splitting up of a complex molecule, such as a protein, into simpler molecules, usually by hydrolysis and mediated by an enzyme.

clinch To set or close a fastener, usually a rivet. Also *clench*.

clingfilm Thin polymer packaging film which quickly bonds to itself or products to be protected.

clinker Incombustible residue, consisting of fused ash, raked out from coal- or coke-fired furnaces; used for road-making and as aggregate for concrete. See **breeze block**.

clinks Internal cracks formed in steel by differential expansion of surface and interior during heating. The tendency for these to occur increases with the hardness and mass of the metal, and with the rate of heating.

clip gauge Type of **extensometer** used in **tensile tests**; relative movement of two arms clipped to the specimen generates an electrical signal proportional to extension.

cloqué Woven or knitted double fabric with a blistered surface pattern.

close annealing The operation of annealing metal products (eg sheets, strip and rod) in closed containers to avoid oxidation. Also *pot* (or *box*) *annealing*.

closed cell See **foam**.

closed loop recycling Material reclamation within factory or industrial system, eg in-house recycling.

closed magnetic circuit The magnetic core of an inductor or transformer without air gap.

closed pore A cavity within a particle of powder which does not communicate with the surface of the powder.

close-packed hexagonal structure See **hexagonal close-packed structures** and panel on p. 62.

close packing of atoms

The atoms of the metallic elements of the periodic table (Appendix 1) can be modelled by perfect spheres. Metals in the solid state usually pack together regularly to give *crystal structures*, most of which are close packed. In other words, the spherical atoms arrange themselves so as to fill the available space to the maximum possible extent. On a flat plane, spheres of equal radius pack together most efficiently in a *hexagonal array*, with each sphere surrounded by six *nearest neighbours* (Fig. 1).

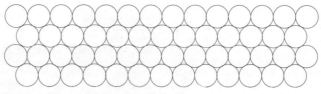

Fig. 1. **A hexagonal array in a flat plane**.

As similar sheets are added successively to form a three-dimensional structure, two types of close packed systems can be created. The second sheet (B) to be added sits in hollows created by the first layer (A), and if the third sits in the hollows created by B in the same way as the first layer, the *hexagonal close packed* (hcp) structure is formed (Fig. 2). The sequence of layers is thus ABABAB... and the unit cell in section shows that each atom is surrounded by 12 similar atoms, ie its co-ordination number is 12.

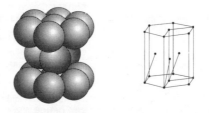

Fig. 2. **Hexagonal close packing**.

Alternatively, the third layer C can be added in a different configuration (C) to either preceding layer (Fig. 3), giving *cubic close packing* (ccp). The sequence of layers is ABCABCABC... and the unit cell is oriented with its diagonal at right angles to the layers of close packed atoms.

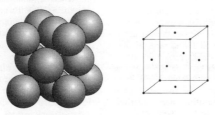

Fig. 3. **Cubic close packing**.

continued on next page

close packing of atoms (contd)

The cubic unit cell possesses a sphere at the centre of each face: hence the alternative term *face-centred cubic* (fcc). Both hcp and fcc structures possess a coordination number of 12 and occupy 74% of the space available.

A third type of close packed structure is created by square arrays, which can pack in only one way.

Fig. 4. **Body-centred cubic packing**.

The unit cell possess a co-ordination number of 8 owing to the less efficiently packed structure (68% of space filled). It comprises a cube with a sphere at the centre and is thus known as *body-centred cubic* (bcc). Other atomic crystals possess unit cells of lower symmetry (eg rhombic, tetragonal) due to distortion of the close packed structures by bonding effects (see Appendix 2, column 5). The close packed models of metal structures are particularly useful for explaining alloy formation (eg solid solutions). In addition, the structure of ionic solids (which are the basis of useful ceramics and refractories) can be explained by close packed lattices of anions into which the smaller cations can be fitted.

close packing of atoms See panel on p. 62.

closures Devices used to effect final sealing of containers, eg screw tops for bottles, lids for boxes. Often polymeric, where design freedom allows discrimination between users, such as child-proof plastic tops for medicine bottles.

cloth Generic term embracing all fabrics woven, felted, knitted, or non-woven etc using any material or man-made fibre or continuous filament or blends of these. Includes apparel, furnishing and industrial fabrics.

cluster A molecular compound containing four or more metal atoms bonded to one another without intervening ligands. Clusters have been prepared with more than 40 metal atoms so bonded, and the internal structure of these approximates to that of metals themselves.

cluster mill Rolling mill in which two small diameter working rolls are each backed up by two or more larger rolls.

cm³ Abbrev. for *cubic centimetre*.

CMOS See **complementary metal-oxide-silicon, CMOS**.

CN Abbrev. for **Coordination Number**.

coacervation The reversible aggregation of particles of an emulsoid into liquid droplets preceding **flocculation**.

coagulation The precipitation of **colloids** from solutions, particularly of proteins. See **gels** p. 143.

coal-tar The distillation products of the high- or low-temperature carbonization of coal. Coal-tar consists of hydrocarbon oils (benzene, toluene, xylene and higher homologues), phenols (carbolic acid, cresols, xylenols and higher homologues), and bases, such as pyridine, quinoline, pyrrole, and their derivatives.

coarsening Increase in the grain size of metals usually by heating for a time and at a temperature where grain growth is rapid.

coated cathode One sprayed or dipped with a compound having a lower **work function** than the base metal, in order to enhance electron emission, which may be thermionic or photo-emission.

coated fabric A fabric coated on one or both sides with a layer of a coating material such as a plastic, esp. polyvinyl chloride or rubber.

coated lens The amount of light reflected from a glass surface can be considerably reduced by coating it with a thin film of transparent material (blooming). The film has a thickness of a quarter of a wavelength and its refractive index is the geometric mean of that of air and glass. Used to increase the light transmission through an optical system by reducing internal reflections, thus also reducing flare.

cobalt Hard, grey metallic element in the eighth group of the periodic table. It is magnetic below 1075°C, and can take a high polish. Symbol Co. See Appendix 1, 2. Electrical resistivity at 20°C is 0.0635 $\mu\Omega$ m. Tensile strength (commercial grade containing carbon) 450 MN m^{-2}. Similar properties to iron but harder; used extensively in alloys, and as a source for radiography or industrial irradiation. Forms a hexagonal structured compound with samarium ($SmCo_5$) which has high anisotropy and coercivity and is used to make short permanent magnets.

cobalt (II) oxide CoO. Used to produce a deep blue colour in glass, and in small quantity to counteract the green tinge in glass caused by the presence of iron.

cobalt steels Steels containing 5–12% cobalt, 14–20% tungsten, about 4% chromium, 1–2% vanadium, 0.8% carbon and a trace of molybdenum used for tools. They have high hardness, and retain this at red heat. See **steels** p. 296.

cockling Local wrinkling of the surface of paper generally due to the release of dried-in strains as the result of moisture take-up.

cocobolo A very durable C American **hardwood** (*Dalbergia*) whose **heartwood** varies from a rich red to a streaked, mellow orange-red, and has an irregular grain and fine texture. Used for small decorative items and for very attractive veneers. Density 990–1200 kg m^{-3} (average 1100 kg m^{-3}) .

cocoon The envelope of silk thread spun by the fully grown silkworm round itself as a protective covering when entering the chrysalid state. If intended for silk manufacture, the cocoon is heated to destroy the pupa and the thread is subsequently reeled.

cocuswood A tropical American **hardwood** of the genus *Brya*, available only in the form of small logs. Typical uses include inlaying, brush backs, turnery, parquet and musical instruments. Also *Jamaica ebony*, *West Indian ebony*.

coefficient of equivalence A factor used in converting amounts of aluminium, iron and manganese into equivalent amounts of zinc, in relation to their effect on the constitution of brass, principally the solubility limit of the alpha solid solution.

coefficient of expansion See **thermal expansion coefficient**.

coefficient of friction See **friction**.

coefficient of restitution The ratio of the relative velocity of two elastic bodies after direct impact to that before impact. If a sphere is dropped from a height on to a fixed horizontal elastic plane, the coefficient of restitution is equal to the square root of the ratio of the height of rebound to the height from which the sphere was dropped. See **impact**.

coefficient of thermal expansion See **thermal expansion coefficient**.

coefficient of viscosity See **viscosity**.

coercive force See **characterizing magnetic materials** p. 57.

coercivity See **characterizing magnetic materials** p. 57.

co-extrusion Process in which duplex or multiplex film made of two or more polymers is extruded simultaneously.

cogging The operation of rolling or forging a metal ingot to reduce it to a **bloom** or **billet**.

coherent units A system of coherent units is one where no constants appear when units are derived from base units.

cohesion Attraction forces by which particles are held together to form a body, generally imparted by the introduction of a temporary binder, or by compaction. It is measured as green strength.

cohesive energy density Measure of the forces of attraction holding molecules together. Defined as

$$\frac{(\lambda - RT)}{M/\rho}$$

where λ is the latent heat of vaporization, R the gas constant, T the absolute temperature, ρ is the density and M the molecular mass of the substance. Easy to determine experimentally for small molecule compounds (eg solvents, plasticizers), more difficult with polymers which cannot vaporize normally. Can however, be inferred from structural formula. See **solubility parameter**.

coiled-coil filament A spiral filament for an electric lamp which is coiled into a further helix to reduce radiation losses and enable it to be run at a higher temperature.

coinage metals Those **transition metals** renowned for their relative inertness and hence widely used in metal coins. Specifically copper, silver and gold.

coining The impressing or repressing of a component in a die and tool set, in which all surfaces of the part are confined, to impart final shape and accuracy of dimensions.

coir The reddish-brown coarse fibre obtained from the coconuts of *Cocos nucifera*. The finer fibres are used for making mats, the coarser ones for brushes, and the shortest for filling mattresses and upholstery. Of interest as a substitute for peat based compost.

cokes Originally, tin plates made from **wrought-iron** produced in a coke furnace. The term is now applied to plates with a thinner tin coating.

cold bend A test of the **ductility** of a metal; it consists of bending a bar when cold through a certain specified angle.

cold casting A shape made by or the process of pouring a mix of thermosetting resin and metal powder into as mould at ambient temperature and allowing it to cure. Often used for pseudo-bronze sculptures where the product has the appearance of the metal.

cold cathode Electrode from which electron emission results from high-potential gradient at the surface at normal temperatures.

cold chisel A chisel for chipping or cutting away surplus metal; it is used with a hand hammer. Different forms of cutting edges (eg flat, cross-cut, half-round) are used for various purposes.

cold drawing (1) Process of producing bar or wire by drawing through a steel die without heating the material. See **wire-drawing**. (2) Tensile deformation of polymer fibre, rod or bar at ambient temperatures (ca 25°C) to produce a stiffer product. Caused by chain **orientation** along tensile axis.

cold galvanizing The coating of iron and steel articles with zinc by suspending them in an organic liquid, subsequently evaporated to leave a zinc film on the article. Now also applied to **electroplating** with zinc, it is alternatively known as *electrogalvanizing*.

cold heading The process of forming the heads of bolts or rivets by **upsetting** the end of the bar without heating the material.

cold junction Junction of **thermocouple** wires with conductors leading to a thermoelectric pyrometer or other temperature indicator or recorder.

cold moulding The use of resin, filler and accelerator to fill the mould, which then polymerizes to form the component.

cold riveting The process of closing a rivet without previous heating; confined to small rivets.

cold rolled Metal that has been rolled to sheet or section at a temperature close to atmospheric. The cold-rolling of metal sheets results in a smooth surface finish.

cold sett, set, sate See **set**.

cold short Metals brittle below their **recrystallization temperature**.

cold slug well Cylindrical, deep hole in runner system designed to aid sprue removal in **injection moulding**.

cold welding The forcing together of like or unlike metals at ambient temperature, often in a shearing manner, so that normal oxide surface films are ruptured allowing such intimate metal contact that **adhesion** takes place.

cold working The operation of shaping metals at temperatures below their recrystallization temperature (ie below $0.5\ T_m$) so as to produce strain-hardening. See **work-hardening**.

Cole–Cole plot Graph of real against imaginary part of complex permittivity, theoretically a semicircle, from which the relaxation time of polar dielectrics can be determined. See **dielectrics and ferroelectrics** p. 92.

collagen Family of structural fibrous proteins abundant in animal tissues, especially **bone** (with **hydroxyapatite**), tendon and skin (with **elastin**). It has the general amino acid sequence –Gly–Pro–Hyp–Gly–x– (x = any amino acid) arranged in a crystalline triple α-helix. When degraded by strong alkali, collagen yields **gelatin**.

collapsible drum Device on which a green tyre is built, needed to make a hollow, cutaway doughnut shape on the tyre-building machine.

collective electron theory Assumption that ferromagnetism arises from free electrons; **Fermi Dirac statistics** identify the **Curie point** with the transition from the ferro- to the paramagnetic state. See **ferromagnetics and ferrimagnetics** p. 128.

collector (1) Any electrode which collects charges which have already completed and fulfilled their function, eg screen grid. (2) Outer section of a transistor which delivers a primary flow of carriers.

collector capacitance The capacitance of the depletion layer forming the collector-base junction of a bipolar **transistor**.

collector current The current which flows at the collector of a transistor on applying a suitable bias.

collector-current runaway The continued increase of the collector current arising from an increase of temperature in the collector junction when the current grows.

colligative properties Those properties of solutions which depend only on the concentration of dissolved particles, ions and molecules, and not on their nature. They include depression of freezing point, elevation of boiling point and osmotic pressure.

collision An interaction between particles in which momentum is conserved. If the kinetic energy of the particles is also conserved, the collision is said to be *elastic*, if not then the collision is *inelastic*. With particles in nuclear physics, there is no contact unless there is *capture*. Collision then means a nearness of approach such that there is mutual interaction due to the forces associated with the particles. Cf **impact**.

collodion A **cellulose** tetranitrate, soluble in a mixture of ethanol and ethoxyethane (1:7); the solution is used for coating materials and, in medicine, for sealing wounds and dressings.

colloid From Gk *kolla*, glue. Name originally given by Graham to amorphous solids, like gelatine and rubber, which spontaneously disperse in suitable solvents to form lyophilic sols. Contrasted with crystalloids on the one hand and with lyophobic sols on the other. The term currently denotes any colloidal system. See **gels** p. 143.

colloidal graphite Extremely fine dispersion of ground graphite in oil. Graphite lowers the surface tension of oil without lowering the viscosity; the oil spreads more easily, taking the graphite to rough surfaces where it can build up a smoothness.

colloidal state A state of subdivision of matter in which the particle size varies from that of true 'molecular' solutions to that of coarse suspensions, the diameter of the particles lying between 1 nm and 100 nm. The particles are charged and can be subjected to electrophoresis, except at the iso-electric point. They are subject to Brownian movement and have a large amount of surface activity. See **gels** p. 143.

colloidal suspension See **gels** p. 143.

colophony, colophonium See **rosin**.

colour fastness The resistance of a dyed textile to change of colour when exposed to specific agents such as water, light or rubbing.

column A vertical pillar or shaft of cast-iron, forged steel, steel plate in box section, stone, timber etc, used to support a compressive load. See **strut**.

columnar crystals Elongated crystals formed by growth taking place at right angles to the temperature gradient within a mould, usually at right angles to the mould wall. They initiate from the layer of **chill crystals** at the surface and the extent to which they grow before solidification is completed is determined by the formation of equi-axed crystals in the interior portions of the casting.

combed yarns Highest quality yarns prepared from carded and combed fibres that have been mechanically straightened and freed from **neps** and short fibres.

combination Formation of a compound.

combination mill A continuous rolling mill in which the shaping mills follow the roughing mill directly.

combined carbon In cast-iron, the carbon present as iron carbide as distinct from that present as graphite. See **graphitic carbon**.

combing The process of further separating and straightening *carded slivers* of fibres and removing impurities and fibres below a specified length.

combining weight See **equivalent weight**.

combustion Chemical union of oxygen with any organic material accompanied by the evolution of light and rapid production of heat (exothermic).

comminution The process of reducing a material to a powder by eg attrition, impact, crushing, grinding, abrasion, milling or chemical methods.

commodity polymer Commercial polymers produced in high tonnage quantities, esp. polystyrene, low-density polyethylene, high-density polyethylene, polypropylene and polyvinyl chloride.

common lead Lead of lower purity than chemical or corroding lead (about 99.85%).

compact The solid produced by confining a powder, with or without a binder, and compressing it in a die. See **green compact**.

compacted graphite cast iron Made by a method like that for ductile cast iron but without allowing the formation of completely spherulitic graphite nodules so that the graphite shape is between that of grey and ductile cast iron. Abbrev. *CG iron*. Also *vermicular iron*.

comparative tracking resistance Abbrev. *CTR*. See **tracking resistance**.

compatibility (1) General term describing state of mixture of materials whether in liquid or solid state. If two chemically distinct substances mix completely in the liquid state, then they are compatible and miscible, and form a homogeneous fluid. See **phase diagrams** p. 236. If some degree of phase separation occurs, then they are partially miscible, but if totally incompatible. then they form immiscible phases. Also a term applied to the solid state, where kinetics of formation may be important. See **isothermal transformation diagrams** p. 177. Origin of all such effects lies in type and sizes of atoms or molecules and type and magnitude of bonding. (2) Tendency of different materials esp. polymers to mix homogeneously at a molecular level. Relatively rare effect in polymers, but see **Noryl**.

compensated semiconductor Material in which there is a balanced relation between **donors** and **acceptors**, by which their opposing electrical effects are partially cancelled.

compensation point The temperature (T_{comp}), below the Curie temperature (T_c), at which the magnetization of certain ferrimagnetic materials vanishes. It arises because of differences in the temperature variations of saturation magnetization of the two opposed sublattices in the ferrimagnetic. Such ferrimagnetics can be magnetized below T_{comp} and between T_{comp} and T_c. See **ferromagnetics and ferrimagnetics** p. 128.

complementary metal-oxide-silicon Abbrev. *CMOS*. A major integrated circuit technology based on combinations of p-channel and n-channel **field effect transistors** fabricated on the same silicon substrate. Especially attractive in low-power applications since the basic CMOS logic gate only consumes significant power during switching.

complete reaction A reaction which proceeds until one of the reactants has effectively disappeared.

complex modulus Result from a dynamic mechanical test, where for example, **tensile modulus** E^* is measured. It is related to the real ($'$) and imaginary ($''$) moduli by the equation: $E^* = E' + iE''$. The ratio E''/E' is *tan delta* or the **loss factor**.

compliance The ease with which a body can be deformed elastically. Equals deflection/force, or reciprocal **stiffness**.

components The individual chemical substances present in a system. See **phase rule**.

composite beam A beam composed of two or more materials bonded together and having different moduli of elasticity, eg reinforced concrete beams, **sandwich beams**.

composite material Structural material made of two or more different materials eg **cermets** or **carbon-** or **glass-fibre reinforced plastics**.

composite resistor One formed of a solid rod of carbon compound.

composite structure Any structure made by bonding two or more different materials, such as metal, plastic, composite material etc.

composite yarn Yarn made from a combination of staple fibres and continuous filaments.

composition The nature of the elements present in a substance and the proportions in which they occur, eg mole fraction.

compound (1) See **chemical compound**. (2) Mixture of polymer(s) with various additives such as fillers, pigments, carbon black, dyes, chopped fibre and speciality chemicals (eg anti-oxidant, accelerator, cross-linking agent). Also *blend*.

compounding Process of mixing polymer compounds, often using **extrusion** and usually performed by trade compounders. But see **masterbatch**.

compound magnet A permanent magnet made up of several laminations.

compressed-air capacitor An electric capacitor in which air at several atmospheres' pressure is used as the dielectric, on account of its high dielectric strength at these pressures.

compression moulding Simple polymer process method where powder, granules or semifinished product are put directly between heated tool faces, the tool faces brought together under pressure and the material thus shaped. Widely used for products of both simple (gramophone records formerly) and complex shape (tyres). And also for thermoset polymers, but being increasingly replaced by injection moulding etc.

compression ratio Injection moulding or **extrusion** term characterizing the screw dimensions, ratio of volume of one screw flight at entry to that at discharge. See **injection moulding** p. 169.

compression set Term used to describe permanent **creep** of rubbers under a compressive load.

compression spring A helical spring with separated coils, or a conical coil spring, with plain, squared or ground ends, made of round, oblong, or square-section wire.

compression test A test in which specimens are subjected to an increasing compressive force, usually until they fail by cracking, buckling or disintegration. A stress-strain curve may be plotted to determine mechanical properties, as in a tensile test. Compression tests are often applied to materials of high compressive but low tensile strength, such as concrete. See **strength measures** p. 301.

compression wood Form of **reaction wood** developed in **softwoods** with a higher lignin content, hence darker colour, and more brittle, with lower tensile but higher compressive strength than normal. Cf **tension wood**.

Compton's rule An empirical rule that the melting point of an element in kelvins is equal to half the product of the **relative atomic mass** and the specific latent heat of fusion.

computer A device or set of devices that can store data and a program that operates on the data. A general purpose computer can be programmed to solve any reasonable problem expressed in logical and arithmetical terms.

computer aided design The use of the computer particularly with high-resolution graphics in a wide range of design activities from the design of cars to the layout of **chips**. The designs can be modified and evaluated rapidly and precisely. Abbrev. *CAD*.

computer aided engineering The application of computers to manufacturing processes in which manual control of machine tools is replaced by automatic control resulting in increased accuracy and efficiency. Abbrev. *CAE*.

computer aided machining, manufacture General terms used to describe manufacturing processes which are computer controlled. The data from a **computer aided design** system (CAD) is used directly to produce the program needed for the machine control. Formerly developed as a separate system is now often integrated with CAD to form a complete design and manufacturing facility. Abbrev. *CAM*.

concentration (1) The amount of eg ions or molecules in a given volume, generally expressed as moles per cubic metre or decimetre, or the number of particles in a given volume, expressed as number per cubic metre (or centimetre). (2) The process by which the concentration of a substance is increased, eg the evaporation of a solvent from a solution.

concrete An artificial stone which can be moulded and then allowed to set. Made from a mixture of cement, aggregate and water. See **cement and concrete** p. 52.

condensation The union of two or more molecules with the elimination of a simpler group, such as H_2O, NH_3 etc.

condensed system One in which there is no vapour phase. The effect of pressure is then practically negligible, and the **phase rule** may be written $P + F = C + 1$.

condenser (1) A chamber into which the exhaust steam from a steam engine or turbine is delivered, to be condensed by the circulation or the introduction of cooling water; in it a high degree of vacuum is maintained by an air pump. (2) The part of a refrigeration system in which the refrigerant is liquefied by transferring heat to the cool-

ing medium, usually water or air. (3) Glass chamber cooled by an external water circuit for collecting volatile liquids. (4) An electrical capacitor (now obsolete).

condenser spinning Fibres are carded and the resultant web is divided into narrow strips which are rubbed together on oscillating rubber or leather aprons to form twistless **slubbings**. These are then spun into *condenser yarns*.

condenser tissue Thin rag paper used as a **capacitor** dielectric. Also *capacitor tissue*.

condenser tubes The tubes through which the cooling water is circulated in a surface condenser, and on whose outer surfaces the steam is condensed.

condenser yarn Yarns spun from clean soft waste material; suitable for cotton blankets, quiltings and towellings.

conditioning Allowing materials to reach equilibrium with the surrounding atmosphere. Samples are frequently conditioned in a standard atmosphere before testing. Cf **seasoning**.

conductance The ratio of the current in the conductor to the potential difference between its ends; reciprocal of **resistance**. SI unit is *siemens*, abbrev. *S*. Also *reciprocal ohms* or *mhos*.

conduction band In the **band theory of solids**, a band which is only partially filled, so that electrons can move freely in it, hence permitting conduction of current.

conduction current The current resulting from the flow of charge carriers in a medium in response to a local electric field (same as *drift current*) as distinct from a **displacement current**.

conduction electrons The electrons situated in the **conduction band** of a solid, which are free to move under the influence of an electric field.

conductivity Inverse of **resistivity**. Symbol σ. The ratio of the *current density* in a conductor to the electric field causing the current to flow. It is the **conductance** between opposite faces of a cube of the material of one metre edge. Unit $\Omega^{-1}\,m^{-1}$.

conductivity modulation That effected in a semiconductor by varying a charge carrier density.

conductor (1) A material which offers a low resistance to the passage of an electric current. See **conductors and insulators** p. 70. (2) That part of an electric circuit which actually carries the current. (3) A material which offers a low resistance to the passage of heat. See **thermal conductivity**.

conductors and insulators See panel on p. 70.

configuration Shape of molecules determined by covalent bonds, so invariant unless bonds are broken. In polymers with asymmetric carbon atoms, it is fixed by polymerization, eg atactic or isotactic. See **polymers** p. 247, **stereoregular polymers** p. 299.

conformation Shape of molecules determined by rotation about single bonds, esp. in polymer chains about carbon-carbon links, eg random coils, oriented chains and zig-zag chains. See **elastomers** p. 106.

conformational analysis The study of the relative spatial arrangements of atoms in molecules, in particular in saturated organic molecules. Viewed along a C–C bond, the three other substituents of the nearer carbon atom are said to be eclipsed with respect to those of the further atom if they cover them. Rotating the nearer atom of the bond by 60° with respect to the further will give the more stable staggered conformation.

conformer Particular conformation of a polymer chain or small group of repeat units.

congruent melting Melting at a constant temperature or pressure of a material in which both phases retain the same composition.

connectivity The way in which atoms are bound together in a molecule, ie the configurations of all the atoms. Stereoisomers have the same connectivity, while structural isomers do not.

conservation laws Usually refers to the classical laws of the separate conservation of mass, of energy, of momentum and of atomic species, which are sufficiently accurate for most physical and chemical processes.

conservation of matter See **law of conservation of matter**.

Considère's construction A construction on a graph of **true stress** versus **engineering strain** which allows the prediction of the onset of **necking** in a material under tensile test. The tangent to the stress-strain curve is drawn that passes through the point (true stress = 0, strain = –1). The point where it touches the curve locates the limit of uniform strain. Important in avoiding necking in metal-forming processes.

constantan An alloy of about 40% nickel and 60% copper, having a high volume resistivity and almost negligible temperature coefficient; used as the resistance wire in resistance boxes etc. Also *Eureka*.

constant boiling mixture See **azeotropic mixture**.

constant proportions See **law of constant (definite) proportions**.

constituent Component of alloy or other compound or of a mixture. It may be present

as an element or in chemical, physical or intermediate combination.

constituents All the substances present in a system.

constitution Structural distribution of atoms and/or ions composing a regularly coordinated substance. Includes percentage of each constituent and its regularity of occurrence through the material.

constitutional formula A formula which shows the arrangement of the atoms in a molecule.

constitutional water See **water of hydration**.

constitution changes Changes in solid alloys which involve the transformation of one constituent to another (as when pearlite is formed from austenite), or a change in the relative proportions of two constituents.

constitution diagram Alternative name for **phase diagram** p. 236.

contact adhesive Type of polymer glue which only activates when pressed against the surfaces to be joined.

contact angle The angle between the liquid and the solid at the liquid-solid-gas interface. It is acute for wetting (eg water on glass) and obtuse for non-wetting (eg water on paraffin wax).

contact pad An area of metallization or conducting track in eg a transistor to which an external electrical connection is made.

contact potential The electrical potential which arises in equilibrium at the contact of dissimilar metals. A related phenomenon arises when one or both materials are semiconductors, in which case there is a finite region over which the potential is developed. See **built-in voltage**. A contact potential can also be formed at junctions betweeen similar materials at different temperatures.

contact-potential barrier Potential barrier formed at the junction between regions with different energy gap or carrier concentration.

contact resistance The resistance at the surface of contact between two conductors, influenced by the nature of the materials, the state of the surfaces and the pattern of current flow.

contact scanning Ultrasonic testing procedures in which the ultrasonic head is acoustically coupled to the material being scanned.

containers Reservoirs for materials, solid or fluid, which must be made of materials themselves inert to both the contents and external environment. Thermoplastics like polyolefins have replaced many traditional materials (eg glass, mild steel) for their ease

conductors and insulators

The primary electrical classification of materials is in terms of conductivity or its reciprocal, resistivity. Values of electrical conductivity range from around 10^8 S m^{-1} for silver to 10^{-17} S m^{-1} for polystyrene. Materials with values towards the extreme ends of this range are generally described as *conductors* and *non-conductors* (or *insulators*), as appropriate. Materials with intermediate values of conductivity are said to be *semiconductors*. See figure below.

The conductivity of a material is dependent on its purity and temperature and also the frequency of the electrical stimulus. The figure illustrates the range with various categories and examples corresponding with room temperature, direct current conditions. The conductivity of metals and extrinsic semiconductors tends to decrease with increasing temperature while that of insulators and intrinsic semiconductors tends to increase.

Direct current conductivity is associated with the drifting of charge in a steady gradient of electric potential. The charge carrier may be described as a delocalized electron (metals), an electron in a conduction band or a hole in a valence band (semiconductors), a hopping electron (from one ion to another in ceramics) or else cations or anions (solid and liquid electrolytes). Alternating current effects are discussed under **dielectrics and ferroelectrics** p. 92.

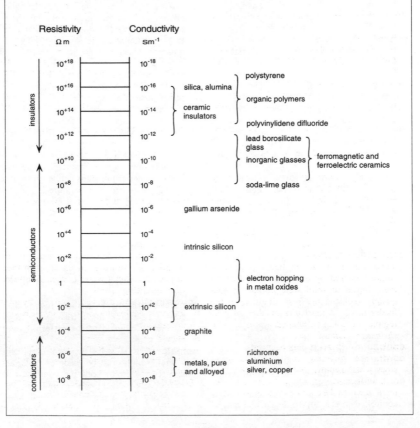

of shaping, low density and chemical resistance. Some grades are however susceptible to **environmental stress cracking**.

continuous beam A **beam** supported at a number of points and continuous over the supports, as distinct from a series of simple independent beams. Also *continuous girder*.

continuous casting Method in which the molten metal is added at the top of the mould while the externally solidified material is withdrawn from the bottom, the mould and emerging product being cooled by water or air jets.

continuous cooling transformation diagram See panel on p. 72.

continuous filament yarn Yarn composed of one or more unbroken filaments. Man-made fibres are usually made as continuous filaments although these may be cut or broken into staple fibres for subsequent manufacture into fabric. Silk is a natural continuous filament. Cf **staple fibres**.

continuous furnace Furnace in which the charge enters at one end, moves through continuously, and is discharged at the other.

continuous mill A rolling-mill consisting of a series of pairs of rolls in which the stock undergoes successive reductions as it passes from one end to the other end of the mill. See **pull-over mill, reversing mill, three-high mill**.

continuous processing Method of producing an article continuously and, in theory, indefinitely. See **extrusion, continuous casting**.

continuous vulcanization Method of processing thermoset rubber product so that it is produced continuously in final cross-linked form. Needs great care to ensure optimum cure, since overcured product is difficult if not impossible to recycle. Simple examples include extruded pipe, more complex that for making conveyor belting. See **Rotocure**.

contraction The percentage shrinkage after processing or heat treatment.

contraction cavities Porous zones in metal castings, usually in the last portions to solidify, caused by the volume contraction from liquid to solid state not being adequately replaced by fresh liquid from the feeder head. Almost unavoidable in some regions of complex shaped castings. Also *shrinkage cavities*.

contraction in area See **necking**.

contraries Anything in the pulp or paper stock which is unwanted in the paper.

controlled cooling Methods of heat treatment in which the cooling cycle is accurately controlled so as to impart the desired properties or structure.

controlled degradation Chemical type of analysis for helping to identify polymers, esp. thermosets, which often have functional groups susceptible to hydrolysis. Soluble fragments may then be identified.

convection current Current in which the charges are carried by moving masses appreciably heavier than electrons.

conversion The fraction of some key reagent which has been used up, hence a measure of the rate or completeness of a chemical reaction.

conversion coating A coating on a metal produced by chemical treatment eg chromatization and phosphating, which impart properties like corrosion resistance. See **passivation, activation**.

converting (1) Removal of impurities from molten metal by blowing air through the melt in eg the **Bessemer process**. (2) Producing a **sliver** of staple textile fibres from a continuous filament tow by cutting or breaking.

cooling analysis Method of analysing cooling time in moulding of polymers, important because it often forms the larger part of the total cycle time. Uses cooling curves and material data (thermal diffusivity, heat distortion temperature) plus product dimensions (usually greatest thickness) to calculate cooling time from estimated **Fourier number**. Product redesign can then be undertaken to reduce maximum thickness, and so increase productivity. At the same time, care is needed to ensure that product stiffness and strength remain within specification. Also needed for used rubber products, eg. tyres, where cure kinetics are critical. May be backed up by direct temperature measurement within tyre using thermocouples. See **injection moulding** p. 169.

cooling circuit System of water tubes within mould tool for maintaining it at a constant pre-set temperature. It thus ensures product reproducibility. Chilled water is usually used, but some engineering plastics demand high temperatures to minimize orientation, so hot water is used. See **injection moulding** p. 169.

cooling curves Curves obtained by plotting time against temperature for a material cooling under constant conditions. The curves show the evolutions of heat which accompany solidification, polymorphic changes in pure metals and various transformations in alloys.

Cooper pairs In a *superconducting* material below its critical temperature, the electrons do not act independently but in

continuous cooling transformation diagrams

Also *CCT diagrams*. These represent transformation characteristics for materials which undergo solid state transformations. They carry time temperature transformation data (*TTT diagrams*) but are presented with start and end of transformation lines relating to the cooling of round bars of varying diameter. Fig. 1 below is a typical example for a medium-carbon low alloy steel. See **isothermal transformation diagrams** p. 177.

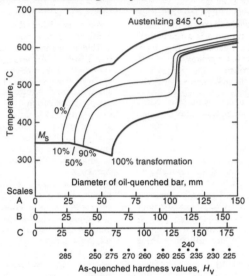

Fig. 1. **Transformation characteristics of a steel bar as a function of diameter and position in the bar**. Scales A, B and C refer to axial, mid-radial and near-surface positions respectively.

Fig. 2. **Structures present in 'as-quenched' bars**.

The abscissa is bar diameter, but with three separate scales indicating the near-surface, mid-radius and axis positions for particular sizes of bar when quenched in oil. The ordinate is a linear scale for reading off temperatures at which transformation will begin or end. Thus referring to the axis position of a 125 mm diameter bar, transformation during the quench will begin at 650°C, while the near surface of the same size bar will begin at about 630°C and finish at 420°C.

In steels, the various transformation products have vastly different hardnesses and mechanical properties so, below the bar diameter, is another scale giving *as-quenched* hardness values associated with the transformation products produced in various positions within a bar. The shaded diagram in Fig. 2 indicates the microstructural constituents which give rise to these hardnesses.

CCT diagrams thus summarize in a practical way the transformation behaviour represented by a TTT diagram. They are extremely useful in selecting for *hardenability* those low alloy steels which are to be hardened and tempered in order to achieve specified levels of mechanical properties in various sizes of bar.

dynamic pairs. The electrons are weakly bound together and form *Cooper pairs*. The BCS (Bardeen, Cooper, Schrieffer) theory uses this concept to give a detailed microscopic theory of superconductivity. See **superconductors** p. 308.

co-ordinate bond See **covalent bond**.

co-ordination compound A compound generally described from the point of view of the central atom to which other atoms are bound or co-ordinated and are called ligands. Normally, the central atom is a (transition) metal ion, and the ligands are negatively charged or strongly polar groups.

co-ordination number The number of atoms or groups (ligands) surrounding the central (nuclear) atom of a complex ion or molecule.

co-ordination polymerization See **chain polymerization** p. 56.

copolymer See panel on p. 75.

copolymer equation Equation which relates structure of copolymers to propagation rate constants etc. See **chain polymerization** p. 56, **copolymer** p. 75.

oopolymerization See **copolymer** p. 75.

copper Bright, reddish metallic element, symbol Cu. Crystallizes in the face centred cubic system. There are several grades of commercially pure copper, all of which are ductile, with high electrical and thermal conductivity, good resistance to corrosion; it has many uses, notably as an electrical conductor and in plumbing and water systems. Basis of brass, bronze, aluminium bronze and other alloys. See **copper alloys** p. 76.

copper alloys See panel on p. 76.

copper loss The loss occurring in electric machinery or other apparatus due to the current flowing in the windings; it is proportional to the product of $(\text{current})^2 \times$ resistance.

copper (II) sulphate $CuSO_4$. Bluestone; a salt, soluble in water, used in copper-plating baths; formed by the action of sulphuric acid on copper; crystallizes as hydrous copper sulphate, $CuSO_4.5H_2O$, in deep-blue triclinic crystals.

coppersmith's hammer A hammer having a long curved **ball-pein** head, used in dishing copper plates.

coquille Glass in thin curved form used in the manufacture of sun glasses. The radius of curvature is traditionally 3.5 in (90 mm). Similar glass of 7 in (180 mm) radius is called *micoquille*.

cord Non-metric timber measure, 128 ft^3 ($8 \times 4 \times 4$ ft, about 3.6 m^3).

corduroy Strong, hard-wearing cloth having a rounded or flattened cord or rib of weft pile running longitudinally; made entirely from cotton, or cotton warp and spun-rayon pile.

cordwood Tree trunks of medium diameter sawn into uniform lengths.

core (1) Region associated with a coil: may be air or a magnetic material to increase inductance. Typical materials are ferrites and punched laminations of soft iron. Construction may be as a complete magnetic circuit (divided to accommodate the coil) or simply a rod inserted into the coil. (2) Assembly consisting of the conductor and insulation of a cable but not including the external protective covering. An arrangement comprising many such assemblies is termed *multicore*. (3) In an atom, the nucleus and all complete shells of electrons. In the atoms of the alkali metals, the nucleus, together with all but the outermost of the planetary electrons, may be considered to be a core, around which the valency electron revolves in a manner analogous to the revolution of the single electron in the hydrogen atom around the nucleus. In this manner, the simple **Bohr model** may be made to give an approximate representation of the alkali spectra. See **atomic structure**. (4) A solid mass of specially prepared sand or loam placed in a mould to provide a hole or cavity in the casting. (5) A steel form in an **injection moulding** tool which creates a cavity in the final product. Side cores are retractable to allow removal of the moulding. Also *fusible core*, *rotating core*.

cored hole A hole formed in a casting by the use of a core, as distinct from a hole that has been drilled.

cored solder Hollow **solder** wire containing a **flux** paste, which allows flux and solder to be applied to the work simultaneously.

core-spun yarn Yarn made with a core (usually of continuous filaments) surrounded by a sheath generally made of staple fibres.

Corfam TN for **poromeric** material formerly used in shoe uppers.

cork The naturally occurring, long exploited, **cellular solid**, density approx. 170 kg m^{-3}, which occurs as a thin layer in the bark of all trees. The cork oak (*Quercus suber*) is unique in having this layer several cm thick. Used for thermal and acoustic insulation, flooring, packaging, as elastic and chemically inert closures for bottles, and for its buoyancy.

cork rubber Rubber to which cork granules have been added to increase bulk and insulation, much used for gaskets.

corona Phenomenon of air breakdown when electric stress at the surface of a conductor exceeds a certain value. At higher values, stress results in luminous discharge.

corona discharge Method of etching polymer surfaces (esp. polyolefins) by electrical discharge. Used for providing chemically active surface ready for printing etc. Other thermoplastics such as acrylonitrile-butadiene-styrene (ABS) can be etched chemically with, for example, chromic acid.

corrosion Chemical degradation of metals and alloys due to reaction with agent(s) in the service environment, eg the rusting of steel in moist air. Eventual failure results from the wasting away of cross-section. May take the form of uniform attack over the whole surface or as highly localized pitting. Sometimes attack may be intergranular and very rapid. See **rusting** p. 271, **stress corrosion**.

corrosion fatigue Acceleration of weakening of structure exposed to cyclic stress by a combination of chemical penetration and fatigue. See **fatigue** p. 123.

corrugated board A layered packaging material produced by sticking a suitable liner to both sides of a fluted paper or papers.

cotton The seed hairs of the cotton plant of which there are many varieties (*Gossypium* spp.). The fibres vary in length according to variety and country of origin but on average are about 20–30 mm. Sea Island and Egyptian cottons are longer and are used for making high-quality fine fabrics and sewing threads.

Cotton–Mouton effect Effect occurring when a dielectric becomes double-refracting on being placed in a magnetizing field H. The retardation δ of the ordinary over the extraordinary ray in traversing a distance l in the dielectric is given by $\delta = C_m \lambda l H^2$ where λ is the light wavelength and C_m is the Cotton–Mouton constant.

cottonwood A wide girthed, N American species of **poplar**.

cotton wool Loose cotton or viscose rayon fibres which have been bleached and pressed into a sheet; used as an absorbent or as a protective agent. Medicated cotton wool sometimes has a distinguishing colour to indicate its special property.

couching Separating the newly formed wet sheet or web of paper from the forming surface and transferring it to a felt.

coulomb SI unit of electric charge, defined as that charge which is transported when a current of one ampere flows for one second. Symbol C.

coulomb force Electrostatic attraction or repulsion between two charged particles.

coulomb scattering Scattering of particles by action of coulomb force.

Coulomb's law Fundamental law which states that the electric force of attraction or repulsion between two point charges is proportional to the product of the charges and inversely proportional to the square of the distance between them. The force also depends on the **permittivity** of the medium in which the charges are placed. In SI units, if Q_1 and Q_2 are the point charges a distance d apart, the force is

$$F = \frac{1}{4\pi\varepsilon}\frac{Q_1 Q_2}{d^2}$$

where ε is the permittivity of the medium. The force is attractive for charges of opposite sign and repulsive for charges of the same sign.

count, count of yarn See **yarn count**.

counter An instrument for recording the number of operations performed by a machine, or the revolutions of a shaft.

counterions See **gegenions**.

couple A system of two equal but oppositely directed parallel forces. The perpendicular distance between the two forces is called its arm and any line perpendicular to the plane of the two forces its axis. The *moment* of a couple is the product of the magnitude of one of its two forces and its arm. A couple can be regarded as a single statical element (analogous to force); it is then uniquely specified by a vector along its axis having a magnitude equal to its moment. Couples so specified combine in accordance with the parallelogram law of addition of vectors.

coupling (1) A device for connecting two lengths of hose etc. (2) A device for connecting two vehicles. (3) A connection between two co-axial shafts, conveying a drive from one to the other.

coupon An extra piece, attached to a forging or casting, from which a test piece can be prepared.

courbaril Hardwood (*Hymenaea courbaril*) from the American tropics whose **heartwood** is reddish- to orange-brown with darker brown and red streaks and a medium texture. The timber is hard, tough, shock-resistant and moderately durable, but is not easy to work. Typical uses are in furniture, cabinet-making, tool handles and ship-building. Mean density 910 kg m^{-3}.

course A row of loops across the width of a knitted fabric.

course density The number of courses per unit length measured along a wale of the knitted fabric.

course length The length of yarn in one course of a knitted fabric.

Courtelle TN for synthetic fibres based on

copolymer

Commercial polymeric materials, whether **thermoplastic** or **thermoset** are frequently copolymers because this is a way of modifying the physical properties of the parent **homopolymer**. Thus styrene-acrylonitrile (SAN) possesses a **glass transition** temperature (T_g) above the boiling point of water and so will remain rigid when used eg for holding hot coffee. Polystyrene possesses a T_g of about 97°C, so will collapse when exposed to boiling water at 100°C. Polystyrene is a brittle material at ambient temperatures, but copolymerization of styrene with polybutadiene gives high-impact polystyrene, a tougher material which can be used for stressed applications. An even tougher terpolymer, ABS plastic, is created by copolymerizing styrene and acrylonitrile with poly-

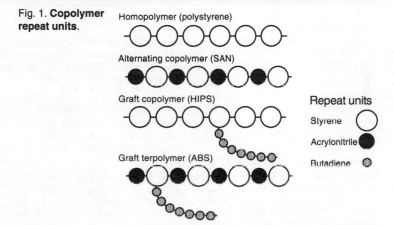

Fig. 1. **Copolymer repeat units**.

Homopolymer (polystyrene)

Alternating copolymer (SAN)

Graft copolymer (HIPS)

Graft terpolymer (ABS)

Repeat units

Styrene ◯

Acrylonitrile ●

Butadiene ◉

butadiene.

The structure of copolymers is described by the sequence of repeat units along the backbone chain (Fig. 1).

By increasing the proportion of butadiene to styrene units, elastomers will be produced. At about 75% by weight of polybutadiene, styrene-butadiene rubber is the most common synthetic rubber and widely used in car tyre treads. It is a *random copolymer* where the repeat units are dispersed irregularly along the linear chains (Fig. 2).

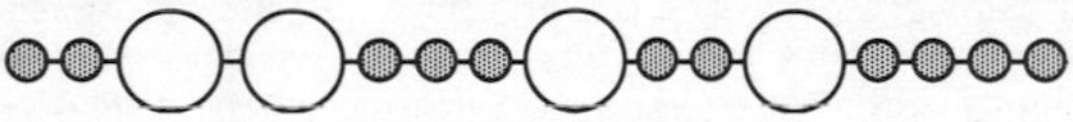

Fig. 2. **Random copolymer**.

The material is vulcanized in the conventional way to produce a network polymer. Alternatively, a different kind of material not requiring vulcanization can be produced by anionic polymerization where a much more regular *block copolymer* such as styrene-butadiene-styrene is produced (Fig. 3).

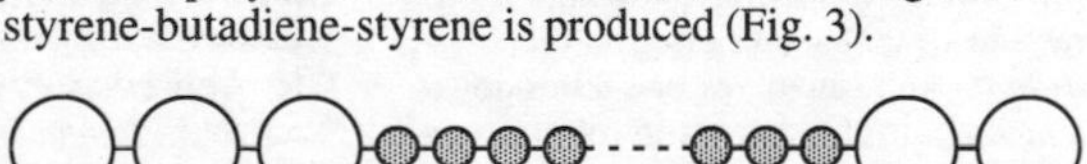

Fig. 3. **Block copolymer**.

Because most polymers are incompatible with one another, the polystyrene chains segregate together to form domains within the elastomeric matrix of polybutadiene. Every polybutadiene chain is anchored in a rigid domain, therefore the material behaves like a crosslinked rubber yet can be processed by conventional injection moulding. As chemical crosslinks are absent, such materials are known as *thermoplastic elastomers* (TPE).

copper alloys

The use of copper alloys dates back to 3000 BC, so not surprisingly there have evolved numerous names for the alloys needed for specific applications together with a wide range of proprietary names for various coppers, brasses and bronzes many of which are essentially similar.

The principal fields of application are:

Electrical conductors, where the alloy content has to be kept low in order to preserve copper's inherently high electrical conductivity, second only to silver.

Chemical applications and heat exchangers exploiting corrosion resistance and high thermal conductivity.

Ornamental and architectural purposes, where a pleasing range of colours may be obtained by appropriate alloying, ranging from salmon pink, through the reds and yellows of gilding metals and bronzes, to the green and silvery tinges of the high zinc and high tin alloys. Additionally, copper cladding develops an attractive greenish-blue patina on exposure to the atmosphere.

In general engineering, copper alloys are used on account of their excellent formability and ductility and as joining alloys for hard soldering and brazing. Copper-tin alloys in powder form are the basis for the manufacture of sintered, self lubricating bearings, extensively used in small electric motors and similar applications.

Current international standards adopt a system based on letters for the main alloying base followed by a number designating the particular alloy, eg CZ stands for copper-zinc (brasses), PB for phosphor bronze, LG for leaded gunmetal, CT for copper-tin (bronzes), AB for aluminium bronze, CN for copper-nickel, and so on.

All the traditional copper alloys are represented in the modern specification system, but many are still referred to by their application names as shown on the next page. This list includes the most common names and their principal alloying ingredients. Proprietary and trade names are omitted.

continued on next page

---

polyacrylonitrile.

coutil Strong, closely woven, herringbone cloth, often of cotton and used for corsetry.

covalency The union of two or more atoms by the sharing of one or more pairs of electrons

covalent bond A chemical bond in which two or more atoms are held together by the interaction of their outer electron clouds. A *co-ordinate bond* involves one atom donating a spare pair of electrons to form a covalent bond. See **bonding** p. 33.

covalent radius Half the internuclear separation of two bonded like atoms. For bonds between unlike atoms, approximate bond lengths may be derived from the sum of the covalent radii for the two bonded atoms.

cover paper A heavy paper or board, generally of distinctive appearance eg coloured and/or embossed and intended for use as the cover of booklets, pamphlets, menus etc.

Cowles dissolver Type of blender for polymers using chamber enclosing rotating impellor blade.

CPVC Abbrev. for **Chlorinated PolyVinyl Chloride.**

CR Abbrev. for **Chloroprene Rubber** or *polychloroprene.*

CR 39 Abbrev. for *Columbia Resin 39.* TN for tough, aliphatic polycarbonate used mainly for spectacle lenses and sunglasses. Thermoset based on allylic resin with high scratch-resistance.

crabwood See **andiroba.**

crack driving force Same as **toughness** or

Nominal compositions (wt%) of some copper alloys					
Name	Zinc %	Tin %	Lead %	Nickel %	Other %
Admiralty gunmetal	2	10			
Admiralty brass	30	1			
Aluminium brass	30				2 Al
Aluminium bronze				2	10 Al, 2 Fe
American brass	35		3		
Architectural bronze	44				Fe + Mn
Arsenical copper					0.5 As
Bearing bronze		11	2		
Bell metal		22.5			
Beryllium copper					2 Be
Binding brass	35		1.5		
Bobiere's metal	37				
Cadmium copper					0.6 Cd
Cap copper	3.5				
Cathode copper (99.5% +)					
Cartridge brass	30				
Chinese art metal	10	1	17.5		
Chinese bronze		22.5			
Chromium copper					0.6 Cr
Clock brass	37		2		
Commercial bronze		10			
Common brass	35–38				
Coinage bronze	1	4			
Constantan					45 Ni
Copper lead			50		
Cupro-nickel					5–30 Ni
Ferry					45 Ni
German silver	18–24				
Gilding metals	5–15				
Hardware bronze	13		2		
Heusler alloy (magnetic)					15 Al, 30 Mn
High-tensile brass	40				Al+Mn+Fe+Sn
Jewellery bronze	12.5				
Leaded gunmetal	2	10	4		
Low brass	20				
Manganese bronze	30				4 Al, 3 Mn, 3 Fe
Manganin					4 Ni, 12 Mn
Muntz metal	38		3		
Naval brass	40	1.25			
Nickel silver	18–25				8–12 Ni
Phosphor bronze		12	0.5		0.5 P
Post Office bronze		1			
Red brass	15				
Sifbronze (brazing alloy)	39				0.25 Si
Silicon brass	40				0.3 Si
Silver copper					0.5 Ag
Spinning brass	37		1		
Tellurium copper					0.5 Te
Tobin brass	40	1			
Tobin bronze	39	0.75	1		
Tough pitch copper (99.5%)					0.5 O
Valve bronze	6	6	4.5		
Yale bronze	8	1	1		
Yellow brass	40				

critical strain energy release rate. See **strength measures** p. 301.

cracking Controlled breakdown of naphtha to give light olefines such as ethylene, propylene, butylenes by heat and pressure (thermal cracking). Such products can be polymerized after purification to give polymeric materials. Term also used for catalysed process (catalytic cracking), which gives different yield. Ethane, propane etc from liquefied petroleum gas and natural gas can be similarly processed to give monomers.

crackled Glassware whose surface has been intentionally cracked by water immersion and partially healed by reheating before final shaping.

crank An arm attached to a shaft, carrying at its outer end a pin parallel to the shaft; used either to give reciprocating motion to a member attached to the pin, or in order to transform such motion into rotary motion of the shaft.

crankshaft The main shaft of an engine or other machine which carries a crank or cranks for the attachment of connecting-rods.

crankshaft motion Form of short chain movement in polyethylene polymers, where pairs of repeat units collaborate in rotation like eponymous engine part. Detected as an absorption peak in the infrared.

cràpe See **crêpe**.

craze Microfeature associated with fracture and failure of polymers. Consists of voided and oriented material formed at crack tip or rubber particles. See **rubber toughening** p. 270. Polymers of high molecular mass form stronger crazes than low molecular mass polymer, so explaining the strength of ultra-high molecular mass polyethylene (UHMPE), for example. Also gives rise to the **strain-whitening** of rubber-toughened polymers.

crazing (1) Pattern of hair cracks that can develop on the surfaces of concrete (esp. precast), paints, varnishes and glazed ceramics (where it may be introduced deliberately for decoration). (2) Effect in polymers, esp. thermoplastics, where surfaces exposed to certain fluids (crazing agents) show crack-like defects. Driving force supplied by **frozen-in strain**. Also associated with cracks formed prior to fracture or failure in absence of fluids. See **craze.**

cream-laid White writing paper made with a laid **watermark**.

cream-wove White writing-paper, in the manufacture of which a wove **dandy roll** has been used.

creep The continuously increasing deformation of materials with time under steady load. It can either be due to irreversible plastic deformation such as exhibited by oure metals above about $0.3 \, T_m$, or due to reversible, viscoelastic deformation as exhibited esp. by polymeric materials. See panel on p. 79.

creep curve Presentation of creep data in the form of strain versus time graph, at various different stress levels for a given temperature. Isochronous stress-strain and isometric creep stress curves can be derived by interpolation. Usually commonly used for viscoelastic polymers. See **creep** p. 79, **viscoelasticity** p. 332.

creep modulus Measure of modulus of material, esp. polymeric, determined from elongation E of specimen (for tensile creep modulus) under a constant applied load. Given by equation: $E_c(t) = \sigma/\mathrm{strain}(t)$ where σ is the constant applied stress. Since it is a viscoelastic quantity, the time-scale of measurement must be quoted if it is to be used in **quasi-elastic calculations**. See **viscoelasticity** p. 332.

creep rupture Type of failure in materials esp. polymers, where constant applied load causes sample to elongate and finally fail by parting. Often involves slow crack growth. See **static fatigue**.

creep strength The ability of a material to resist deformation under constant stress, measured as the amount of creep induced by a constant stress acting for a given time and temperature. See **creep** p. 79.

creep tests Methods for measuring the resistance of materials to creep by determining time-extension curves under constant loads.

crêpe A woven fabric with a distinctive rough, crinkled appearance because of the special high-twisted yarns from which it is made. Similar fabrics are also made by warp- and weft-knitting. Also *cràpe*.

crêpe de chine A light crêpe fabric made from continuous filament yarns. Also *cràpe de chine*.

crêpe paper Crinkled paper produced by doctoring the moist web from a supporting cylinder, so increasing elongation in the machine direction. Used principally for packaging and industrial applications but also for decorative purposes. Also *cràpe paper*.

crêpe rubber Raw, unvulcanized sheet rubber, not chemically treated in any way.

cresols A common name for the hydroxytoluenes, $CH_3.C_6H_4.OH$, monohydric phenols.

cretonne A heavy printed cotton fabric

creep

Represented by the curve plotted to display the variation of plastic strain with time at constant temperature and under constant force, usually in tension. The curve has three distinct stages numbered from one to three and referring to *primary*, *secondary* and *tertiary creep* (see Fig. 1a). Families of such curves are often plotted on the same axes in order to show the effect of temperature.

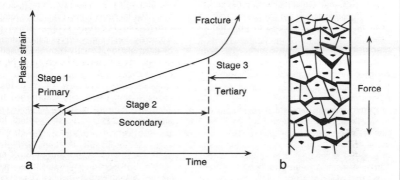

Fig. 1. (a) **The three stages of creep**. (b) **Grain separation and void formation in the tertiary stage**.

All such creep is permanent plastic deformation of the material, although the test piece will contract elastically if the load is removed. Stage 1 creep is *transient* and starts at a high rate that quickly diminishes with time. This leads into the *secondary* or *steady state creep* in stage 2, where the rate remains constant with time. Stage 2 may be maintained for the remainder of the test but, more usually, the rate increases as stage 3 is reached and then continues to accelerate to the point of fracture.

Each stage of the creep curve is associated with different mechanisms of deformation within the material. Essentially stage 1 is a redistribution of dislocations assisted by thermally activated climb and recovery; stage 2 represents a running balance between work hardening and recovery, with diffusion controlled creep and grain boundary slidings involved. In stage 3 grain separation and the formation of voids (Fig. 1b) reduce the effective cross-sectional area supporting the load which then leads to the rate accelerating up to fracture. Unlike a tensile fracture where the cross-section visibly reduces by necking prior to the final break, creep fractures show much less change in the original area because the reduction in effective cross-section is accounted for by microscopic internal voids. Oxidation may also play a part in certain types of material.

Increasing temperature speeds up the above mechanisms and thus leads to strain effects presenting at higher rates and in shorter times. At temperatures below $0.3\ T_m$ (see **homologous temperature**), stage 3 may never appear, whereas at very high temperatures (above $0.9\ T_m$) all three stages of the curve may overlap to a degree where none is separately distinguishable. See **stress-rupture curves**.

Creep also occurs in viscoelastic materials (eg polymers) but different mechanisms apply and the creep strains are partially or wholly recoverable on unloading. See **viscoelasticity** p. 332.

often used as a furnishing fabric.

crevice corrosion In a liquid-containing system, the acceleration of corrosive attack encountered in crevices and cracks which are partly segregated from the main flow and where build up of ions and salts or oxygen deficiency may occur. See **rusting** p. 271.

crimp The waviness of a fibre, measured as the difference between the straightened and crimped fibre expressed as a percentage of the straightened length. The crimp of a yarn in a woven fabric is measured similarly by comparing the length of the fabric with the length of the yarn removed from it and straightened.

Crimplene TN for polyester fibre with artificial crimp made by heat treating duplex co-extrudate.

crimping Pressing into small regular folds or ridges in: (1) the reduction of cross-section of a bar material by progressively corrugating it along its surface to give an increase in length; (2) bending or moulding to a required shape; (3) folding or bending sheet metal to provide stiffness.

critical cooling rate In the heat treatment of steels, the rate of cooling required to prevent nucleation of **pearlite** and to secure the formation of **martensite** in steel. With carbon steel, this means cooling in cold water, but the critical rate is reduced by the addition of other elements, hence the use of **oil-** and **air-hardening steels**.

critical range In steels, the range of temperature in which the reversible change from **austenite** (stable at high temperature) to **ferrite**, **pearlite** and **cementite** (stable at low temperature) occurs. The upper limit varies with carbon content; the lower limit for slow heating and cooling is about 700°C.

critical strain energy release rate Same as **toughness**. See **strength measures** p. 301.

critical stress intensity factor Same as **fracture toughness**. See **strength measures** p. 301.

critical temperature Temperature at which a sudden transition occurs such as a **phase change** or the onset of **superconductivity**.

crizzling Fine cracks in the surface of glass, due to thermal stresses set up by localized chilling during manufacture. See **crazing**.

CRO Abbrev. for **cathode ray oscilloscope**.

cropping The operation of cutting off the end or ends of an ingot to remove the **pipe** and other defects.

cross direction In a paper sheet or web, the direction at right angles to the machine direction in the plane of the sheet.

crosslink density Density of chemical crosslinks in a polymer. In rubbers, the **kinetic theory of elasticity** predicts that the **shear modulus**, G, is directly proportional to the crosslink density. Also *degree of crosslinking*.

cross-linking The formation of side bonds between different chains in a polymer, thus increasing its rigidity. Usually achieved in rubbers with sulphur to give disulphidic or polysulphidic links between vicinal carbon atoms. Physical cross-links are used in **thermoplastic elastomers** to provide elastomeric properties. See **elastomers** p. 106, **thermosets** p. 318.

cross-ply Tyre in which reinforcement belting layers are inclined to one another, and composed only of polymer fibre, eg rayon. Generally of lower life than radial tyre, in which steel belting is also used. See **tyre technology** p. 325.

cross-section (1) In atomic or nuclear physics, the probability that a particular interaction will take place between particles. The value of the cross-section for any process will depend on the particles under bombardment and upon the nature and energy of the bombarding particles. Suppose N_0 particles per second are incident on a target area A containing N particles and $N_{0'}$ of the incident particles produce a given reaction, then if $N_{0'} \ll N_0$, $N_{0'} = N_0 N\sigma/A$ where σ is the cross-section for the reaction; σ can be imagined as a disk of area σ surrounding each target particle. Measured in **barns**. (2) The section of a body (eg a girder or moulding) at right angles to its length; a drawing showing such a section.

cross slip The transfer of **glide** of a **dislocation** from one slip plane to another during deformation or thermal recovery.

crown See **paper sizes**.

crown glass (1) Glass made in disk form by blowing and spinning, having a natural fire-finished surface but varying in thickness with slight convexity, giving a degree of distortion of vision and reflection. Cf **bullion point**. (2) That class of optical glasses with $v_d > 50$ if $n_d > 1.60$ and $v_d > 55$ if $n_d < 1.60$, where v_d is the Abbé number, or reciprocal of the **dispersivity**, and n_d is the **refractive index** at 587.6 nm. Cf **flint glass**. (3) Loosely used for **soda-lime-silica glass**.

CRT Abbrev. for *Cathode Ray Tube*.

crucible A refractory vessel or pot in which metals are melted. In chemical analysis, smaller crucibles, made of porcelain, nickel or platinum, are used for igniting precipitates, fusing alkalis etc.

crucible furnace A furnace, fired with coal, coke, oil, or gas, in which metal contained in crucibles is melted.

crucible steel Now obsolete. Steel made by melting **blister bar** or **wrought iron**, charcoal and ferro-alloys in small (45 kg) crucibles This was the first process to produce steel in a molten condition, hence the product was called *cast steel*. Now replaced by electric-furnace steelmaking.

crucible tongs Tongs used for handling crucibles.

crude oil See **petroleum**.

cryoscope Instrument for the determination of freezing or melting points.

cryoscopic method The determination of the **molecular mass** of a polymer by observing the lowering of the freezing point of a suitable solvent. See **GPC**.

crypto- Prefix from Gk *kryptos*, hidden.

crystal Solid substance showing some marked form of geometrical pattern, to which certain physical properties, angle and distance between planes, refractive index etc can be attributed.

crystal diamagnetism Property of negative susceptibility shown by silver, bismuth etc. See **characterizing magnetic materials** p. 57.

crystal glass A colourless, highly transparent glass of high **refractive index**, which may be 'lead crystal', a somewhat misleading term as it means different things in different glassmaking districts. See **lead glass**.

crystal growing Technique of forming semiconductors by extracting crystal slowly from molten state. Also *crystal pulling*. See **Czochralski process**.

crystal indices See **Miller indices**.

crystal lattice Three-dimensional repeating array of points used to represent the structure of a crystal, and classified into 14 groups by Bravais.

crystalline solid A solid in which the atoms or molecules are arranged in a regular manner, the values of certain physical properties depending on the direction in which they are measured. When formed freely, a crystalline mass is bounded by plane surfaces (faces) intersecting at definite angles. See **X-ray diffraction**.

crystallinity See **degree of crystallinity**.

crystallites Very small, often imperfectly formed crystals.

crystallizable polymers Those polymers capable of crystallizing due to stereoregular chains. See **crystallization of polymers** p. 82.

crystallization Formation of crystals from melt or solution.

crystallization of polymers See panel on p. 82.

crystallizing rubbers Those elastomers with stereoregular chains (eg isotactic or cis-repeat units) which are capable of crystallizing. They include natural rubber and cis-polybutadiene but not styrene butadiene rubber. Crystallization may occur simply in storage (see **stark rubber**) or by strain-crystallization.

crystallography Study of internal arrangements (ionic and molecular) and external morphology of crystal species, and their classification into types.

crystal melting point Phase transition where crystals melt to form liquid, symbol T_m. For small molecule substances, usually identical with freezing or crystallization point. For polymers it can be substantially different. See **crystallization of polymers** p. 82, **polymer melting**.

crystal momentum The product of the Dirac constant, $\hbar$ ($= h/2\pi$) and the wavevector q of a *phonon* in a crystal. For a *photon* the product $\hbar k$, where k is its wavevector, is the momentum it carries, as its frequency is proportional to the magnitude of k. This is not so for a phonon, so the crystal momentum is just a useful fiction used in the discussion of scattering processes. Also *pseudo-momentum*.

crystal nuclei The minute crystals whose formation is the beginning of crystallization.

crystal oscillator Valve or transistor oscillator in which frequency is held within very close limits by rapid change of mechanical impedance (coupled piezoelectrically) when passing through resonance.

crystal pick-up See **piezoelectric crystal**.

crystal pulling See **crystal growing**.

crystal rectifier One which depends on differential conduction in semiconducting crystals, suitably 'doped', such as Ge or Si.

crystal structure The whole assemblage of rows and patterns of atoms, which have a definite arrangement in each crystal. The arrangement in most pure metals may be imitated by packing spheres, and the same applies to many of the constituents of alloys. See **close packing of atoms** p. 62.

C-stage Final stage in curing process of phenol formaldehyde resin, characterized by infusibility and insolubility in alcohol or acetone. See **thermosets** p. 318.

CTR Abbrev. for **Comparative Tracking Resistance**. See **tracking resistance**.

Cuban mahogany See **American mahogany**.

cubeoctohedron A solid body made up of eight square faces joined at the corners, connected by eight equilateral triangles. It can also be thought of as a cube with each corner

crystallization of polymers

Polymer materials are not intrinsically crystallizable: chain irregularity found in random copolymer units or *atacticity* prevent the close approach of chains needed to form crystallites. But when stereoregular chains are found, as in isotactic polystyrene or polypropylene, then crystallization can occur, resulting in an improvement in thermal and mechanical properties that is usually offset by a loss in optical properties, higher density and greater contraction after moulding. The *crystal melting point* (T_m) is *always* higher than the *glass transition temperature* (T_g), and correlation of the two transitions suggests that T_g (kelvin)~ 2/3 T_m (kelvin) although there is a wide divergence.

The rate of crystallization depends on molecular mass and the temperature of crystallization, reaching a maximum about halfway between T_g and T_m and being greatest for lower mass polymers. Thus polyethylene terephthalate ($T_g = 65°C$, $T_m = 276°C$) crystallizes most rapidly at about 175°C, a feature which controls the injection blow moulding process for drink containers. High rates must be avoided to ensure clarity of the container walls.

At a molecular level, polymers crystallize to form chain folded *lamellae*, which themselves twist and branch to form *fibrils*, then a *wheatsheaf*, which in turn grows into a *spherulite* (see Fig. 1 below). Impingement of spherulites finally occurs, although amorphous material between the lamellae and fibrils prevents 100% crystallization. But see **high performance fibres** p. 158. Under certain conditions eg high pressure, extended chain crystals can be formed in crystallizable polymers.

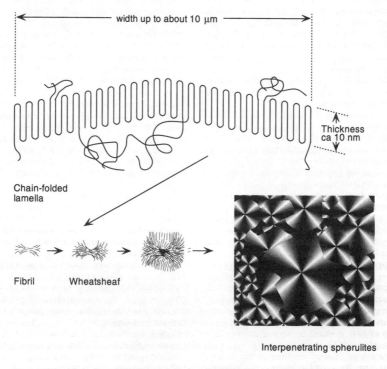

width up to about 10 μm

Thickness ca 10 nm

Chain-folded lamella

Fibril   Wheatsheaf

Interpenetrating spherulites

Fig. 1. Stages in the crystallization of polymers.

chamfered. When 12 equal spheres are packed such that their centres are at the vertices of a cuboctahedron the internal void will just accommodate one further such sphere.

cubic close packing The stacking of spheres formed by stacking close packed layers in the sequence ABCABC. The unit cell of such an arrangement is a face-centred cube, with four atoms per cell. This structure is adopted by many metals, eg Cu, Ag and Au. See **close packing of atoms** p. 62, Appendix 2. Abbrevs. *CCP*, *FCC*.

cullet Waste glass added with the raw materials to accelerate the rate of melting of the batch. See **glasses and glassmaking** p. 144.

cumulative distribution In an assembly of particles, the fraction having less than a certain value of a common property, eg size or energy. Cf **fractional distribution**.

cup-and-cone fracture A characteristic form of tensile fracture of ductile metals in which one side of the break is flat topped and cone shaped, while the other side is a matching cup. Caused by the fracture starting as a plane crack in the mid section of the test piece and at 90° to the tensile axis, which then develops as a 45° shear fracture as it approaches the outer surface.

cupel A thick-bottomed shallow dish made of bone ash; used in the **cupellation** of lead beads containing gold and silver, in the assay of these metals.

cupellation The operation employed in recovering gold and silver from lead. It involves the melting of the lead containing these metals and its oxidation by means of an air-blast.

cup flow figure The time in seconds taken by a test mould of standard design to close completely under pressure, when loaded with a charge of moulding material, used esp. for **thermosets**.

cupola furnace A shaft furnace used in melting pig-iron (with or without iron or steel scrap) for iron castings. The lining is firebrick. Metal, coke and flux (if used) are charged at the top, and air is blown in near the bottom.

cupped wire Wire in which internal cavities have been formed during drawings.

cupro-nickel An alloy of copper and nickel; usually contains 15, 20, or 30% of nickel; is very ductile and has high resistance to corrosion. See **copper alloys** p. 76.

cup shake A **shake** between concentric layers of wood. Also *ring shake*.

cure curve Plot of viscosity or torque versus time for crosslinking polymer system, usually referring to rubbers. Viscosity or torque usually measured with oscillating disc rheometer or Mooney viscometer at standard temperature. Curve usually shows an initial drop, then rises gradually to a plateau where **vulcanization** is complete. Important for assessing vulcanizing agents etc. See **overcure, undercure, cure rate index, scorch.**

cure cycle Usually a rubber technology term for the sequence of steps during **vulcanization**, from initiation of cross-linking to final set of product. Also more widely applied to all **thermosets**.

cure rate index Term applied to average gradient of main part of cure curve of **vulcanizing** rubber system.

Curie point (1) Temperature above which a **ferromagnetic** (or ferrimagnetic) material becomes paramagnetic. (2) Temperature (*upper Curie point*) above which a **ferroelectric** material loses its polarization. (3) Temperature (*lower Curie point*) below which some ferroelectric materials lose their polarization. Also *Curie temperature, magnetic transition temperature*.

Curie's law For paramagnetic substances, the magnetic susceptibility is inversely proportional to the absolute temperature.

Curie temperature See **Curie point**.

Curie–Weiss law At the **Curie point**, θ, a ferromagnetic material becomes paramagnetic. Well above this temperature its paramagnetic susceptibility is $\chi = C/(T-\theta)$, where T is the absolute temperature and C is the Curie constant.

curing The chemical process undergone by a thermosetting plastic by which the liquid resin cross-links to form a solid. This may be initiated, or accelerated, by heat. Curing generally takes place during the moulding operation, and may require from a few seconds to several hours for its completion. Cf **C-stage, setting**.

curl (1) A roughly hewn block of timber cut from a crotch and intended for cutting into veneers. (2) A paper defect caused by unequal movement of the two surfaces of the sheet due to changes in the ambient moisture or temperature.

current A flow of eg electric charge, water, air.

Cushyfoot TN for type of inclined shear mount using a rubber and steel plate laminate.

cut The thickness of the metal shaving removed by a cutting tool.

cutback **Bitumen** that has been diluted with suitable solvents, eg kerosine, to make it liquid and easier to handle.

cuticle Flat overlapping scales that lie on the surface of animal hair and wool. They cover the internal core or cortex.

cut-off wheel Thin abrasive wheel made of flexible material which is used to cut metals, concrete etc.

cutter Any tool used for severing, often more specifically a milling cutter. See **milling-machine**.

cutting compound A mixture of water, oil, and soft soap, etc, used for lubricating and cooling the cutting tool in machining operations. Also *coolant*.

cutting tools Tools used for the machining of metals etc, eg broach, cutter, lathe tools, milling-cutter, planer tools, reamer, screwing die, shaper tools, slotting tools, tap, twist drill.

CV Abbrev. for **Continuous Vulcanization**.

CVD See **chemical vapour deposition**.

cyanidation vat A large tank, with a filter bottom, in which sands are treated with sodium cyanide solution to dissolve out gold.

cyanide hardening Case-hardening in which the carbon content of the surface of the steel is increased by heating in a bath of molten sodium cyanide.

cyanoacrylate Methyl and ethyl 2-cyanoacrylate monomer, well-known for its use as an adhesive, with no gap-filling, but excellent penetration of rough surfaces due to low viscosity. Anionic polymerization is initiated by traces of water on the surfaces to be joined, and occurs rapidly to form a strong bond.

cyanobiphenyls Aromatic molecules with linked benzene nuclei, $R–C_6H_5–C_6H_5–CN$, which shows **liquid crystal** or **nematic** properties. See **liquid crystal phases** p. 194.

cycle A series of occurrences in which conditions at the end of the series are the same as they were at the beginning. Usually, but not invariably a cycle of events is recurrent.

cycle time Overall time to make a shaped product. Applied esp. to eg injection moulding process.

cyclic compounds Closed-chain or ring compounds consisting either of carbon atoms only (carbocyclic compounds), or of carbon atoms linked with one or more other atoms (heterocyclic compounds).

cyclo- (1) Prefix from Gk *kyklos*, circle. (2) In chemistry, containing a closed carbon chain or ring.

cycloalkanes Hydrocarbons containing saturated carbon rings. Also *cyclanes*.

cyclohexane C_6H_{11}, mp 2°C, bp 81°C, rel.d. 0.78. A colourless liquid, of mild ethereal odour.

cyclotron resonance The resonant coupling of electromagnetic power into a system of charged particles undergoing orbital movement in a uniform magnetic field. Used for the quantitative determination of the band parameters in semiconductors. See **Landau levels**.

Cycolac TN for **acrylonitrile-butadiene-styrene** (ABS) polymers (US).

cysteine *2-amino-3-mercaptopropanoic acid.* $HS.CH_2CH(NH_2).COOH$. The L- or D-form of this amino acid is found in proteins, often in its oxidized form, cystine. Symbol Cys, short form C.

cystine The dimer resulting from the oxidation of cysteine. The resulting disulphide bridge is an important structural element in proteins, as it often connects groups otherwise distant in the protein chain.

Czochralski process Single crystal growth process, especially for semiconductor applications, in which a crystal is grown by slowly withdrawing (pulling) a seed from the melt contained in a crucible. Crystal and melt are continuosly counter rotated to minimize thermal and compositional fluctuation effects. Cf **Bridgman process, float zone.**

D

d- Abbrev. for **dextrorotatory**.

δ- Substituted on the fourth carbon atom of a chain.

D Symbol for **deuterium**.

D- Abbrev. for **dextrorotatory**.

Λ Prefixed symbol for a double bond beginning on the carbon atom indicated.

Dacron TN for a polyester fibre (US).

dahoma A hardwood (*Piptadeniastrum*) from the African tropical rainforests, whose **heartwood** is a uniform yellow-orange to golden brown, interlocked grain with a coarse texture, and is durable. **Sapwood** attacked by powder post beetle. Used in heavy duty applications, eg buildings, flooring, piling. Density 560–780 kg m^{-3} (average 690 kg m^{-3}).

dalton Alternative name for the **atomic mass unit**.

Dalton's atomic theory States that matter consists ultimately of indivisible, discrete particles (*atoms*) and atoms of the same element are identical, chemical action takes place as a result of attraction between these atoms, which combine in simple proportions. It has since been found that atoms of the chemical elements are not the ultimate particles of matter, and that atoms of different mass can have the same chemical properties (isotopes). Nevertheless, this theory of 1808 is fundamental to chemistry. See **atomic structure**.

Dalton's law See **law of multiple proportions**.

Dalton's law of partial pressures The pressure of a gas in a mixture is equal to the pressure which it would exert if it occupied the same volume alone at the same temperature.

damask (1) A figured fabric made with satin and sateen weaves, in which background and figure have a contrasting effect; used mainly for furnishing. (2) Linen cloth of damask texture, used for tablecloths and towellings; also a cotton cloth of similar nature, used for tablecloths; both fabrics are reversible.

damping Commonly seen in polymeric materials, where viscoelastic effects cause energy dissipation, which appears as heat. Exploited in vibration isolation devices, such as rubber engine mounts etc. Also found in elastic materials. See **internal friction**, **Snoek effect**.

damping capacity The ability to absorb energy from external source, eg sound waves or vibrations. It is a measure of **internal friction** associated with the atomic or molecular disturbances induced by the external energy.

dandy roll Hollow cylinder covered with wire cloth situated on top of the machine wire of a papermaking machine so that the surface of the roll makes contact with the upper surface of the wet web. The wire cloth may be such that a wove or laid pattern is imparted to the paper or names or other designs secured to it to produce corresponding **watermarks** in the paper.

DAP Abbrev for **DiAlphanylPhthalate**.

DAP resins Type of polyester resin made using diallyl phthalate. Compares favourably with phenol formaldehyde resins for electrical insulation. Diallyl isophthalate, abbrev. *DAIP*, is also used.

dark red heat Glow emitted by metal at temperatures between 550° and 630°C.

dark resistance The resistance of a photocell in the dark.

Darvic TN for unplasticized polyvinyl chloride sheet (UK).

Davisson–Germer experiment The first demonstration (1927) of wave-like diffraction patterns from electrons by passing them through a nickel crystal.

d.c. Abbrev. for **direct current**, also *dc*.

d.c. bias (1) In an electronic amplifier, the direct signal applied to an active component which sets the quiescent conditions for the device. Thereafter, an a.c. signal may be applied. (2) In a magnetic tape recorder, the addition of a polarizing direct current in the signal recording to stabilize magnetic saturation.

d.c. resistance The resistance which a circuit offers to the flow of a direct current. Also *true (ohmic) resistance*.

deactivation The return of an activated atom, molecule, or substance to the normal state. See **activation** (2).

dead burnt Descriptive of such carbonates as limestone, dolomite, magnesite, when they have been so kilned that the associated clay is vitrified, part or all of the volatile matter removed and the slaking quality lowered.

dead knot A **knot** which is partially or wholly separated from the surrounding wood.

deal (1) A piece of timber of cross-section roughly 250×75 mm (10×3 in). (2) See **red deal**, **white deal**.

de Broglie wavelength That associated with a particle by virtue of its motion, ie $\lambda = h/p$ where λ is the wavelength, h is **Planck's constant** and p the particle's relativistic momentum. Only for electrons and other elementary particles can the de Broglie wavelength be large enough to produce observable diffraction effects. See **electron diffraction**.

Debye and Scherrer method A method of **X-ray crystallography** applicable to powders of crystalline substances or aggregates of crystals.

Debye-Hückel theory A theory of electrolytic conduction which assumes complete ionization and attributes deviations from ideal behaviour to inter-ionic attraction.

Debye length Maximum distance at which coulomb fields of charged particles in a plasma may be expected to interact.

Debye temperature The Debye theory of specific heats of solids can be usefully applied to many solids using only one fitting parameter, the Debye temperature.

Debye theory of specific heats of solids Based on the assumption that the thermal vibrations of the atom of a solid can be represented by harmonic oscillators whose energies can be quantized. The oscillator frequencies are distributed, up to a maximum (cut-off) frequency, according to the normal modes of vibration of a *continuous* medium. For many substances the theory gives a satisfactory agreement with experiment over a wide range of temperature. See **Blackman theory of specific heats of solids**.

Debye unit Unit of electric dipole moment equal to 3.34×10^{-30} C m or 10^{-18} e.s.u.

decalcomania paper A transfer paper for conveying a design on to pottery etc.

decalescence The absorption of heat that occurs when iron or steel is heated through the **arrest points**. See **recalescence**.

decarburization Removal of carbon from the surface of steel by heating in an atmosphere in which the concentration of decarburizing gases exceeds a certain value.

deci- Prefix with physical unit, meaning one-tenth.

decimetre One-tenth of a **metre**.

decitex See **tex**.

deckle edge A rough feather edge on the four sides of a hand-made sheet of paper due to stock seeping beneath the deckle. Similar effects can be simulated artificially. Also the irregular edges of a web of paper before trimming.

decomposition The breaking down of a substance into simpler molecules or atoms.

decorative laminate Laminates with a highly resistant, frequently decorative, surface based on **melamine resins** coated on resin impregnated paper. See **kraft paper**.

decrement Ratio of successive amplitudes in a damped harmonic motion.

decrepitation (1) The crackling sound made when crystals are heated, caused by internal stresses and cracking. (2) Breakdown in size of the particles of a powder due to internal forces, generally induced by heating.

deep drawing The process of cold working or drawing sheet or strip metal by means of dies into shapes involving considerable plastic deformation of the metal.

defect (1) Anything which can cause a product to fail in its specified function. Usefully classified as minor, major, serious and very serious, with the latter two potentially able to cause injury and severe injury and corresponding degrees of economic loss. (2) Crystal lattice imperfection which may be due to the introduction of a minute proportion of a different element into a perfect lattice, eg indium into germanium crystal, to form an intrinsic semiconductor for a transistor. A 'point' defect is a **vacancy** or an 'interstitial atom', while a 'line' defect relates to a **dislocation** in the lattice.

defect sintering Sintering whereby particles are introduced as a fine dispersion in a sintered body, or by chemical action during sintering. Mobility in heat treatment, with or without working, enables the introduced atoms or molecules to migrate through defects giving marked modification of properties.

defect structure Intense localized misalignment or gap in the crystal lattice, due to migration of ions or to slight departures from stoichiometry. The resulting opportunity for mobility is important for semiconductors, catalysis, photography, rectifiers, corrosion etc. Cf **dislocation**.

definite proportions See **law of constant (definite) proportions**.

deflagration Sudden combustion, generally accompanied by a flame and a crackling sound.

deflection (1) The amount of bending or twisting of a structure or machine part under load. (2) The movement of the pointer or pen of an indicating or a recording instrument.

deflocculate The break up agglomerates to form a stable colloidal dispersion.

defoaming agent Substance added to a boiling liquid to prevent or diminish foaming. Usually hydrophobic and of low surface tension, eg silicone oils.

deformation map See panel on p. 87.

degaussing Neutralization of the magnetization of a mass of magnetic material, eg a ship, by an encircling current.

degeneracy Two or more quantum states are said to be degenerate if they have the same energy. If the energy level can be realized in n different ways the energy has an n-fold degeneracy.

deformation map

A two-dimensional representation of temperature and stress which shows the deformation behaviour expected in a given metallic or ceramic material. They are useful in predicting **creep** behaviour in service at moderately elevated temperatures. Fig. 1 is such a map for aluminium of small grain size.

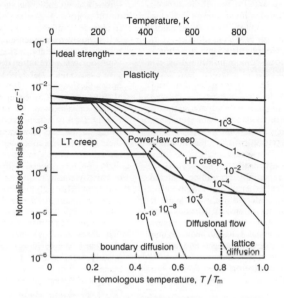

Fig. 1. **Deformation map for aluminium of small grain size**.

Temperature is plotted as *homologous temperature*, T/T_m, though for practical purposes the Celsius scale may be more useful. The stress axis is normalized by expressing the applied stress as a ratio of the *elastic modulus E*.

The 'ideal strength' line across the top of the map represents the stress at which plastic deformation takes place by the creation of new dislocations and their *avalanche* movement, and is therefore at a level greater than would be permitted in the design of a component for service. At somewhat lower levels of applied stress, deformation occurs by movement of pre-existing dislocations, but much more gradually as the latter are released. This is labelled the *plasticity zone* in the upper part of the map.

Below the plasticity zone is the region where *power law creep* takes place. Here the mechanisms are governed by *dislocation climb*, where obstacles to glide are overcome by the dislocations being able to move out of one slip plane and into another by thermal activation. In this region contours are drawn of equal strain rate in units of fractional strain per second.

Finally, at the bottom of the diagram, is the region of *diffusional flow*, which does not involve dislocation movement. Atomic diffusion allows changes of grain shape to take place over long periods with the result that grains elongate in the direction of the applied tensile force, giving an overall extension of the material. If diffusion through the crystal lattice predominates, this movement is called *Nabarro-Herring creep* but, if boundary diffusion predominates, it is called *Coble creep* after their respective discoverers.

degenerate gas (1) That which is so concentrated, eg electrons in the crystal lattice of a conductor, that the **Maxwell–Boltzmann law** is inapplicable. (2) Gas at very high temperature in which most of the electrons are stripped from the atoms. (3) An electron gas which is far below its **Fermi level** so that a large fraction of the electrons completely fills the lower energy levels and has to be excited out of these levels in order to take part in any physical processes.

degenerate semiconductor One in which the conduction approaches that of a simple metal.

degradation General term for reactions which cause loss of integrity of polymer properties. It covers depolymerization as well as chain oxidation, ozone cracking, ultraviolet degradation etc.

degree of crosslinking See **crosslink density**.

degree of crystallinity Total crystalline content of a partially crystalline material, esp. polymers. Experimentally found using density tests or inferred from wide-angle X-ray scattering measurements. Kinetics of development described by **Avrami equation**. Depends on polymer type and chain regularity as well as rate of cooling from melt. See **crystallization of polymers** p. 82.

degree of dissociation The fraction of the total number of molecules which are dissociated.

degree of ionization The proportion of the molecules or 'ion-pairs' of a dissolved substance dissociated into charged particles or ions.

degree of polymerization Number of repeat units in a polymer chain (usually an average). Equals molecular mass of chain divided by molecular mass of repeat unit. Abbrev. *DP*.

degree of swelling Extent of swelling in rubbers, judged either by change in linear dimensions or by volumetric change.

degrees of freedom (1) The number of variables defining the state of a system (eg pressure, temperature) which may be fixed at will. See **phase rule**. (2) Number of independent capacities of a molecule for holding energy, translational, rotational and vibrational. A molecule may have a total of $3n$ of these, where n is the number of atoms in the molecule.

de Haas–van Alphen effect Oscillations in the magnetic moment of a metal as a function of $1/B$ where B is the magnetic flux density. Interpretation of this effect gives important information about the shape of the **Fermi surface**.

dehydration (1) The removal of H_2O from a molecule by the action of heat, often in the presence of a catalyst, or by the action of a dehydrating agent, eg concentrated sulphuric acid. (2) The removal of water from crystals, tars, oils etc, by heating, distillation or by chemical action.

de-ionized water Water from which ionic impurities have been removed by passing it through cation and anion exchange columns.

delamination (1) Peeling of surface layers in moulded products, a type of mould defect often caused by contamination from mould release agents or foreign polymer. (2) Separation of fabric or reinforced layers in a composite material, caused by water ingress or poor adhesion etc.

deliquescence The change undergone by certain substances which become damp and finally liquefy when exposed to the air, owing to the very low vapour pressure of their saturated solutions; eg calcium chloride.

Delrin TN for an acetal resin (US). See **polyoxymethylene**.

delta iron The polymorphic form of iron stable between 1403°C and the melting point (about 1532°C). The crystal lattice is the similar to that of α-iron and different from that of γ-iron.

delustrant Dense inorganic material, frequently titanium dioxide, added to a manmade fibre before it is extruded. In this way a range of fibres may be obtained with different lustres and opacities.

demagnetization (1) *Removal* of magnetization of ferromagnetic materials by the use of diminishing saturating alternating magnetizing forces. (2) *Reduction* of magnetic flux by the internal field of a magnet, arising from the distribution of the primary magnetization of the parts of the magnet. (3) *Removal* by heating above the **Curie point**. (4) *Reduction* by vibration.

demagnetization factor Diminution factor (N) applied to the intensity of magnetization (M) of a ferromagnetic material, to obtain the demagnetizing field (ΔH), ie $\Delta H = NM$. N depends primarily on the geometry of the body concerned.

demijohn Narrow-necked, glass wine or spirit container of more than 2 l capacity.

demoulding temperature Temperature at which moulding can be safely removed from tool without permanent distortion. Also freeze-off temperature. It is closely related to the crystallization temperature or glass-transition temperature (for non-crystalline polymers).

demy See **paper sizes**.

dendrite Many crystals grow in the first instance by branches developing in certain directions from the nuclei. Secondary branches are later thrown out at periodic intervals by the primary ones and in this way a skeleton crystal, or *dendrite*, is formed. The interstices between the branches are finally filled with solid which in a pure material is indistinguishable from the skeleton. In many alloys, however, the final structure consists of skeletons of one composition merging gradually into a matrix of another.

denier A unit used for the thickness of yarns; the mass in grams of 9000 m of yarn. Superseded by **tex**. See **yarn count**.

denim A strong cotton twill fabric made from yarn-dyed warp (often blue) and undyed (sometimes even unbleached) weft yarn. Used for the manufacture of overalls, boiler suits and jeans.

dense Of optical glass, having a high **refractive index**.

densification All modes of increasing the density of materials, including the effect of sintering contraction on powder density.

Densithene TN for polyethylene loaded with lead powder. Used for radioactive shielding.

density Materials property; ratio of the mass of a material to its volume. Symbol, ρ; units, kg m^{-3}. Its reciprocal is the **specific volume**.

density of states The number of electronic states per unit volume having energies in the range from E to $E + dE$; an important concept of the **band theory of solids**.

density test Analytical method for determining material density. Crudest method involves flotation in fluids of different densities, so only giving density limits to an unknown sample. This test often combined with flame test and pencil hardness test to give guide to polymer type. A better method uses calibrated density columns with fluid of regularly varying density, so that the position of the unknown in the tube gives the density precisely. Care is needed in interpretation due to variation of polymer density with added fillers etc.

dental materials Synthetic materials replacing or restoring function of teeth eg polymethyl methacrylate for dentures. Other materials include gold, **amalgams**, **porcelain**, **ionomer resins**.

dentine Structural **biocomposite** of very fine (20–40 nm long) needle crystals of **hydroxy apatite** embedded in a **collagen** fibre matrix, very similar to compact bone.

deodar See **cedar**.

deoxidant See **deoxidizer**.

deoxidation The process of reduction or elimination of oxygen from molten metal before casting by adding elements with a high oxygen affinity, which form oxides that tend to rise to the surface.

deoxidized copper Copper from which dissolved oxygen remaining after refining has been removed by the addition of a deoxidizer. Residual amounts of deoxidizer in solid solution, lower the electrical conductivity below that of tough-pitch copper, but the product is more suitable for working operations.

deoxidizer A substance which will eliminate or modify the effect of the presence of oxygen, particularly in metals. Also *deoxidant*.

dephosphorization Elimination, partial or complete, of phosphorus from steel, in basic steel-making processes. Accomplished by forming a slag rich in lime.

depletion In semiconductors, the local reduction of the density of charge carriers. Cf **enhancement**.

depletion layer In semiconductors, a non-neutral region in which the majority carrier density is reduced below that of the local dopant concentration. Such layers arise naturally at the junction between p-type and n-type material and can be induced in the semiconductor adjacent to an insulator in eg **metal-oxide-silicon** (MOS) structures.

depoling See **poling**.

depolymerization Reverse of polymerization which is induced by heat, free radicals, photons, radiation etc to produce monomer molecules. Only a few polymers degrade to monomer alone (eg polymethyl methacrylate, polytetrafluoroethylene). Most leave a complex mixture of degradation products.

depth of fusion The depth to which a new weld has extended into the underlying metal or a previous weld.

de-scaling The process of (1) removing scale or metallic oxide from metallic surfaces by pickling, or by mechanical means; (2) removing scale from the inner surfaces of boiler plates and water tubes.

de-seaming Process of removing surface blemishes, or superficial slag inclusions from ingots or blooms.

design See **product design**, **industrial design**, **engineering design**.

design engineering Wide term implying forethought and analysis applied to practical ends. Encompasses all practical methods used in industrial, product and engineering design.

desilverization The process of removing silver (and gold) from lead after softening.

Desmodur TN for isocyanates used to produce polyurethanes in conjunction with polyester resins.

desoxyribose nucleic acid The natural polymer present in the cell nucleus of all living things and which ultimately controls all biosynthesis and reproduction. Consists of linked perfectly alternating phosphate (PO_4) and ribose sugar ($C_5H_{10}O_5$) groups in a copolymer chain of molecular mass varying from 4×10^5 (viruses) to well in excess of 10^7 (animals, plants). To each sugar group is attached a pendant base group, adenine (A), cytosine (C), guanine (G) or thymine (T) linked with hydrogen bonds to similar base groups in a parallel chain. The double stranded molecule adopts a helical conformation with the base groups stacked like a pile of coins inside the entwining phosphate-sugar chains. Abbrev. *DNA*.

Destriau effect A form of electroluminescence arising from localized regions of very intense electric field associated with impurity centres in the phosphor.

desulphurizer Chemical agent added to liquid steel before casting in order to combine with sulphur and transfer to a **basic slag**, rich in lime. Similar to *dephosphorizer*.

detail paper A translucent tracing paper, usually unoiled.

detergent cracking Effect of aggressive detergents (eg Teepol) on thermoplastics, esp. polyolefins, causing deep, isolated cracks. Solved by using higher molecular mass materials. See **environmental stress cracking**.

de-tinning Chlorine treatment to remove tin coating from metal scrap.

deuteration The addition of or replacement by deuterium atoms in molecules.

deuterium Isotope of element hydrogen having one neutron and one proton in nucleus. Symbol D.

deutero- Prefix, in general denoting second in order, derived from Gk *deuteros*, second. Particularly, in chemistry, containing heavy hydrogen (**deuterium**).

devitrification A physical process in silicate glasses which causes a change from the glassy state to a minutely crystalline state. The change can occur over time, and the glass becomes turbid and more brittle. It is employed in a controlled way to produce **glass-ceramics**.

dextrorotatory Said of an optically active substance which rotates the plane of polarization in a clockwise direction when looking against the incoming light.

dextrose See **glucose**.

diagonal A tie or strut joining opposite corners of a rectangular panel in a framed structure.

Diakon TN for polymethyl methacrylate moulding powder (UK).

dial gauge A sensitive measuring instrument in which small displacements of a plunger are indicated in 1/1000 mm, or similar length units, by a pointer moving over a circular scale.

dialphanyl phthalate A common plasticizer for polyvinyl chloride. Abbrev. *DAP*.

See **polyvinyl chloride** p. 249.

diamagnetism Phenomenon in some materials in which the susceptibility is negative, ie the magnetization opposes the magnetizing force, the permeability being less then unity. It arises from the precession of spinning charges in a magnetic field. The susceptibility is generally one or two orders of magnitude weaker than typical paramagnetic susceptibility. See **characterizing magnetic materials** p. 57.

diamond One of the crystalline forms of carbon; it crystallizes in the cubic system, rarely in cubes, commonly in forms resembling an octahedron, and less commonly in the tetrahedron. Curved faces are characteristic. It is the hardest mineral; hence its value as an abrasive, and for arming rock-boring tools etc; its high dispersion and birefringence makes it valuable as a gemstone. Occurs in blue ground, in river gravels and in shore sands.

diamond die A wire drawing die containing a diamond insert which reduces wear. Cf **bushing**.

diamond-like carbon A material containing carbon and hydrogen deposited by plasma-enhanced chemical vapour deposition from hydrocarbon gases. Amorphous but with substantial quantities of sp^3-bonded carbon atoms, it has high mechanical hardness (approaching that of diamond), high electrical resistivity and high thermal conductivity. Its optical properties are also similar to those of pure diamond.

diamond tool A diamond of specified shape, mounted in a holder which is used for precision machining of non-ferrous metals and ceramics.

diamond wheel A rotating wheel for a grinding machine in which small diamonds are embedded, and which is used for grinding and cutting very hard materials.

diatomite, diatomaceous earth A siliceous deposit occurring as a whitish powder consisting essentially of the frustules of diatoms. It is resistant to heat and chemical action, and is used in fireproof cements, insulating materials, as an absorbent in the manufacture of explosives and as a filter. Also *infusorial earth, kieselguhr, tripolite*.

dia-stereoisomers Stereoisomers which are not simple mirror images (enantiomers) of one another. For example, in a molecule with two chiral centres, the L,L- and D,D-forms are enantiomers, while the L,D- and D,L-forms are dia-stereoisomers.

dibasic acids Acids containing two replaceable hydrogen atoms in the molecule.

dibenzoyl-peroxide $C_6H_5.CO.OO.CO.C_6H_5$. Relatively stable organic peroxide used mainly as a catalyst in polymerization reactions.

dichloroethylenes *Dichloroethenes*, $C_2H_2Cl_2$. Exist in cis-form, bp 48°C and trans-form, bp 60°C. Prepared industrially from tetrachloroethane. Used as source material for vinyl chloride, the monomer of polyvinyl chloride.

dichroism The property possessed by some crystals (eg tourmaline) of absorbing the ordinary and extraordinary ray to different extents; this has the effect of giving to the crystal different colours according to the direction of the incident light. Cf **Polaroid**.

die (1) A metal block used in stamping operations. It is pressed down on to a blank of sheet-metal, on which the pattern or contour of the die surface is reproduced. (2) The element complementary to the punch in press tool for piercing, blanking etc. (3) An internally threaded steel block provided with cutting edges, for producing screw threads by hand or machine. (4) A tool made of very hard material, often tungsten carbide or diamond, with a (bell-mouthed) hole, usually circular, used to reduce the product cross-section by plastic flow, in wire or tube drawing. (5) The steel tool which shapes the extrudate (see **extrusion** p. 121).

diecasting Casting of metals or plastics into permanent moulds, made of suitably resistant non-deforming metal. See **gravity diecasting**, **pressure diecasting**.

diecasting alloys Alloys suitable for diecasting, which can be relied on for accuracy and resistance to corrosion when cast. Aluminium-base, copper-base, tin-base, zinc-base and lead-base alloys are those generally used.

die-fill ratio The ratio of uncompacted powder volume to the volume of the **green compact**.

dielectric Substance, solid, liquid or gas, which can sustain a steady electric field, and hence an insulator. It can be used for cables, terminals, capacitors etc. See **dielectrics and ferroelectrics** p. 92.

dielectric breakdown Passage of large current through normally non-conducting medium at sufficiently intense field strengths accompanied by a relative reduction of resistance and, in solids, mechanical damage.

dielectric constant See **dielectrics and ferroelectrics** p. 92, **permittivity**.

dielectric current A changing electric field applied to a perfect dielectric gives rise to a displacement current. For a real dielectric there will also be a conduction current or absorption current giving rise to energy loss in the dielectric.

dielectric diode A capacitor whose negative plate can emit electrons into, eg CdS crystals, so that current flows in one direction.

dielectric dispersion Variation of permittivity with frequency.

dielectric fatigue Breakdown of a dielectric subjected to a repeatedly applied electric stress, insufficient to break down dielectric if applied once or a few times.

dielectric heating Radio frequency heating in which power is dissipated in a non-conducting medium through dielectric hysteresis. It is proportional to $V^2fS\,t^{-1}$, where V is applied voltage, f the frequency, S the area of the heated specimen and t its thickness. This is the principle used in microwave ovens.

dielectric loss Dissipation of power in a dielectric under alternating electric stress;

$$W = \omega CV^2 \tan \delta,$$

where W = power loss, V = r.m.s. voltage, C = capacitance, tan δ = **loss tangent**, ω = $2\pi \times$ frequency.

dielectric polarization Phenomenon explained by formation of doublets (dipoles) of elements of dielectric under electric stress.

dielectric relaxation Time delay, arising from dipole moments in a dielectric when an applied electric field varies.

dielectrics and ferroelectrics See panel on p. 92.

dielectric and ferroelectric materials See panel on p. 94.

dielectric strength Electric stress necessary to break down a dielectric. It is generally expressed in kV per mm of thickness. The stress, steady or alternating, is normally maintained for 1 minute when testing. See **dielectrics and ferroelectrics** p. 92.

die lubricant Lubricant applied to reduce friction between the material and the die walls during working processes; it is usually a solid such as a soap. It may be incorporated in powders when interparticle friction is also reduced.

diene Organic compound containing two double bonds between carbon atoms in its structure.

die sinking The engraving of dies for coining, paper embossing and similar operations.

die swell Ratio of extrudate diameter to die diameter, caused by elastic recovery of polymer chains after **extrusion**. Due to **melt memory**. Accommodated by careful die design. See **melt elasticity**.

diethylene glycol *2,2-dihydroxyethoxyethane*, $(C_2H_4OH)_2O$. Colourless liquid; bp 245°C. Used as a solvent, eg for cellulose nitrate. Its monoethyl ether is known as carbitol, also used widely as a solvent. Derivatives (esters and ethers) used as plasticizers.

differential permeability Ratio of a small change in magnetic flux density of magnetic

dielectrics and ferroelectrics

The application of a steady electric field to a material without free charges, ie all charges are bound, will displace positive and negative charges in opposite directions to some degree, without leading to the passage of a steady current. This dielectric response results in such a material becoming electrically polarized. The net polarization (P) depends in general upon the strength of the applied field (E) and upon the disposition and polarizability of the constituent ions or molecules of the material.

The dielectric constant (or relative permittivity) measures the effect of polarization and is conveniently illustrated by reference to a parallel plate capacitor (Fig. 1).

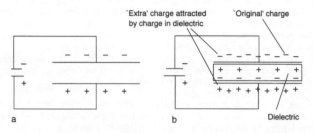

Fig. 1. **A parallel plate condenser with free space** (a) **and a dielectric** (b) **between the plates**.

When there is a dielectric between the plates, the electric field polarizes it, inducing a surface charge. This is neutralized by the extra charge supplied by the battery with the result that for the same battery voltage, more charge is stored on the capacitor plates than is the case without the dielectric. The dielectric constant (ε_r) is the factor by which the capacitance is thereby increased and is a property of the dielectric material.

In a sufficiently large *dc* electric field, a dielectric will break down because it is no longer able to resist the passage of a steady current. The limiting field of the bulk material is termed the dielectric strength.

In *ac* fields, energy is absorbed if the small displacements of charge associated with the changing polarization lead to inelastic interactions with the host structure. When there is absorption of energy from the field there is in effect a component of current density which is locally in phase with the macroscopic electric field. The ratio of in-phase to quadrature current is known as tan δ (or *loss tangent*) and is a characteristic of a particular material. Equivalently tan δ is the ratio of the imaginary to real parts of a complex dielectric constant.

In paraelectric materials, the distribution of electric charge is not uniform within the constituent molecular and ionic structures in the absence of any external field. In this case there are permanent electric dipoles within the material. In an external field these dipoles enhance the contribution of the induced polarization, giving rise to substantially higher dielectric constants. At high frequencies this interaction is essentially between a polarizing field and the refractive index of the material, leading to an electro-optic effect.

There are three classes of polarization which contribute to the dielectric constant: (i) electronic polarization arising from the displacement of electrons around an atom; (ii) ionic polarization arising from the displacement of ions; (iii) orientational polarization arising from the partial alignment of any induced and permanent dipoles which are free to rotate. At high frequencies only the first is able to contribute. The last is only significant at low frequencies. Polarization is disrupted by thermal vibrations of the material.

continued on next page

dielectrics and ferroelectrics (contd)

If the internal dipoles in a material remain aligned when the external field is removed (the material has a remanent polarization), it is said to be a ferroelectric, by loose analogy with ferromagnetics. Ferroelectrics possess a domain structure below a critical temperature known as the Curie temperature, T_c. Above T_c such materials are *paraelectric*.

The polarization (P) of a ferroelectric follows the external polarizing field (E) as illustrated in Fig. 2a. As with magnetic materials, the changing of domain size and orientation is not perfectly reversible and the polarization curve reveals hysteresis. When the polarizing field is removed the material is left with a *remanent polarization*.

In ferroelectrics, electrical polarization is closely coupled to the crystal structure so that all ferroelectrics are to some degree piezoelectric. Fig. 2b shows a schematic polarization-strain (P-s) curve (sometimes called a *butterfly curve*). Similarly, the interaction between polarization and thermal energy gives rise to temperature-sensitive remanent polarization, the *pyroelectric effect*.

Fig 2. **Schematic curves for** (a) **polarization and** (b) **strain vs the applied polarizing field.**

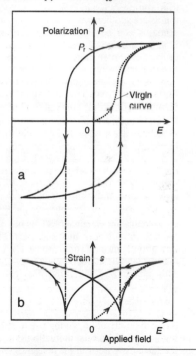

material to change in the magnetizing force producing it, ie slope of the $B(H)$ **curve** at the point in question, often expressed in dimensionless form as $dB/\mu_0 dH$.

differential scanning calorimetry Method of thermal analysis of materials similar to differential thermal analysis, but with enhanced accuracy. Used for studying crystallization effects in materials, **glass transitions**, oxidation and degradation processes, **curing cycles** etc. Abbrev. *DSC*.

differential resistance Ratio of a small change in the voltage drop across a resistance to the change in current producing the drop, ie the slope of the voltage-current characteristic for the material. See **dynamic resistance**.

differential scanning calorimetry Type of thermal analysis similar to **differential thermal analysis**. Abbrev. *DSC*.

differential susceptibility Ratio of a small change in the intensity of magnetiza-

dielectric and ferroelectric materials

All materials contain electronic charges and so respond, to varying degrees, to the application of an electric field. In metals these charges are free to move so an applied field induces a macroscopic transport current in the specimen. Insulators, on the other hand, contain mostly bound charges and therefore offer more resistance to the flow of current. The extent to which a material can resist the flow of charge in an electric field determines its *permittivity*, a measure of its dielectric character.

An externally applied field induces electric dipoles in a predominantly ionic material such as lithium fluoride.This behaviour is termed *paraelectric*. The internal field produced by the collective effect of these dipoles (the *polarization*) opposes that applied externally, reducing the magnitude of the field in the interior of the sample. Although the net polarization tends to be small in paraelectric materials, they offer significantly better dielectric properties than covalently bonded insulators such as glass, which are less polarizable.

Ferroelectric materials exhibit the highest permittivities because they belong to a *non-centrosymmetric polar crystal* class. This means that individual domains are spontaneously polarized below their **Curie temperature** (T_c). These domains can be aligned reversibly by an externally applied electric field. Non-polar centrosymmetric materials are characterized by their **piezoelectric** properties. Polar, non-centrosymmetric materials on the other hand exhibit *pyroelectricity* and non-linear *electro-optic* properties. Ferroelectrics fall into both categories and hence feature predominantly in many electrical and electronic applications. The properties of some of the more commonly used centrosymmetric materials are described briefly below.

Ceramic compounds in the lead zirconate titanate (PZT) family of ferroelectrics have large piezoelectric coupling coefficients and are widely used in high-field electroacoustic applications such as underwater sound generators and receivers. Furthermore PZT can be made optically transparent by substituting 5–10% of the lead cations with lanthanum to form PLZT without compromising its ferroelectric properties. PLZT is therefore the basis of a variety of electro-optic and birefringent devices, such as *flash goggles* which can be voltage switched to become opaque when exposed to intense optical radiation.

Lead scandium tantalate (PST) is becoming an important ferroelectric material for use in dielectric **bolometers**. PST undergoes a ferroelectric-paraelectric transition at around room temperature. This results in a high permittivity that may be stabilized over a wide temperature range by the application of a field. This makes PST particularly useful for high-resolution, uncooled, thermal imaging arrays.

continued on next page

tion of a magnetic material to the change of magnetizing force producing it, ie the slope of the $M(H)$ **curve**.

differential thermal analysis The detection and measurement of changes of state and heats of reaction, esp. in solids and melts, by simultaneously heating two samples of identical heat capacities and noting the difference in temperature between them, which becomes very marked when one of the two samples passes through a **transition temperature** with evolution or absorption of heat but the other does not. Abbrev. *DTA*.

diffusion The general transport of matter (atoms, molecules, ions) through thermal agitation. A net flux results from diffusion when there is a concentration gradient. In a crystalline solid, interstitial atom, lattice vacancy and impurity atom diffusion are thermally activated. Diffusion is often used to introduce controlled quantities of impurities into the surfaces of semiconductors (for doping) and metals (for carburizing and nitriding etc). See **diffusion coefficient, Fick's laws of diffusion**.

diffusion capacitance The rate of change

Lithium niobate (LN) is an important ferroelectric material because it can be grown in large single crystals, making it suitable for surface acoustic wave (SAW) devices that require a smooth uniform piezoelectric surface for their operation. In addition LN exhibits a significant electro-optic effect (ie its refractive index is anisotropic and can be varied by the application of an electric field). This makes it useful in a variety of optical waveguide switching applications.

A number of organic polymers have useful dielectric properties. Polyvinylidene fluoride (PVDF) is piezoelectric and mechanically conformable. It is used in noise-cancelling microphones and *flexural bimorph strips*.

Triglycine sulphate (TGS) has the highest recorded pyroelectric coefficient and is used in infrared detectors. A major disadvantage is its moisture sensitivity which limits its application. Single-crystal lithium tantalate (LT) is atmospherically stable and available in large areas and is therefore often used in preference to TGS in single-element thermal detectors.

Tables 1–3 summarise properties typical of selected dielectric and ferroelectric materials although the values are sensitive to composition and processing.

Table 1. **Dielectric materials**.

Material	ε_r	tan $\delta/10^{-4}$
Alumina	10	5–20
Porcelain	5	75
Mica	5	3
Quartz	4	2
Barium titanate, BT	500	150
Lead zirconate titanate	1000	40
Polycarbonate	3	10
Polystyrene	3	1
Polytetrafluoroethylene	2	2
Polyvinyl chloride	3	160

Table 2. **Piezoelectric materials**.

Material	Type	T_c (°C)	ε_r	Piezo. coeff. d_{33} (pC N^{-1})
Potassium titanate, KTN	Single crystal	49	14 000	200
Polyvinylidine fluoride	Polymer	> mp	12	28
Barium titanate, BT	Single crystal	125	3000	86
Lead zirconate titanate	Polycrystal	25	2000	450

Table 3. **Pyroelectric materials**.

Material	Type	T_c (°C)	ε_r	Pyro. coeff. (μC m^{-2}K^{-1})
Triglycine sulphate, TGS	Single crystal	69	55	550
Polyvinylidine fluoride	Polymer	> mp	12	27
Lead titanate, LT	Single crystal	665	47	230
Lead scandium titanate	Polycrystal	25	1500	2300*

* with 5 V μm^{-1} bias field

of an injected charge concentration with variation in the applied voltage in a semi-conductor diode or similar structure.

diffusion coating Methods by which an alloy or metal is allowed to diffuse into the surface of an underlying metal. They can involve heating and exposing the metal to a solution of the coating material.

diffusion coefficient In the diffusion equation (Fick's first law), the coefficient of proportionality between molecular flux and concentration gradient. Symbol D. Units, $m^2 s^{-1}$, as for **thermal diffusivity**. The approximate distance, x, moved by a diffusing species in time, t, is:

$$x = \sqrt{Dt} .$$

Since diffusion is a *thermally activated process*, the Arrhenius rate equation applies,

$$D = D_0 \exp(-E_a / kT) .$$

Also *diffusivity*. See **Fick's laws of diffusion**.

diffusion current The net flux of particles down a density gradient. In particular, that current resulting from the diffusion of charge carriers in a concentration gradient as distinct from a **drift current**. In electrolysis, the maximum current at which a given bulk concentration of ionic species can be discharged, being limited by the rate of migration of the ions through the **diffusion layer**.

diffusion law See **Fick's laws of diffusion**, **Graham's law**.

diffusion layer In electrolysis, the layer of solution adjacent to the electrode, in which the concentration gradient of electrolyte occurs.

diffusion length Average distance travelled by carriers in semiconductor between generation and recombination.

diffusion potential The potential difference across the boundary of an electrical double-layer in a liquid.

diffusion welding A method in which high temperature and pressure cause a permanent bond between two metallic surfaces without melting or large-scale deformation. A solid metal filler may be sandwiched between the surfaces to aid the process.

diffusivity See **diffusion coefficient, thermal diffusivity**

di-iso-octyl phthalate Plasticizer used in flexible polyvinyl chloride. Abbrev. *DIOP*.

dilatancy The behaviour of a fluid when stressed in which it increases its resistance to further stress by increasing its shear rate (eg wet beach sand or polyvinyl chloride plastisol). Pseudoplastic fluids behave in the opposite way.

dilatometer An apparatus for the determination of transition points of solids by measuring changes in length or volume with temperature.

dimer Molecular species formed by the union of two like molecules, the simplest **oligomer**. Adj. *dimeric*.

dimity Strong cotton fabric that appears striped because of a corded pattern; used chiefly for mattress coverings.

DIN Abbrev. for **Deutsche Institut für Normung**. The German national standards organization; in particular, their system of photographic speed rating with logarithmic increments, plus many material and product specifications.

dioctylphthalate A common plasticizer for polyvinyl chloride. Abbrev. *DOP*. See **polyvinyl chloride** p. 249.

diode (1) Simplest electron tube, with a heated cathode and anode; used because of unidirectional and hence rectification properties. (2) Semiconductor device with similar properties, evolved from primitive crystal **rectifiers** for radio reception.

diode characteristic A graph showing the current-voltage characteristics of a vacuum-tube or semiconductor diode. In particular it will show marked differences between currents in the forward and reverse directions, and, in the case of **avalanche** or **Zener diodes**, sudden increases in reverse current when the applied voltage reaches a critical value.

diode isolation Isolation of the circuit elements in a micro-electronic circuit by using the very high resistance of a reverse-biased p-n junction.

diode laser See **optoelectronics**.

diols Dihydric alcohols, chiefly represented by the glycols in which the hydroxyl groups are attached to adjacent carbon atoms.

DIOP Abbrev. for **Di-Iso-Octyl Phthalate**.

diphenyl-methane diisocyanate A type of aromatic isocyanate monomer used to make polyurethanes. Abbrev. *MDI*.

dipolar molecule One which has a permanent moment due to the permanent separation of the effective centres of the positive and negative charges.

dipole Equal and opposite charges separated by a close distance constitute an electric dipole. A bar magnet or a coil carrying a steady current produce a magnetic dipole.

dipping The immersion of pieces of material in a liquid bath for surface treatment such as pickling or galvanizing.

dip soldering The method of soldering previously fluxed components by immersing them in a bath of molten solder. Ideal for bulky assemblies with complicated or multiple joints, and for fast automatic operation.

Dirac's theory Using the same postulates as the **Schrödinger equation**, plus the requirement that quantum mechanics conforms with the theory of relativity, an electron must have an inherent angular momentum and magnetic moment (1928).

direct chill casting A method like **continuous casting** but for larger cross-sections in which the hollow mould is closed at the bottom by a platform. This is gradually lowered as the metal becomes solid on the outside and therefore able to contain the melt, platform, mould and metal being appropriately cooled.

direct-fired Furnace in which the fuel is delivered into the heating chambers.

direct process The method originally used for obtaining from ore a form of iron similar to **wrought iron** in one operation, ie without first making pig-iron.

direct stress The stress produced at a section of a body by a force whose resultant passes through the centre of gravity of the section.

disappearing-filament pyrometer An instrument used for estimating the temperature of a furnace by observing a glowing electric-lamp filament against an image of the interior of the furnace formed in a small telescope. The current in the filament is varied until it is no longer visible against the glowing background. From a previous calibration the required temperature is derived from the current value.

discard Portion of material which has to be rejected by virtue of the nature of a working process, eg in direct extrusion the bulk of the billet can be forced through the die orifice but a small fraction always remains inside the die chamber due to frictional effects, this portion is separated from the extruded product and becomes the discard.

discharge (1) The abstraction of energy from a cell by allowing current to flow through a load. (2) Reduction of the potential difference at the terminals (plates) of a capacitor to zero. (3) Flow of electric charge through gas or air due to ionization, eg lightning, or at reduced pressure, as in fluorescent tubes. See **field discharge**.

discharge bridge Measurement of the ionization or discharge, in dielectrics or cables, depending on the amplification of the high-frequency components of the discharge.

discharge lamp One in which luminous output arises from ionization in gaseous discharge. See **discharge** (3).

disintegration A process in which a nucleus ejects one or more particles, applied esp., but not only, to spontaneous radioactive decay.

disintegration constant The probability of radioactive decay of a given unstable nucleus per unit time. Statistically, it is the constant λ, expressing the exponential decay

$\exp(-\lambda t)$ of activity of a quantity of this isotope with time. It is also the reciprocal of the mean life of an unstable nucleus. Also *decay constant, transformation constant*.

disk friction The force resisting the rotation of a disk in a fluid. It is important in the design of centrifugal machinery as it decreases efficiency and causes a rise in the pressure of the fluid being pumped.

dislocation A lattice imperfection in a crystal structure which exerts a profound effect on structure sensitive properties such as strength, hardness, ductility and toughness. Has a configuration of an extra half plane of atoms inserted in the crystal stacking and the associated structural displacements near the end of the half plane result in atomic movement, eg slip, at much lower applied stresses than would occur in a perfect crystal. Plays a fundamental role in accounting for deformation and strengthening phenomena in metals. In an annealed crystal the density of dislocation lines is of the order of 10^9 per square millimetre which rises to 10^{13} when the material is heavily cold worked, owing to interactions during deformation. There are two types, edge and screw, both of which are characterized by a **Burgers vector** which represents the amount and direction of slip when the dislocation moves. See **deformation map** p. 87.

dislocation glide The movement of dislocations along slip planes during the process of deformation.

dispersion (1) See **molecular mass distribution** p. 212. (2) The dependence of wave velocity on the frequency of wave motion; a property of the medium in which the wave is propagated. In the visible region of the electromagnetic spectrum, dispersion manifests itself as the variation of **refractive index** of a substance with wavelength (or colour) of the light. A prism forms a spectrum by dispersion.

dispersion curve A plot of frequency against wavelength for a wave in a dispersive medium. See **phonon dispersion curve, acoustic branch**.

dispersion forces Old name for weak intermolecular forces. See **van der Waals forces**. Also *London forces*.

dispersion hardening Hardening of a material by introducing a fine dispersion of particles into a ductile matrix which increases the applied stress necessary to move **dislocations**. See **precipitation hardening**.

dispersion medium A substance in which another is colloidally dispersed.

dispersive mixing Process where particle clumps, eg carbon black, are broken down during mixing with polymer by high shear forces. Essential to combine with **distribu-**

tive mixing. Also *intensive mixing*.

dispersive power See **dispersivity**.

dispersivity The ratio of the difference in the refractive indices (*n*) of a medium for specified wavelengths in the red (R) and violet (V) to the mean refractive index diminished by one. This may be written as follows:

$$v = \frac{n_V - n_R}{\bar{n} - 1} \, .$$

Also *dispersive power*.

displacement (1) Vector representing the electric flux in a medium and given by $D = \varepsilon E$, where ε is the **permittivity** and E is the electric field. Also *dielectric strain, electric flux density*. (2) In mechanics, the vector of distance moved by a body.

displacement current Integral of the displacement current density through a surface. The time rate of change of the electric flux. Current postulated in a dielectric when electric stress or potential gradient is varied. Distinguished from a normal or conduction current in that it is not accompanied by motion of current carriers in the dielectric. Concept introduced by Maxwell for the completion of his electromagnetic equations.

displacement series See **electrochemical potential series**.

display Device for presenting information from an electronic system. See **liquid crystal displays** p. 192.

dispersoid A particle of a second phase material distributed through a host solid by means eg of precipitation.

disproportionation (1) A reaction in which a single compound is simultaneously oxidized and reduced, eg the spontaneous reaction in water of soluble copper (I) salts to form equal amounts of copper (0) and copper (II). (2) A chain termination reaction where two active free radical chain ends transfer electrons to form two dead chains. See **chain polymerization** p. 56.

disruptive discharge The discharge arising from the breakdown (puncture) of the dielectric of a **capacitor** caused by an electric field which it cannot withstand.

disruptive strength Obsolete term for the stress at which a material fractures under tension. See **strength measures** p. 301.

disruptive voltage That which is just sufficient to puncture the dielectric of a capacitor. A test voltage is normally applied for one minute. See **breakdown voltage**.

dissipation factor The tangent of the phase angle (δ) for an inductor or capacitor.

$$\tan \delta \approx \frac{\sigma}{\omega \varepsilon} \, ,$$

σ is the electrical conductivity, ε the **permittivity** of the medium, and ω is $2\pi \times$ the

frequency. For low loss components, the dissipation factor is approximately equal to the power factor. Also *loss tangent*.

dissociation The reversible or temporary breaking-down of a molecule into simpler molecules or atoms.

dissociation constant The equilibrium constant for a process considered to be a dissociation. Commonly it is applied to the dissociation of acids in water.

dissolution The taking up of a substance by a liquid, with the formation of a homogeneous solution.

dissymmetrical See **asymmetrical**.

distortion (1) Any departure from an intended or original shape because of internal stress or the release of **residual stress** in the material. (2) The permanent change in shape of a moulding or shaped product caused by relief of **frozen-in strain**, often by a rise in temperature.

distributive mixing Process where additives and fillers are mixed with polymer to produce a homogeneous material. Must be combined with **dispersive mixing**. Also *extensive mixing, blending*.

Dittus-Boelter equation An equation for the transfer of heat from tubes to viscous fluids flowing through them.

$$u \propto c \left(\frac{k \Delta s}{d} \right)^{1/3} \left(\frac{vd}{u/s} \right)^{1/2},$$

where u = film transfer factor, k = thermal conductivity, Δ = logarithmic mean temperature difference between tube and liquid, d = thickness of fluid stream, s = relative density of fluid, c = specific heat capacity of fluid, v = mean velocity of fluid in tube, and u/s = kinematic viscosity of the fluid.

divalent Capable of combining with two atoms of hydrogen or their equivalent. Also having an oxidation state of two.

DLC Abbrev. for **diamond-like carbon**.

DMC Abbrev. for **Dough Moulding Compound**.

DMOS transistor See **double-diffused metal-oxide semiconductor**. Also *double-diffused transistor*.

DMTA Abbrev. for **Dynamic Mechanical Thermal Analysis**.

DNA Abbrev. for **Desoxyribose Nucleic Acid**.

dobby Mechanism over the top or at the side of a loom. Operated by punched cards, it lifts and lowers the healds to move the warp threads in timed sequence to form the design in the cloth.

doctor A blade-like device resting at a shallow angle on the down-running surface of a roll or cylinder to remove unwanted material. Used in printing and papermaking.

doeskin cloth Similar to **beaver cloth** but finer and lighter, usually made from merino wool.

dolly (1) A heavy hammer-shaped tool for supporting the head of a rivet during the forming of a head on the other end. (2) A shaped block of lead used by panel beaters when hammering out dents.

dolomite The double carbonate of calcium and magnesium, occurring as cream-coloured crystals or masses with a distinctive pearly lustre. Calcined dolomite is used as a basic refractory for withstanding high temperatures and attack by basic slags in metallurgical furnaces. The name is also used to describe refractories made from magnesian limestone, which does not necessarily contain the mineral dolomite. See **magnesium oxide**.

domain (1) In ferroelectric, ferromagnetic and ferrimagnetic materials, a region where there is saturated polarization, depending only on temperature. The transition layer between adjacent domains is the **Bloch wall**, and the average size of the domain depends on the constituents of the material and its heat treatment. (2) Microstructural unit formed in polymeric materials, esp. by segregation of different chain segments in copolymers. In styrene butadiene styrene block copolymers, they may form a regular array, and typically have a diameter of 10 nm. See **copolymer** p. 75. In toughened polymers, they may form within the rubber particles. See **rubber toughening** p. 270.

donor Impurity atom which adds to the conductivity of a semiconductor by contributing electrons to a nearly empty conduction band, so making n-type conduction possible. See **acceptor**, **impurity**.

donor level See **energy levels**.

DOP Abbrev for **DiOctylPhthalate**

doped junction One with a semiconductor crystal which has had impurity added during a melt.

doping Addition of known impurities to a semiconductor or other material, to achieve the desired extrinsic properties.

Dorn effect The production of a potential difference when particles suspended in a liquid migrate under the influence of mechanical forces eg gravity; the converse of **electrophoresis**.

dor silver Silver bullion, ie ingots or bars, containing gold.

dose Quantity of material introduced; eg in ion implantation, the number of ions per square metre which are implanted into a surface.

double-action press A press fitted with two slides, permitting multiple operations, such as blanking and drawing, to be performed.

double bar and yoke method A ballistic method of magnetic testing, in which two test specimens are arranged parallel to each other and clamped to yokes at the ends to form a closed magnetic circuit. A correction for the effect of the yokes is made by altering their position on the bars and repeating the test.

double bond Covalent bond involving the sharing of two pairs of electrons.

double-cover butt joint A butt joint with a cover plate on both sides of the main plates.

double decomposition A reaction between two substances in which the atoms are rearranged to give two other substances. In general it may be written $AB + CD \rightarrow AC + BD$.

double-diffused metal-oxide semiconductor A metal-oxide semiconductor manufacturing process involving two stages of diffusion of impurities through a single **mask**, enabling depletion-mode or enhancement-mode transistors to be produced on a chip. The technique keeps **channels** short so that the devices are suitable for high-speed logic or microwave applications. Abbrev. *DMOS*.

double-diffused transistor One which is produced by using **double-diffused metal-oxide semiconductor** techniques.

doubled yarns See folded yarns.

double jersey A range of weft-knitted fabrics (rib or interlock) made on fine gauge machines. The construction is usually chosen so that the fabric has reduced extensibility.

double refraction Division of an electromagnetic wave in an anisotropic medium into two components propagated with different velocities, depending on the direction of propagation, causing two images to appear (eg in calcite). In uniaxial crystals the components are called the *ordinary ray* where the wavefronts are spherical and the *extraordinary ray* where the wavefronts are ellipsoidal. In biaxial crystals both wavefronts are ellipsoidal. Also *birefringence*.

double-row ball journal bearing A rolling bearing which has two rows of caged balls running in separate tracks. There are rigid, self-aligning, and angular contact double-row ball journal bearings.

double salts Compounds having two normal salts crystallizing together in definite molar ratios.

dough moulding compound Unsaturated polyester, thermosetting moulding materials in the form of dough or putty, and usually reinforced with chopped glass fibres. Abbrev. *DMC*.

Douglas fir *Pseudotsuga menziesii*, the

most important timber of the N American continent, and one of the best known **softwoods** in the world. It is not, however, a true fir (genus *Abies*). Light reddish-brown in colour, with prominent growth rings and fairly straight grain, it is moderately durable and is unaffected by powder-post beetle attack. More veneer and plywood produced than from any other species; it is also used extensively in heavy construction, laminated timber, joinery, paper pulp etc. Mean density 530 kg m^{-3}. Also *British Columbian pine, Oregon pine, red pine, false hemlock*.

dowel (1) A pin fixed in one part which, by accurately fitting in a hole in another attached part, locates the two, thus facilitating accurate reassembly. (2) A pin similarly used for locating divided patterns.

Down's process Electrolytic method of producing sodium metal and chlorine from fused salt at 600°C.

down time Time during which a machine, eg a computer or printing press (or series of machines), is idle because of adjustment, cleaning, reloading or other maintenance.

Dow process Extraction of magnesium from sea-water by precipitation as hydroxide with lime; also, electrolytic production of magnesium metal from fused chloride.

draft angle Slight angle between core and tool to aid the withdrawal of the product from mould at the end of the moulding cycle.

drafting The drawing-out or attenuation of the web of textile fibres passing through a card, drawframe, speedframe, or spinning machine. Measured by the ratio of the linear density of input to output materials. Cf **draw ratio**.

drag angle The angle between the welding rod and the normal to the surface of the weld.

drain In a field-effect transistor, the region into which majority **carriers** flow from the **channel**; comparable with the collector in a conventional bipolar transistor.

Dralon TN for German polyacrylonitrile staple fibre.

drawability A measure of the ability of a material to be drawn, as in forming a cuplike object from a flat metal blank.

drawing (1) The process of producing wire, or giving rods a good finish and accurate dimensions, by pulling through one or a series of tapered dies. (2) The process of hot or cold plastic stretching of polymeric fibre, tape, rod or sheet to orient the molecules and so increase the tensile strength and elastic modulus in the draw direction. (3) Running together and attenuation of a number of slivers or tops of staple fibres, preparatory to making **slubbings** and **rovings**. (4) Alternative term for **tempering** applied to the softening of steels which have been hardened by quenching.

drawing-down The operation of reducing the diameter of a metal bar, and increasing its length, by forging. Cf **cold drawing**.

drawing temper The operation of tempering hardened steel by heating to some specific temperature and quenching to obtain some definite degree of hardness. See **tempering, steels** p. 296.

draw ratio (1) Ratio of deformed to undeformed length in a polymer sheet during eg vacuum forming. Usually about 2 : 1 overall, but may reach 5 : 1 at corners of product. (2) Longitudinal extension of polymer fibre, tape or rod in post-production operation to increase modulus of product. Normally conducted while material is still hot. Draw ratios up to 72 times may be used in gel-spun polymers like Dyneema fibre. (3) The ratio of the linear density of undrawn yarn to that of the drawn yarn. See **drawing**.

dresser (1) An iron block used in forging bent work on an anvil. (2) A mallet for flattening sheet-lead. (3) A tool for facing and grooving millstones, or for trueing grinding wheels.

dress-face finish Lustrous nap on woollen cloths achieved by milling, raising and cropping the fibres at the cloth surface, then laying them in the same direction. Used in **beaver** and **doeskin** cloths.

dressing Fettling of castings and mouldings, the removal of **flash** and **runners**.

drier Furnace used to de-water ore products without changing their composition.

driers Substances accelerating the drying of vegetable oils, eg linseed oil in paints. The most important representatives of this group are the naphthenates, resinates and oleates of lead, manganese, cobalt, calcium and zinc.

drift (1) A tapered steel bar used to draw rivet holes into line. (2) A brass or copper bar used as a punch.

drift current That resulting from the drift of charge carriers in a local electric field as distinct from a **diffusion current**.

drifting test A workshop test for ductility; a hole is drilled near the edge of a plate and opened by a conical **drift** until cracking occurs.

drift mobility See **mobility**.

drill (1) A revolving tool with cutting edges at one end, and having flats or flutes for the release of chips; used for making cylindrical holes in metal. (2) Also the **drilling machine** which turns the drill. (3) The operation of making a hole in a workpiece. (4) A heavy woven fabric (often cotton) with diagonal lines on the surface.

drilling machine A machine tool for drilling holes, consisting generally of a vertical standard, carrying a table for supporting the work and an arm provided with bearings for the drilling spindle.

driving fit A degree of fit between two mating pieces such that the inner member, being slightly larger than the outer, must be driven in by a hammer or press.

drop forging The process of shaping metal parts by forging between two dies, one fixed to the hammer and the other to the anvil of a pneumatic or mechanical hammer. The dies are expensive, and the process is used for the mass-production of parts such as connecting-rods, crankshafts etc. Also *drop stamping*.

drop hammer A gravity-fall hammer or a double acting stamping hammer used to produce **drop forgings** by stamping hot metal between pairs of matching dies secured to the anvil block and to the top of the drop hammer respectively.

drop hammer test Product impact test of the compression resistance of a paper or cardboard box. See **impact tests** p. 165.

dropping mercury electrode A half-element consisting of mercury dropping in a fine stream through a solution. Used in *polarography*, a continuously renewed mercury surface being formed at the tip of a glass capillary, the accumulating impurities being swept away with the detaching drops of mercury.

drop stamping See **drop forging**.

dross Metallic oxides that rise to the surface of molten metal in metallurgical processes.

drossing Removal of scums, oxidized films and solidified metals from molten metals.

Drude law A law relating the specific rotation for polarized light of an optically active material to the wavelength of the incident light:

$$\alpha = \frac{k}{\lambda^2 - \lambda_0^2},$$

where α = specific rotation, k = rotation constant for material, λ_0^2 = dispersion constant for material, λ = wavelength of incident light. The law does not apply near absorption bands.

dry assay The determination of a given constituent in ores, metallurgical residues and alloys, by methods which do not involve liquid means of separation.

dry battery A battery composed of **dry cells**.

dry brushing The process of gently brushing a fabric to raise the fibres on the surface (eg with a teazle).

dry cell A **primary cell** in which the contents are in the form of a paste. See **Leclanché cell**.

dry copper Copper containing oxygen in excess of that required to give **tough pitch**. Such metal is liable to be brittle in hot- and cold-working operations.

dry electrolytic capacitor One in which the negative pole takes the form of a sticky paste, which is sufficiently conducting to maintain a gas and oxide film on the positive aluminium electrode.

dry etching In the processing of semiconductors, the use of gas discharge media to etch features, as opposed to the use of wet chemicals. See **semiconductor device processing** p. 280.

dry ice Solid (frozen) carbon dioxide, used in refrigeration (storage) and engineering. At ordinary atmospheric pressure it sublimes slowly.

dry indicator test Method of determining the resistance of a paper to penetration by aqueous media. A powder that intensifies in colour when wet, is sprinkled on a test sample floating on water and the time measured until an agreed degree of colour change has taken place.

dry joint A faulty solder joint giving high-resistance contact due to residual oxide film.

dry laying The formation of a web of fibres by carding or air-laying preparatory to the manufacture of a non-woven fabric.

dry spinning The production of polymeric filaments by the extrusion of a solution of the polymer in a volatile liquid which is then removed by evaporation.

dry-spun flax The coarse flax yarn obtained from a dry roving. See **wet-spinning**.

DSC Abbrev. for **Differential Scanning Calorimetry**.

DTA Abbrev. for **Differential Thermal Analysis**.

dual in-line package A common integrated-circuit package having two parallel rows of connectors at right angles to the body, as required for insertion into pre-drilled holes in a printed-circuit board. Abbrev. *DIL*.

du Bois balance An instrument used for measuring the permeability of iron or steel rods. The magnetic attraction across an air gap in a magnetic circuit, of which the sample forms a part, is balanced against the gravitational force due to a sliding weight on a beam.

duck A plain, bleached cotton or linen cloth, used for tropical suitings. Heavier makes are used for sails, tents and conveyor belting. Similar to **canvas**.

duct (1) A hole, pipe or channel carrying a fluid, eg for lubricating, heating or cooling. (2) A large sheet metal tube or casing through which air is passed for forced-draught, ventilating or conditioning purposes.

ductile-brittle transition temperature
That at which the failure mode of a material, esp. metals and plastics, changes from ductile, higher energy, to brittle, lower energy, as the temperature is reduced. Symbol T_b. Transition often mapped by **impact tests** p. 165. Also *brittle temperature*.

ductile cast iron Cast iron in which the free graphite has been induced to form as nodules by adding cerium or magnesium in the molten state which gives a marked increase in ductility. Abbrev. *SG iron* for **spherulitic graphite cast iron**.

ductile fracture Type of fracture in any material where substantial deformation has occurred away from fracture surfaces. Usually associated with yielding in materials.

ductility Ability of materials to be deformed by working processes and to retain strength and freedom from cracks when their shape is altered. See **work-hardening**.

duffel, duffle Heavy, low-quality, woven woollen fabric raised on both sides. Short coats made from this fabric.

duke A notepaper size, 178 × 143 mm (7⅝ in).

Dulong and Petit's law See **law of Dulong and Petit**.

dummying The preliminary rough-shaping of the heated metal before placing between the dies for drop-forging.

dungaree A cotton cloth, with a twill weave, made from coloured warp and weft yarns, generally used for overalls.

Dural, Duralumin TNs for a precipitation hardenable aluminium alloy containing nominally 4% copper and 0.5% manganese. See **aluminium alloys** p. 8.

dust Particulate material which is or has been airborne and which passes a 200 mesh BS test sieve (76 μm).

dust core Magnetic circuit embracing or threading a high-frequency coil, that is made of ferromagnetic particles compressed into an insulating matrix binder, thus reducing losses at high frequency because of eddy currents.

dust explosion An explosion resulting from the ignition of small concentrations of flammable dust (eg finely divided metal particles, coal dust, sugar or flour) in the air.

Dutch gold A cheap alternative to gold-leaf, consisting of copper-leaf, which, by exposure to the fumes from molten zinc, acquires a yellow colour.

dwell Term used to describe the pause or hold in the application of pressure in a moulding process.

dyeline base paper Paper with controlled chemical and physical properties to enable it to be coated satisfactorily with a diazo compound and thereafter used to make a dyeline print.

dyestuffs Groups of aromatic compounds having the property of dyeing textile fibres or transparent polymers.

dynamic friction See **friction**.

dynamic mechanical test Type of test which seeks to measure mechanical properties, eg tensile modulus E, under dynamic conditions, such as regular vibrations. A common apparatus for polymeric materials is the **torsion pendulum**. Similar tests give a **complex modulus**, which can be divided into real and imaginary parts.

dynamic mechanical thermal analysis Method of measuring complex moduli of materials as a function of temperature, using torsion pendulum, vibrating reed etc. Abbrev. *DMTA*.

dynamics That branch of applied mathematics which studies the way in which force produces motion.

dynamic resistance The relationship between voltage and current at a given position on the non-linear static characteristic of a device in an electrical circuit, eg a diode. It may be regarded as the tangent to the characteristic curve at that point and is often assumed to be constant over a small range of voltage and current. See **differential resistance**.

dynamic viscosity See **coefficient of viscosity**.

Dyneema TN for gel-spun polyethylene fibre (Europe). See **high-performance fibres** p. 158.

dys- Prefix from Gk *dys-*, in English mis-, un-.

dysprosium A metallic element, a member of the rare-earth group, symbol Dy. See Appendix 1, 2.

E

e Symbol for the electron (e⁻) or positron (e⁺).

e Symbol for the elementary charge; 1.6022×10^{-19} coulomb.

ε Symbol for: (1) emissivity; (2) linear strain; (3) permittivity.

E Symbol for: (1) **Young's modulus**; (2) **electric field**; (3) **internal energy** (US); (4) **electrochemical potential** or *electrochemical force*.

ear A cast or forged projection integral with, or attached to, an object, for supporting it, or attaching another part to it pivotally. Also *lug*.

earing Excessive elongation along edges and folds of metals being shaped by deep-drawing or rolling.

Early effect Variation of junction capacitance and effective base thickness of transistor with the supply potentials.

earth Connection to main mass of Earth by means of a conductor having a very low impedance.

earthenware General term for vessels and other products made from opaque, permeable ceramic whiteware, which is fired at a lower temperature (1100–1150°C), and so has a lower proportion of glass phase, and is more porous and less strong than eg **porcelain**. Must be glazed to make it impermeable. Cf **bone china, faience, stoneware, terracotta**.

earthquake mount Laminated bearing of steel and rubber for buildings, designed to resist vibrations from earthquakes. Widely used in California and Japan.

East African olive See **olive**.

East Indian satinwood See **Ceylon satinwood**.

easy axis Crystal axis in magnetic materials along which magnetization is energetically preferred in the absence of external fields. See **magnetocrystalline anisotropy**.

EBM Abbrev. for **Electron Beam Machining**.

ebonite Highly crosslinked and filled natural or synthetic rubber, vulcanized with up to about 40% sulphur. A hard brittle material, it was formerly much used for battery cases etc but has now been largely displaced by polypropylene, polycarbonate etc. Also *hard rubber, vulcanite*.

ebony Hardwood from a tree of the genus *Diospyros*, a native of W Africa (*D. crassiflora*), India and The Celebes (*D. celebica*), very dense, dark brown to black, straight grained and fine textured. It is naturally resistant to most diseases and insects.

Typical uses are in musical instruments, tool and cutlery handles, inlay work, veneers, etc. Density 1000–1090 kg m⁻³.

ebullioscopy The determination of the molecular mass of a material by observing the elevation of the boiling point of a suitable solvent.

EBW Abbrev. for **Electron Beam Welding**.

eccentric (1) Displaced with reference to a centre; not concentric. (2) A crank in which the pin diameter exceeds the stroke, resulting in a disk eccentric to the shaft; used as a crank, particularly for operating steam-engine valves, pump plungers etc.

ECM Abbrev. for **ElectroChemical Machining**.

ecobalance A description of an industrial process in terms of the materials and energy inputs and the outputs of solid, liquid and gaseous wastes. See **life cycle analysis, life cycle inventory**. Also *ecoprofile*.

ecoprofile See **ecobalance**.

ecru Unbleached knitted fabrics and their colour.

EDAX Abbrev. for **Energy Dispersive Analysis of X-rays**.

eddy An interruption in the steady flow of a fluid, caused by an obstacle situated in the line of flow; the *vortex* so formed.

eddy current heating See **Induction heating**.

eddy currents Those arising through varying electromotive forces consequent on varying magnetic fields, resulting in diminution of the latter and dissipation of power. They are one of the main causes of heating in motors, transformers, etc (known also as *iron loss*, cf **copper loss**). To minimize them the iron cores of such machines are composed of many layers of thin iron sheet (laminations) which are insulated from one another to reduce the currents' ability to flow in the direction in which the field tries to induce them. This in turn slightly increases the reluctance of the magnetic circuit, thereby reducing the overall efficiency of the machine, hence a compromise must be struck. Used for mechanical damping and braking (as in electricity meters) and for induction heating as applied in case-hardening. Also *Foucault current(s)*.

eddy current testing Non-destructive test using an electromagnetic field to induce eddy currents in a component. Discontinuities, flaws, internal cracks and changes in shape in a metal component are detected by variations in the signal produced in pick-up coils located nearby.

edge dislocation Line defect within a crystal in which the **Burgers vector** is perpendicular to the line of the **dislocation**.

edge sealed The edge of a fabric that has been made without the usual **selvedge** and has been cut and sealed (usually by the heating and melting of a thermoplastic fibre) to prevent fraying.

edge tool A hand-worked, mallet-struck or machine-operated cutting tool with one or a regular pattern of cutting edges, ie excluding grinding wheels.

effective mass For electrons and/or holes in a semiconductor, effective mass is a parameter which may differ appreciably from the mass of a free electron, and which depends to some extent on the position of the particle in its energy band. This modifies the mobility and hence the resulting current.

effective particle density The mass of a particle divided by its volume, including opened and closed pores.

effective value That of a simple parameter which has the same effect as a more complex one, eg r.m.s. value of a.c. euals the d.c. value for many purposes.

effervescence The vigorous escape of small gas bubbles from a liquid, esp. as a result of chemical action.

efficiency A non-dimensional measure of the performance of a piece of apparatus, eg an engine, obtained from the ratio of the output of a quantity, eg power, energy, to its input, often expressed as a percentage. The power efficiency of an *IC engine* is the ratio of the shaft- or brake-horsepower to the rate of intake of fuel, expressed in units of energy content per unit time. It must always be less than 100% which would imply perpetual motion. Not to be confused with *efficacy*, which takes account only of the output of the apparatus, and is not given an exact quantitative definition.

efflorescence The loss of water from a crystalline hydrate on exposure to air, shown by the formation of a powder on the crystal surface.

effusion The passage of a gas through a small aperture. See **Graham's law**.

eggshell finish A soft dull finish on paint or paper.

E-glass Designation for a glass of composition (wt %): SiO_2 52.9, CaO 17.4, Al_2O_3 14.5, B_2O_3 9.2, MgO 4.4, Na_2O 1.0, K_2O 1.0. It is the most widely used glass fibre for the reinforcement of **composite materials**, with a chemical resistance between those of **A-** and **C-glass**.

EIA Abbrev. for **Environmental Impact Assessment**.

eighteen-electron rule The structures of most stable organometallic compounds of the **transition metals**, eg ferrocene, can be rationalized by showing that 18 valence electrons can be associated with each metal atom. See **electron octet**.

Einstein–de Haas effect When a magnetic field is applied to a body the precessional motion of the electrons produce a mechanical moment that is transferred to the body as a whole. See **Barnett effect**.

Einstein diffusion equation As a result of **Brownian motion**, molecules or colloidal particles migrate an average distance, δ, in each small time interval, τ. The equation for the diffusion of a spherical particle of radius, r, through a fluid of viscosity, η is:

$$D = \frac{\delta^2}{2\tau} = \frac{RT}{6\pi\eta r N_A},$$

where D is the diffusion coefficient, N_A is **Avogadro's number**, R is the **gas constant** and T the absolute temperature.

Einstein equation for the specific heat of a solid One mole of the solid consists of N_A atoms, each vibrating with frequency, v, in three dimensions. From quantum theory, the molar specific heat is:

$$C_v = \frac{3R x^2 e^x}{(e^x - 1)^2},$$

where $x = hv(kT)^{-1}$, R is the gas constant, h is **Planck's constant**, k is **Boltzmann's** constant and T the absolute temperature.

Einstein law of photochemical equivalence Each quantum of radiation absorbed in a photochemical process causes the decomposition of one molecule.

Einstein photoelectric equation That which gives the energy of an electron, just ejected photoelectrically from a surface by a photon, ie $E = hv - \phi$, where $E =$ kinetic energy, $h =$ **Planck's constant**, $v =$ frequency of photon, and $\phi =$ **work function**.

Einstein theory of specific heat of solids Based on the assumption that the thermal vibrations of the atoms in a solid can be represented by harmonic oscillators of one frequency, whose energy is quantized. See **Debye theory, Blackman theory**.

EIS Abbrev. for **Environmental Impact Statement**.

ejector pin Steel pin which is activated at end of moulding cycle to push moulding from tool cavity.

eka- A prefix, coined by Mendeleev, denoting the element occupying the next lower position in the same group in the periodic table, used in the naming of new elements and unstable radio-elements. See Appendix 1.

elastic constant Materials property which is the ratio of stress to strain for different modes of **elastic deformation**. In linear, **iso-**

tropic elastic materials, only four constants are needed to characterize fully the elastic behaviour of the material: **Young's modulus**, *E*, **shear** (or *rigidity*) **modulus**, *G*, **bulk modulus**, *K*, which have the units of stress (N m^{-2}) and Poisson's ratio, ν, which is dimensionless. These are related through:

$$E = 2G(1 + \nu) = 3K(1 - 2\nu)$$

so that either *E* and ν, or *K* and *G*, are sufficient to allow the other pair to be determined. In anisotropic elastic materials, up to 36 elastic constants are required. Also *elastic modulus* or *modulus of elasticity*. See **Lamé constants, specific modulus, tensile modulus.**

elastic deformation Any change in shape in response to an applied force in which the initial shape is recoverable with no sensible time delay when the applied force is removed.

elastic fatigue Obsolete and confusing term for time-dependent recovery from 'elastic' deformation. Now recognized as an aspect of **viscoelasticity** p. 332.

elasticity The tendency of matter, whether gaseous, liquid or solid, to return to its original size or shape, after having been stretched, compressed, or otherwise deformed. See **elastic constant, Hooke's law.** Cf **plasticity, viscoelasticity** p. 332.

elastic limit The highest stress that can be applied to a material without producing a measurable amount of plastic (ie permanent) deformation. Usually assumed to coincide with the **proportional limit.** See **strength measures** p. 301.

elastic liquid Polymer melt or concentrated solution exhibiting memory of previous condition, usually behaving like an uncrosslinked elastomer. In one demonstration, a viscous fluid snaps back into its container after cutting the pouring liquid with scissors! See **Weissenberg effect, die swell, melt fracture, Kaye effect.**

elastic modulus See **bulk modulus, complex modulus, elastic constant, Poisson's ratio, shear modulus, storage modulus, tensile modulus, Young's modulus.**

elastic strain The recoverable strain or fractional deformation undergone by a material, ie that which disappears as the straining force is removed.

elastin A structural protein abundant in elastic body tissues such as skin and internal membranes. A heteropolymer of some sixteen different amino acids, it is largely noncrystalline and exhibits long-range reversible elasticity like natural rubber. Permanent crosslinks occur through the lysine residues in the main chain. See **resilin.**

elastomers See panel on p. 106.

elastomeric state General condition of amorphous thermoplastic polymers above glass-transition temperature. See **viscoelasticity** p. 332.

electret Permanently polarized dielectric material, formed by cooling a **ferroelectric** (eg barium titanate) from above a **Curie point** or waxes (eg carnauba) in a strong electric field.

electrical conductivity See **conductivity.**

electrical discharge machining Spark erosion technique in which metal is removed as sparks pass between a shaped electrode and the work. Can be used for machining irregular holes etc. Abbrev. *EDM.*

electrical resistivity See **resistivity.**

electric field Region in which forces are exerted on any electric charge present.

electric field strength The strength of an electric field is measured by the force exerted on a unit charge at a given point. Expressed in volts/metre. Symbol *E.*

electric polarization The dipole moment per unit volume of a dielectric.

electric potential That measured by the energy of a unit positive charge at a point, expressed relative to zero potential.

electric strength Maximum voltage which can be applied to an insulator or insulating material without sparkover or breakdown taking place. The latter arises when the applied voltage gradient coincides with a breakdown strength at a temperature which is attained through normal heat dissipation.

electric susceptibility The amount by which the relative permittivity of a dielectric exceeds unity, or the ratio of the polarization produced by unit field to the permittivity of free space. See **dielectrics and ferroelectrics** p. 92.

electroactive polymer composites See panel on p. 107.

electrobrightening See **electrolytic polishing.**

electroceramic processing See panel on p. 109.

electrochemical constant See **faraday.**

electrochemical machining Removing material from a metal by anodic dissolution in a bath in which electrolyte is pumped rapidly through the gap between the shaped electrode and the stock. Abbrev. *ECM.*

electrochemical potential Difference in voltage between anode and cathode in an electrochemical cell.

electrochemical potential series The classification of redox half-reactions, written as reductions, in order of decreasing reducing strength. Thus the combination of any half reaction with the reverse of one further down the series will give a spontaneous reaction. For reference, the half reaction

elastomers

Any material in the elastomeric state, specifically polymers which are elastomeric at ambient temperatures and are lightly crosslinked for stability. Most linear polymers become elastomeric as the temperature rises above T_g (see **viscoelasticity** p. 332) but it is those polymers which are elastomeric at ambient temperature or below which have become important for engineering applications, whether *general purpose rubbers* (GP) or *speciality materials*. Historically, the most important elastomer was natural rubber crosslinked with sulphur (vulcanization) and often compounded with carbon black to reinforce and toughen the material. A wide range of synthetic rubbers is now available, many of which compete directly with the natural material.

All these materials possess a double bond in the main chain, a feature which reduces **steric hindrance**, making the polymer chains highly flexible and able to adopt a large number of different conformations at ambient temperature (see **rotational isomerism**). This is the origin of both long-range elasticity and the retractive force. Provided chain slippage (ie viscous flow) is prevented by a small degree of crosslinking (1–2 % typically), then elastomers will extend by several hundred percent, when stressed, and return to their original shape when released.

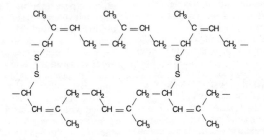

Fig. 1. **Crosslinking of rubbers**.

Natural rubber and butadiene polymers are crosslinked (vulcanized) with sulphur, the chains being crosslinked between sites adjacent to the double bond. See Fig. 1. Accelerators are normally used to improve the efficiency of the reaction. The relatively small stress needed to extend elastomers is entropic rather than energetic in origin as it is in most rigid materials like glass and steel, reflecting the tendency of chains to return to their equilibrium random coil conformation. It also helps to explain why the modulus increases with rising temperature, rather than falling as in glass and steel. Thermal vibration increases with temperature, so increasing the number of possible conformations of the random coils. See **kinetic theory of elasticity**.

Another phenomenon shown by elastomers is the rise in temperature when strained quickly, a direct analogue of the **Joule–Thomson effect** in gases. The temperature rise in natural rubber, eg common elastic bands when stressed on the lips, is even greater due to strain crystallization. The release of the heat of fusion (ΔH_f) heats the sample up by about one degree C, but on release, the crystallites melt and the rubber cools by roughly the same amount.

For most engineering applications (bushes, engine mounts, seals, bearings) the high extensibility is not needed and it is usually desirable to increase the modulus. Although increasing the degree of crosslinking will help, the material loses its toughness (eg ebonite). The preferred alternative is to reinforce with carbon black, as in the ubiquitous car tyre. See **tyre technology** p. 325.

electroactive polymer composites

Electroactive polymer composites are a class of materials developed specifically to extend the properties and range of application of established single phase piezoelectric materials. Materials in common use tend to be limited to either acoustic generation or reception applications, but not both. Lead titanate zirconate (PZT), for example, has a relatively high hydrostatic strain coupling constant, d_h, but suffers from a reduced hydrostatic voltage coefficient, g_h, by virtue of its high dielectric constant. The practical applications of PZT are further limited by its stiffness and high density. Polyvinylidene fluoride (PVDF), on the other hand, has a conformal structure and low dielectric constant which gives rise to a high g_h, but is difficult to fabricate. In addition it has a relatively low **Curie point** (temperature) which can lead to problems with depoling and a low d_h, making it suitable for sound generation applications.

The properties of both PZT and PVDF are inhibited further by the tensor nature of the piezoelectric effect. An applied hydrostatic strain to either material, for example, will couple via the individual piezoelectric strain coefficients d_{31} and d_{33} to produce two significant, but opposing voltages across the specimen. These cancel to reduce d_h and so limit the overall performance of the material.

Electroactive composites are an attempt to combine the desirable properties of piezoelectric ceramics and polymers without suffering their intrinsic limitations. They consist of a mixture of processed piezoelectric ceramic and an electrically inert polymer such as polypropylene. The non-piezoelectric polymer gives the composite the desired low permittivity and density whilst its compliant nature enables it to be shaped into specialized geometries and to resist thermal shock. Finally, the connectivity of the ceramic phase can be adjusted to optimize its piezoelectric response.

Two main factors govern the choice of electroactive ceramic and polymer components: (1) the way the individual phases couple under the application of an electric field and (2) the overall physical and compositional symmetry of the product material. The physical connection between the component phases is the key to the piezoelectric properties of the composite. Each phase may be connected in 0, 1, 2 or 3 dimensions, corresponding to 10 possible diphasic connectivities (*binary combinations*). Only composites with 1–3 or 0–3 connectivity, however, are in common use for piezoelectric transduction applications. In this notation the first digit identifies the dimensionality of the electroactive phase of the composite, 0 refers to an unconnected, or discrete point phase, 1 to fibres or rods, 2 to sheets or planes and 3 to a full 3D connectivity. Modelled and practical 0–3 and 1–3 connected structures are illustrated in Fig. 1.

Fig. 1. **Modelled and practical structures for piezoelectric polymer complexes.**

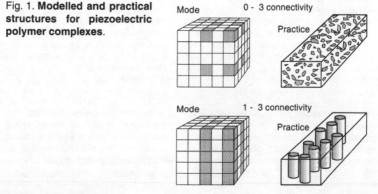

Mode   0 - 3 connectivity

Practice

Mode   1 - 3 connectivity

Practice

continued on next page

electroactive polymer composites (contd)

A major advantage of the 1–3 material is that most of the applied stress is taken across the ceramic phase which amplifies the piezoelectric properties of the composite. In addition, d_{31} is greatly reduced in this connectivity so the voltage cancellation effect associated with the 3D piezoelectric effect is less significant.

Although 0–3 composites do not have the same stress concentration factor as 1–3 materials, they offer the advantage of being much easier to fabricate. They can be produced in the form of thin sheets or moulded into shapes for specific applications. This is particularly important for large area, conformal applications such as underwater noise cancelling devices, for which composites offer significantly better impedance matching properties than ceramics.

The properties of 0–3 electroactive composites are compared with common single-phase piezoelectric materials in Table 1.

Table 1. **Comparison of properties of electroactive composites**.

Property	Ceramic (PZT)	Polymer (PVDF)	Composite (0–3)
d_h (pC N^{-1})	60	–10	30
g_h (mV mN^{-1})	2	–100	110
ε_r	1000–2400	12	30
Density (kg m^{-3})	7600	1800	4000

$2H^+ + e^- = H_2$ is taken as having an energy of zero. Also *electrode potential series*.

electrochemistry That branch of chemistry which deals with the electronic and electrical aspects of processes in a liquid or solid phase.

electrode Conductor whereby an electric current is led into or out of a liquid (as in an electrolytic cell), or a gas (as in an electric discharge lamp or gas tube), or a vacuum (as in a valve).

electrodeposition Deposition electrolytically of a substance on an electrode, as in **electroplating** or **electroforming**.

electrode potential Voltage between electrode and electrolyte in electrochemical cell.

electro-extraction The recovery of a metal from a solution of its salts by electrolysis, the metal depositing on the cathode. Also *electrowinning*.

electrofluor A transparent material which has the property of storing electrical energy and releasing it as visible light.

electrofluorescence See **electroluminescence**.

electroforming A primary process of forming metals, in which parts are produced by electrolytic deposition of metal on a conductive removable mould or matrix.

electrogalvanizing Electrodepositing zinc on metals for corrosion protection. Thinner than hot-dipped coating and consequently it has a much reduced life outdoors.

electrographite Carbon transformed into graphite by heating to temperatures in the range 2200°C to 2800°C.

electrohydraulic forming The discharge of electrical energy across a small gap between electrodes immersed in water, which produces a shock wave. The wave travels through the water, until it hits and forces a metal workpiece into the particular shape of the die which surrounds it, thus, eg piercing holes, rock-crushing etc. Sometimes called *explosive forming*.

electroluminescence The luminescence produced by the application of an electric field to a dielectric **phosphor** (such as Mn doped ZnS). Also *electrofluorescence*. See **Gudden–Pohl effect**.

electrolysis Chemical change, generally decomposition, effected by a flow of current through a solution of the chemical, or its molten state, based on ionization.

electrolyte Chemical, or its solution in water, which conducts current through ionization. Molten salts are also electrolytes.

electrolyte strength Extent towards complete ionization in a dilute solution. When concentrated, the ions join in groups, as indi-

electroceramic processing

Ceramic is the term given to a broad class of materials that are composed of inorganic, polycrystalline compounds. They are commonly complex oxides or nitrides of metals or semimetals and are characteristically refractory in nature; ceramics are brittle and difficult to form into specialized geometries.

Ceramics play an important role in a number of electronic devices such as those used in piezoelectric, pyroelectric, electro-optic and electromagnetic applications. Their most common use is as dielectric fillings for capacitors in passive components although more specialized applications such as ferro-electric memories and transducers are becoming increasingly common.

Bulk ceramics for electronic applications are formulated to a specific composition by a number of distinct chemical and mechanical processes. Typical examples are the high-temperature superconductor containing yttrium, barium and copper, $YBa_2Cu_3O_{7-x}$ (*YBCO*), where x denotes a departure from stoichiometry, and the *ferroelectric*, barium titanate, $BaTiO_3$ (*BT*). Powders of the oxides of the various metal, or cation, components are mixed thoroughly in the appropriate proportions for the composition. Alternatively carbonates or nitrates, which decompose readily to oxides, may be used. For example, $BaCO_3$ and TiO_2 are mixed in a 1:1 cation ratio in the fabrication of BT. The oxide/carbonate/nitrate mix is heated to a temperature at which the individual powders partially react to form a single compound. This process is called *calcination* and is typically performed in the range 500–800°C. The calcined powder is then finely ground, or milled, before being compacted, typically in cylindrical dies, in pressures up to 80 MPa.

In a further heat treatment, the pressed pellet is sintered for several hours at a temperature usually a few hundred degrees above that needed for calcination. This process completes the reaction of the constituent cations and transforms the compacted powder into a single ceramic body with grains typically 5–20 µm in diameter. The sintered block is usually sliced into wafers for further processing such as lapping, polishing, electroding and, for ferroelectric devices, **poling** by the application of an electric field. These steps are summarized in Fig. 1.

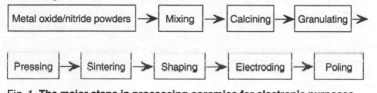

Fig. 1. **The major steps in processing ceramics for electronic purposes**.

continued on next page

cated by lowered **mobility**.

electrolytic capacitor An electrolytic cell in which a very thin layer of non-conducting material has been deposited on one of the electrodes by an electric current. This is known as 'forming' the capacitor, the deposited layer providing the dielectric. Because of its thinness a larger capacitance is achieved in a smaller volume than in the normal construction of a capacitor. In the so-called *dry electrolytic capacitor* the dielectric layer is a gas which however is actually 'formed' from a moist paste within the capacitor.

electrolytic cell An electrochemical cell in which an externally applied voltage causes a chemical change to occur, such as the breakdown of water into hydrogen and oxygen. Opposite of **galvanic cell**.

electrolytic copper Copper refined by electrolysis. This gives metal of high purity (over 99.94% copper), and enables precious metals, such as gold and silver, to be recovered.

electrolytic corrosion Corrosion produced by contact of two different metals *electroceramic processing*

A variety of methods has been developed to improve the chemical uniformity and mechanical properties of the processed ceramics. A homogeneous mixture of the component cations, for example, can be achieved by dissolving soluble forms of the base compounds (usually nitrates) in deionized water and spray drying them at 200°C. Again, sintering under pressure (*hot isostatic pressing*) in a controlled oxidizing environment may give fully dense, pore-free ceramic. This process is aided significantly for most ceramics if the pre-sintered powder consists of fine, spherical particles at around 1 μm diameter. Calcination and sintering times and temperatures, post-sintering cooling rates and gaseous firing environment are other key variables that may be optimized empirically to control the properties of the processed material. The control of impurities and/or minority ceramic phases, formed or liberated during the processing, is particularly important since these tend to concentrate at grain boundaries and can significantly alter the electrical properties of the product.

The use of thin film ceramics for small-area electronic devices is becoming increasingly common, particularly in multi-layer ceramic capacitors. Two of the most successful manufacturing techniques involve deposition of the precursor onto a substrate from either a solution in **sol-gel processes** or a vapour in metallo-organic chemical vapour deposition (MOCVD) processes. See **epitaxial growth of semiconductors** p. 117. Both are multiple step: thin sub-micron layers are deposited and fired repeatedly in each process to build a film about 5 μm thick. Thin film ceramics are commonly applied in pyroelectric transducers, where their low mass gives a high thermal response, and in ferroelectric devices that require the application of an electric field.

For applications which require a homogeneous film over a large surface area, such as pyroelectric infrared imaging arrays, wafers of bulk ceramic are usually thinned mechanically by lapping. This is time consuming and wasteful and cannot produce 'films' thinner than about 20 μm.

The choice of electrode for the processed ceramic and the method of application depends largely on the shape and composition of the ceramic surface. In high *permittivity* applications, such as piezoelectric transducers, the electrode must adhere closely to the ceramic. This can be achieved by evaporating an electrode (usually gold) under vacuum onto the polished surface (no pretreatment of a grown thin film is usually necessary). Alternatively a silver-loaded epoxy can be fired on the ceramic if adhesion is difficult. A simple silver paint deposited on an as-cut, rough ceramic is often suitable for low-permittivity applications such as positive temperature coefficient (PTCR) devices.

when an electrolyte is present and current flows.

electrolytic dissociation The reversible splitting of substances into oppositely charged ions.

electrolytic grinding A metal bonded and diamond impregnated grinding wheel removes metal from the stock but in addition the abradant is an insulator between the wheel acting as anode and the stock which has electrolyte flowing over it. This allows **electrochemical machining** to occur.

electrolytic lead Lead refined by the Betts process; has purity of about 99.995–99.998% lead.

electrolytic machining An electrochemical process based on the same principles as

electroplating, except that the workpiece is the anode and the tool is the cathode, resulting in a deplating operation. Sometimes, eg in **electrolytic grinding**, combined with some abrasive action.

electrolytic polishing By making the metal surface an anode and passing a current under certain conditions, there is a preferential solution so the microscopic irregularities vanish, leaving a smoother surface. Also *anode brightening, anode polishing, electrobrightening, electropolishing.*

electrolytic refining The method of producing pure metals, by making the impure metal the anode in an electrolytic cell and depositing a pure cathode. The impurities either remain undissolved at the anode or pass into solution in the electrolyte. Also *electrofining*.

electrolytic zinc Zinc produced from its ores by roasting (to convert sulphide to oxide), solution of oxide in sulphuric acid, precipitation of impurities by adding zinc dust, and final electrolytic deposition of zinc on aluminium cathodes. Product has purity over 99.9%.

electromagnet Soft iron core, embraced by a current-carrying coil, which exhibits appreciable magnetic effects only when current passes.

electromagnetic pick-up One in which the motion of the stylus, in following the recorded track, causes a fluctuation in the magnetic flux carried in any part of a magnetic circuit, with consequent electromotive forces in any coil embracing such magnetic circuit.

electromagnetic pole-piece In a U-shaped core, the pole-pieces are attached at the free end and are often conical in shape to concentrate the magnetic field in the air gap.

electromagnetic radiation The emission and propagation of electromagnetic energy from a source including long (radio) waves, heat rays, light, X-rays and γ-rays.

electromagnetics See **electromagnetism**.

electromagnetism Science of the properties of, and relations between, magnetism and electric currents. Also *electromagnetics*.

electrometallization The electrodeposition of a metal on a non-conducting base, either for decorative purposes or to give a protective covering.

electrometallurgy A term covering the various electrical processes for the industrial working of metals, eg electrodeposition, electrorefining and operations in electric furnaces.

electron A fundamental particle with negative electric charge of 1.602×10^{-19} C and mass 9.109×10^{-31} kg. Electrons are a basic constituent of the atom and are distributed around the nucleus in shells and the electronic structure is responsible for the chemical properties of the atom. Electrons also exist independently and are responsible for many electric effects in materials. Due to their small mass, the wave properties and relativistic effects of electrons are marked. The positron, the antiparticle of the electron, is an equivalent particle but with a positive charge. Either electrons or positrons may be emitted in β-decay. Electrons, muons and neutrinos form a group of fundamental particles called *leptons*.

electron affinity (1) The energy required to remove an electron from a negatively charged ion to form a neutral atom. (2) Tendency of certain substances, notably oxidiz-ing agents, to capture an electron. (3) See **work function**.

electron-beam analysis Scanning a microbeam of electrons over a surface in vacuo and analysing the secondary emissions to determine the distribution of selected elements.

electron-beam welding Heating components to be welded by a concentrated beam of high-velocity electrons in vacuo. Abbrev. *EBW*.

electron binding energy Same as **ionization potential**.

electron cloud (1) The density of electrons in a volume of space, as the position and velocity of an electron cannot be simultaneously specified. (2) The nature of the valence electrons in a metal, where their non-attachment to specific nuclei gives rise to electrical conduction.

electron compound A solid phase appearing in metallic systems whose composition is largely determined by a particular ratio of valency electrons to number of atoms, eg CuZn occurring at a ratio of 3 : 2. Usually associated with extensive solubility on either side of the ideal ratio.

electron conduction That which arises from the drift of conduction electrons in metallic materials when an electric field is applied. See **n-type** and **p-type semiconductor**.

electron-deficient Of a substance which does not have enough valence electrons to form 'normal' chemical bonds. Usually, the term is used for the compounds of boron, but most metallic bonding is of this type.

electron density The number of electrons per unit volume, particularly conduction electrons in semiconductors, measured as the number per cubic metre (or per cubic centimetre).

electron diffraction Investigation of crystal structure by the patterns obtained on a screen from electrons diffracted from the surface of crystals or as a result of transmission through thin metal films.

electron dispersion curve A curve showing the electron energy as a function of the wave vector under the influence of the periodic potential of a crystal lattice. Experiments and calculations which determine such curves give important information about the energy gaps, the electron velocities and the **density of states**.

electronegativity The relative ability of an atom to retain or gain electrons. There are several definitions, and the term is quantitative on a scale between about one and four. It is useful in predicting the strengths and the polarities of bonds, both of which are greater when there is a significant electronegativity difference between the atoms forming the bond. See Appendix 2.

electron emission The liberation of electrons from a surface.

electron gas The 'atmosphere' of free electrons in vacuo, in a gas or in a conducting solid. The laws obeyed by an electron gas are governed by **Fermi–Dirac statistics**, unlike ordinary gases to which Maxwell–Boltzmann statistics apply.

electronic Pertaining to devices or systems which depend on the flow of electrons; the term covers most branches of electrical science other than electric power generation and distribution. Telecommunications, radar and computers all use electronic components and techniques. Electronic engineering is a field which encompasses the application of electronic devices, as opposed to physical electronics which is the study of electronic phenomena in vacuo, in gases, or in solids.

electronic charge The unit in which all atomic and subatomic charges are expressed. It is equal to 1.602×10^{-19} coulombs; that on the electron is negative.

electronic configuration The descriptions of the electrons of an atom or a molecule in terms of orbitals.

electronic theory of valency Valency forces arise from the transfer or sharing of the electrons in the outer shells of the atoms in a molecule. The two extremes are complete transfer (ionic bond) and close sharing (covalent bond), but there are intermediate degrees of bond strength and distance.

electron mass A result of relativity theory, that mass can be ascribed to kinetic energy, is that the effective mass (m) of the electron should vary with its velocity according to the experimentally confirmed expression:

$$m = \frac{m_0}{\sqrt{1 - (v/c)^2}},$$

where m_0 is the mass for small velocities, c is the velocity of light, and v that of the electron.

electron mobility See **mobility**.

electron octet The (up to) eight valency electrons in an outer shell of an atom or molecule. Characterized by great stability, in so far as the complete shell round an atom makes it chemically inert, and round a molecule (by sharing) makes a stable chemical compound.

electron pair Two valence electrons shared by adjacent nuclei, so forming a bond.

electron paramagnetic resonance See **electron spin resonance**.

electron-phonon scattering An important process that contributes to the cause of electrical resistivity; the electrons are scattered by the thermal vibrations of the crystal lattice.

electron probe analysis A beam of electrons is focused on to a point on the surface of the sample, the elements being detected both qualitatively and quantitatively by their resultant X-ray spectra. An accurate (1%) non-destructive method needing only small quantities (micron size) of sample. See **energy dispersive analysis of X-rays**.

electron spin See **spin**.

electron spin resonance A branch of microwave spectroscopy in which there is resonant absorption of radiation by a paramagnetic substance, possessing unpaired electrons, when the energy levels are split by the application of a strong magnetic field. The difference in energy levels is modified by the environment of the atoms. Information on impurity centres in crystals, the nature of the chemical bond and the effect of radiation damage can be found. Also *electron paramagnetic resonance*. Abbrev. *ESR*.

electron transfer chain A series of chemical components, starting with glucose, each successively oxidized in *aerobic respiration* in living organisms to release energy and synthesize ATP by the stepwise transfer of electrons to their ultimate acceptor, oxygen (reduced to water). The process involves three large multiprotein complexes (*flavoproteins*, *cytochromes* and other metalloproteins), with the smaller components of *ubiquinone and cytochrome c*.

electron trap An acceptor impurity in a semiconductor.

electron-volt General unit of energy of moving particles, equal to the kinetic energy acquired by an electron losing one volt of potential, equal to 1.602×10^{-19} J. Abbrev. *eV*.

electro-optical effect See **Kerr effects** (1).

electrophoresis Motion of charged particles under an electric field in a fluid, positive groups to the cathode and negative groups to the anode.

electroplating Deposition of one metal on another by electrolytic action on passing a current through a cell, for decoration or for protection from corrosion etc. Metal is taken from the anode and deposited on the cathode, through a solution containing the metal as an ion.

electrochemical series The classification of redox half-reactions, written as reductions, in order of decreasing reducing strength. Thus the combination of any half reaction with the reverse of one further down the series will give a spontaneous reaction. For reference, the half reaction $2H^+ + e^- = H_2$ is taken as having an energy of zero. Also *electrode potential series, electromotive series.*

electropolishing See **electrolytic polishing**.

electropositive Carrying a positive charge of electricity. Tending to form positive ions, ie having a relatively negative electrode potential.

electroradiescence Emission of ultraviolet or infrared radiation from dielectric **phosphors** on the application of an electric field.

electrorefining See **electrolytic refining**.

electrostatic bonding See **electrovalence**.

electrostatic field Electric field associated with stationary electric charges.

electrostatic flocking The application of a coloured flock directed by an electrostatic field on to a fabric pretreated with an adhesive. The fibres of the flock protrude from the surface of the fabric giving it a characteristically prickly feel. The products are often used as wall-hangings.

electrostatic Kerr effect See **Kerr effects** (2).

electrostatics Section of science of electricity which deals with the phenomena of electric charges substantially at rest.

electrostriction Change in the dimensions of a dielectric accompanying the application of an electric field.

electrothermoluminescence Changes in **electroluminescence** radiation resulting from changes of dielectric temperature. (Some dielectrics show a series of maxima and minima when heated.) The complementary arrangement of observing changes in thermoluminescence radiation when an electric field is applied is termed **thermoelectroluminescence**.

electrovalence Chemical bond in which an electron is transferred from one atom to another, the resulting ions being held together by electrostatic attraction.

electrowinning See **electro-extraction**.

electrum (1) An alloy of gold and silver (55–88% of gold) used for jewellery and ornaments. (2) Nickel-silver (copper 52%, nickel 26% and zinc 22%); it has the same uses as other nickel-silvers.

Elektron alloys TN for magnesium-based light alloys with up to 4.5% copper, up to 12% aluminium and perhaps some manganese and zinc.

element Simple substance which cannot be resolved into simpler substances by normal chemical means. Because of the existence of isotopes of elements, an element cannot be regarded as a substance which has identical atoms, but as one which has atoms of the same **atomic number**.

elemental analysis Quantitative analysis of a substance to determine the relative amounts of the elements that make it up.

elephant UK size of paper (23×28 in). Cf ISO sizes.

elevation A view (eg side or end elevation) of a component or assembly drawn in projection on a vertical plane.

elimination The removal of a simple molecule (eg of water, ammonia etc) from two or more molecules, or from different parts of the same molecule. See **condensation** (2).

Elinvar A nickel-chromium steel alloy with variable proportions of manganese and tungsten. Used for eg watch hairsprings because of its constant **elastic modulus** at different temperatures.

Ellingham diagram Graph of standard free energy of formation of esp. metallic oxides plotted against temperature. Enables prediction of the stability of oxides and the ease of reduction to metal using carbon, hydrogen or light metals such as aluminium in the **aluminothermic process**.

elm Trees of the genus *Ulmus*, native to N Europe (*U. hollandica*) and N America (*U. americana*), the wood is coarse-textured and light- to reddish-brown. It is noted for its durability under water, and is used for wagon making, coffins, agricultural implements, gymnasium equipment, furniture, pulley blocks, ship building etc. Susceptible to Dutch elm disease, caused by a fungus *Ceratocystis ulmi* which, spread by bark beetles carrying infected spores, defoliates and kills the tree within weeks. Mean density 560 kg m^{-3} (rock elm, *U. thomasii*, 700 kg m^{-3}). Also *American white elm, Dutch elm, English elm, smooth leaved elm.*

elongation The percentage plastic extension produced in a **tensile test**.

elongation ratio (1) The ratio of the length of a powder particle to its breadth. See **aspect ratio**. (2) See **extension ratio**.

embossed paper A paper surface, to which a pattern has been imparted by passing the sheet or web through the nip of suitable rolls in an embossing **calender**. The upper roll is of steel engraved with the appropriate design and the other, backing, roll is of compressible fibrous material.

embrittlement Reduction in **toughness** developing after heat treatment or over a period of service. Some metals and plastics exhibit reduced impact toughness at subzero temperatures or may degrade at ordinary temperatures in ways which reduce their ability to absorb energy when stressed to the point of fracture.

embroidery (1) Lace work consisting of a ground of net on which an ornamental design has been stitched. (2) Ornamental work done by needle or machine on a cloth, canvas, or other ground.

emery An abrasive powder consisting of a mixture of **carborundum** with either **magnetite** or **haematite**.

emery paper, emery cloth, emery buff Paper, or more often cloth, surfaced with emery powder, held on by an adhesive solution; used for polishing and cleaning metal. See **emery**.

emf, e.m.f. Abbrevs. for **ElectroMotive Force**.

emission Release of electrons from parent atoms on absorption of energy in excess of normal average. This can arise from (1) *thermal* (thermionic) agitation, as in valves, Coolidge X-ray tubes, cathode-ray tubes; (2) *secondary* emission of electrons, which are ejected by impact of higher energy primary electrons or ions; (3) *photoelectric* release on absorption of quanta above a certain energy level; (4) *field* emission by actual stripping from parent atoms by high electric field.

emission spectrum Wavelength distribution of **electromagnetic radiation** emitted by self-luminous source.

emitter In a **transistor**, the region from which charge carriers, that are to be minority carriers in the base, are injected into the base.

empirical formula (1) Chemical formula deduced from the results of analysis which is merely the simplest expression of the ratio of the atoms in a substance. In molecular materials it may, or may not, show how many atoms of each element the molecule contains: eg methanal, CH_2O, ethanoic acid, $C_2H_4O_2$, and lactic acid, $C_3H_6O_3$, have the same percentage composition, and consequently, on analysis, they would all be found to have the same empirical formula. (2) A relationship founded on experience or experimental data only, not deduced in form from purely theoretical considerations.

empiricism The regular scientific procedure whereby scientific laws are proposed by inductive reasoning from relevant observations. Critical phenomena are deduced from such laws for experimental observation, as a check on the assumptions or hypotheses inherent in the theory correlating such laws. Scientific procedure, described by empiricism, is not complete without the experimental checking of deductions from theory.

empty band See **energy band**.

emulsifier An apparatus with a rotating, stirring or other device used for making emulsions.

emulsifying agents Substances whose presence in small quantities stabilizes an emulsion, eg ammonium linoleate, certain benzene-sulphonic acids etc.

emulsion A colloidal suspension of one liquid in another, eg milk. See **gels** p. 143.

emulsion polymerization See **chain polymerization** p. 56.

enamel (1) The hard external layer of a tooth consisting almost entirely of large elongated apatite (calcium phosphate) crystals set vertically on the surface of the underlying dentine. (2) Surface coating of opaque glass fused onto metal articles for decoration or to provide a hard, inert and impermeable layer on eg cooking vessels. Also *vitreous enamel*. See **frit**. (3) General term for hard surface coatings, eg oil-based paints containing resin and used for wire insulation and in building.

enantiomerism See **enantiomorphism**.

enantiomorphism Mirror image isomerism. A classical example is that of the crystals of sodium ammonium tartrate which Pasteur showed to exist in mirror image forms. Adj. *enantiomorphous*. Also *enantiomerism*.

encapsulation Provision of a tightly fitted envelope of material to protect, eg a metal during treatment or use in an environment with which it would otherwise react in a detrimental manner. Similarly, the coating of, eg an electronic component in a resin to protect it against the environment.

end Applied variously during textile making: (1) in spinning, an individual strand; (2) in weaving, a warp thread; (3) in finishing, each passage of a fibre through a machine; (4) a length of finished fabric shorter than the normal piece.

end group analysis Method of determining molecular mass of polymers by titration of eg acidic groups at chain ends. Accuracy limited to relatively low-mass polymers such as prepolymers.

endoscope Instrument for inspecting and photographing (1) otherwise inaccessible sites in eg machinery, nuclear installations etc, or (2) internal cavities of the body in medicine. Fibre-optics are normally used to both illuminate and observe the remote site from outside.

endothermic An endothermic chemical reaction is one that absorbs thermal energy (heat) ie ΔH positive. Without an input of thermal energy the reaction will not proceed. Cf **exothermic**.

end-quench test See **Jominy test**.

endurance limit In **fatigue testing**, the number of cycles which may be withstood without failure at a particular level of stress. See **fatigue strength** p. 124.

energetics The abstract study of the energy relations of physical and chemical changes. See **thermodynamics**.

energy The capacity of a body for doing

work. Mechanical energy may be of two kinds: *potential energy*, by virtue of the position of the body, and *kinetic energy*, by virtue of its motion. Energy can take a wide variety of forms. Both *mechanical* and *electrical* energy can be converted into *heat* which is itself a form of energy. Electrical energy can be stored in a a capacitor to be recovered at discharge of the capacitor. *Elastic potential energy* is stored when a body is deformed or changes its configuration, eg in a compressed spring. All forms of wave motion have energy; in electromagnetic waves it is stored in the electric and magnetic fields. In any closed system, the total energy is constant – the *conservation of energy*. Units of energy: SI unit is the **joule** (symbol J) and is the work done by a force of 1 newton moving through a distance of 1 metre in the direction of the force. The CGS unit, the *erg* is equal to 10^{-7} joules and is the work done by a force of 1 dyne moving through 1 cm in the direction of the force. The foot-pound force (ft-lb f) of the British system equals 1.356 J. See **kinetic energy, potential energy, mechanical equivalent of heat, kilowatt-hour, electron-volt, BTU**.

energy band In a solid the energy levels of the individual atoms combine to form bands of allowed energies separated by forbidden regions. The individual electrons are considered to belong to the crystal as a whole rather than to a particular atom. The energy bands are the consequence of the motion of the electron in the periodic potential of the crystal lattice. A solid for which a number of the bands are completely filled and the others empty, is an insulator provided the energy gaps are large. If one band is incompletely filled or the bands overlap then metallic conduction is possible. For semiconductors, there is a small energy gap between the filled and empty band and *intrinsic conduction* occurs when some electrons acquire sufficient energy to surmount the gap.

energy barrier The minimum amount of free energy which must be attained by a chemical entity in order to undergo a given reaction. See **activation energy**.

energy dispersive analysis of X-rays Method of elemental analysis of materials by scanning back-scattered X-rays from high-voltage electron bombardment, usually in a **scanning electron microscope**. Characteristic emission peaks enable identification of most elements. Useful for analysing inclusions, impurities etc. Abbrev. *EDAX*.

energy gap Range of forbidden energy levels between two permitted bands. See **energy band**.

energy levels In semiconductors, a *donor level* is an intermediate level close to the

conduction band; being filled at absolute zero, electrons in this level can acquire energies corresponding to the conduction level at other temperatures. An *acceptor level* is an intermediate level close to the valence band, but empty at absolute zero; electrons in the valence band can acquire energies corresponding to the intermediate level at other temperatures. See **Fermi level**. Electron energies in atoms are limited to a fixed range of values, termed *permitted energy levels*, and represented by horizontal lines drawn against a vertical energy scale. Cf **band theory of solids**.

energy product See **characterizing magnetic materials** p. 57.

eng See **keuring**.

Engel process Patented process for the production of large vessels, as in **rotomoulding**.

engine Generally, a machine in which energy is applied to do work: particularly, a machine for converting heat energy into mechanical work; loosely, a locomotive.

engineering design Methods used to plan, test and produce engineering devices and machines, such as CAD, failure modes and effects analysis, materials selection. See **industrial design, product design**.

engineering polymer Term rather loosely applied to speciality polymers offering improved properties over commodity polymers, eg polysulphones for heat resistance or polycarbonate for toughness.

engineering strain See **strain**.

engineers' bending theory Simple relationship between the parameters in the bending of beams, encapsulated in the double equality:

$$\frac{M}{I} = \frac{\sigma}{y} = \frac{E}{R},$$

where M is the **bending moment**, I is the **second moment of area**, σ is the **stress** developed a distance y from the **neutral axis**, E is the **Young's modulus** of the beam material and R is the radius of curvature to which it is bent.

engine sized paper Paper made from stock to which appropriate chemicals have been added to confer resistance to penetration by aqueous liquids.

enhancement In metal-oxide-silicon devices, the creation of a conducting channel as a consequence of an externally applied gate-substrate bias voltage. Semiconductor beneath the gate accumulates an excess of charges which are nominally the minority carriers and is said to have undergone inversion.

entanglement molecular mass See **molecular mass distribution** p. 212.

entanglements Knots and loops formed between polymer chains above a critical molecular mass. See **molecular mass distribution** p. 212.

enthalpy Thermodynamic property of a working substance defined as $H = U + PV$ where U is the internal energy, P the pressure and V the volume of a system. Associated with the study of heat of reaction, heat capacity and flow processes. SI unit is the joule.

ento- Prefix from Gk *entos*, within.

entropy In thermal processes, a quantity which measures the extent to which the energy of a system is available for conversion to work. If a system undergoing an infinitesimal reversible change takes in a quantity of heat dQ at absolute temperature T, its entropy is increased by $dS = dQ/T$. The area under the absolute temperature-entropy graph for a reversible process represents the heat transferred in the process. For an adiabatic process, there is no heat transfer and the temperature-entropy graph is a straight line, the entropy remaining constant during the process. When a thermodynamic system is considered on the microscopic scale, equilibrium is associated with the distribution of molecules that has the greatest probability of occurring, ie the state with the greatest degree of disorder. *Statistical mechanics* interprets the increase in entropy in a closed system to a maximum at equilibrium as the consequence of the trend from a less probable to a more probable state. Any process in which no change in entropy occurs is said to be *isentropic*.

entropy of polymerization Change of entropy during polymerization. Large and negative due to ordering of monomer molecules into chains. See **ceiling temperature**.

environmental impact assessment The European Economic Community's equivalent of the US **environmental impact statement**. Abbrev. *EIA*.

environmental impact statement In the US, a detailed analysis, required of any agency undertaking a major federal project, of its effects on the environment. Abbrev. *EIS*.

environmental stress (corrosion) cracking Describes a variety of phenomena and mechanisms in which the initiation and propagation of cracks in materials subject to stress is accelerated by environmental chemicals. Their **fracture toughness** is thereby reduced and therefore their structural integrity. In polymers, the **residual stresses** (or *frozen-in strains*) resulting from processing can be sufficient to cause cracking in the presence of many organic liquids (cf **crazing**) and the term *environmental*

stress cracking (abbrev. *ESC*) tends to be used. This also affects the response of polymers to external loading eg **creep rupture** and **fatigue**. In other materials (metals, ceramics, glasses etc) the term *environmental stress corrosion cracking* (abbrev. *ESCC*) is preferred and the phenomenon is more usually associated with external loading eg **fatigue** and **static fatigue**. Cf **stress corrosion**.

EPDM Abbrev. for **Ethylene-Propylene-Diene rubber**.

epi- (1) Containing a condensed double aromatic nucleus substituted in the 1,6-positions. (2) Containing an intramolecular bridge.

epi Abbrev. for an epitaxially grown layer, especially silicon semiconductor.

epichlorhydrin *1-Chloro-2,3-epoxy propane*, C_3H_5ClO, bp 117°C. A liquid derivative of glycerol formed by reaction with hydrogen chloride to give dichlorohydrin, which in turn is treated with concentrated potassium hydroxide solution. Used in the production of **epoxy resins**. See **thermosets** p. 318.

epichlorhydrin rubber A speciality rubber made with ethylene oxide as co-monomer. Abbrev. *CHR*.

Epikote TN for a range of **epoxy resins**, used for castings, encapsulation (potting) and surface coatings.

epitaxial Having the same crystal axes. Used to describe extensions grown or deposited onto a single crystal substrate; the material in the epitaxial layer must have a lattice spacing and structure close to that of the substrate.

epitaxial growth of semiconductors See panel on p. 117.

epitaxial transistor One in which the collector consists of a high-resistivity epitaxial layer deposited on a low-resistivity substrate, the emitter and base regions being formed in or on this layer by diffusion techniques. This type of construction results in a very thin base region, important for effective high-frequency operation, and a relatively large collector, which ensures good heat dissipation.

epitaxy The growth or deposition of epitaxial layers. See **epitaxial growth of semiconductors**.

epitropic fibre A synthetic fibre whose surface contains particles (eg of carbon) that are electrically conducting.

EPOS TN for database of properties of polymers supplied by large UK manufacturer.

epoxy resins Polymers derived from epichlorhydrin and bisphenol-A. Widely used as structural plastics, surface coatings and adhesives, and for encapsulating and

---

epitaxial growth of semiconductors

Semiconducting material is extremely sensitive to chemical impurities and structural imperfections. Many high performance semiconductor devices are fabricated in single crystal material which has been deposited on single crystal substrates. In this way the crystalline orientation of the deposited layer is determined by the substrate but the purity or composition of the active layer is independently controlled. The growth of such layers is termed *epitaxy*. Epitaxial layers can be grown (with simultaneous doping) by a number of means.

In *liquid phase epitaxy* (LPE), layers are grown from a molten source on an appropriate substrate. For example, gallium arsenide (GaAs) layers can be grown by the controlled cooling of a liquid solution of arsenic in gallium on a gallium arsenide substrate.

Vapour phase epitaxy (VPE) introduces material for the growing layer via a gas or vapour. For example, silicon is deposited by chemical vapour deposition (CVD) from a mixture of silicon tetrachloride and hydrogen onto a silicon substrate held at 1200°C, significantly below the melting temperature (1412°C). Layers grown too cool or too quickly are polycrystalline.

In *metallo-organic chemical vapour deposition* (MOCVD), metallic species are introduced as gaseous organic complexes. For indium phosphide (InP) deposition, for example, trimethyl indium vapour is mixed with hydrogen and phosphine (PH_3) and passed over a heated substrate where epitaxial growth takes place.

In *molecular beam epitaxy* (MBE), each species to be deposited is supplied as a molecular beam effusing from an oven containing the pure material. The technique is particularly suited to the growth of compound semiconductors, especially when complex compositional variations such as **heterostructures** are required. Deposition is carried out in ultra-high vacuum apparatus which assures the highest levels of purity and permits *in situ* analysis of the growing material.

---

embedding electronic components. Characterized by low shrinkage on polymerization, good adhesion, mechanical and electrical strength and chemical resistance. Also *epoxide resins*. See **thermosets** p. 318.

equations of motion The set of relations between distance *s*, time *t*, initial velocity *u*, final velocity *v* and acceleration *a*:

$$v = u + a\,t,$$

$$v^2 = u^2 + 2\,a\,s,$$

$$s = \tfrac{1}{2}\,(u + v)\,t,$$

$$s = u\,t + \tfrac{1}{2}\,a\,t^2.$$

equation of state (1) Any equation relating thermodynamic state functions, ie those system variables whose value is independent of the path taken to reach a particular state of the system. Examples include the **free energy equation** and the **ideal gas** equation. See **thermodynamics**. equi- Prefix from L. *aequus*, equal.

equilibrium (1) The state reached in a reversible chemical reaction when the reaction velocities in the two opposing directions are equal, so that the system has no further tendency to change. (2) The state of a body at rest or moving with constant velocity. A body on which forces are acting can be in equilibrium only if the resultant force is zero and the resultant torque is zero. (3) Thermal state of a system at which no further heat flow occurs and all components of the system are at the same temperature.

equilibrium constant The ratio, at equilibrium, of the product of the active masses of the molecules on the right side of the equation representing a reversible chemical reaction to that of the active masses of the molecules on the left side. See **law of mass action**.

equilibrium diagram See **phase diagram** p. 236.

equipartition of energy The **Maxwell–Boltzmann law**, which states that the available energy in a closed system eventually distributes itself equally among the degrees of freedom present.

equivalence, photochemical See **Ein-**

stein law of photochemical equivalence.

equivalent See **equivalent weight**.

equivalent conductance Electrical conductance of a solution which contains 1 gram-equivalent weight of solute at a specified concentration, measured when placed between two plane parallel electrodes, 1 m apart. *Molar conductance* is more often used.

equivalent proportions See **law of equivalent** (or **reciprocal**) **proportions**.

equivalent surface diameter The diameter of a sphere which has the same effective surface area as that of an observed particle when determined under the same conditions.

equivalent volume diameter The diameter of a sphere which has the same effective volume as that of an observed particle when determined under the same conditions.

equivalent weight That quantity of one substance which reacts chemically with a given amount of a standard. In particular, the equivalent weight of an acid will react with one mole of hydroxide ions, while the equivalent weight of an oxidizing agent will react with one mole of electrons.

erbium A metallic member of the group of **rare earth elements**. Symbol Er. Found in the same minerals as dysprosium. See Appendix 1, 2.

erbium laser Laser using erbium in YAG (*Yttrium-Aluminium-Garnet*) glass. It has the advantage of operating between 1.53 and 1.64 μm, a range in which there is a high attenuation in water. This feature is of particular importance in laser applications to eye investigations, since a great deal of energy absorption occurs in the cornea and aqueous humour before reaching the delicate retina.

ergonomics Methods intended to improve product performance by matching its design and materials of construction to the user. It includes study of human physiology and biomechanics (eg muscle strength), variability in human form (eg hand size) on the one side and product properties on the other (eg mechanical strength and stiffness). Handtool design, for example, has improved dramatically through use of lightweight plastics carefully shaped to fit the user's limbs. US *Human factors*.

Erichsen test A test in which a piece of metal sheet is pressed into a cup by means of a plunger; used to estimate the suitability of sheet for pressing or drawing operations.

erosion The removal of material from components subject to fluid flow, particularly when the fluid contains solid particles.

ES(C)C Abbrev. for **Environmental Stress (Corrosion) Cracking**.

esparto A rush (*Macrochloa tenacissima*) native to N Africa and southern Spain; the main raw material for paper pulp before wood pulp, and yielding paper with excellent printing qualities. Also *esparto grass*, *Spanish grass*.

ESR Abbrev. for **Electron Spin Resonance**.

ester Esters are derivatives of acids obtained by substitution of the replaceable hydrogen by alkyl radicals. See **polyesters**.

esterification The direct action of an acid on an alcohol, resulting in the formation of esters. See **step polymerization** p. 298.

Et A symbol for the ethyl radical C_2H_5-.

etching (1) Method of showing the structure of materials by attacking a highly polished surface with a reagent that has a differential effect on different crystals or different constituents. (2) Removing films from the surface of materials to facilitate the subsequent deposition of another coating eg paint. See **dry etching, semiconductor device processing** p. 280.

etching pits Small cavities formed on the surface of materials during etching.

ethane $CH_3.CH_3$, a colourless, odourless gas of the alkane series. The second member of the alkane series of hydrocarbons. Chemical properties similar to those of methane.

ethene The IUPAC name for ethylene, $H_2C=CH_2$, mp $-169°C$, bp $-103°C$, a gas of the alkene series, obtained from the cracking of petroleum. Used for making polyethylene and polyethylene oxide.

ethenoid resins Resins made from compounds containing a double bond between two carbon atoms, ie the acrylic, vinyl and styrene groups of plastics.

ether *Alkoxyalkane*. (1) Any compound of the type R–O–R, containing two identical, or different, alkyl groups united to an oxygen atom; they form a homologous series C_nH_{2n+2} O. (2) Specifically, diethyl ether.

ethyl acrylate $CH_2=C.COOC_2H_5$. Colourless liquid, bp 101°C. Used in the manufacture of plastics.

ethylene See **ethene**.

ethylene diamine tartrate Chemical in crystal form, exhibiting marked piezoelectric phenomena; used in narrow-band carrier filters. Abbrev. *EDT*.

ethylene oxide C_2H_4O, bp 13.5°C, a mobile colourless liquid, monomer for polyethylene oxide.

ethylene polymers Group of common polymers made from ethylene monomer. Includes low-density polyethylene, high-density polyethylene and medium-density polyethylene plus copolymers like **ethylene-propylene-diene rubber**. See **chain polymerization** p. 56.

ethylene-propylene-diene rubber A **terpolymer** which is elastomeric due to low crystallinity from random copolymer structure. A small amount of diene is added to aid **vulcanization**. Abbrev. *EPDM*. Also *Ethylene-proplyene rubber, EPM*. See **elastomers** p. 106.

ethylene-propylene rubber See **ethylene-propylene-diene rubber**.

ethylene-vinylacetate copolymer The relatively low-cost thermoplastic elastomer used as a **hot-melt adhesive** and for small injection mouldings (eg stapler base-pads). Abbrev. *EVA*.

ethyl group The monovalent radical —C_2H_5.

Ettinghausen effect A difference in temperature established between the edges of a metal strip carrying a current longitudinally, when a magnetic field is applied perpendicular to the plane of the strip. Effect is very small and is analogous to **Hall effect**.

Euler buckling limit Criterion for the elastic **buckling** of a structural member under compressive loading where the loading is along the centroid of the member's cross-section. The buckling load, P, is given by:

$$P = \frac{n^2 \pi^2 E I}{L^2}$$

where E is **Young's modulus**, I is the minimum **second moment of area** of the section, L is the loaded length of the member and n is the half wavelength of the buckled shape.

Eureka See **constantan**.

European plane Hardwood (*Platanus acerifolia*), yielding a useful general purpose, but perishable, timber. Light reddish-brown with characteristic fleck markings, straight-grained and fine-textured. Density about 620 kg m^{-3}. Used in furniture, cabinet-making and in decorative applications. Selected logs are cut at about 45° to the growth rings (quarter sawn) to accentuate the flecked appearance. This is **lacewood** and used for decorative panelling, etc. The related *P. occidentalis* is known as sycamore in the US, but should not be confused with the true **sycamore**, *Acer pseudoplatanus*.

europium Metallic element of the rare earth group. Symbol Eu. Contained in black monazite sand. See Appendix 1, 2.

eutectic The isothermal transformation of a liquid solution simultaneously to different solid phases. It represents the lowest temperature for solidification of any mixture in that part of the system and involves simultaneous crystallization of two constituents in a binary system (or of three in a ternary sys-

tem, four in a quaternary, and so on) and occurs at a unique temperature. Eutectic liquids also exhibit high fluidity compared with compositions on either side, which freeze over a range of temperature and pass through a pasty region during solidification. See **phase diagram** p. 236.

eutectic change The transformation from the liquid to the solid state in a eutectic alloy. It involves the simultaneous crystallization of two constituents in a binary system and of three in a ternary system.

eutectic point The point in the binary or ternary constitutional diagram indicating the composition of the eutectic liquid and the temperature at which it solidifies.

eutectic structure The particular arrangement of the constituents in a eutectic alloy which arises from their simultaneous crystallization from the melt.

eutectic system A binary or ternary alloy system in which one particular alloy solidifies at a constant temperature which is lower than the beginning of solidification in any other alloy.

eutectic welding A welding or brazing process which is carried out below the melting points of the components joined by exploiting the eutectic properties of the materials involved.

eutectoid Similar to a eutectic except that it involves the simultaneous formation of two or three constituents from another solid constituent instead of from a liquid. Eutectoid point and eutectoid structure have similar meanings to those given for eutectic.

eutectoid steel Steel having the same composition as the eutectoid point in the iron-carbon system (0.87%C), and which therefore consists entirely of the eutectoid at temperatures below 723°C. See **pearlite**.

eutexia The property of being easily melted, ie at a minimum melting point.

eV Abbrev. for **electron-volt**.

EVA Abbrev. for **Ethylene-VinylAcetate copolymer**.

evaporation The conversion of liquid into vapour.

Ewing curve tracer An instrument for throwing a curve representing the hysteresis loop of a sample of iron on to a screen. A mirror is deflected horizontally in proportion to the magnetizing force and vertically in proportion to the flux produced.

Ewing permeability bridge A measuring device in which the flux produced in a sample of iron is balanced against that produced in a standard bar of the same dimensions. The magnetizing force on the bar under test is varied until balance is obtained, and from the value of the force so found the permeability can be estimated.

exchange energy Term, of quantum mechanical origin, in the energy balance of magnetic domains which accounts for the interaction between neighbouring dipole moments. Cf **super exchange energy**.

excitation (1) Current in a coil which gives rise to a magnetomotive force in a magnetic circuit, esp. in a generator or motor. (2) The magnetomotive force itself. (3) The magnetizing current of a transformer.

exciting coil A coil on a field magnet, or any other electromagnet, which carries the current for producing the magnetic field.

exciton Bound hole-electron pair in semiconductor. These have a definite half-life during which they migrate through the crystal. Their eventual recombination energy releases as a photon or, less often, several phonons.

excluded volume Property of real polymer chains, where they cannot pass through one another. Leads directly to problem of entanglements in chains. See **statistical chain.**

exclusion principle See **Pauli exclusion principle**.

exfoliation Lifting away of the surface of a metal due to the formation of corrosion products beneath the surface, a common result of rusting. Also *spalling*.

exo-electron One, emitted from the surface of a metal or semiconductor, which comes from a metastable trap with very low binding energy under conditions such that electrons in their ground state could not be emitted.

exothermic An exothermic chemical reaction is one that evolves thermal energy (heat), ie ΔH negative. Cf **endothermic**.

expanded plastics Foamed plastic materials, eg polyvinyl chloride, polystyrene, polyurethane, polythene etc, created by the introduction of pockets or cells of inert gas (air, carbon dioxide, nitrogen etc) at some stage of manufacture. Used for heat insulation purposes or as core materials in **sandwich beam** construction, because of their low densities; also, because of lightness, for packaging, and (as foam rubber) for upholstery cushioning, artificial sponges etc. Often highly flammable. See **foam.**

expanding metals Metals or alloys, eg two parts antimony to one part bismuth, which expand in final stage of cooling from liquid; used in type-founding.

expansion joint A special pipe joint used in long pipelines to allow for expansion, eg a horseshoe bend, a corrugated pipe acting as a bellows, a sliding socket joint with a stuffing box.

expansion rollers Rollers on which one end of a large girder or bridge is often carried, to allow for movement resulting from thermal expansion; the other end of the girder etc is fixed. See **bridge bearing**.

explosion, explosive welding Welding of two components brought into forceful contact in a controlled explosion.

explosive forming One of a range of high-energy rate-forming processes by which parts are formed at a rapid rate by extremely high pressures. Low and high explosives are used in variations of the explosive forming process; with the former, known as the cartridge system, the expanding gas is confined; with the latter, the gas need not be confined and pressure of up to one million atmospheres may be attained. See **electrohydraulic forming.**

explosive rivet A type of **blind rivet** which is clinched or set by exploding, electrically, a small charge placed in the hollow end of the shank.

expression Residual liquid left in a fabric after squeezing (eg on a mangle), calculated as a percentage of the dry fabric.

extended-chain crystal Conformation of polymers where individual chains are fully extended in the crystalline state. In polyethylene, the extended chain is a linear zig-zag and occurs in high-density polyethylene crystallized at high pressure (ca 5 kbar) or gel-spun, oriented fibre. Chain-folding is the most common state of polymer single crystals, with the exception of PTFE and natural cellulose.

extender (1) A substance added to paint as an adulterant or to give it body, eg barytes, china clay, French chalk, gypsum, whiting. (2) In synthetic resin adhesives, a substance (eg rye flour) added to reduce the cost of gluing or to adjust viscosity. (3) A non-compatible plasticizer used as an additive to increase the effectiveness of the compatible plasticizer in the manufacture of elastomers. (4) Low molecular mass substance which may be used to replace plasticizer, esp. in polyvinyl chloride compounds. See **polyvinyl chloride** p. 249.

extension ratio The ratio of extended length to original length of rubber samples. Symbol λ (= new length / original length). Also *elongation ratio*. See **strain.**

extension spring A helical spring with looped ends which stores energy when in the stretched condition.

extensometer Instrument for measuring dimensional changes of a material esp. during a mechanical test such as the **tensile test**. Many different principles of operation, eg optical lever, **moiré fringes**, optical interferometry, various transducers etc. See **clip gauge**.

extrusion

The operation of producing rods, tubes and various solid and hollow sections, by forcing suitable material through a die by means of a ram. Applied to numerous non-ferrous metals, alloys and other substances, notably plastics (for which a screw-drive is frequently used). In addition to rods and tubes, extruded plastics include sheets, film and wire-coating.

Shaping polymer products of constant cross-section is achieved by extrusion of the melt through a die. The method is also widely applied in conjunction with other shaping methods (eg *blown film*, *blow moulding*), for mixing purposes (*eg cavity transfer mixing*) and for preparing polymer granules (eg lace). Like **injection moulding**, melt transport utilizes a screw within a heated barrel, often arranged in parallel as in twin-screw extruders (Fig. 1).

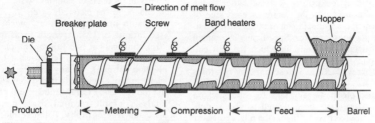

Fig. 1. **Extrusion moulding machine for plastics**.

Pipe forms a particularly large market eg medium-density polyethylene gas pipes, unplasticized polyvinyl chloride (uPVC) water pipes, high- and low-density polyethylene water tubing, plasticized polyvinyl chloride tubing. Special cooling jigs are needed to maintain pipe quality and the pipe is pressurized to maintain its shape. Rates of extrusion are relatively low (shear rate ~10 to 10^3 s^{-1}) and so relatively high molecular mass thermoplastic is normally used; a feature which produces higher strength in the products than injection moulding. An upper limit to rate is created by **melt fracture** or **sharkskin** and all extruders are affected by **die swell**. A defect analogous to the weld lines found in injection mouldings can occur if melt temperatures are too low (so-called *spider lines*). They are caused by division of the melt front head of the die by the three screws holding the mandrel in position.

High-modulus polyethylene rod can be produced by hydrostatic extrusion and many other materials can be extruded in a similar way (eg green clay pipes). Metal products such as copper tubing and aluminium sections, on the other hand, are always solid-state extruded.

extinction coefficient Spectroscopic term applied to molecular group, determining absorption at a particular wavelength, esp. infrared. See **Beer's law**.

extraordinary ray See **double refraction**.

extra thirds A former size of cut card, 44.5 × 76 mm (1¾ × 3 in).

extreme pressure lubricant A solid lubricant, such as graphite, or a liquid lubricant with additives which form oxide or sulphide coatings on metal surfaces exposed where, under very heavy loading, the liquid film is interrupted, thus mitigating the effects of dry friction.

extrinsic semiconductor See **intrinsic and extrinsic semiconductors** p. 173.

extrusion See panel on p. 121.

extrusion blow moulding Conventional route for making plastic bottles etc. See **blow moulding**. Cf **injection blow moulding**.

extrusion zones Parts of barrel where different processes occur, eg feed section. See **extrusion** p. 121.

F

F Symbol for **farad**.

F Symbol for: (1) **faraday**; (2) **Helmholtz free energy** (US); (3) **force**.

fabric A coherent assembly of fibres and/or yarns that is long and wide but relatively thin and strong; a cloth. See **fibre assemblies** p. 130.

face The working surface of any part; as the sole of a carpenter's plane, the striking surface of a hammer, the surface of a slide-valve, or the surface of the steam chest on which it slides, the seating surface of a valve, the flank of a gear-tooth etc.

face-centred cubic A crystal lattice with a cubic unit cell, the centre of each face of which is identical in environment and orientation to its vertices. Specifically a common structure of metals, in which the unit cell contains four atoms, based on this lattice. Abbrev. *FCC*. See **close packing of atoms** p. 62.

faced cloth General term for fabrics whose surfaces have been treated to give a rich, luxuriant effect by laying the pile, eg **beaver** and doeskin cloths. See **dress-faced cloth**.

facing sand Moulding sand with admixed coal dust, used near pattern in foundry flask to give the casting a smooth surface.

faconné Figured fabric with a **Jacquard** or **dobby pattern**.

factor of safety Design factor which is the ratio of the failure stress of the material in a structural member or structure to the safe permissible stress in it. Abbrev. *FS*.

fadometer An instrument used to determine the resistance of a dye or pigment to fading.

faggot Made by forming a box with four long flat bars of **wrought iron** and filling the interior with scrap and short lengths of bar.

faggoted iron Wrought iron bar made by heating a **faggot** to welding heat and rolling down to a solid bar. If the process is repeated double-faggoted iron is obtained. Now obsolete.

faience Originally the 16th-century French name for **earthenware** glazed with an opaque, tin-based glaze, from Faenza, Italy. Now used in the UK for large glazed blocks and slabs used in buildings, in France for any glazed, porous earthenware, and in the US for a decorated earthenware with a transparent glaze. Cf **bone china**, **earthenware**, **porcelain**, **stoneware**, **terra-cotta**.

failsafe Design of component or product to provide alternative support or function in the case of failure eg by redundancy.

failure modes and effects analysis The design activity for process and product which aims to eliminate defects before production or launch of product in marketplace. Involves a committee of designers and engineers who review systematically potential defects in terms of ease of detection, seriousness and ease of correction. Each defect is assigned a number (scale of 1–10) by mutual agreement, and the product number matched against the RPN (risk priority number). If above a critical value, the defect is corrected. Of particular importance in product liability. Abbrevs. *FMEA, FMECA*.

failure of materials Loss of structural integrity in products or samples by some form of change in the material, eg chemical (see **corrosion and degradation**), mechanical (see **fatigue** p. 123, **strength measures** p. 301), dimensional (see **tolerance**), physical (see **demagnetization**), electrical (see **dielectric fatigue**), or any combination of such kinds of change. Studied systematically in products using **failure modes and effects analysis** or **fault tree analysis**, or in materials using **fractography**, **fracture mechanics** etc.

Fajans rule for ionic bonding The conditions which favour the formation of ionic (as opposed to covalent) bonds are (a) large cation, (b) small anion, (c) small ionic charge, (d) the possession by the cation of an inert gas electronic structure.

false bottom (1) A removable bottom placed in a vessel to facilitate cleaning; a casting placed in a grate to raise the fire bars and reduce the size of the fire. (2) Any secondary bottom plate or member used to reduce the volume of a container or to create a secondary container.

false hemlock See **Douglas fir**.

false-twist Method of producing textured, continuous filament, yarn from thermoplastic fibres, in which it is twisted, heated, allowed to cool and set, and then twisted in the opposite direction. See **textured yarn**.

family tool Injection moulding tool possessing cavities of different shape, so that several parts can be made in one operation. They are usually sub-components of one final product, such as a telephone handset.

farad The practical and absolute SI unit of electrostatic capacitance, defined as that which, when charged by a potential difference of one volt, carries a charge of one coulomb. Equal to 10^{-9} electromagnetic units and 9×10^{11} electrostatic units. Symbol F. This unit is in practice too large, and the subdivisions, *microfarad* (μF), *nanofarad* (nF) and *picofarad* (pF), are in more

fatigue

A phenomenon which results in the sudden fracture of a component after a period of cyclic loading in the elastic regime. Failure is the end result of a process involving the initiation and growth of a crack, usually at the site of a **stress concentration** on the surface. It can also be sub-surface. Both eventually so reduce the effective cross-sectional area that the component ruptures under a normal sevice load, but one at a level which has been satisfactorily withstood on many previous occasions before the crack propagated. The final fracture may occur in a *ductile* or *brittle* mode depending on the characteristics of the material.

Fatigue fractures have a characteristic appearance which reflects the initiation site and the progressive development of the crack front, culminating in an area of final overload fracture. See Fig. 1a which illustrates fatigue failure in a circular shaft. The initiation site is shown and the shell-like markings, often referred to as *beach markings* because of their resemblance to the ridges left in the sand by retreating waves, are caused by arrests in the crack front as it propagates through the section. The hatched region on the opposite side to the initiation site is the final region of ductile fracture.

Sometimes there may be more than one initiation point and two or more cracks propagate. This produces features as in Fig. 1b with the the final area of ductile fracture being a band across the middle. This type of fracture is typical of double bending where a component is cyclically strained in one plane or where a second fatigue crack initiates at the opposite side to a developing crack in a component subject to reverse bending. Some stress-induced fatigue failures may show multiple initiation sites from which separate cracks spread towards a common meeting point within the section.

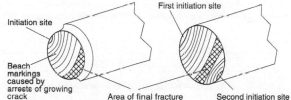

Fig. 1. **Schematic illustrations of fatigue fractures in circular shafts**. *Left*, a single site initiated at a change in cross section. *Right*, two initiation sites on opposite sides.

general use.

faraday Quantity of electric charge carried by one mole of singly charged ions, ie 9.6487×10^4 coulombs. Symbol F.

Faraday effect The rotation of the plane of polarization of electromagnetic radiation in propagating through a material parallel to a magnetic field.

-farious Suffix meaning *arranged in so many rows*.

fastener An article designed to fasten together two or more other articles, usually in the form of a shaft passing through the articles to be fastened, eg nails, screws, rivets, pins.

fastness The ability of a colour in a textile to remain unchanged when exposed to a specified agency including light, rubbing and washing.

fathom A cubic timber measure, $(6 \text{ ft})^3$ or 216 ft^3. (NB The fathom is also a nautical unit for depth = 6 ft.)

fatigue See panel on p. 123.

fatigue limit The upper limit of the range of stress that a metal can withstand indefinitely. If this limit is exceeded, failure will eventually occur.

fatigue-testing machine A machine for subjecting a test piece to rapidly alternating or fluctuating stress, in order to determine its

fatigue strength

Fatigue strength is determined by applying different levels of cyclic stress to individual test specimens and measuring the number of cycles to failure. Standard laboratory tests use various methods for applying the cyclic loading, eg rotating bend, cantilever bend, axial push-pull and torsion. The data are plotted in the form of a *stress-number of cycles to failure (S-N) curve*. See Fig. 1. Owing to the statistical nature of the failure, several specimens have to be tested at each stress level. Some materials, notably low-carbon steels, show a flattening off at a particular stress level as at (a) in Fig. 1 which is referred to as the *fatigue limit*. In principle, components designed so that the applied stresses do not exceed this level should not fail in service. The difficulty is that a localized stress concentration may be be present or introduced during service which leads to initiation, despite the design stress being nominally below the 'safe' limit.

Most materials, however, exhibit a continually falling curve as at (b) in Fig. 1 and the usual indicator of fatigue strength is to quote the stress below which failure will not be expected in less than a given number (usually 10^8) cycles, which is referred to as the *endurance limit*.

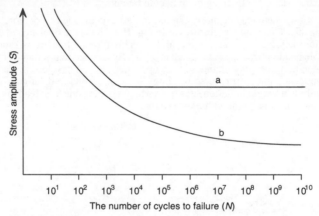

Fig. 1. *S-N* **curves for two classes of material**. See text above.

In practice, loading is seldom if ever uniform and sinusoidal, as in the tests used for *S-N* measurements. Often complete components or assemblies, eg a railway bogie frame or aircraft fuselage, will be tested by subjecting them to an accelerated loading spectrum reproducing what they are likely to experience over their entire service lifetime.

continued on next page

fatigue limit.

fault-finding General description of locating and diagnosing faults, according to a pre-arranged schedule, generally arranged in a chart or table, with or without special instruments. US *trouble-shooting*.

fault tree analysis Design method which aims to track a specific product failure back to the original cause(s) using a network. Complementary to **failure modes and ef-**

fects analysis. Abbrev. *FTA*.

FCC Abbrev. for **Face-Centred Cubic**.

F-centre An electron trapped at a negative ion vacancy in an ionic crystal. *F*-centres can be formed by the release of electrons by irradiation with X-rays or by producing stoichiometric excess of anions in the crystal. F^1-centres consist of *F*-centres with a further electron trapped in the same vacancy. F_A-centres are *F*-centres modified by one *fatigue strength (contd)*

Fatigue damage which leads to initiation and growth of the crack is cumulative. Several methods are used for predicting service lifetimes to take account of a foreseeable loading spectrum comprising wide ranging and complex life cycles, some levels of which may be severely damaging.

Miner's rule

One of the most useful is Miner's rule, which requires the calculation of all the fractional contributions of fatigue damage for the number of cycles at each particular stress level in the spectrum; failure is to be expected when the summation of these fractions reaches one. That is, if N_1 cycles are required to produce complete failure at stress level S_1, then n_1 cycles at that stress is considered to have used up n_1/N_1 of the fatigue life; n_2/N_2 at stress level S_2 and so on with similar terms n_x/N_x for the entire loading spectrum. The moment of expected failure is reached after the time when all the fractional contributions in that particular loading spectrum have built up to 1.0. By applying such a method it is possible to remove a component when it approaches the end of its fatigue life and replace it with another, thus guarding against catastrophic failure.

Introduced compressive stresses

The nature and level of static stress a component experiences also influence fatigue life because the applied cyclic loads are superimposed on any which already exist. Fatigue cracking only advances under the tensile part of the cycle. Hence, if a static load is compressive, it will reduce the tensile peak of each cycle and extend the fatigue life. A tensile static load on the other hand, will reinforce the tensile part of the cycle and accelerate failure. This effect is often exploited advantageously by introducing residual *compressive* stresses at the surfaces of components subject to cyclic loading in service, by techniques like **peening** and **surface hardening**. Male screw threads are preferably rolled to form rather than machined from bar, because plastic deformation at the thread root and the residual compressive stress significantly increase fatigue resistance.

neighbouring cation being different from the cations of the lattice. *F*-centre aggregates can be formed by arrays of nearest neighbour centres. *F*-centres give rise to broad optical absorption bands.

F-class insulation A class of insulating material which will withstand a temperature of 155°C.

feasibility diagram Plot of flow length versus cavity thickness in an injection moulding tool, with curves for shot volume, clamp force and melt pressure. Together, they define a feasibility envelope, which together with mouldability index curves, defines the window within which moulding can occur.

feed (1) The rate at which the cutting tool is advanced. (2) Fluid pumped into a vessel, eg feed-water to a boiler. (3) Mechanism for advancing material or components into a machine for processing.

feeder head See **hot top**.

feed screw A screw used for supplying motion to the feed mechanism of a machine

tool.

feedstock energy Energy of oil gas or coal used as feedstock for polymer production. Part of the feedstock energy can be recouped by incineration of polymer products following use.

feel Term describing the physical character of a cloth or paper when handled. Cf **handle**.

feeler gauge A thin strip of metal of known and accurate thickness, usually one of a set, used to measure the distance between surfaces or temporarily placed between working parts while setting them an accurate distance apart.

fell The edge of the cloth in a loom, where the picks of weft are beaten up by the reed.

fellmongering Obtaining wool from the skins of slaughtered sheep. Cf **pulling, shearing, skin wool.**

felt (1) A densely matted non-woven fabric containing wool or hair that has passed through a felting process of heat, steam and pressure. (2) Heavily milled woven fabric with a matted fibrous surface. (3) In papermaking, a woven blanket or a synthetic fabric in the form of an endless band to give support to the web at various points on the paper machine eg in a wet press, **MG** dryer or drying cylinders.

felting (1) The formation of a felt during processing (eg by **milling**) or during wear; a property esp. of woollen fabrics resulting from the structure of the cuticle of the fibres. (2) The natural action by which fibres adhere in papermaking.

fent Damaged pieces of cloth or off-cuts sold by weight to wholesalers or market traders for sale to consumer; if suitable, recycled within factory or system. See **waste**.

FEP Abbrev. for **Fluorinated Ethylene Propylene**.

Fermi decay Theory of ejection of electrons as β-particles.

Fermi–Dirac gas An assembly of particles which obey **Fermi–Dirac statistics** and the **Pauli exclusion principle**. For an extremely dense Fermi gas, such as electrons in a metal, all energy levels up to a value E_F, the Fermi energy, are occupied at absolute zero.

Fermi–Dirac statistics Statistical mechanics laws obeyed by system of particles whose wave function changes sign when two particles are interchanged, ie the **Pauli exclusion principle** applies.

Fermi level The energy level at which the probability of occupancy by an electron is 0.5; it depends on the distribution of energy levels and the number of electrons available. In semiconductors, the number of electrons is relatively small and the Fermi level is affected by donor and acceptor impurities.

Fermi surface A constant energy surface in k-space which encloses all occupied electron states at absolute zero in a crystal.

ferric iron (III) chloride $FeCl_3$. Brown solid. Deliquescent, soluble in ethanol. Uses: coagulant in sewage and industrial wastes, mordant, photo-engraving etching of copper, chlorination and condensation catalyst, disinfectant, pharmaceutical, analytical reagent.

ferric iron (III) oxide Fe_2O_3. The common red oxide of iron. Used in metallurgy, pigments, polishing and theatrical rouge, gas purification, and as a catalyst.

ferric iron (III) sulphate $Fe_2(SO_4)_3$. A yellowish-white powder which dissolves slowly in water. Uses: pigments, water purification, dyeing, disinfectant, medicine.

ferricyanide The complex ion $Fe(CN)_6^{3-}$ Also *hexacyano ferrate* (III).

ferri-, ferro-. (1) Prefixes from L. *ferrum*, iron. (2) Denoting trivalent and divalent iron respectively.

ferrimagnetic A material, or of a material, which exhibits ferrimagnetism. See ferro-

magnetics and ferrimagnetics p. 128.

ferrimagnetism Phenomenon in some magnetically ordered materials in which there is incomplete cancellation of the *antiferromagnetically* arranged spins giving a net magnetic moment; observed in ferrites and similar materials. See **ferromagnetics and ferrimagnetics** p. 128.

ferrite (1) The **body-centered cubic** form of iron and of solid solutions based on it. In pure iron, α-ferrite is stable up to 1183 K, whilst δ-ferrite occurs between 1663 K and the melting point (1811 K). (2) A ceramic iron oxide compound having **ferrimagnetic** properties. Those with **inverse spinel** (cubic) structure and the general formula $MO.Fe_2O_3$, where M is a divalent **transition metal** (such as Mn, Fe, Co, Ni, Cu) tend to have high permeability and low coercivity. Being ceramics they have high resistivity and are much used in high-frequency applications where eddy current losses are critical. The hexagonal structured ferrites such as barium and strontium ferrites, $MO(Fe_2O_3)_6$ have a very high coercivity. Useful as short, permanent magnets. (3) tetracalcium aluminoferrite, $4CaO.Al_2O_3$. Fe_2O_3; a constituent (~8 wt%) of Portland cement, it hydrates at a slow-to-medium rate during the setting reaction. See **cement and concrete** p. 52.

ferrite core A magnetic core, usually in the form of a small toroid, made of **ferrite** material such as nickel ferrite, nickel-cobalt ferrite, manganese-magnesium ferrite, yttrium-iron garnet etc These materials have high resistance and make eddy-current losses very low at high frequencies.

ferro- Prefix. See **ferri-**.

ferrochromium A master alloy of iron and chromium (60–72% chromium) used in making additions of chromium to steel and cast-iron.

ferrocyanide The complex ion $Fe(CN)_6^{4-}$ Also *hexacyano ferrate* (II).

ferroelectric materials Dielectric materials (usually ceramics) with domain structure, which exhibit spontaneous electric polarization. Analogous to ferromagnetic materials (see ferromagnetism). They have relative permittivities of up to 105, and show dielectric hysteresis. See **dielectrics and ferroelectrics** p. 92.

ferromagnetic A material, or of a material, which exhibits **ferromagnetism**. See **ferromagnetics and ferrimagnetics** p. 128.

ferromagnetic resonance A special case of *paramagnetic resonance*, exhibited by ferromagnetic materials and often termed *ferroresonance*. Explained by simultaneous existence of two different pseudo-stable states for the magnetic material B-H curve, each associated with a different magnetization current for the material. Oscillation between these two states leads to large currents in associated circuitry.

ferromagnetics and ferrimagnetics See panel on p. 128.

ferromagnetism Phenomenon in some magnetically ordered materials in which there is a bulk magnetic moment and the magnetization is large. See **ferromagnetics and ferrimagnetics** p. 128.

ferromanganese A master alloy of iron and manganese, used in making additions of manganese to steel or cast-iron.

ferromolybdenum A master alloy of iron and molybdenum (55–65% molybdenum) used in adding molybdenum to steel and cast-iron.

ferronickel Alloy of iron and nickel containing more than 30% nickel. Lower nickel alloys are known as nickel steel. See **Elinvar, Invar, Mumetal, permalloy.**

ferroprussiate paper   See **blueprint paper**.

ferrosilicon A master alloy of iron and silicon, used as a de-oxidant in steelmaking or for making silicon steels and cast-irons.

ferrospinel A crystalline material which has the equivalent function to that of the M in a ferrite. See **ferrite** (2).

ferrous [iron(II)] oxide FeO. Black oxide of iron.

ferrous [iron(II)] sulphate $FeSO_4.7H_2O$. Also *mineral copperas*.

ferrule (1) A short length of tube. (2) A circular gland nut used for making a joint. (3) A slotted metal tube into the ends of which the conductors of a joint are inserted. The whole is soldered solid. When the conductors are oval, the ferrule is in two parts to allow for the fact that the major axes of the oval sections may not coincide.

FET Abbrev. for **Field-Effect Transistor**.

fettling 'Making good', eg: (1) of hearth and walls of a furnace, where erosion has damaged the refractory lining; (2) the trimming of feeders and excess material from a moulding or casting

fibre (1) Any type of vegetable, animal, regenerated, synthetic, or mineral filament which is long in relation to its thickness and is fine and flexible. Yarns and fabrics are manufactured from them, by spinning and weaving, or knitting, felting, bonding etc and they are also used directly as reinforcement in **composite materials**. (2) Term for arrangement of the constituents of metals parallel to the direction of working. It is applied to the elongation of the crystals in severely cold-worked metals, to the elongation and stringing out of the inclusions in hot-worked metal, and to preferred orientations.

fibre assemblies See panel on p. 130.

fibre-board Building or insulating board made from fibrous material such as wood pulp, waste paper and other waste vegetable fibre. May be homogeneous, bitumenbonded or laminated. See **hardboard, insulating board**.

fibre, man-made See **man-made fibre**.

fibre optics Branch of optics based on the properties and use of **optical fibre**.

fibre, regenerated See **regenerated fibre**.

fibre, synthetic See **synthetic fibre**.

fibril (1) Bundle of aligned, crystalline polymer chains, as in cellulose. (2) Form of growth of aligned polymer lamellae. See **crystallization of polymers** p. 82.

fibrillation The process by which a film (eg of polypropylene) is converted into fibres. The film is deliberately stretched in order to orient the molecules. Further manipulation such as twisting, splits the film longitudinally into a yarn comprising an interconnected mass of fibres.

fibroin Structural protein of silk fibre. See **biological polymers** p. 27.

Fick's laws of diffusion Model of the **diffusion** process expressed as two laws. The first states that the rate of diffusion of a species, or molar flux, J, in a given direction is proportional to the concentration gradient in that direction, ie

$$J = -D\frac{\partial C}{\partial x},$$

where J is expressed as (number) m^{-2} s^{-1}, C as (number) m^{-3} and D is the **diffusion coefficient** (m^2 s^{-1}). The negative sign indicates flow down the gradient. The second law combines the first with a continuity equation and the assumption that D is a constant, and states that the rate of change of concentration with time, t, is proportional to the change in concentration gradient with distance in a given direction, ie in one dimension:

$$\frac{\partial C}{\partial t} = D\frac{\partial^2 C}{\partial x^2}.$$

fiddleback See **sycamore**.

field effect transistor A transistor in which the field associated with the voltage applied to a gate electrode creates/destroys/modifies a conducting channel between source and drain electrodes. Abbrev. *FET*.

field emission That arising, at normal temperature, through a high-voltage gradient causing an intense electric field at a metallic surface and stripping electrons from surface atoms.

field-emission microscope One in which

ferromagnetics and ferrimagnetics

Ferromagnetic and ferrimagnetic materials, by virtue of the behaviour of electric charge within the constituent atoms, together with the spatial disposition of the atoms themselves, are able to interact strongly with external *magnetizing fields* (H). The effects are described in terms of the *magnetization* (M).

The magnetism of individual atoms (their **magnetic moment**) is associated with electron orbits and electron spin but the principal source of magnetism in solids is the latter as the basic orbital effects get cancelled out owing to the nature of atomic bonding. Solids in which the constituent atoms have no unpaired electrons exhibit **diamagnetism**, a small effect arising from the reaction of electron orbits to the application of an external magnetic field, according to **Lenz's law**. The diamagnetic *susceptibility* (M/H) is small and negative. When the constituent atoms have unpaired spins, such as arise in the *d* shells of the transition elements, a material shows a *paramagnetic* response. Atoms with unpaired electron spins behave like tiny, independent bar magnets, aligning with and enhancing an external magnetizing field. Paramagnetic susceptibilities are small, positive and diminished by increasing temperature.

In the magnetic materials of technological importance there is a long-range magnetic order. In effect the atomic bar magnets interact co-operatively so that large groups of atoms within a sructure have a common orientation of their magnetism. Only a very few elements, notably iron, cobalt and nickel show this effect of spontaneous magnetization, which is termed *ferromagnetism*. In the quantum mechanical description, the alignment of magnet moments is ascribed to the exchange interaction, which energetically favours magnetic order. The ordered state is destroyed by thermal agitation above a characteristic, critical temperature known as the **Curie point** or temperature (T_c). Above the Curie temperature, paramagnetic behaviour is displayed. Many alloys involving one or more ferromagnetic elements are also ferromagnetic.

In some materials the magnetic order is antiferromagnetic, favouring oppositely aligned (rather than co-aligned) neighbours so that the whole exhibits no net magnetization. In various ferrites, neighbouring sites are similarly ordered but where unequal magnetic moments are oppositely aligned a net magnetization persists. This effect is termed *ferrimagnetism*. See Fig. 1. In ferrites the magnetic moments are due to unpaired spins of electrons in the d shell of transition metal ions.

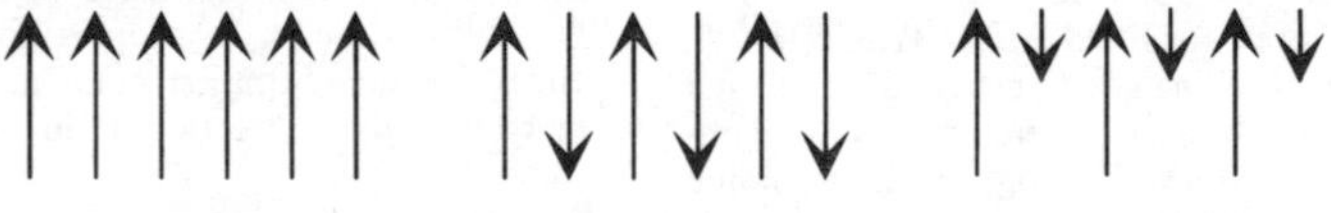

ferromagnetic   antiferromagnetic   ferrimagnetic

Fig. 1. **Magnetic order**. (a) ferromagnetic; (b) antiferromagnetic; (c) ferrimagnetic.

continued on next page

the positions of the atoms in a surface are made visible by means of the electric field emitted on making the surface the positive electrode in a high-voltage discharge tube containing argon at very low pressure. When an argon atom passes over a charged surface atom, it is stripped of an electron, and thus is drawn toward the negative electrode, where it hits a fluorescent screen in a position corresponding to that of the surface atom.

field oxide Silicon dioxide grown on a wafer for the purpose of isolating the active regions where devices are fabricated.

figured fabric Any fabric with a complex woven pattern produced by the **dobby** or **Jacquard** mechanisms.

---

ferromagnetics and ferrimagnetics (contd)

Ferromagnetics and ferrimagnetics are generally referred to simply as magnetic materials. The large groups of magnetically ordered atoms or ions within such magnetic materials are called **domains**. The orientation of the magnetization of a domain is the result of a complex interaction between the magnetic ordering, the crystal structure, the atomic spacing, the magnetization of adjacent domains and any external magnetizing field. The bulk magnetization of a single crystal or polycrystalline specimen is due to the net effect of the various internal domains and may be anywhere between zero and a saturation value which corresponds with the parallel alignment of the magnetization of every domain. During magnetization of a bulk sample, domains grow, contract and rotate the orientation of their magnetization with respect to the crystal axes.

The domain structure can be controlled through several processing strategies, such as grain texturing, grain size control and annealing in a magnetic field. The ease with which domains can be nucleated or change their shape and orientation is strongly influenced by the presence and distribution of non-magnetic inclusions and other defects such as grain boundaries. As a result it is possible to engineer magnetically hard materials which can retain their magnetization even in the presence of a demagnetizing field (**permanent magnets**) or else magnetically soft materials which readily respond to the magnetizing influence of an external field. See **characterizing magnetic materials** p. 57.

---

filament See **continuous filament yarn**.

filet net Any type of square ground-mesh of lace; a woven net.

filled band An energy-level band in which there are no vacancies. Its electrons do not contribute to valence or conduction processes.

filler (1) Any inert solid substance added to plastics either for economy (eg wood-flour, clay) or to modify its properties (eg mica, aluminium or other metal powder). (2) White mineral matter in finely divided form added to the stock to improve opacity, dimensional stability and reduce costs of paper. In some circumstances it may assist in achieving a smooth finish. Certain loadings such as titanium dioxide are expensive but have special properties such as preventing undue loss of opacity when paper is waxed. Also *loading*.

filler metal The metal required to be added at the weld in welding processes in which the fusion temperature is reached. In metal-electrode welding, the electrode, usually fluxed and coated, is melted down to provide the filler metal.

fillet (1) A narrow strip raised above the general level of a surface. (2) A rounded internal corner in a plastic or metal article, to avoid a possible weakness arising from an abrupt change in cross-section. See **stress concentration** p. 303.

fillet weld A weld at the junction of two parts, eg plates, at right angles to each other, in which a fillet of welding metal is laid

down in the angle created by the intersection of the surfaces of the parts.

filling (1) Insoluble materials such as China clay added with starch or gum to increase the weight of fabrics or to alter their appearance. Also *loading*. (2) Used in Canada and US for weft yarns.

film (1) Any thin layer of substance, eg that which carries a light-sensitive emulsion for photography, that which carries iron (III) oxide particles in a matrix for sound recording. (2) Thin layer of material deposited, formed or adsorbed on another, down to mono-molecular dimensions, eg electroplated films, oxide on aluminium, sputtered depositions on glass or microcomponents. (3) Packaging material, esp. from a thermoplastic polymer.

filter Material with pore or mesh sizes within a defined range and used to separate particles or macromolecules from a suspension or solution. See **filter paper, membrane filter**.

filter cake The layer of precipitate which builds up on the cloth of a filter press.

filter paper Paper, consisting of pure cellulose, which is used for separating solids from liquids by filtration. Filter paper for quantitative purposes is treated with acids to remove all or most inorganic substances, and has a definite ash content.

filter press An apparatus used for filtrations; it consists of a set of frames covered with filter cloths into which the mixture which is to be filtered is pumped.

filtrate The liquid freed from solid matter

fibre assemblies

Fibre assemblies can be made from either **staple fibre** or continuous *fibre* in several distinctly different ways as shown in Fig. 1. Both can be twisted together to form *yarn*, but staple fibres can be used directly for *felt* and *non-woven* fabrics.

Continuous fibres   Staple fibres

Yarns

Fabrics   Fabrics

Woven   Knitted   Braided   Lace and net   Felt   Non-wovens

Fig. 1. **Types of fibre assembly**.

Woollen *felts* for hats, billiard tables, blankets, carpet underlay etc are formed by repeatedly compressing woollen fibres togther so that the scales on the fibre surface can interlock and bind the material. Smoother fibres, such as the synthetics, have to be entangled by repeatedy punching with barbed needles to create a felt. Such materials are used as *geotextiles* to provide a porous medium which will stabilize the underlying soil, sand etc. Another familiar form of non-woven fabric is formed by a more or less random, planar array of cellulose fibres hydrogen bonded together at their points of contact, namely paper. See **paper and papermaking** p. 231.

Fig. 2. **The S-twist and Z-twist**.

Spinning fibres to make yarn is the basis of one of our oldest technologies. The main variables governing the integrity of the yarn are the type of fibre, which affects the frictional force between fibres, the number of fibres in the cross-section (a minimum of 30 is usually required) and the degree of twist. Twist direction is descibed as *S-twist* or *Z-twist* (Fig. 2). The maximum strength of a staple fibre is half that of the individual fibres, while in one based on continuous fibres, the maximum strength is in its untwisted state.

continued on next page

after having passed through a filter.

filtration The separation of solids from liquids by passing the mixture through a suitable medium, eg filter paper, cloth, glass wool, which retains the solid matter on its surface and allows the liquid to pass through.

fin (1) One of several thin projecting strips of metal formed integral with an air-cooled engine cylinder or a pump body or gear-box to increase the cooling area. (2) A thin project-ing edge on a casting or stamping, formed by metal extruded between the halves of the die; any similar projection.

fine gold Pure 24-carat gold.

fineness (1) The state of subdivision of a substance. (2) The purity of a gold or silver alloy; stated as the number of parts per thousand that are gold (or silver).

fine papers Papers of high quality, used for graphic purposes.

fibre assemblies (contd)

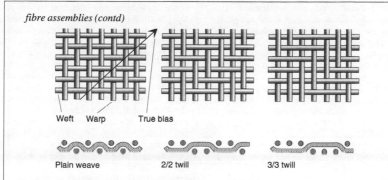

Weft   Warp   True bias

Plain weave   2/2 twill   3/3 twill

Fig. 3. **Simple weaving patterns**.

Weaving is the process of creating a fabric by building up a parallel array of yarns (the *weft*) that pass over and under a perpendicular array (the *warp*) in a repeating pattern. Different repeat sequences plus the use of different yarn materials, colours, textures and sizes lead to the huge variety of woven cloth types (eg Fig. 3), a selection of which is listed in the Dictionary.

In contrast to weaving, *knitting* forms a fabric by the intermeshing of loops of yarn in repeated sequences, in effect creating a series of knots (Fig. 4). Each horizontal row of loops is called a *course* and each vertical line, a *wale*. In an analogous way to weaving, a large variety of different effects can be obtained.

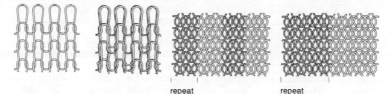

repeat   repeat

Plain knitted fabric   Weft-knitted plush   1 × 1 rib   2 × 2 rib

Fig. 4. **Some knitting sequences**.

Another important class of fibre assembly is represented by *rope* and the related string and *cable*. Rope is distinguished from string in having a minimum diameter of 4 mm. The traditional construction is known as *hawser-laid*, in which three strands are twisted together in a helical array. Each strand consists of *base yarns* twisted into a *primary yarn*, which in turn is spun with other primaries into a *roping yarn*. Roping yarns are then spun to form the strand. A structure more resistent to untwisting is the *plaited* rope in which the strands are woven together rather than just twisted.

fines That portion of a powder composed of particles under a specified size.

fingering Combed, soft-twisted, worsted yarn of the type generally used for hand knitting; usually 2, 3 or 4 ply.

fingerprinting By analogy with the eponymous forensic tool, application of analytical methods (esp. **spectroscopy**) to materials which allow identification of an unknown sample by comparison with standards. Macromolecules like DNA can be cut specifically to give a characteristic pattern of oligomers of different lengths. See **materials matching**.

fining The melting operation in which molten glass is made almost free from undissolved gases. Also *founding*, *plaining*, *refining*. See **glasses and glassmaking** p. 144.

fir Trees of the genus *Abies*, giving a valuable

structural **softwood**. See **whitewood**. The species *A. balsamea*, that is found in eastern N America, yields **Canada balsam** and is used for **pulp**.

fire bars Cast-iron bars forming a grate on which fuel is burnt, as in domestic fires, boiler furnaces etc.

fire-box That part of a locomotive-type boiler containing the fire; the grate is at the bottom, the walls and top being surrounded by water.

fireclay Clay consisting of minerals predominantly of SiO_2 and Al_2O_3, low in Fe_2O_3, CaO, MgO etc. Clays which soften only at high temperatures are used widely as refractories in metallurgical and other furnaces.

fire foam Mixture of foaming but non-flammable substances used to seal off oxygen and to extinguish fire without use of water.

fire polishing The polishing of silicate glass and glassware by localized melting of the surface in a flame.

fire refining The refining of blister copper by oxidizing the impurities in a reverberatory furnace and subsequently removing the excess oxygen. May be used as an alternative to electrolytic refining but is in any case carried out as a necessary preliminary to this.

fire retardant adhesive Heavy duty adhesives intended to fasten lagging around hot surfaces. These are usually alkyd based, but may contain chlorinated paraffins and antimony oxide to render the film noncombustible.

first-order reaction One in which the rate of reaction is proportional to the concentration of a single reactant, ie

$$\frac{dc}{dt} = -kc,$$

where c is the concentration of the reagent.

fish glue (1) **Isinglass**. (2) Any glue prepared from the skins of fish (esp. sole, plaice), fish-bladders and offal.

fission Spontaneous or induced fragmentation of heavy elements into two or more light atoms of comparable mass together with energetic neutrons. Used in the latter half of the 20th century as the basis for nuclear power generation.

fixed carbon Residual carbon in coke after removal of hydrocarbons by distillation in inert atmosphere.

fixture A device used in the manufacture of (interchangeable) parts to locate and hold the work without guiding the cutting tool.

flame cutting Cutting of ferrous metals by oxidation, using a stream of oxygen from a blow pipe or torch on metal preheated to about 800°C by fuel gas jets in the cutting torch.

flame hardening The use of an intense

flame from eg an oxy-acetylene burner for local heating of the surface of a hardenable ferrous alloy which is then immediately cooled.

flame retarders Speciality chemicals added to normally flammable materials, esp. polymers and fibres, to increase resistance to fire and flame. Solids include $CaCO_3$ (powdered chalk, whiting etc), antimony oxide (Sb_2O_3) and hydrated alumina. Fluids include tritolylphosphate and trixylylphosphate plasticizers, esp. for polyvinyl chloride conveyor belting.

flame temperature The temperature at the hottest spot of a flame.

flame test (1) The detection of the presence of an element in a substance by the colouration imparted to eg a Bunsen flame. (2) Simple way of identifying polymers by smelling (with care) odour emitted by burning material. Systematic methods like infrared spectroscopy, are much more accurate and safer. Such tests are usually combined with pencil hardness and density tests, etc to give a rough guide to polymer identity.

flammability Tendency of a material to ignite and continue to burn. Since most polymers contain carbon and hydrogen, their heats of combustion are large, so will tend to be flammable (esp. if finely divided, as in fibres or particles). Burning behaviour can be measured in several ways, as in ASTM D635-74, where a rod is held horizontally in a flame. See **UL 94 flammability** and **oxygen index**.

flannel A soft all-wool fabric, the weave being either plain or twill. The cloth is preshrunk and lightly raised.

flannelette A cotton fabric of plain or twill weave, raised on both sides and used for pyjamas, nightdresses, sheets and working shirts.

flash The excess material forced out of a mould during moulding. Sometimes referred to as a *fin*, if the mould is in two halves.

flashed glass A term sometimes applied to glass coloured by the application of a thin layer of densely coloured glass to a thicker, colourless, base layer.

flashover An electric discharge over the surface of an insulator.

flashover voltage The highest value of a voltage impulse which just produces flashover.

flash photolysis **Photolysis** induced by light flashes of short duration but high intensity, eg from a laser.

flash welding An electric welding process similar to butt welding, in which the parts are first brought into very light contact. A high voltage starts a flashing action between the two surfaces, which continues while sufficient forging pressure is applied to the parts to complete the weld.

flat glass Generic term for **sheet**, **plate** and **float glass** suitable for eg windows. Made from soda-lime-silica glass of typical composition (wt %): SiO_2 72, Na_2O 14, CaO 8, MgO 3, Al_2O_3 2.

flats (1) Iron or steel bars of rectangular section. (2) The sides of a (hexagonal) unit.

flatter (1) Smith's tool resembling a flat-faced hammer, which is placed on forged work and struck by a sledge hammer. (2) Draw-plate for producing flat wire such as hair springs.

flatting mill Rolling mill which produces strip metal or sheet.

flat yarn (1) Man-made continuous filaments that have not been twisted or textured. (2) A straw-like filament.

flax Plants of *Linum usitatissimum* grown in temperate regions and used for the fibres obtained from their stems and for their seeds (**linseed oil** and animal food). The cellulosic fibres are prepared by **retting** followed by cleaning (including **hackling** and **scutching**).

fleece wool Wool obtained by shearing the living sheep.

flexible manufacturing system An arrangement of computer-controlled machines which can easily be adapted and used in different sequence to modify the manufacturing route for a particulr component or to restructure operations for a different product. Abbrev. *FMS*.

flexural strength To avoid problems of failure at the grips with brittle materials, their strength is often measured in bending rather than in the tensile test; the flexural strength tends to be somewhat higher than the tensile strength. See **strength measures** p. 301.

flicker noise Noise with an associated power spectrum that is inversely proportional to frequency, associated with a variety of phenomena such as fluctuations in the value of the resistivity of a resistor material.

flight angle The property of a screw helix in injection moulding or extrusion. Defined as angle subtended by screw flight to the line orthogonal to screw axis. Optimal value of 30° for conveying powders and 10–15° for granules.

flint glass (1) Originally **lead glass**; the good quality silica needed to ensure freedom from colour was obtained from crushed flint. (2) Now that class of optical glasses with v_d < 50 if n_d < 1.60 and v_d < 55 if n_d > 1.60, where v_d is the *Abbé number*, or reciprocal of the **dispersivity**, and n_d is the **refractive index** at 587.6 nm. Cf **crown glass**. (3) Also loosely used for all colourless glass other

than **flat glass**.

flip chip Mounting method to maximize heat transfer and minimize lead inductance in which **chips** are inverted and directly bonded to contacts in a **hybrid** or **printed circuit**.

flitch A piece of timber of greater size than 4 in × 12 in (100 mm × 300 mm).

float (1) A measure of timber, equalling 18 loads. (2) In a woven fabric, the length of yarn between adjacent intersections. (3) A defect in a fabric caused by a thread passing over other threads with which it is designed to interweave (4) A pattern thread in a lace.

float glass The vast majority of **flat glass** is now produced by the float process. A broad stream of molten glass at about 1050°C is spread across the surface of a bath of molten tin (in a reducing atmosphere to avoid oxidation of the tin) and is drawn off at the other end when rigid, at about 600°C. The two glass surfaces emerge flat, parallel and fire polished, so that they give clear, undistorted vision without requiring any further working. Cf **plate glass**, **sheet glass**.

float zone The process technique for growing single crystals without the use of a crucible. A seed is fused to a bar of purified stock material. Stock and seed ends are counter-rotated about a vertical axis while a local molten zone (created by a radio frequency coil around the rod) is swept slowly away from the seed. Cf **Bridgman process**, **Czochralski process**.

flocculation The coalescence of a finely divided precipitate into larger particles. See **gels** p. 143.

flocculent Existing in the form of cloud-like tufts or flocs.

flock (1) Waste fibres produced in the processes of finishing woollen cloths, used for budding and upholstery purposes. (2) Short cut or ground wool, cotton or man-made fibres for spraying on to adhesive coated backings for furniture and upholstery purposes. See **electrostatic flocking**.

flong A board made of **papier-mâché** used for making moulds from an original typesetting for casting duplicate metal *stereotype* printing plates.

floor temperature Critical temperature below which polymerization cannot occur. Unlike ceiling temperature, a relatively rare phenomenon but encountered in plastic sulphur.

flow box A compartment on a papermaking machine immediately before the machine wire or other forming unit, supplied with stock and designed to ensure uniform mixing within the stock and the means to control its flow on to the wire and even distribution.

flow forming A metal-forming operation, in

which thick blanks of aluminium, copper, brass, mild steel or titanium are made to flow plastically by rolling them under pressure in the direction of roller travel to result in components, often conical, having a wall thickness much less than the original blank thickness. Also *flospinning*.

flow length ratio Ratio of greatest distance from gate to end of cavity, to wall thickness in an injection moulding tool. Used to choose best grade of polymer for mouldfill.

flow lines (1) Witness marking or lines which appear on the surface of manufactured components to reveal the direction of material flow in the shaping operation. (2) Bands or structural features within a sectioned component which reveal the direction of material flow during working. (3) Lines found on exterior of moulded polymer products, due to the poor mixing of polymer melt.

fluid Any substance such as a gas, liquid or powder which flows. It differs from a solid in that it can offer no permanent resistance to change of shape.

fluidics The science of liquid flow in tubes etc which strongly simulates electron flow in conductors and conducting plasma. The interaction of streams of fluid can thus be used for the control of instruments or industrial processes without the use of moving parts.

fluid lubrication A state of perfect lubrication in which the bearing surfaces are completely separated by a fluid or viscous oil film which is induced and sustained by the relative motion of the surfaces.

Fluon TN for polytetrafluoroethylene.

fluorescence Emission of radiation, generally light, from a material during illumination by radiation of usually higher frequency, or from the impact of electrons. See **phosphor**.

fluorescent lamp A mercury-vapour electric-discharge lamp with the inside of the bulb or tube coated with fluorescent material so that ultraviolet radiation from the discharge is converted to visible light.

fluorescent penetrant inspection The use of a fluorescent dye which will penetrate any minute crack in a component. After immersion the component is wiped dry and any subsequent seepage from fissures is detected by irradiation at the exciting wavelength. See **penetrant flaw detection**.

fluorescent whitening agents Special dyes widely used to 'whiten' textiles, paper etc, and sometimes incorporated in detergents. Their effect is based on ability to convert invisible ultraviolet light into visible blue light, giving fabrics greater uniformity of reflectance over the visible part of the spectrum. Also *optical bleaches*, *optical whites*.

fluorescent yield Probability of a specific excited atom emitting a photon in preference to an Auger electron. See **Auger effect**.

fluorinated ethylene propylene Plastic with many of the properties of polytetrafluoroethylene (PTFE), having very good chemical resistance; it is unaffected by moisture and has a wide temperature range of application from -260 to $+200°C$. Abbrev. *FEP*.

fluorine The lightest halogen (see Appendix 1) and the most electronegative (non-metallic) element, symbol F. Molecular fluorine is a pale greenish yellow diatomic gas and chemically highly reactive and thus is never found as the free element. Combines with carbon to form inert polymers with low coefficient of friction, eg polytetrafluoroethylene.

fluoropolymers All those organic polymers containing fluorine atoms and hence showing some degree of heat and solvent resistance, polytrichlorofluoroethylene, polytetrafluoroethylene, Viton rubber etc.

fluoroscopy Examination of objects by observing their X-ray shadow on a fluorescent screen; used to examine contents of luggage, packages without unwrapping, quality of welding etc.

flux (1) Material added to a furnace charge, which combines with those constituents not wanted in the final product and improves the fluidity of the resultant slag, eg limestone or fluorspar in ironmaking. (2) A substance added to a joint prior to welding, soldering or brazing which improves wetting by the filler, prevents oxidation of the heated surfaces and dissolves existing infusible oxide films. (3) A vector field quantity eg **magnetic flux, electric flux**.

flux density See **characterizing magnetic materials** p. 57.

flux gate Magnetic reproducing head in which a magnetic flux (due to flux leakage from signals recorded on magnetic tape) is modulated by a high-frequency saturating magnetic flux in another part of the magnetic circuit.

flux quantization A magnetic effect in superconductors. A ring of material is placed in a uniform magnetic field and then cooled below its critical temperature so that it becomes superconducting. When the external field is removed it is found that the ring has trapped the field in its hole. If the flux of the trapped field is measured as a function of the strength of the applied field, it is found to be quantized in steps of $(h/2e)$ where h is **Planck's constant** and e is the electronic charge. This shows superconductivity to be a quantum effect. See **superconductors** p. 308.

fly shuttle Mechanism, invented by John

Kay (Bury, Lancashire, UK) in 1733, for propelling the shuttle across the loom. It superseded hand-shuttling.

flywheel A heavy wheel attached to a shaft (eg an engine crankshaft) either to reduce the speed fluctuation resulting from uneven torque, or to store up kinetic energy to be used in driving a punch, shears etc, during a short interval.

FMEA Abbrev. for **Failure Modes and Effects Analysis**.

FMS Abbrev. for **Flexible Manufacturing System**.

foam A class of **cellular solid** which can be thought of as a type of **composite** material whose matrix incorporates a gas rather than a solid. Polymers (see **expanded plastics**), metals, ceramics and glass (see **cellular glass**) can all be foamed, variously by blowing or beating the gas into the melt, by chemical reactions evolving a gas, by using hollow fillers (*syntactic foam*) or by dissolving or melting out a particulate second phase. See **reticulated foam**. Used for eg buoyancy, shock absorption, cushioning, thermal insulation and, in sandwich panels, for enhanced flexural rigidity with small weight penalty.

foambacked fabric Dress and furnishing fabrics that are bonded on the back to polyether or polyester foams by adhesive or flame treatment.

foamed plastics See **expanded plastics**.

foam moulding Any polymer process which gives a partly or fully foamed product, esp. **structural foam moulding, sandwich moulding**.

folded yarns Yarns formed from two or more single yarns twisted together for strength or special appearance. The products are known as 2-, 3-, 4- etc fold or ply yarns.

folding strength The number of double folds needed to break a test strip of the paper, under prescribed conditions. See **strength measures** p. 301.

folk weave A loosely woven rough fabric made from coarse yarns including coloured ones.

foolscap A superseded size of writing and printing paper 13 in × 17 in. (US, 13 in × 16 in).

foot super 1 ft^2 or 144 in^2 of timber.

force That which, when acting on a body which is free to move, produces an acceleration in the motion of the body, measured by rate of change of momentum of body. The unit of force is that which produces unit acceleration in unit mass. See **newton**. Extended to denote loosely any operating agency. Electromotive force, magnetomotive force, magnetizing force etc are strictly misnomers.

force constant Bond stiffness defined as

force needed to deform a specific covalent bond divided by the deformation produced by that force (k_1 for stretching and k_2 for bending). For the single carbon-carbon bond, $k_1 = 436$ N m^{-1} and $k_2 = 0.35$ N m^{-1}. See **theoretical stiffness**.

force diagram A diagram in which the internal forces in a framed structure, assumed pin-jointed, are shown to scale by lines drawn parallel to the members themselves. Also *reciprocal* or *stress diagram*.

force on a moving charge If a charge q is moving with a velocity **v** in a magnetic field **B**, then the force on the charge is

$$F = q(v \times B).$$

If the moving charges are within a conductor, the force of a short length **l** is

$$F = i(l \times B),$$

where i is the current.

forehand welding That in which the palm of the torch or electrode hand faces the direction of travel so that the metal ahead of the weld position is preheated.

forehearth Bay in front of a furnace into which molten products can be run.

forensic engineering Application of engineering methods to help determine facts at issue in civil and criminal cases. Typical civil cases involve personal injury, product liability, contract and intellectual property. Methods include those of forensic sciences as applied to practical situations, fractography applied to broken products, design analysis in patent cases, stress analysis of products etc.

forensic science Application of scientific methods to legal problems, in order to determine facts at issue in both criminal and civil cases. Methods include materials identification and materials matching using chemical (eg elemental analysis) or physical (eg X-ray analysis) techniques.

forge Open hearth or furnace with forced draught; place where metal is heated and shaped by hammering.

forging The operation of shaping malleable materials by means of hammers or presses.

forging machines Power hammers and presses used for forging and drop forging. See **drop forging**.

formability See **drawability**.

formaldehyde *Methanal*, H.CHO, bp −21°C, a gas of pungent odour, readily soluble in water, and usually used in aqueous solution. Formaldehyde easily polymerizes to give polyoxymethylene. See **acetal resin**.

formaldehyde resins Synthetic resins which are condensation products of methanal with hydroxybenzenes, urea etc. Also *methanal resins*. See **thermosets** p. 318.

formation The pattern of the fibres in paper when viewed by transmitted light.

forming Changing the shape of a metal component without in general altering its thickness. Cf **drawing**.

form tool, forming cutter Any cutting tool which produces a desired contour on the work-piece by being merely fed into the work, the cutting edge having a profile similar to, but not necessarily identical with, the shape produced.

formula (1) The representation of the types and relative numbers of atoms in a compound, or the actual number of atoms in a molecule of a compound. It uses chemical symbols and subscripts, eg H_2SO_4, C_6H_6. (2) A rule or law expressed in algebraic symbols.

forward-bias Said of a semiconductor diode, p-n junction, or the emitter-base junction of a transistor, when the polarity of the applied emf is such as to allow substantial conduction to take place.

Foucault currents See **eddy currents**.

foulard A lightweight dress fabric with a printed pattern, made either of silk or man-made filament yarns or of high-quality Sea Island or Egyptian cotton.

fouling (1) Coming into accidental contact. (2) Deposition or incrustation of foreign matter on a surface, as of carbon in an engine cylinder, or marine growth on the bottom of a ship or on structures subject to the action of sea-water.

founding See **fining**.

foundry A workshop in which metal objects are made by casting in moulds.

fourdrinier The standard type of papermaking machine characterized by a machine wire, part of which (the upper, forming, surface) is horizontal or nearly so. See **paper and papermaking** p. 231.

Fourier number A dimensionless parameter used for studying heat flow problems. It is defined by $\lambda t / C_p \rho l^2$ where λ is the thermal conductivity, t is the time, C_p the specific heat at constant pressure, ρ the density and l a linear dimension.

Fourier transform infrared Most recent form of infrared spectroscopy involving interferometric method giving enhanced resolution etc. Abbrev. *FTIR*.

FPS The system of measuring in *feet, pounds* and *seconds*.

fractional crystallization The separation of substances by the repeated partial crystallization of a solution.

fractional distillation Distillation process for the separation of the various components of liquid mixtures. An effective separation can only be achieved by the use of fractionating columns attached to the still. See **petroleum**.

fractional distribution In an assembly of particles having different values of some common property such as size or energy, the fraction of particles in each range of values is called the fractional distribution in that range. Cf **cumulative distribution**.

fractionation See **fractional distillation**.

fractography Study of fracture surfaces of materials to determine nature and origin of product failure, eg whether brittle or ductile, single or multiple origins, association with stress-concentrations, nature of crack propagation. For example, stop-start mechanism may be indicated by beach marks on surface, so indicating **fatigue** p. 123. Analysis usually starts with macrography, then micrography, often using **scanning electron microscopy**. Important investigative method in **forensic engineering**. See **beach marks**, **fracture lances**, **hackle**, **mirror**, **mist**, **striations**, **Wallner lines**.

fracture energy Energy required to form unit area of fractured surface, numerically equal to half the **toughness**. See **strength measures** p. 301.

fracture lances Characteristic, spearlike, fracture surface markings arising from fracture under torsional loading.

fracture mechanics Stress analysis of the conditions and criteria for crack propagation in materials in terms of such properties as **toughness** and **fracture toughness**. See **fracture mode**, **Griffith equation**, **strength measures** p. 301, **stress-intensity factor**.

fracture mode Three fundamental ways of loading a crack in a body; the opening-mode, forward shear (or sliding)-mode, and transverse shear (or tearing)-mode, designated as modes I, II and III respectively. See Fig. 3 in **strength measures** p. 301.

fracture of materials Loss of structural integrity by propagation of cracks in products or test samples. Often analysed using **fracture mechanics** and **fractography**. May be brittle or ductile, depending on state of material, stress concentrations, rate of test etc.

fracture test Any test to determine strength or crack propagation resistance of a material. Additional information is provided by **fractography** of the fracture surfaces. See **strength measures** p. 301.

fracture toughness A materials property derived from **fracture mechanics** and providing a measure of the material's resistance to crack propagation. It is the critical value of K_{IC}, the mode I **stress intensity factor**, for a crack in a material to propagate to failure. It is related to the **toughness** G_C by

$$K_{IC} = \sqrt{E' \, G_C} \, .$$

For **plane strain**, $E' = E$, the **Young's modulus** of the material, while for **plane stress**

$$E' = \frac{E}{1 - v^2},$$

where v is **Poisson's ratio**. The plane strain value for K_{IC} is lower than the plane stress value, so is normally taken as the conservative measure of resistance to cracking. See **fracture mode, strength measures** p. 301.

francium The heaviest alkali metal. Symbol Fr. See Appendix 1, 2.

free atom Unattached atom assumed to exist during reactions. See **free radicals**.

free beaten stuff Lightly beaten stuff with minimum hydration. The resultant paper is low in strength, bulky, porous and opaque. Such stock permits easy drainage of water on the machine wire.

free cementite Iron carbide in cast-iron or steel not associated with the ferrite in pearlite.

free charge One not bound to an atomic nucleus.

free-cutting See free machining.

free-cutting brass α-brass containing about 2–3% of lead, to improve the machining properties. Used for engraving and screw machine work. See **copper alloys** p. 76.

free electron theory Early theory of metallic conduction based on concept that outer valence electrons, which are not localized in forming bonds, are free to migrate through a crystal, so forming **electron gas**. Now superseded by **energy band** theory.

free energy The capacity of a system to perform work, a change in free energy being measured by the maximum work obtainable from a given process. See **Gibbs free energy, Helmholtz free energy**.

free energy equation See **Gibbs–Helmholtz equation**.

free ferrite Ferrite in steel or cast-iron not associated with the cementite in pearlite.

free machining An alloy with additions to make it easier to machine and so reduce machining time and power consumption, usually effected by causing chips to break up rather than absorb energy by plastically deforming, eg the addition of lead to steels and copper alloys.

free radical polymerization Polymerization of doubly bonded monomers using a free radical catalyst such as benzoyl peroxide. See **chain polymerization** p. 56.

free radicals Groups of atoms in particular combinations capable of free existence under special conditions, usually for only very short periods (sometimes only microseconds). Because they contain unpaired electrons, they are paramagnetic, and this fact has been used in determining the degree of dissociation of compounds into free radicals. See **chain polymerization** p. 56.

free volume Term for the empty space surrounding atoms and molecules in solid or fluid materials. Symbol $v_f = v-v_0$ where v is the **specific volume** (reciprocal of density) and v_0 the specific volume extrapolated to 0 K. Used in molecular theories of the glass transition and of viscosity.

freeze line Line formed on **blown film** showing onset of crystallization, so transparent below but translucent above.

freezing mixture A mixture of two substances, generally of ice and a salt, or solid carbon dioxide with an alcohol, used to produce a temperature below 0°C.

freezing point The temperature at which a material solidifies. Pure materials, eutectics and some intermediate constituents freeze at constant temperature; alloys generally solidify over a range. See **phase diagram** p. 236.

French gold A copper-based alloy with about 16.5% zinc, 0.5% tin and 0.3% iron. Also *oroide*.

Frenkel defect Disorder in the crystal lattice, due to some of the ions (usually the cations) having entered interstitial positions, leaving a corresponding number of normal lattice sites vacant. Likely to occur if one ion (in practice the cation) is much smaller than the other, eg in silver chloride and bromide.

frequency factor The pre-exponential factor in **Arrhenius's equation** when applied to chemical reactions, expressing the frequency of successful collisions between the reactant molecules.

fretting corrosion Corrosion due to slight movements of unprotected metal surfaces, left in contact either in a corroding atmosphere or under heavy stress.

friction The resistance to sliding motion between two surfaces in contact. The frictional force opposing the motion is equal to the applied force up to the onset of motion when its value is known as the *limiting friction*. Any increase in the applied force will then cause slipping. *Static friction* is the value of the limiting friction just before slipping occurs. *Dynamic friction* is the value of the frictional force after slipping has occurred, is smaller than the static friction, and can depend on sliding speed. The *coefficient of friction*, symbol μ, is the ratio of the limiting friction to the normal reaction between the sliding surfaces. Under normal conditions μ is a constant between a given pair of materials of specified surface quality and lubrication conditions. With polymers, μ depends on the normal reaction and on the duration of loading or the sliding speed.

frictional electricity Static electricity produced by rubbing bodies or materials together, eg an ebonite rod with fur. See tribo-electrification.

friction calendering Passing fabric between two rolls of a **calender** designed so that one of the rolls is highly polished and rotates faster than the other. This produces a glaze on the fabric surfaces.

friction glazing A method of glazing paper in which one of the **calender** rolls revolves at a greater peripheral speed than that of the others. A very high polish is obtained.

friction rollers See **anti-friction bearing**.

friction spinning A method for converting staple fibres into a yarn by feeding a sliver onto a rotating perforating roller through which air is being sucked. Another roller is set near so that the yarn is formed from the rapidly twisting fibres at the nip of the rollers.

friction welding Welding in which the necessary heat is produced frictionally, eg by rotation, and forcing the parts together.

Friedel and Crafts' synthesis The synthesis of alkyl substituted benzene hydrocarbons and aromatic ketones, by the action of halogenoalkanes or acyl halides on aromatic hydrocarbons in the presence of anhydrous aluminium chloride.

Friedel–Crafts polymers Heat-resistant aromatic polymers made by alkylation of phenols, structurally intermediate between PF resins and polyphenylenes.

frieze Woven woollen overcoating fabric whose surface has been heavily milled and raised.

fringed micelle model Model of polymer crystallites, where chains wander between crystallites rather than folding into lamellae. See **crystallization of polymers** p. 82.

frit The pulverized **glaze** used esp. for making **enamel** ware.

froth Liquid foam. A gas-liquid continuum in which bubbles of gas are contained in a much smaller volume of liquid, which is expanded to form bubble walls. The system is stabilized by oil, soaps or emulsifying agents which form a binding network in the bubble walls.

frozen equilibrium The state of a solid at low temperature, which is prevented from attaining the theoretically possible thermodynamic equilibrium, because its molecular motion has become too slow.

frozen-in strain Non-equilibrium state of polymer moulding where chain orientation is high. Can be relieved by annealing, which may result in distortion of shape. Detected using polarized light in transparent polymers. Also *residual strain*.

frozen-in stress See **residual stress**.

FSD Abbrev. for **Full-Scale Deflection**.

FTA Abbrev. for **Fault Tree Analysis**.

FTIR Abbrev. for **Fourier Transform Infra-Red**.

fuel cell A galvanic cell in which the oxidation of a fuel (eg methanol) is utilized to produce electricity.

fuel oils Oils obtained as residues in the distillation of petroleum; used, either alone or mixed with other oils, for domestic heating and for furnace firing (particularly marine furnaces); also as fuel for internal combustion engines.

fugacity The tendency of a gas to expand or escape; substituted for pressure in the thermodynamic equations of a real gas. Analogous to *activity*. See **ideal gas**.

fugitometer An apparatus for testing the fastness of dyed materials to light.

full annealing Of steel, heating above the critical range, followed by slow cooling, as distinguished from (a) annealing below the critical range, and (b) normalizing, which involves air-cooling. See **annealing** p. 13.

fullerene See **buckminsterfullerene**.

fuller's earth A non-plastic clay consisting essentially of the mineral montmorillonite, and similar in this respect to bentonite. Used originally in 'fulling', ie absorbing fats from wool, hence the name. The fuller's earth of English stratigraphy is a small division of the Jurassic System in the S Cotswolds.

full hard The stage in the tempering or work hardening of some ferrous and non-ferrous alloys just below that at which the metal cannot be further deformed by bending.

fulling See **milling of textiles**.

fully fashioned Knitted fabrics and garments that are shaped by increasing or decreasing the number of wales. This ensures that the garment fits more closely.

fumaric acid *Ethane 1,2-dicarboxylic acid*, HOOC.CH=CH.COOH, small prisms which do not melt, but sublime at about 200°C, with the formation of maleic anhydride. Fumaric acid being the trans-form. Used in polyester resins.

fume Cloud of airborne particles, generally visible, of low volatility and less than a micrometre in size, arising from condensation of vapours or from chemical reaction.

functional group Small cluster of linked atoms with chemically active bonds, of esp. interest in step polymerization. Examples in monomers include hydroxyl (–OH), carboxyl (–COOH), amine (–NH$_2$) and isocyanate (–NCO). Examples in polymers include amide (–CO–NH–), ester (–CO.O–), carbonate (–OCO.O–) and urethane (–NH–CO.O–). See **step polymerization** p. 298.

furan group A group of heterocyclic compounds derived from furan, C_4H_4O, a compound containing a ring of four carbon atoms and one oxygen atom.

furan resins A group of speciality thermoset polymers derived from the partial polymerization of furfuryl alcohol, or from condensation of furfuryl alcohol with either furfural or methanal, or of furfural with ketones, and used as baked plastic coatings on metal, as adhesives and as resin binders for stoneware.

furnace atmosphere There are three main classes: (1) *oxidizing*, produced when air volumes are in excess of fuel requirements; (2) *neutral*, when the air to fuel ratios are perfectly proportioned; (3) *reducing*, due to deficiency of combustion air. See **protective furnace atmosphere**.

furnace brazing A high-production method of copper-brazing steel, without flux, in a reducing atmosphere, or of brazing steels, copper and copper alloys with brasses or silver-brazing alloys, in continuous or in batch furnaces.

furnish The ingredients from which paper is manufactured.

fuse A device used for protecting electrical apparatus against the effect of excess current. It consists of a piece of fusible metal, which is connected in the circuit to be protected, and which melts and interrupts the circuit when an excess current flows. The term fuse also includes the necessary mounting and cover (if any).

fused silica See **vitreous silica**.

fusible alloys, metals Alloys of bismuth, lead and tin (and sometimes cadmium or mercury) which melt in the 47–248°C temperature range; used as solders and for safety devices in fire extinguishers, boilers etc.

fusible core Low-melting alloy core in injection moulding tool. Used to create products with complex shapes and many re-entrant angles, and removed at end of cycle by melting out (eg sports racquet, inlet manifolds).

fusible plug A plug containing a metal of low melting point used, for example, in the crown of a boiler fire box to prevent serious overheating of the plates if the water level falls below them.

fusing point See **melting point**.

fusion (1) The solid to liquid phase change. (2) The nuclear reaction in which light nuclei combine to form a heavier one (plus fragments), usually with substantial release of energy. The controlled release of energy from the thermonuclear fusion reaction between deuterium and tritium nuclei (requiring very high temperature and pressure) has been attained for short periods in experimental reactors and has been proposed as the basis for power generation.

fusion cones See **Seger cones**.

fustian A term including a number of hard-wearing fabrics usually of cotton but differing widely in structure and appearance, but all heavily wefted; they are used for clothing and furnishings. See **corduroy, moleskin, velveteen**.

G

g Abbrev. for gram(me).

g A symbol for: (1) acceleration due to gravity; (2) specified efficiency; (2) osmotic coefficient.

γ Symbol for: (1) substituted on the carbon atom of a chain next but two to the functional group; (2) substituted on one of the central carbon atoms of an anthracene nucleus; (3) substituted on the carbon atom next but two to the hetero-atom in a heterocyclic compound; (4) a stereoisomer of a sugar.

G Symbol for **giga**, ie 10^9.

G Symbol for: (1) **free energy**; (2) **toughness** or strain energy release rate; (3) Gibbs function; (4) thermodynamic potential; (5) **shear modulus**.

gaberdine A firm twill fabric (eg with worsted warp and cotton weft), with the warp predominating on the surface; used for dress and suiting cloths and light showerproof overcoatings. All-cotton gaberdine is also used for similar purposes.

gaboon Mahogany-like wood from a **hardwood** tree (*Aucoumea*), found in parts of C and W Africa. Planed surfaces of the wood are silky. Not very durable. Used for veneer and plywood manufacture, furniture and fittings. Average density 430 kg m^{-3}.

gadolinium A rare metallic element; trivalent; one of the **rare earth elements**. Symbol Gd. See Appendix 1, 2.

Gal Abbrev. for **galactose**.

galactans The anhydrides of galactose. They comprise many gums, agar and fruit pectins, and occur in algae, lichens, mosses.

galactose A hexose of the formula $CH_2OH.(CHOH)_4.CHO$; thin needles; mp 166°C; dextrorotatory. It is formed together with D-glucose by the hydrolysis of milk-sugar with dilute acids. Stereoisomeric with glucose, which it strongly resembles in properties. Present in certain gums and seaweeds as a polysaccharide galactan and as a normal constituent of milk.

gallium A metallic element in the third group of the periodic table. Symbol Ga. Used in fusible alloys and high temperature thermometry. Gallium arsenide it is an important semiconductor. See Appendix 1, 2.

gallium arsenide. Compound semiconductor (GaAs) in near stoichiometric proportions that is more difficult to process than silicon but with a higher band gap and higher electron mobility. It has **sphalerite** (zinc blende) structure.

gallon Liquid measure. One imperial gallon is the volume occupied by 10 lb avoirdupois

of water. One imperial gallon = 4.54609 litres, = 6/5 US gallon. One US gallon = 3.785 43 litres = 5/6 imperial gallon.

galvanic cell An electrochemical cell from which energy is drawn. Cf **electrolytic cell**.

galvanic corrosion Corrosion resulting from the current flow between two dissimilar metals in contact with an electrolyte. For example if zinc and copper are in electrical contact in a damp atmosphere, a current will flow under a potential of 1.14 V due to the zinc becoming anodic and corroding away while the copper will form the cathode of the cell and remain unaffected.

galvanic series Electrochemical series for different metals and alloys in specific electrolytes, eg sea water.

galvanizing The coating of steel or iron with zinc, generally by immersion in a bath of zinc, covered with a **flux**, at a temperature of 425–500°C. The zinc may alternatively be electrodeposited from cold sulphate solutions. The zinc is capable of protecting the iron from atmospheric corrosion even when the coating is scratched, since the zinc is preferentially attacked by carbonic acid, forming a protective coating of basic zinc carbonates. Cf **electrogalvanizing**.

galvanoluminescence Feeble light emitted from the anode in some electrolytic cells.

gamma brass The γ-constituent in brass is hard and brittle and is stable between 60 and 68% of zinc at room temperature. γ-brass is an alloy consisting of this constituent.

gamma iron The polymorphic form of iron which is stable between 906 and 1403°C. It has a **face-centred cubic** lattice and is non-magnetic. Its range of stability is lowered by carbon, nickel and manganese, and it is the basis of the solid solutions known as **austenite**.

gangue The portion of an ore which contains no metal; valueless minerals in a lode.

gap Range of energy levels between the lowest of conduction electrons and the highest of valence electrons.

gap-filling Term applied to property of adhesives in adequately packing the free space in a joint, implies a high-viscosity material such as epoxy resin rather than monomeric cyanoacrylate.

garnet The natural mineral, $Mn_3Al_2Si_3O_2$, in which the aluminium ion is tetrahedrally co-ordinated and the silicon is octahedrally co-ordinated. Also a related class of **ferrimagnetic materials**, having a similar basic structure, but containing three trivalent cations.

gas carburizing The introduction of carbon into the surface layers of low carbon steels by heating in a reducing atmosphere of gas high in carbon, usually hydrocarbons or hydrocarbons and carbon monoxide.

gas constant The constant of proportionality R in the equation of state for 1 mole of an **ideal gas**, $pV = RT$, where p is the pressure, V the volume and T the absolute temperature. $R = 8.314$ J K^{-1} mol^{-1}. Also *molar gas constant.*

gas evolution The liberation of gas bubbles during the solidification of metals. It may be due to the fact that the solubility of a gas is less in the solid than in the molten metal, as when hydrogen is evolved by aluminium and its alloys, or to the promotion of a gas-forming reaction, as when iron oxide and carbon in molten steel react to form carbon monoxide. See **blowholes, unsoundness.**

gas-filled filament lamp One in which the bulb contains an inert gas.

gas-liquid chromatography Form of partition or adsorption chromatography in which the mobile phase is a gas and the stationary phase a liquid. Solid and liquid samples are vaporized before introduction on to the column. The use of very sensitive detectors has enabled this form of chromatography to be applied to submicrogram amounts of material. Abbrev. *GLC.*

gas moulding See **hollow moulding.**

gas porosity Small voids with the body of a moulding or casting caused by the evolution and/or entrapment of gas.

gas tar Coal tar condensed from coal gas, consisting mainly of hydrocarbons. Distillation of tar provides many substances, eg ammoniacal liquor, 'benzole', naphtha and creosote oils, with a residue of pitch. Dehydrated, it is known as 'road tar', and used as a binder in road-making.

gas tungsten-arc welding Welding processes in which a tungsten electrode is used to form an arc without being consumed and contributing to the weld pool. The weld pool and arc are protected by an inert gas (usually argon). Commonly *TIG welding.*

gas turbine A simple, high-speed machine used for converting heat energy into mechanical work in which stationary nozzles discharge jets of expanded gas (usually products of combustion) against the blades of a turbine wheel. Used in stationary power and other plants, locomotives, marine (esp. naval) craft, jet aero-engines, and experimentally in road vehicles.

gas welding Any metal welding process in which gases are used in a combination to obtain a hot flame. The most commonly used gas welding process employs the oxy-acetylene combination which develops a flame temperature of 3200°C. Some plastics, esp. polyethylene, may be fusion-jointed by a form of low-temperature gas welding without flame.

gate (1) In a mould, the channel or channels through which molten material is led from the runner, down-gate or pouring-gate to the mould cavity. (2) The term used to denote the small restricted space in an injection mould between the mould cavity and the passage carrying the plastic moulding material. See **injection moulding** p. 169. Also *geat, git, sprue.* Also *tab, fan-, diaphragm-, submarine-gates,* by reference to their shape. (3) The control electrode in a **field effect transistor** which electrically modifies the nature of a conducting channel between source and drain electrodes.

gate voltage The control voltage for an electronic gate. The voltage applied to the 'gate' electrode of a field-effect transistor.

gatherer A person who gathers a charge of glass (a *gather*) on a blowpipe or gathering iron for the purpose of forming it into ware or feeding a charge to a machine for that purpose.

gauche A form of staggered molecular conformation in which the substituents being considered make a dihedral angle of 60° to one another. The most stable conformation of hydrogen peroxide has the hydrogen atoms in this relationship.

gauge (1) An object or instrument for the measurement of dimensions, pressure, volume etc. (2) An accurately dimensioned piece of metal for checking the dimensions of work or of less precisely made gauges. (3) A tool used for measuring lengths, as a *micrometer gauge.* (4) The diameter of wires and rods, eg *Birmingham Wire Gauge, Brown & Sharpe Wire Gauge.* (5) That portion of a test piece over which some property such as strain or elongation is to be measured. (5) In textiles it relates to the fineness of a knitted fabric and generally indicates the number of needles per cm in warp or weft knitting machines.

gauge number An arbitrary number or letter denoting the gauge or thickness of sheet metal or the diameter of wire, rod or twist drills in one of the many non-metric gauge number systems.

gaussmeter Instrument measuring magnetic flux density. This term is most widely used in US.

gauze Lightweight woven fabric of open texture.

Gay-Lussac's law Of volumes: when gases react, they do so in volumes which bear a simple ratio to one another and to the volumes of the resulting substances in the gaseous state, all volumes being measured at the same temperature and pressure. Also *Charles' law*.

gear (1) Any system of moving parts transmitting motion, eg levers, gear-wheels etc. (2) A set of tools for performing some particular work. (3) A mechanism built to perform some special purpose, eg steering gear, valve gear. (4) The position of the links of a steam-engine valve motion, as astern gear, mid-gear etc. (5) The actual gear ratio in use, or the gear-wheels involved in transmitting that ratio, in an automobile gearbox, as first gear, third gear etc.

gearing Any set of gear-wheels transmitting motion. See **gear**.

gegenions The simple ions, of opposite sign to the colloidal ions, produced by the dissociation of a colloidal electrolyte. Also *counterions*.

gel The apparently solid, often jellylike, material formed from a colloidal solution on standing. A gel offers little resistance to liquid diffusion and may contain as little as 0.5% of solid matter. Some gels, eg gelatin, may contain as much as 90% water, yet in their properties are more like solids than liquids. See **gels** p. 143.

gelatin(e) A colourless, odourless and tasteless glue, prepared from albuminous substances, eg bones and hides. Used for foodstuffs, photographic films, glues etc.

gelation The process whereby plasticized polyvinyl chloride compounds by the application of heat undergo an irreversible change to soft, rubbery thermoplastic materials. See **plastisols**.

gel-permeation chromatography Method for analysing molecular mass distribution of non-crosslinked polymers, by selective elution of polymer solution through a microporous gel (crosslinked polystyrene). Common solvent is tetrahydrofuran. For insoluble polymers, high-temperature gel-permeation chromatography in decalin etc is available. Abbrev. *GPC*. See **molecular mass distribution** p. 212.

gels See panel on p. 143.

gel spinning Fibre-making process using swollen gel of high molecular mass polymer, esp. high-density polyethylene. See **high-performance fibres** p. 158.

generation rate Rate of production of *electron-hole pairs* in semiconductors.

geochemistry The study of the chemical composition of the Earth's crust.

geometrical isomerism A form of stereo-isomerism in which the difference arises because of hindered rotation about a double bond or a bond that is part of a ring. Thus but-2-ene has two isomers, depending on whether the methyl groups are on the same

(cis) or opposite (trans) sides of the double bond. See **polymers** p. 247.

Georgian (wired) glass See **wired glass**.

geotextile A textile material used in civil engineering. Strong fabrics made of synthetic fibres are frequently used in road-making and for the stabilization of embankments.

germanium A metalloid element in the fourth group of the periodic table. Greyish-white in appearance. Symbol Ge. A semiconductor, used in early devices before the development of silicon processing. See Appendix 1, 2.

German silver A series of alloys containing copper, zinc and nickel within the limits; copper, 25–50%; zinc, 10–35%; and nickel 5–35%. Also *nickel silver*. See **copper alloys** p. 76.

gettering The removal of harmful impurities (or defects in a crystalline solid) by the scavenging action of other impurities (or defects). Strategy used in semiconductor technology to improve crystal purity in active regions.

GFRP Abbrev. for **glass-fibre reinforced plastic**.

ghost In steel, a band in which the carbon content is less than that in the adjacent metal and which therefore consists mainly of ferrite. Also *ghost line*.

ghost line See **ghost**.

Gibbs' adsorption theorem Solutes which lower the surface tension of a solvent tend to be concentrated at the surface, and conversely.

Gibbs–Duhem equation For binary solutions at constant pressure and temperature, the chemical potentials (μ_1, μ_2) vary with the mole fractions (x_1, x_2) of the two components as follows:

$$\frac{\partial \mu_1}{\partial \ln x_1} = \frac{\partial \mu_2}{\partial \ln x_2}.$$

Gibbs' free energy The difference of the enthalpy and the product of the entropy and the temperature of a system. Usual symbol G $(G = H - TS)$. A calculated negative change in G indicates a spontaneous process in a closed system at constant pressure.

Gibbs–Helmholtz equation An equation of **thermodynamics**,

$$\Delta G = \Delta H - T\Delta S$$

where ΔG = change in free energy, ΔH = change in enthalpy, ΔS = entropy change and T = the absolute temperature. It is applied to electrochemical cells in the form:

$$\Delta G = -zFE,$$

where ΔG = the free energy change in the cell reaction, z = the number of electrons

gels

Gels have properties intermediate between the liquid and the solid states. Thus, they deform elastically and recover, yet can often be induced to flow at higher stresses. They have extended three-dimensional network structures and are highly porous, so many gels contain a very high proportion of liquid to solid (one description of them is a highly concentrated solution of a liquid in a solid). The networks can be permanent or temporary and are based on polymeric molecules.

Gels occur in many fields. Examples include: *vulcanized rubber* when it absorbs a liquid and swells; *plasticized PVC, thixotropic paint*, photographic emulsion, jam and mayonnaise; *gel permeation chromatography* (GPC), for measuring the molecular mass distribution in polymers; finings for beers and wines (eg *isinglass* and *gelatin*); the drying agent, *silica gel; gel spinning* to produce highly oriented polymer fibres. They form the basis of the **sol-gel process** for making inorganic glasses and ceramics. The gel network is formed by polymerization in solution; eliminating the interstitial liquid and collapsing the residue by sintering produces a glass without melting.

The classical starting point for gels is a suspension of a colloid (a collection of particles with sizes in the range 1 nm–0.1 μm) in a liquid. The stability of such suspensions is governed by interactions between the particles and between the particles and the liquid. Some eg inks, can exist indefinitely, whilst others, eg milk, are unstable. Their stability is largely governed by the electric charges on the particles, which might be due to surface ionization in a polar liquid, or to absorption of dissolved ions. The type of ions present in the liquid determine whether the particles are positively or negatively charged; in a colloidal suspension in water, the charges on the particles change from positive to negative as the concentration of OH^- ions is increased (ie the pH rises). The pH at which the changeover occurs is known as the *isoelectric point*. At the isoelectric point the particles are uncharged, therefore there is no electrostatic repulsion between them. *Brownian motion* can bring them close enough to combine through eg *van der Waals forces*, forming larger and larger particles which fall out of the suspension, a process known as *coagulation*. If polymeric molecules are present, these can provide bridges between the particles, creating the open three-dimensional network required of a gel; this is known as *flocculation*.

transferred, F = one faraday (96 496 coulombs) and E = the reversible emf of the cell. See **Nernst equation**.

Gibbs–Konowalow rule For the phase equilibrium of binary solutions. At constant pressure the equilibrium temperature is a maximum or minimum when the compositions of the two phases are identical, and vice versa, eg in eutectics or azeotropes. The corresponding statements hold for pressure at constant temperature.

Gibbs' phase rule See **phase rule**.

giga- Prefix used to denote 10^9 times, eg a gigawatt is 10^9 watts. Abbrev. G.

gilding metal Copper-zinc alloy containing zinc up to 15%. See **copper alloys** p. 76.

gillion Rarely term for 10^9; preferable term *giga-* or G.

gin A machine that is used to remove cotton fibres from the seeds: the process is known as *ginning*.

gingham Lightweight, plain weave cotton cloth woven with coloured yarns to produce a check pattern.

girder A beam, usually steel, to bridge an open space. Girders may be rolled sections, built up from plates or of lattice construction. See **continuous beam**.

glarimeter Instrument for measuring the gloss of a paper surface based on light reflectance.

glass See **glasses and glassmaking** p. 144.

glass-ceramic Ceramic processed to final shape as a glass and then induced to crystalize by controlled heat treatment. Has improved thermal and mechanical properties over parent silicate glass. Used in eg cooker hobs. See **glasses and glassmaking** p. 144, **silica and silicates** p. 285.

glasses and glassmaking See panel on p. 144.

glass fibre Glass melted and then drawn out

glasses and glassmaking

The word 'glass' is most commonly associated with the family of hard, brittle, non-crystalline solids based on silica (see **silica and silicates** p. 285), fused togther with various other oxides (eg those of calcium, sodium, boron, phosphorous, magnesium and potassium). *Polymeric glasses* are also well-known; less so are those based on the *chalcogenides* (sulphur, selenium and tellurium), the halides, certain organic compounds and on carbon. More recently, *metallic glasses* have been produced. Their common characteristic is that their structures are *amorphous*, ie with none of the regularity of crystals.

Glasses are sometimes defined as those non-crystalline solids which exhibit a transition in behaviour, the *glass transition*, with temperature; the remainder are classified as amorphous. The location of the glass transition temperature, T_g, is not as well-defined as the crystalline melting temperature, T_m, and depends on the rate of cooling (Fig. 1). The most common route to a glass is the cooling of a liquid at a sufficiently high rate to avoid nucleation and growth of crystals (in metals this requires cooling rates of around 10^6 K s^{-1}!). The resulting solid retains the structural disorder of the liquid. Other methods exist, including condensing a gas, creating disorder in solids by shock waves or radiation damage, and chemical reactions in solution to produce a gel precursor (the **sol-gel process**).

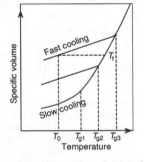

Fig. 1.   **Specific volume (= density^{-1}) plotted against temperature for a glass**.

continued on next page

continuously through platinum bushings (almost always in multiples of 204) into fine fibres (diameters range from 3–25 μm). See **A-, C-, E-, S-** and **Z-glass.**

glass-fibre reinforced plastic   Class of **composite materials** comprising a polymeric **matrix** reinforced with glass fibre. The matrix can be thermoplastic (used chiefly with chopped fibres) or thermosetting (allowing use with continuous fibres and/or glass textiles). Widely used as a structural material in eg aircraft, boats and ships, road and rail vehicles, sports equipment, printed circuit boards. Abbrev. *GFRP.*

glassine   Transparent glazed greaseproof paper. Produced by long beating of the stock and high glazing of the finished paper.

glasspaper   Paper coated with glue on

which is sprinkled broken glass of a definite grain size; used for rubbing down surfaces. Cf **emery paper, sandpaper.**

glass transition   Characterized by a change in slope of the volume vs temperature and enthalpy vs temperature curves as glass-forming materials are cooled or heated, the transition is general to all glass materials whether inorganic, polymeric or metallic. Its location, the *glass transition temperature*, T_g, varies with the rate of cooling or heating. Theories and models of the transition abound, but none is generally applicable or accepted. See **glasses and glassmaking** p. 144, **viscoelasticity** p. 332.

glass transition temperature   See **glass transition.**

glaze   Glass or glass-like surface coating of

The manufacture of silicate glasses is based on the readily available ingredients, silica and sodium carbonate. Glasses made from these alone are vulnerable to attack by water. To reduce this effect, some of the sodium carbonate is replaced by calcium carbonate to make soda-lime-silica glass as used for such products as windows, bottles and jars, light bulbs and spectacle lenses. Since the carbonates decompose to form oxides and CO_2 during glass-making, the glass is formed from a mixture of oxides (as its name suggests).

The first stage in making the glass is to melt the pulverized raw materials in a furnace. One of these (SiO_2) has a very high melting point (~2000 K). To avoid the need (and expense) of heating to this temperature, a quantity of powdered scrap glass, known as *cullet*, is added to the raw materials. This makes up 15–30% of the total and by melting before any of the raw constituents, it conducts heat and dissolves the other ingredients together. This significantly reduces the time required to melt the mixture and hence it saves energy. Fig. 2 shows a type of continuous glass melting furnace used, and a sketch of the temperature profile along its length. The processes taking place are complex and interdependent. They fall into the three categories of melting, refining (or *fining*) and homogenization. The furnace is either gas or oil-fired and is usually arranged so that the maximum temperature (about 1750 K for soda-lime-silica glass) is reached approximately one third of the way along the melting chamber where the refining process takes place.

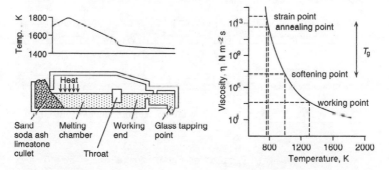

Fig. 2. **Glassmaking furnace and temperature profile (above).**

Fig. 3. **Plot of viscosity versus temperature for a soda-lime silica glass.**

Associated with the melting is the reaction of the silica with the sodium and calcium carbonates which leads to the evolution of carbon dioxide gas. The process of refining is the clearing away of these gas bubbles. To aid this process, fining agents such as sodium sulphate are sometimes used to sweep out the bubbles, and this movement also aids the homogenization of the melt, together with convection and the flow of the melt through the tank.

Beyond the throat of the furnace is the working end where the fining and homogenization occur, as the temperature is reduced to the working range of the glass. Since there are no abrupt changes in the viscosity of glass as a function of temperature, glasses are frequently characterized by a set of reference temperatures at which their viscosities have defined values. These are shown in Fig. 3 for a soda-lime-silica glass. In the melting chamber the viscosity η is of the order of $10 \text{ N m}^{-2} \text{ s}$ (similar to that of thick treacle). When

continued on next page

glasses and glassmaking (contd)

the glass is being worked (either moulded or drawn) η is between 10^2 and 10^5 N m^{-2} s, so the temperature at which η is 10^3 N m^{-2} s is arbitrarily called the *working point*. The *softening point* is defined to be the temperature at which η = $10^{6.6}$ N m^{-2} s, which is a viscosity associated with the creep at a prescribed rate of a given length of a glass rod or fibre under its own weight. The *annealing point* ($\eta = 10^{12.4}$ N m^{-2} s) is the temperature at which any internal stresses are relieved after several minutes. The *glass transition temperature*, T_g, lies somewhere between the softening and annealing points. Finally, the *strain point* ($\eta = 10^{13.5}$ N m^{-2} s) is the highest temperature from which the glass can be cooled rapidly without causing significant levels of internal stress to be developed.

The structure of the resulting material is based on a random network of predominantly covalently bonded SiO$_4$ tetrahedra, in which each oxygen atom is shared between two tetrahedra (*bridging*). The sodium and calcium oxides ionize and donate oxide ions to the network, partially breaking up the network and forming gaps. The sodium and calcium ions locate at these gaps, so providing ionic bonds between unbridged tetrahedra (Fig. 4). Due to their lower energy and non-directionality, these ionic bonds reduce the viscosity of the melt, enabling glass processing at lower temperatures. Ions such as Na$^+$ and Ca^{2+} are known as *network modifiers*.

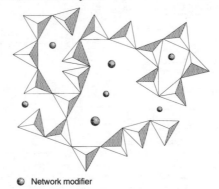

● Network modifier

Fig. 4. **Schematic structure of soda-lime-silica glass**.

finish applied to materials. See **frit**. Cf **enamel**.

glazed imitation parchment   A highly glazed packaging paper, generally bleached and frequently opacified, suitable for waxing or coating for use in food wrapping applications. Abbrev. *GIP*.

glide   The movement of dislocations along slip planes. See **dislocation**.

globular cementite   In steel, cementite occurring in the form of globules instead of in lamellae (as in **pearlite**) or as envelopes round the crystal boundaries (as in **hyper-eutectoid steel**). Produced by very slow cooling, or by heating to between 600 and 700°C.

globular pearlite   See **granular pearlite**.

glory-hole   (1) A subsidiary furnace, in which articles may be reheated during manufacture. (2) An opening exposing the hot interior of a furnace and used for eg fire polishing of glass.

glow discharge   An electrical discharge through a low-pressure gas, producing a **plasma**. The ionized media of glow discharges are much use in surface engineering and microfabrication.

Glu   Abbrev., depending on the context, for **glucose** or *glutamic acid*.

glucans   The condensation polymers of glucose, eg cellulose, starch, dextrin, glycogen etc.

gluco-   See terms prefixed **glyc-**.

glucose   The commonest aldohexose, with

the formula $C_6H_{12}O_6$ and the major source of energy in animals. Starch and cellulose are condensation polymers of glucose. Also *dextrose, grape sugar*.

glue See **adhesive**.

Gly Symbol for **glycine**.

glycerine   *Glycerol, propan-1,2,3-triol*, $CH_2OH.CHOH.CH_2OH$, a syrupy hygroscopic liquid. Colourless. It is a raw material for alkyd resins, nitroglycerine, printing inks, foodstuff preparations etc. Also *glycerin*.

glycerine litharge cement A mixture of litharge (lead (II) oxide) and glycerine which rapidly sets to a hard mass.

glycerol-phthalic resins See **alkyd resins**.

glycine Aminoethanoic acid, the simplest of the naturally occurring amino acids. Symbol Gly, short form G.

glycols Dihydric alcohols, of the general formula $C_nH_{2n}(OH)_2$. They are viscous liquids.

glyoxal *Ethan 1,2-dial*, $CHO.CHO$. A dialdehyde, a yellow liquid, mp 15°C, bp 51°C (776 mm), forming green vapours, but it is not stable, and polymerizes to insoluble paraglyoxal, $(CHO.CHO)_3$.

glyptal resins See **alkyd resins**.

gob (1) A measured portion of molten glass as fed to machines making glass articles. (2) A lump of hot glass gathered on a *punty* or **blowing iron**.

gob process One for making glass hollow ware, in which glass is delivered by an automatic feeder in the form of soft gobs of suitable shape to a forming unit. Also *flow process, gravity process*.

godets (1) Small rollers used to regulate the speed of extruded filaments during manmade fibre production. (2) An insert of material used to shape a garment being assembled from fabric.

gold A heavy, yellow, metallic element, occurring in the free state in nature. Symbol Au. Has electrical resistivity of about 0.02 $\mu\Omega$ m. Most of the metal is retained in gold reserves but some is used in jewellery, dentistry and for decorating pottery and china. In coinage and jewellery, the gold is alloyed with varying amounts of copper and silver. White gold is usually an alloy with nickel, but as used in dentistry this alloy contains platinum or palladium. See Appendix 1, 2.

gold filled The agreed term for a coating of gold over a base metal which should be stamped with a fraction giving the proportion by weight of the gold present and the carat fineness of the gold alloy.

gold-leaf Pure gold beaten out into extremely thin sheets, so that it may be applied to surfaces which are to be gilded.

Goldschmidt process (1) See **aluminothermic process**. (2) Detinning of coated iron by use of chlorine.

Gore-tex TN for microporous polytetrafluoroethylene, made by stretching polymer in controlled way so as to create fine pores, which allow diffusion of air and water vapour but prevent liquid water ingress. Used in waterproof clothing.

government rubber-styrene The chief synthetic rubber developed in the US during World War II. Based on styrene and butadiene. Abbrev. *GR-S*. See **SBR**.

GPC Abbrev. for **Gel Permeation Chromatography**.

G–P zones Abbrev. for **Guinier–Preston zones**.

grade Individual specification of a particular polymer. Includes details of composition (additives, fillers etc), polymer type (homopolymer, copolymer etc) and molecular mass distribution. Usually given a proprietory code by the manufacturer. Applies *mutatis mutandis* to other materials.

Grafil TN for carbon fibre (Japan).

graft polymer Polymer where chain branches occur, esp. where branches are different to main chain. See **copolymer** p. 75.

Graham's law The velocity of effusion of a gas is inversely proportional to the square root of its density.

grain (1) Small particles of a crystalline substance. May be discrete as in a single grain of mineral or one in a polycrystalline array, as in a phase mixture or a recrystallized metal (see Fig. 1 in **annealing** p. 13). The crystal structure is continuous across each individual grain and boundaries are formed where one grain meets its neighbours because the orientation of the crystal lattice is different. Grains may sometimes be visible to the unaided eye without any preparation, but more usually a section needs to be polished and etched to reveal their size, form and distribution. (2) Feature of the texture of wood determined by the direction of the long axes of the **tracheid** cells (ie parallel to the long axes of trunk and branches) and showing the marked **anisotropic** nature of wood. See **structure of wood** p. 305. Cf **banding** in metals. (3) Unit of mass, the basis of the obsolete *apothecaries' weight* system. One grain = 15.4 g.

grain boundary Zone formed at the junction of single crystals in a polycrystalline material. Impurities tend to accumulate here by being excluded from normal growth of each crystal.

grain growth Stage in the annealing process of cold-worked metals, in which holding the metal at above about 0.4–0.5 T_m (melting temperature) after recrystallization has taken place allows the average grain size of the metal crystals to increase. See **annealing** p. 13.

grain refining Production of small closely knit grains, resulting in improved mechanical properties. Particularly with aluminium alloys it is achieved by small additions to melts of substances such as boron which cause fine nucleation on casting.

grain size Particle size. The average diameter or expressed dimension of the grains or crystals in a sample of metal or rock.

grain-size analysis US term for particle-size analysis. It is also the literal translation for the German phrase used to specify particle-size analysis. Confusion arises because grain-size analysis is also used to describe procedures for measuring the size of crystallites in a cast or sintered metal.

grain-size control Control of the rate at which grains grow when metal is heated above the recrystallization range; the control may be effected by the addition of elements which anchor grain boundaries (eg aluminium to steel) or by regulation of the temperature and time of the recrystallization process.

grain structure Size, shape and orientation of crystallites forming the microstructure of most metals, alloys and ceramics.

grammage Preferred term for the mass in grams of one square metre of the paper or board under the prescribed conditions of test. See **basis weight, substance**.

gram(me)-atom The quantity of an element whose mass in grams is equal to its relative atomic mass. A *mole of atoms*.

gram(me) calorie See **calorie**.

gram(me)-molecular volume See **molar volume**.

gram(me)-molecule See **mole**.

granular Said of powder particles having an approximately equidimensional but irregular shape. .

granular, globular pearlite Pearlite in which the cementite occurs as globules instead of as lamellae. Produced by very slow cooling through the critical range, or by subsequent heating just below the critical range.

granules Main form in which a compounded polymer is supplied for further processing to shape.

graphical methods The name given to those methods in which items, such as forces in structures, are determined by drawing diagrams to scale, as distinct from calculation.

graphic formula A chemical formula in which every atom is represented by the appropriate symbol, valency bonds being indicated by dashes; eg H–O–H, the graphic formula for water.

graphic statics A method of finding the

stresses in a framed structure, in which the magnitude and direction of the forces are represented by lines drawn to a common scale.

graphite One of the two naturally occurring forms of crystalline carbon, the other being diamond. It occurs as black, soft masses and, rarely, as shiny crystals (of flaky structure and apparently hexagonal) in igneous rocks; in larger quantities in schists particularly in metamorphosed carbonaceous clays and shales, and in marbles; also in contact metamorphosed coals and in meteorites. Graphite has numerous applications in trade and industry now much overshadowing its use in 'lead' pencils. Much graphite is now produced artificially in electric furnaces using petroleum as a starting material. Widely used in refractories and as a moderator in nuclear power plants. Also *black lead*, *plumbago*. See **colloidal graphite**.

graphite resistance A resistance unit consisting of a rod of graphite, which has a high ohmic value; also a variable resistance made up of piles of graphitized disks of cloth under a variable pressure.

graphitic carbon In cast-iron, carbon occurring as **graphite** instead of as cementite.

graphitization The transformation of amorphous carbon to graphite brought about by heat. It results in a volume change due to the alteration in atomic lattice layer spacing. It is reversible under bombardment by high-energy neutrons and other particles. See **Wigner effect**.

grate That part of a furnace which supports the fuel. It consists of fire bars or bricks so spaced as to admit the necessary air.

gravimetric analysis The chemical analysis of materials by the separation of the constituents and their estimation by weight.

gravitation The name given to that force of nature which manifests itself as a mutual attraction between masses, and whose mathematical expression was first given by Newton, in the law which states: 'Any two particles of matter attract one another with a force directly proportional to the product of their masses and inversely proportional to the square of the distance between them'. This may be expressed by the equation:

$$F = G\frac{m_1 m_2}{d^2},$$

where F is the force of gravitational attraction between bodies of mass m_1 and m_2, separated by a distance d. G is the constant of gravitation = 6.670×10^{-11} N m^2 kg^{-2}.

gravity diecasting A process by means of which castings of various alloys are made in steel or cast-iron moulds, the molten metal being poured by hand, and where the pressure is solely due to the hydrostatic head of the metal in the mould. See **diecasting, pressure diecasting**.

gravity drop hammer A type of machine hammer used for **drop forging**, in which the impact pressure is obtained from the kinetic energy of the falling ram and die. Also *board hammer*.

greaseproof A quality of paper possessing grease resistant characteristics brought about by heavy beating of a suitable fibrous furnish.

greaseproof paper Any paper that in its natural state or as the result of coating or other treatment resists penetration by oils or greases.

green compact Ceramic *preforms* before sintering, which are held together by the cohesive forces resulting from compaction alone or with the assistance of a temporary binder.

greenheart Very strong, dense, yellowish-green to black timber of an evergreen **hardwood** tree (*Ocotea*) from S America; largely used for piles and underwater work because of its considerable resistance to the attack of the teredo, a wood-boring mollusc. Also used for bridges, fishing rods, flooring and longbows. Density 1030 kg m^{-3}.

green strength See **strength measures** p. 301.

green tyre Tyre in unvulcanized state ready for final shaping. See **tyre technology** p. 325.

grey Fabrics in the state they leave the loom or knitting machine before any scouring or bleaching has been carried out.

grey iron Pig- or cast-iron in which nearly all the carbon not included in **pearlite** is present as flakes of graphite carbon. See **mottled iron, white iron**.

Griffith equation Relates the **tensile strength**, σ_t, of a material to the critical flaw size, or crack length, a_c for an edge crack or $2a_c$ for an internal crack, required for fracture ie failure by crack propagation:

$$\sigma_t = \sqrt{E'G_C / \pi a_c}$$

where G_C is the **toughness** of the material. For **plane strain**, $E' = E$, the **Young's modulus** of the material, while for **plane stress**,

$$E' = \frac{E}{1 - v^2},$$

where v is **Poisson's ratio**. See **strength measures** p. 301.

grinding machine A machine tool in which flat, cylindrical, or other surfaces are produced by the abrasive action of a high-speed grinding wheel.

grinding medium The solid charges (balls, pebbles, rods etc) used in suitable mills for grinding certain materials, eg cement, pigments etc, to a fine powder.

grinding wheel An abrasive wheel for cutting and finishing metal and other materials. It is composed of abrasive particles, such as silicon carbide or emery, held together by a bond or binding agent, which may be either a vitrified material or a softer material, such as shellac or rubber.

grit (1) Hard particles, usually mineral, of natural or industrial origin, retained on a 200 mesh BS test sieve (76 μm). (2) A measure indicating the sizes of the abrasive particles in a grinding wheel, usually expressed by a figure denoting the number of meshes per linear inch in a sieve through which the particles will pass completely.

grit blasting A process used in preparation for metal spraying, which cleans the surface and gives it the roughness required to retain the sprayed metal particles.

grommet A ring or collar used to line a sharp-edged hole through which a cable or similar material passes.

gross energy requirement Total energy expended in manufacture of different materials. Highly electropositive metals (eg aluminium, titanium) possess high energy contents and recycling after use is widely practised. Polymers possess intermediate energy contents (process energy + feedstock energy), which are greater than that of glass or paper.

ground (1) In **lace** manufacture, the mesh which forms a foundation for a pattern. (2) The base fabric to which is secured the figuring, threads, eg pile or loops in carpet or terry cloths.

ground state State of nuclear system, atoms etc, when at their lowest energy, ie not *excited*. Also *normal state*.

group (1) A vertical column of the periodic table, containing elements of similar properties. (2) Metallic radicals which are precipitated together during the initial separation in qualitative analysis. (3) A number of atoms which occur together in several compounds.

group reaction The reaction by which members of a group are precipitated.

growing The production of semiconductor crystals by slow crystallization from the molten state.

growth Permanent increase in size of a component, leading to distortion, which occurs when materials are in service over a long period. For example, common cast iron firebars, subjected to thermal cycling in the range 700–900° C or held for long periods above 480°C, suffer extensive distortion and cracking due to an increase in **specific vol-**

ume resulting from the breakdown of iron carbides to graphite in the microstructure. The cracking leads to internal oxidation which further adds to the damage.

GR-S Abbrev. for **Government Rubber-Styrene**.

guayule A bush which occurs in the US and Mexico and is a potential source of natural rubber. Other sources of natural rubber considered in the past have included dandelion latex.

Gudden–Pohl effect A form of electroluminescence which follows metastable excitation of a **phosphor** by ultraviolet light.

Guerin process Used in presswork to cut or form sheet metal by placing it between a die made of a cheap material and a thick rubber pad which adapts itself to the die while under pressure.

guide mill A rolling-mill equipped with guides to ensure that the stock enters the mill at the correct point and angle.

Guillemin effect The tendency of a bent magnetostrictive rod to straighten in a longitudinal magnetic field.

Guinier–Preston zones A clustering of solute atoms on certain crystallographic planes in supersaturated solid solutions which occurs during precipitation hardening. Two types are seen, referred to as GP I in the initial stage followed by GP II at an intermediate stage of the hardening process. See **precipitation hardening** p. 254.

Guldberg and Waage's law See **law of mass action**.

gum See **American red gum**, **Red River gum**.

gum arabic A fine, yellow or white powder, soluble in water, rel.d. 1.355. It is obtained from certain varieties of acacia, the world's main supply coming from the Sudan and Senegal. Used in pharmacy for making emulsions and pills; also in glues and pastes. Also *acacia gum*, *Senegal gum*.

gummed paper Paper coated with a moisture activated adhesive, eg dextrin, gum arabic etc.

gums Non-volatile, colloidal plant products

which either dissolve or swell up in contact with water. On hydrolysis, they yield certain complex organic acids in addition to pentoses and hexoses.

gunmetal A copper-tin alloy (ie bronze) either Admiralty gunmetal (copper 88%, tin 10%, zinc 2%) or copper 88%, tin 8% and zinc 4%. Lead and nickel are frequently added, and the alloys are used as cast where resistance to corrosion or wear is required, eg in bearings, steam-pipe fittings, gears. See **copper alloys** p. 76.

Gurley densometer Instrument for measuring the air resistance of a sample of paper. The time in seconds recorded for the passage of a given volume of air under specified conditions eg 100 ml is called the *gurley*.

gusset, gusset plate A bracket or stay, cast or built up from plate and angle, used to strengthen a joint between two plates which meet at a joint, as the junction of a boiler shell with the front and back plates, or between connecting members of a structure.

gutta percha The coagulated **latex** of *Palaquium oblongifolium*. Consists mainly of 60% crystalline trans-1,2-polyisoprene compared to 0–10% crystallinity for natural rubber (the cis isomer). A hard inelastic solid ($T_m = 56-65°C$), it was formerly used for submarine cables before being displaced by polyethylene etc. See **natural rubber**.

gypsum Crystalline mineral of hydrated calcium sulphate, $CaSO_4.2H_2O$. Occurs in bulk form as **alabaster**, in fibrous form as satin spar, and as clear, colourless, monoclinic crystals of selenite. Used in making **plaster of Paris**, plaster and plasterboard, and is an important constituent (~3.5 wt%) of Portland cement. It hydrates very rapidly during the setting reaction, and helps to control the initial setting rate. See **cement and concrete** p. 52.

gyromagnetic ratio The ratio γ of the magnetic moment of a system to its angular momentum. For orbiting electrons $\gamma = e/2m$ where e is the electronic charge and m is the mass of the electron. γ for electron spin is twice this value.

H

h Symbol for: (1) height; (2) **Planck's constant.**

H Symbol for: (1) **magnetic field strength**; (2) **enthalpy.**

HAC Abbrev. for **High Alumina Cement.**

hackle Rough, ridged region of tensile fracture surface in brittle materials which follows the **mirror** and **mist** regions. The surface in the hackle region makes an angle to the previous plane of crack propagation, because the crack has branched while running at maximum fracture velocity, frequently ejecting a piece of material. See **fractography.**

hackling Process of combing scutched flax in the hackling machine, in order to divide and make parallel the long fibres, and to remove the short ones and impurities.

hack-saw (1) A mechanic's hand-saw used for cutting metal. It consists of a steel frame, across which is stretched a narrow saw-blade of hardened steel. (2) A larger saw, similar to the above, but power-driven.

Hadfield's steel See **manganese steel.**

haematite Oxide of iron, Fe_2O_3, crystallizing in the trigonal system. It occurs in a number of different forms: kidney iron-ore massive; specular iron-ore in groups of beautiful, lustrous, rhombohedral crystals as, for example, from Elba; bedded ores of sedimentary origin, as in the Pre-Cambrian throughout the world; and as a cement and pigment in sandstones. Also *haematite.*

hafnium A metallic element in the fourth group of the periodic table. Symbol Hf. It occurs in zirconium minerals, where its chemical similarity but relatively high neutron absorption makes it a troublesome impurity in zirconium metal for nuclear engineering. Used to prevent recrystallization of tungsten filaments. See Appendix 1, 2.

hair Animal fibre of variable diameter and length occurring externally on most mammals. Composed mainly of the protein **keratin.** Formerly used for rope, fabric etc or to reinforce building materials such as plaster and brick clay in composite form. Excludes sheep's wool and the invertebrate product, silk, but the soft shorter fibres from certain animals may also be called wool but qualified by the animal's name, eg angora wool.

hair cloth A material generally composed of coarse hair and cotton-yarn; used as a stiffening for coats and in upholstery.

haircord Cotton fabrics of light weight, in which cords are produced by running two warp threads together at frequent intervals.

haircord carpet A hard carpet made from animal hair.

half-life (1) Time taken in any chemical reaction for half the starting material to be transformed into product(s). (2) Time in which half of the atoms of a given quantity of radioactive nuclide undergo at least one disintegration. Also *half-value period.* The half-life T is related to the *decay constant* λ by $T = (\ln 2)/\lambda$.

half-stuff Paper raw materials which have been converted into pulp but not yet beaten.

halides Fluorides, chlorides, bromides, iodides and astatides.

hali-, halo- Prefixes from Gk *hals,* salt.

Hall coefficient The co-efficient of proportionality (R_H) in the **Hall effect** relation $E_H = R_H \, jB$ where E_H is the resulting transverse electric field, j is the current density and B is magnetic flux density.

Hall effect The generation of a transverse electric field in a conductor or semiconductor when carrying current across a magnetic field.

Hall mobility Mobility (mean drift velocity in unit field) of current carriers in a semiconductor as calculated from the product of the Hall coefficient and the conductivity.

Hall–Petch equation An empirical relation between the yield stress σ_Y of a metal or alloy and its average grain size d which states that

$$\sigma_Y = \sigma_0 + kd^{-1/2},$$

where σ_0 and k are experimentally determined constants. The effect is due to the pinning of dislocations by grain boundaries. The equation shows why grain size should be kept as small as possible to achieve highest strength. Also *Hall–Petch relation.* See **strength measures** p. 301.

Hall probe A small probe which uses the **Hall effect** to compare magnetic fields.

halogen One of the seventh group of elements in the periodic table (see Appendix 1) for which there is one electron vacancy in the outer energy level, viz. fluorine (F), chlorine (Cl), bromine (Br), iodine (I), astatine (At). The main oxidation state is -1.

halogenation The introduction of halogen atoms into an organic molecule by substitution or addition.

haloid acids A group consisting of hydrogen fluoride, hydrogen chloride, hydrogen bromide and hydrogen iodide.

hammer scale The scale of iron oxide which forms on work when it is heated for forging.

hand feed The hand operation of the feed

mechanism of a machine tool. See **feed**.

handle The subjective reaction obtained by feeling a fabric and assessing its roughness, harshness, flexibility, softness etc.

handmade paper Paper made in single sheets by dipping a mould into a vat containing stock so that the requisite amount of stock is picked up and, by skilful shaking, is distributed and formed into the sheet. The wet sheets are couched on to felts, pressed to remove water, dried and, if necessary, tub sized. Handmade papers are characterized by their permanence and durability, appearance of quality and excellent properties for watercolour painting.

hand mould Wooden frame, accompanied by a pair of deckles, covered with a wove or laid pattern woven wire on which a sheet of hand-made paper is formed.

hank General term for a reeled length of yarn.

hard (1) Said of bulk magnetic material which retains its magnetization. See **characterizing magnetic materials** p. 57. (2) Glass having a relatively high **softening point**.

hardboard **Fibre-board** that has been compressed in drying, giving a material of greater density than **insulating board**.

hard bronze Copper-based alloy used for tough or dense castings; based of 88% copper plus tin with either some lead or zinc. See **copper alloys** p. 76.

hard drawn Term applied to wire or tube which has been greatly reduced in cross-section without annealing.

hardenability The propensity of a steel to transform to a hard **martensite** when cooled from the **austenitic** state. A steel with low hardenabilty will only form martensite when cooled rapidly (eg water quenched in thin section) whereas one possessing a high hardenability may be cooled slowly in air and will still transform to a hard martensite. Improvement of hardenability is one of the prime reasons for alloying medium carbon steels, since it allows components of large cross section to be hardened prior to tempering without risk of cracking. Commonly assessed by the **Jominy** (end quench) **test**. See **steels** p. 296.

hardener An **accelerator**.

hardening The process of making steel hard by cooling from above the critical range at a rate that prevents the formation of ferrite and pearlite and results in the formation of martensite. May involve cooling in water, oil or air, according to composition and size of article and the hardenability of the steel. The steel must contain sufficient carbon (above about 0.3%) to achieve a useful hardening response.

hardening, quenching media Liquids

into which steel components are plunged for hardening. They include cold water and brine to increase the cooling power for the fastest quench rates and special mineral oils and polymers for slower and intermediate rates.

hard-facing (1) The application of a surface layer of hard material to impart, in particular, wear resistance. (2) A surface so formed. The composition is generally of high melting-point metals, carbides etc, applied by powder, wire or plasma arc spraying, or by welding.

hard glass **Borosilicate glass**, whose high softening point is principally due to boron compounds. Resistant to thermal shock and to chemical action.

hard head Alloy of tin with iron and arsenic left after refining of tin.

hard lead All antimonial lead; metal in which the high degree of malleability characteristic of pure lead is destroyed by the presence of impurities, of which antimony is the most common.

hardmetal Sintered tungsten carbide. Used for the working tip of high-speed cutting tools. See **sintering**.

hard metals Metallic compounds with high melting-points; typified by refractory carbides of the **transition metals** (4th to 6th groups of the periodic table) Most notable are tungsten, tantalum, titanium and niobium carbides.

hardness (1) See **hardness measurements** p. 153. (2) Degree of vacuum in an evacuated space, esp. of a thermionic valve or X-ray tube. Also penetrating power of X-rays, which is proportional to frequency.

hardness measurements See panel on p. 153.

hard plating Chromium plating deposited in appreciable thickness directly on to the base metal, that is, without a preliminary deposit of copper or nickel. The coating is porous, but offers resistance to corrosion and to wear.

hard segment Term used for rigid parts of block copolymer polymers, esp. polyurethanes and block polyesters, formed from isocyanate or polyester groups respectively. Such rigid chain segments often co-crystallize, so reinforcing the final material.

hard solders For jopining metals, usually copper-silver-zinc alloys, which melt at temperatures between 600°C and 850°C; they have greater strength those based on lead-tin alloys. See **silver solder**.

hardwood Deciduous, broad-leaved trees (ash, balsa, beech, oak, teak etc), one of the two distinct families of tree. Cf **softwood**. See **structure of wood** p. 305.

Hartmann dispersion formula An em-

hardness measurements

There is a bewildering variety of tests purporting to give some measure of the hardness of materials. These range from a scale of what scratches what (**pencil hardness, Mohs scale**) through measuring the size of the impression left by an indenter of prescribed geometry under a known load (**Vickers, Knoop** and **Brinell hardness**) or the depth to which an indenter penetrates under specified conditions (*Rockwell B* and *C, Shore A* or *IRHD*), to the height of rebound of a ball or hammer dropped from a given distance (**Shore scleroscope**). Not surprisingly, perhaps, each test produces a different number (some on arbitrary scales) for the hardness of a given material. The approximate correlation between different scales of hardness is shown in Fig. 1.

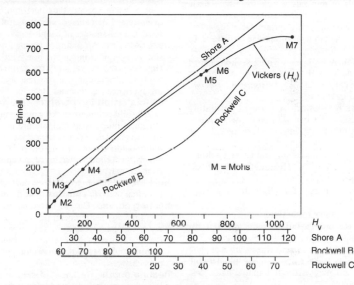

Fig. 1. **Relation between different hardness scales**.

The chief attractions of the various hardness tests are that they are relatively simple and quick to perform and that they are virtually non-destructive. Thus they are very well suited for quality control purposes. However, there is not a well-defined materials property called hardness, and what all these tests measure is differing combinations of the elastic, plastic, and sometimes fracture behaviour of materials.

Relating the results to properties such as **yield stress** and **Young's modulus** is not straightforward. At one extreme, the *International Rubber Hardness Degree* (IRHD or Shore A) measures solely the elastic response of rubbers. It is widely used as a check on the degree of *cure* or crosslinking of a rubber, and there is an approximate correlation between the hardness value and the **shear modulus**. At the other extreme, the size of the plastically deformed impression produced by indenters such as the 136° diamond pyramid in the **Vickers hardness test** p. 331 must obviously bear some relation to the yield stress. It should also be borne in mind that, in those materials which exhibit time-dependence of either yield stress (eg metals above 0.5 T_m) or both yield stress and elastic modulus (eg the majority of plastics), the size of the indentation will increase with time, and thus their hardness value will depend on how long the load is applied.

pirical expression for the variation of the refractive index n of a material with the wavelength of light λ;

$$n = n_0 + \frac{c}{(\lambda - \lambda_0)^a} ,$$

where n_0, c, λ_0 are constant for a given material. For glass a is about 1.2.

Harvey process Toughening treatment for alloy steels, involving superficial carburization, followed by heating to a high temperature and quenching with water.

hazardous materials Elements, compounds or minerals which are highly toxic (eg beryllium metal, asbestos, radioactive materials) or highly inflammable (eg pyrophoric metals) or otherwise injurious to health and safety.

H-beam A steel beam with a section shaped like the letter H, the cross-piece of which is relatively long. Also *H-girder*, *I-beam*, *rolled steel joist*.

H-class insulation A class of insulating material which can withstand a temperature of 180°C. Other letters, which see, denote different temperatures.

HCP Abbrev. for **Hexagonal Close Packing**.

HDI Abbrev. for **Hexamethylene DiIsocyanate**.

HDPE Abbrev. for **High-Density Polyethylene**.

He The symbol for **helium**.

head Recording and reproducing unit for magnetic tape, containing exciting coils, a laminated core, in ring form with a minute gap. Flux leakage across this gap enters the tape and magnetizes it longitudinally. See **flux gate**.

head box See **flow box**.

head-to-head unit Pair of asymmetric repeat units fused at identical ends. Normal structure of chain is head-to-tail. Kind of chain defect in polmers.

heart valve implant Artificial valve grafted to heart tissue to replace diseased valves; usually composed of metal frame with synthetic fabric covering (eg polyethylene terephthalate) and polymer (eg polypropylene) ball or disc.

heartwood The older, inner core of wood in tree trunks and branches, which no longer functions in storage and conduction; in many species it is impregnated with gums and resins which help it to resist decay and make it denser and darker than the surrounding sapwood.

heat Heat is energy in the process of transfer between a system and its surroundings as a result of temperature differences. However, the term is still used also to refer to the energy contained in a sample of matter. Also

for **temperature**, eg forging or welding *heat*. For some of the chief branches in the study of heat, see **calorimetry, heat units, internal energy, latent heat, mechanical equivalent of heat, radiant heat, radiation, specific heat capacity, temperature, thermal conductivity, thermometry.**

heat balance Evaluation of operating efficiency of a furnace or other appliance, the total heat input being apportioned as to heat in the work, heat stored in brickwork, or refractory materials, loss by conduction, radiation, unburnt gases in waste products, sensible heat in dry flue gases, and latent heat of water vapour, thus determining the quantity and percentage of heat usefully applied and the sources of heat losses.

heat detector An indirect-acting thermostat for operation in conjunction with a gas-flow control valve, and for controlling working temperatures in furnaces and heating appliances up to about 1000°C.

heat-distortion temperature The point at which solid polymers sag or cannot maintain structural integrity as temperature is raised, so lies near their T_g. It is important in injection moulding, determining the point below which the product can be removed safely from the machine.

heat drop Colloq. transfer of heat energy, but often used for **enthalpy heat drop**.

heating curves Curves obtained by plotting time against temperature whilst heating under constant conditions. The curves show the absorption of heat which accompanies melting and arrest points marking polymorphic changes.

heating depth Thickness of skin of material which is effectively heated by dielectric or eddy-current induction heating, or radiation.

heat-insulating concrete High-alumina cement and a lightweight aggregate, eg kieselguhr, diatomic earth or vermiculite, to reduce heat transfer through furnace walls etc.

heat of formation Strictly the *enthalpy of formation*. The net quantity of heat evolved or absorbed during the formation of one mole of a substance from its component elements in their standard states. Symbol ΔH_f. See **standard heat of formation**.

heat of polymerization Strictly, enthalpy of polymerization. Heat given out during polymerization, caused by bond formation. Strict control is necessary in large-scale commercial operations. See **chain polymerization** p. 56. Symbol ΔH_p.

heat of solution The thermal energy evolved or absorbed when one mole of a substance is dissolved in a large volume of a solvent. Symbol ΔH_s.

heat-resisting alloy Alloys developed to withstand high stresses at very high temperatures as in the fan blades of aeroengines.

heat setting Stabilizing fibres, yarns, or fabrics by means of heating under controlled conditions. Thus a fabric may be heat set when it is held flat under tension while being extended in length and breadth. The resultant fabric will tend to retain its flatness in use. Pleats may also be heat set so that they remain clearly defined in a garment.

heat treatment Generally, any heating operation performed on a solid material, eg heating for hot-working, or annealing after cold-working. Particularly, the thermal treatment of steel by normalizing, hardening, tempering etc; used also in connection with precipitation-hardening alloys, such as those of aluminium.

heat units See **joule**, **calorie**.

heavy chemicals Those basic chemicals which are manufactured in large quantities, eg sodium hydroxide, chlorine, nitric acid, sulphuric acid.

heavy hydrogen Same as **deuterium**.

heavy water Deuterium oxide, or water containing a substantial proportion of deuterium atoms (D_2O or HDO).

Heisenberg principle See **uncertainty principle**.

heli-arc welding A welding process in which helium is used to shield the weld area from contamination by atmospheric oxygen and nitrogen.

helical gears Gear-wheels in which the teeth are not parallel with the wheel axis, but helical (ie parts of a helix described on the wheel face), being therefore set at an angle with the axis.

helical spring A spring formed by winding wire into a helix along the surface of a cylinder; sometimes erroneously termed a spiral spring.

helium Chemically inert element, symbol He. The gas is monatomic, liquefies at temperatures below 4K, and undergoes a phase change to a form known as liquid helium II at 2.2K. The latter form has many unusual properties believed to be due to a substantial proportion of the molecules existing in the lowest possible quantum energy state; see superfluid. Liquid helium is the standard coolant for devices working at cryogenic temperatures. The atom has an extremely stable nucleus identical to an α-**particle**. See **noble gases**. See Appendix 1, 2.

helix A line, thread, wire or other structure curved into a shape such as it would assume if wound in a single layer round a cylinder; a form like a screw-thread which is very common in biological macromolecules, eg DNA.

Hellesen cell A dry cell with zinc and car-

bon electrodes and a depolarizer of manganese dioxide.

Helmholtz free energy Similar to Gibbs free energy but with internal energy substituted for enthalpy. Symbol A (or F in US). A negative change in A is indicative of a spontaneous change in a closed system at constant volume.

hematite See **haematite**.

hemi- Prefix from Gk *hemi*, half.

hemicelluloses A group of polysaccharides in the matrix of plant cell walls; homo- and hetero-polymers, linear and branched, of xylose, glucose and other sugars.

hemicolloid A particle up to 2.5×10^{-8} m in length; 20–100 molecules.

hemisphere The half of a sphere, obtained by cutting it by a plane passing through the centre. As applied to the Earth, the term usually refers to the *northern* or the *southern hemisphere*, the division being by the equatorial plane.

hemlock A N American **hardwood** (*Tsuga*) whose **heartwood** is a pale brown, with a straight grain and fine texture. Neither durable nor resistant to insects. Used in general construction and joinery, for railway sleepers, broom handles, wood pulp and plywood, and for veneers. Average density 500 kg m^{-3}.

hemp The bast fibre of the hemp (pot) plant, *Cannabis sativa*, generally used for making string and ropes. Certain other fibres such as manila and sisal are sometimes incorrectly called hemps.

HEMT Abbrev. for **High Electron Mobility Transistor**.

henequen Bast fibre obtained from *Agava fourcroydes* and is similar to sisal.

Henry's law The amount of a gas absorbed by a given volume of a liquid at a given temperature is directly proportional to the pressure of the gas.

Henschel mixer Type of dry blender of polymer powders using high-speed rotating disc fitted with sharp blades, which creates circulating vortex.

HERF Abbrev. for **High-Energy Rate Forging** or **Forming**.

herringbone Any cloth made from the eponymous weave.

hertz SI unit of frequency, indicating number of cycles per second (c/s). Symbol Hz.

hesitation line Visual defect formed on suface of hollow moulded plastic product where extra sharp impression of tool cavity surface is formed by onset of gas pressurization.

hessian A strong plain-weave jute fabric, used for packing material, sacks, in tarpaulin manufacture, and as a furnishing fabric and wall covering.

Hess's law The net heat evolved or absorbed in any chemical change depends only on the initial and final states, being independent of the stages by which the final state is reached.

heter-, hetero- Prefixes from Gk *heteros*, other, different.

heterocyclic compounds Cyclic or ring compounds containing carbon atoms and other atoms, eg O, N, S, as part of the ring.

heterogeneous A chemical system consisting of more than one phase.

heterojunction Junction between different semiconductor materials in a heterostructure, such as between GaAs and AlGaAs.

heteropolar Having an unequal distribution of charge, as in covalent bonds between unlike atoms.

heteropolymers See **polymers** p. 247.

heterostructure In semiconductor technology, certain materials grown epitaxially in layers having different composition and properties offering extra degrees of freedom in device engineering. See **epitaxial growth of semiconductors** p. 117.

hex- Prefix from Gk *hex*, six.

hexagonal close-packed structure An arrangement of atoms in crystals which may be imitated by packing spheres; characteristic of many metals. The disposition of atomic centres in space can be related to a system of hexagonal cells. See **close packing of atoms** p. 62.

hexamethylenediamine *1,6-diaminohexane*, $H_2N(CH_2)_6NH_2$. Important as a constituent material of nylon 6.6, which is a step-growth polymer formed from nylon salt. See **step polymerization** p. 298.

hexamethylene diisocyanate Type of aliphatic isocyanate monomer used to make polyurethanes. Abbrev. *HDI*.

Hibbert standard A standard of magnetic flux linkage suitable for fluxmeter or galvanometer calibration. It comprises a stabilized magnet producing a radial field in an annular gap, through which a cylinder carrying a multiturn coil can be dropped.

hickory The product of a **hardwood** tree (*Carya*) common to eastern N America. The **heartwood** is reddish-brown to brown (red hickory) while the **sapwood** is much lighter (white hickory). Fairly straight-grained with a coarse texture. It bends well and is used for making sports goods, chairs, tool handles etc. as well as in plywood and veneers. Density 700–900 kg m^{-3} (average 820 kg m^{-3}).

high alumina cement Cement containing a higher proportion of alumina (30–50%) than ordinary Portland cement, it is faster setting, less affected by low temperature during setting, and more resistant to sea-water and acids when set. But it has been found to degrade in warm, humid environments, eg

swimming pools. Made by fusing a mixture of bauxite and chalk or limestone and grinding the resultant clinker. Abbrev. *HAC*.

high brass Common brass of 65/35 copper-zinc alloying, as distinct from deep-drawing brass with 66–70% copper. See **copper alloys** p. 76.

high-carbon steel Hypereutectoid steels containing more than 0.8% carbon. Such steels consist of iron carbide (*cementite*) and **pearlite** when slow cooled. They are capable of being heat treated to high hardness but tend to be brittle. Used for metal working formers and fine edge cutting tools eg files. See **steels** p. 296.

high-conductivity copper Metal of high purity, having an electrical conductivity not much below that of the international standard, which is a resistance of 0.153 028 ohms for a wire 10 m in length and weighing 10 g.

high-density polyethylene Highly crystalline ethylene polymer made at low pressure using Ziegler–Natta type catalysts. Linear chains with little branching, with high T_m (ca 140°C) and high density (ca 960 kg m^{-3}). Competes with low-density polyethylene for packaging. Abbrev. *HDPE*.

high electron mobility transistor A **heterojunction** device in which electron current is confined to an undoped, high mobility region. Abbrev. *HEMT*.

high-energy rate forging Methods in which the ram is accelerated to very high velocities by the release of compressed gas, usually to complete an operation in one blow. Abbrev. *HERF*. Also *high-velocity forging*.

high-energy rate forming Any of a recently developed family of processes, in which metal parts are rapidly compacted, forged and extruded, by the application of extremely high pressures.

high-frequency heating Heating (induction or dielectric) in which the frequency of the current is above mains frequency; from rotary generators up to ≈ 3000 Hz and from electronic generators 1–100 MHz. Also *radio heating, microwave heating*.

high-frequency induction furnace It is essentially an air transformer, in which the primary is a water-cooled spiral of copper tubing, and the secondary the metal being melted. Currents at a frequency above about 500 Hz are used to induce eddy currents in the charge, thereby setting up enough heat in it to cause melting. Used in melting steel and other metals. Lower frequency (50 Hz) is used for melting non-ferrous metals, where a loop of liquid forms the secondary of the transformer and the furnace is never emptied completely in order to preserve this loop.

Highgate resin Fossil gum-resin occurring in the Tertiary London clay at Highgate in North London.

high-lead bronze Soft matrix metal used for bearings, of copper/tin/lead alloys in approximate proportions 80, 10 and 10.

high-performance fibres See panel on p. 158.

high recombination rate contact The contact region between a metal and semiconductor (or between semiconductors) in which the densities of charge carriers are maintained effectively independent of the current density.

high-speed steel A range of high alloy steels used for metal-cutting tools. They are formulated to retain their hardness at a low red heat, and hence tend not to soften when used at high rates of machining, as would lower alloy steels of similar hardness. High speed steels usually contain 12 to 18% tungsten or 6–8% tungsten plus 5–8% molybdenum, with up to 5% chromium and 5% cobalt. Carbon is in the range 0.7 to 1.2% and small amounts of other elements, eg vanadium are usually included. Abbrev. *HS steel*. See **steels** p. 296.

high-strength brass A type of brass based on the 60% copper-40% zinc composition, to which manganese, iron and aluminium are added to increase the strength. Manganese bronze denotes a variety in which manganese is the principal addition, but most varieties now contain all three elements. See **copper alloys** p. 76.

HIP Abbrev. for *hot isostatic pressing*. See **ceramics processing** p. 54.

hip joint implant Artificial composite joint used to replace diseased hip joint. Typically constructed of ultra-high molecular mass polyethylene socket, ceramic ball attached to metal alloy stem adhesively bonded with polymethyl methacrylate to thigh bone. The technology has been extended to most joints of the human body.

HIPS Abbrev. for *High-Impact Polystyrene*. See **copolymer** p. 75.

HI-PVC See **polyvinyl chloride** p. 249.

Hofmeister series The simple anions and cations arranged in the order of their ability to coagulate solutions of lyophilic colloids.

hol- Prefix. See **holo-**.

Holden permeability bridge A permeability bridge in which the standard bar and the bar under test carry magnetizing coils, and are connected by yokes to form a closed magnetic circuit. The magnetizing currents are varied until there is no magnetic leakage between the yokes.

hole Vacancy in a normally filled energy band, either as result of electron being elevated by thermal energy to the conduction band, and so producing a hole-electron pair; or as a result of one of the crystal lattice sites being occupied by an acceptor impurity atom. Such vacancies are mobile and contribute to electric current as if positive carriers. Analagously, in oxide ceramics, a singly charged oxygen ion can be viewed as a doubly charged ion plus a hole.

hole current That part of the current in a material due to the migration of **holes**.

hole mobility See **mobility**.

hole theory of liquids Interpretation of the fluidity of liquids by regarding them as disordered crystal lattices with mobile vacancies or holes.

hole trap An impurity in a semiconductor which can release electrons to the conduction or valence bands and so trap a 'hole'.

holland A glazed cotton or linen fabric used principally for window blinds and interlinings.

hollander beater In papermaking, an horizontal trough with a dividing wall parallel to its longer side which stops short of the ends to provide a continuous channel around which stuff may circulate under the propulsion of the rotating beater roll. The surface of the latter is fitted with metal bars parallel to the roll shaft which impart a rubbing and cutting action on the fibres in the narrow gap between them and similar fixed bars beneath the roll. The extent of this action largely controls the properties of the finished paper. Now generally superseded by the refiner, which performs a similar function.

Hollofil TN for hollow polyester fibre, with lowered thermal conductivity due to entrapped air, used therefore for insulation in sleeping bags etc (US).

hollow glass microspheres Glass bubbles with diameters in the region 10–150 μm and typical density of 280 kg m^{-3}. Used as fillers for plastics, chiefly for the low density of the resulting composite material. Cf **ballotini**.

hollow moulding Type of **injection moulding** process where nitrogen gas is injected into polymer product while still molten to create void. Also *gas-melt process*.

holly Widespread **hardwood** (*Ilex*) whose **heartwood** is a fine-textured, irregularly grained, creamy white. Used in decorative applications, keys of keyboard instruments, and in veneers. Average density 800 kg m^{-3}.

holmium A metallic element, one of the **rare earth elements**. Symbol Ho. See Appendix 1, 2.

holo- Also *hol-*. Prefixes from Gk. *holos*, whole.

hologram Image produced by **holography**.

holographic interferometry The superimposition of two holograms of the same

high-performance fibres

The intrinsic strength of the carbon-carbon bond (eg as in diamond) suggests that organic polymers should be among the strongest materials known. But the opposite seems to be true: their tensile moduli are among the lowest of engineering materials. In fact, a high-performance organic fibre has been known since the late 1950s, ie carbon fibre made by controlled pyrolysis and orientation of polyacrylonitrile (PAN).

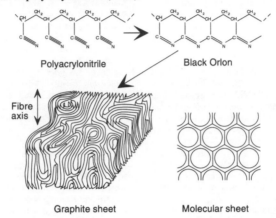

Fig. 1. **The polymerization of polyacrylonitrile to form carbon fibre**.

The sheets of carbon atoms bonded via sp^2 *orbitals* are aligned along the fibre axis and laterally interleaved like a rolled-up newspaper. When strained along the fibre axis, carbon-carbon bonds are stressed directly, resulting in a maximum tensile modulus of 520 GN m^{-2} (over 2.5 times that of steel). With a specific gravity of 1.96, its **specific modulus** ($E\rho^{-1}$) is almost 10 times that of the best steel wire (see Table 1, next page). Its **specific strength** ($\sigma_{TS} \rho^{-1}$) is also extremely high (E is the tensile modulus and σ_{TS} the tensile strength of the fibre). When woven into cloth and impregnated with epoxy or polyester resin, carbon fibres are therefore ideal materials for aerospace applications.

A similar conversion of linear polymers into high-modulus fibres was not, however, achieved until 1968 when workers at du Pont discovered a way of

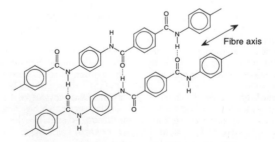

Fig. 2. **Chemical structure of aramid fibres**.

continued on next page

spinning *liquid crystal aramid oligomers* into aramid fibre (TNs Kevlar, Twaron). The material is nearly 100% crystalline, unlike most polymers, and all the chains are aligned along the fibre axis.

The material is stabilized laterally by hydrogen bonds (cf **protein structure**) and the material is also very stable thermally, showing no T_g or T_m but degrading at temperatures in excess of 450°C. Like carbon fibre, several grades of aramid fibre are available commercially for application in composites (see table below). Unlike carbon fibre however, this fibre can be made into rope for engineering applications. To maximize the potential of the fibre without introducing weakening knots etc, the fibre yarn is laid parallel to the cable axis and special terminations have been developed to connect cables to other structures (*Parafil construction*). Such cables (rated loads 10–200 tonnes) have been used for tethering oil rigs as well as reinforcing cracked structures like concrete cooling towers.

More recently (Dutch State Mines, 1984) a method of spinning high-performance fibres from any linear polymer has been developed. The process, known as **gel-spinning**, involves creating a dry gel from **theta-solvents**. Each ultrahigh molecular mass polymer chain is a random coil, not entangled with neighbouring coils. When hot stretched and drawn some 72 times, the chains crystallize and orient along the fibre axis. Although high-density polyethylene can be drawn conventionally, it is impossible to eliminate the chain-folded crystals originally present.

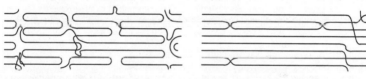

(a) High-density polyethylene fibre   (b) Gel-spun polyethylene fibre

Fig. 3. **The effect of gel-spinning on polyethylene fibre.**

Although this material is not as stable thermally as either aramid or carbon fibre, it possesses a specific gravity of less than one, so finds application for yacht rope etc. Like the other fibres, it is prepared with a diameter of about 10 μm and is routinely available as continuous fibre.

Table 1. (ρ is the density, E the tensile modulus, σ_{TS} the tensile strength)

Fibre	Property units	Density, ρ kg m^{-3}	E GN m^{-2}	σ_{TS} GN m^{-2}	$E\rho^{-1}$ MN m kg^{-1}	$\sigma_{TS}\rho^{-1}$ MN m kg^{-1}
aramid (Kevlar 29)		1440	60	2.8	42.9	2.00
(Kevlar 49)		1440	124	3.1	86.1	2.15
polyethylene (Spectra 900)		970	120	2.6	124	2.68
polypropylene (gel-spun)		910	36	1.0	39.6	1.10
carbon fibre UHM		1960	520	1.9	265	0.91
HM		1850	480	2.0	259	1.08
UHS		1750	270	5.2	154	2.97
HS		1760	265	2.8	151	1.59
steel piano wire		7860	210	3.0	26.7	0.38

object produces a pattern of interference fringes if any changes in surface displacement have occurred. Used as a *non-destructive testing method* for measuring deformations (eg **stress analysis**), vibration analysis and detection and monitoring of cracks.

holography Imaging technique which records and reconstructs the wave front emanating from an illuminated object. Coherent light from a laser is split in two; one is a reference beam and the other illuminates the object. The waves scattered by the object and the reference beam are recombined to form an interference pattern on a photographic plate, the *hologram*; this records both the amplitude and phase of the scattered light. When the hologram is itself illuminated by light from a laser or other point source, two images are produced; one is virtual but the other is real and can be viewed directly. So a three-dimensional image of the object can be produced. Also *ultrasonic holography*.

homespun A coarse tweed hand-woven from handspun wool.

homo- Prefix from Gk *homos*, same.

homogeneous Said of a system consisting of only one phase, ie a system in which the chemical composition and physical state of any physically small portion are the same as those of any other portion.

homogenization See **glasses and glassmaking** p. 144.

homologous series A series of organic compounds, each member of which differs from the next by the insertion of a $-CH_2-$ group in the molecule. Such a series may be represented by a general formula and shows a gradual and regular change of properties with increasing molecular mass, eg alkanes, paraffins.

homopolar Having an equal distribution of charge, as in a covalent bond between like atoms.

homopolar magnet One with concentric pole pieces.

homopolymer Polymer in which all repeat units are identical. See **polymers** p. 247.

Honduras mahogany See **American mahogany**.

honeycomb (1) *Cellular solid*, structural material made by bending and bonding together thin sheets of eg aluminium or paper to give an array of channels of hexagonal cross-section like its eponym. Used as the filling in **sandwich beams** to give lightweight, rigid products such as aircraft flooring. (2) Fabric with the threads forming ridges and hollows to give a cell-like appearance. Generally woven from coarse soft yarns in compact structures and used for

towels and bedspreads.

honing The process of finishing cylinder bores etc to a very high degree of accuracy by the abrasive action of stone or silicon carbide slips held in a head having both a rotatory and axial motion.

Hooke's law The basic statement of linear elasticity, originally formulated by Robert Hooke in 1676 and published as an anagram of '*ut tensio sic vis*' (or, 'as the extension, so the force'); it was Thomas Young who realized over 100 years later that the proportionality was between **stress**, σ, and **strain**, ε, ie $\sigma = E\varepsilon$, where the proportionality constant is **Young's modulus**. This is the constitutive equation for elastic deformation in tension, and analogous equations apply for other deformation modes. Although materials are intrinsically non-linear, the law is a good approximation to the behaviour of most types within the range of recoverable, small strains. The main exceptions are polymers, which are not only viscoelastic, but also significantly non-linear, and elastomers, where strains can reach several hundred per cent.

hoop stress The largest of the three stresses in the wall of a tube under pressure, acting around the circumference of the tube; usual symbol, σ_θ. The other two are the axial stress, which is one-half of the hoop stress, and the radial stress through the thickness of the tube wall, which is negligible when the diameter is greater than 20 times the wall thickness.

Hoppus foot A largely obsolete unit of volume obtained by using the square of the quarter-girth for trees or logs instead of the true sectional area; only applied to timber in the round.

hopsack (matt) weave A variant of the plain weave in which two or more ends and picks weave as one.

horizontal flash tool Tool normally used in injection moulding, where mating parts meet at right angles to main injection direction. Flash polymer may be extruded here during moulding, and must be removed in a post-moulding operation. See **vertical flash tool**.

horn (1) Hard body parts, usually external, of many mammals (eg fingernails); one of the pointed or branched hard projections borne on the head of many animals. Composed largely of the protein **keratin** often together with inorganic minerals in biocomposite. (2) The material derived from natural hard body parts, shaped by cutting and often also thermoformed to make thin sheet (formerly for lantern glazing) and cups etc.

hornbeam European **hardwood** tree of the genus *Carpinus*, yielding a fairly perishable, dull white wood whose uses are chiefly interior, such as flooring, inlaying, marquetry, musical instruments and in veneers. It can also be dyed black to simulate ebony. Average density 750 kg m^{-3}.

horseshoe filament An electric lamp filament in the shape of a single half-turn.

horseshoe magnet Traditional form of an electro- or permanent magnet, as used in many instruments, eg meters, magnetrons.

host Essential crystal, base material or matrix of a luminescent material.

hot crack Crack formed during cooling due to the stresses set up by the volume change on solidification of one or other of the components of an alloy or metal.

hot-die steels Shock- and temperature-resistant alloys used in high-temperature forging. See **steels** p. 296.

hot dip galvanizing The process of producing a corrosion-resistant zinc coating on articles by immersion in molten zinc or spelter, as distinct from electroplating with zinc.

hot-drawn Term describing metal wire, rod or tubing which has been produced by pulling it heated through a constricting orifice.

hot electron An electron in excess of the thermal-equilibrium number and, for metals, having energy greater than the **Fermi level**. In semiconductors, the energy must be a definite amount above the edge of the conduction band. Hot electrons (or holes) can be generated by photo-excitation, tunnelling, minority-carrier injection or Schottky emission over a forward-biased p-n junction, or by abnormal electric fields in non-conductors.

hot-ground pulp Mechanically ground wood pulp in which the minimum of water is used, so allowing the temperature to rise by friction.

hot isostatic pressing Making a compact by application of heat as well as pressure. The combined effect is general softening and/or the liquefaction of a phase to allow sintering. Far greater amounts of pressure and heat would be required separately. Also *hot pressing*. See **ceramics processing** p. 54.

hot melt adhesive Polymeric equivalent of solder. A thermoplastic applied in molten state by glue-gun directly onto surfaces to be joined, which bond together as the melt solidifies. Low molecular mass ethylene vinylacetate is a popular hot-melt adhesive.

hot-plate welding Method of joining similar thermoplastics by heating to above T_g or T_m with a steel platen, and bringing together under load. Used for joining medium density polyethylene gas pipes, battery cases etc. Care needed in control of temperature and load to ensure good fusion at the weld line. **Sacrificial-tape welding** is used for critical

products.

hot-pressed Paper finished by glazing between hot metal plates.

hot runner **Runner** into injection moulding tool cavity which is heated to conserve polymer. See **injection moulding** p. 169.

hot-short, red-short Said of metals that tend to be brittle at temperatures at which hot-working operations are performed.

hot top Feeder head, often heated by exothermic reaction compounds, containing a reservoir of molten metal drawn on by a cooling ingot as it solidifies, thus avoiding porosity. See **Ingot, ingot mould.**

hot-working The process of shaping metals by rolling, extrusion, forging etc, at temperatures above about 0.6 T_m. The hot-working range varies from metal to metal, but it is, in general, a range in which recrystallization proceeds concurrently with the working, so that no strain-hardening occurs.

Hoyt's metal A tin base (91.5%) white metal, containing also antimony (3.4%), lead (0.25%), copper (4.3%) and nickel (0.55%).

HS steel Abbrev. for **High Speed steel**.

huckaback A woven linen or cotton cloth with a rough surface, used for towels and glass-cloths. Different from terry-towelling fabric which is a pile fabric. See **terry fabric**.

hue The perception of colour which discriminates different colours as a result of their wavelengths. Hue, chroma (degree of saturation) and value (brightness) specify a colour on the Munsell scale.

human factors See **ergonomics**.

Hund's rule For electrons of otherwise equal energy, spins are aligned parallel as much as possible and in separate orbitals. This results in the strongly paramagnetic properties of many transition metal compounds which have incomplete inner electron shells.

Hv Symbol for Vickers hardness. See **Vickers hardness test** p. 331. Also *VHN*, *VPN*.

hyal-, hyalo- Prefixes from Gk *hyalos*, clearstone, glass.

hyaluronic acid Natural polysaccharide which exists in body tissues (eg joint synovial fluid) and often binds with proteins. See **chondroitin**.

hybrid circuit An electronic sub-assembly formed by combining different types of individual integrated circuits (and some discrete components). See **printed, hybrid and integrated circuits** p. 257.

hydrated ion Ion surrounded by molecules of water which it holds in a degree of orientation. Hydronium ion. Solvated H ion of formula $[H_2O \rightarrow H]^+$ or H_3O^+.

hydrates Salts which contain water of crystallization. See **water of hydration**.

hydraulic cement A cement which will harden under water.

hydraulic glue A glue which is partially able to resist the action of moisture.

hydraulic press An upstroke, downstroke or horizontal press, with one or more rams, working at approximately constant pressure for deep drawing and extruding operations or at progressively increasing pressure for baling, plastics moulding etc.

hydraulics The science relating to the flow of fluids. adj. *hydraulic*.

hydr-, hydro- Prefixes from Gk *hydor*, gen. *hydatos*, water.

hydrides Compounds formed by the union of hydrogen with other elements. Those of the non-metals are generally molecular liquids or gases, certain of which dissolve in water (oxygen hydride) to form acid (eg hydrogen chloride) or alkaline (eg ammonia) solutions. The alkali and alkaline earth hydrides are crystalline, salt-like compounds, in which hydrogen behaves as the electronegative element. They contain H^- ions and, when electrolysed, give hydrogen at the anode. Transition elements give alloy or interstitial hybrids.

hydro See **hydr-**.

hydrocarbons A general term for organic compounds which contain only carbon and hydrogen. They are divided into saturated and unsaturated hydrocarbons, aliphatic (alkane or fatty) and aromatic (benzene) hydrocarbons.

hydrocelluloses Products obtained from cellulose by treatment with cold concentrated acids. They still retain the fibrous structure of cellulose, but are less hygroscopic.

hydrochloric acid HCl. An aqueous solution of hydrogen chloride gas. Dissolves many metals, forming chlorides and liberating hydrogen. Used in industry for many purposes, eg for the manufacture of chlorine, pickling, tinning, soldering etc.

hydrodynamic lubrication Thick-film lubrication in which the relatively moving surfaces are separated by a substantial distance and the load is supported by the hydrodynamic film pressure.

hydrodynamic process A process for shallow forming and embossing operations, in which high-pressure water presses the blank against a female die, there being no solid punch to conform to the die contour. Similar processes involve explosive charges.

hydrodynamics That branch of dynamics which studies the motion produced in fluids by applied forces.

hydrofluoric acid Aqueous solution of hydrogen fluoride. Dissolves many metals,

with evolution of hydrogen. Etches glass owing to combination with the silica of the glass to form silicon fluoride, hence it is stored in, eg polyethylene vessel.

hydroforming A hydraulic forming process in which the shape is produced by forcing the material by means of a punch against a flexible bag partly filled with hydraulic fluid and acting as a die.

hydrogen The element with the lightest atoms, forming diatomic molecules H_2. Symbol H. Molecular hydrogen is a colourless, odourless, diatomic gas, water being formed when it is burnt. The element is widely distributed as part of the water molecule, occurs in many minerals, eg petroleum, and in living matter. Hydrogen is used as a reducing agent in organic synthesis and metallurgy, oxyhydrogen and atomic hydrogen welding flames. See Appendix 1, 2.

hydrogen bond A weak inter- or intramolecular force resulting from the interaction of a hydrogen atom bonding with an electronegative atom with a lone pair of electrons, eg O or N. Hydrogen bonding is important in polyamides and is responsible for much of the tertiary structure of **proteins**. See **bonding** p. 33.

hydrogen electrode For pH measurement, a platinum-black electrode covered with hydrogen bubbles. Although rarely used in practice, it defines the hydrogen scale of electrode potentials. Also *Hildebrand electrode*. See **electrochemical series**.

hydrogen embrittlement Effect produced on metal by sorption of hydrogen during pickling or electroplating operations.

hydrogen ion An atom of hydrogen carrying a positive charge, ie a proton; in aqueous solution, hydrogen ions are hydrated, H_3O^+ (the **hydroxonium ion**).

hydrogen ion concentration See **pH**.

hydrogen oxide Water.

hydrogen scale A system of relative values of electrode potentials, based on that for hydrogen gas, at standard pressure, against hydrogen ions at unit activity, as zero. See **hydrogen electrode**.

hydrolysis (1) The formation of an acid and a base from a salt by interaction with water; it is caused by the ionic association of water. (2) The decomposition of polymers by interaction with water, either in the cold or on heating, alone or in the presence of acids or alkalis; eg polyesters form alcohols and acids; oligo- and polysaccharides yield monosaccharides on boiling with dilute acids.

hydrolysis of polymers Degradation of step-growth polymers by water. Must be guarded against during heat treatment or moulding. See **step polymerization** p. 298.

hydronium ion See **hydroxonium ion**.

hydroperoxides Intermediate compounds formed during the oxidation of unsaturated organic substances, eg fatty oils, such as linseed oil. They contain the group –OOH.

hydrophilic colloid A colloid which readily forms a solution in water.

hydrophobic colloid A colloid which forms a solution in water only with difficulty.

hydrosol A colloidal solution in water.

hydrostatic extrusion Form of extrusion in which the material to be shaped is preshaped to fit a die which forms the lower end of a high-pressure container. The container is filled with the pressure-transmitting liquid, and pressure is built up in the liquid by a plunger until the metal is forced through the die orifice. See **extrusion**.

hydrostatic pressing Equivalent to isostatic pressing. See **ceramics processing** p. 54.

hydroxides Compounds of the basic oxides with water. The term hydroxide (a contraction of hydrated oxide) is applied to compounds that contain the –OH or hydroxyl group.

hydroxonium ion The hydrogen ion, normally present in hydrated form as H_3O^+. Also *hydronium ion*

hydroxyapatite Hydrated calcium phosphate occurring widely natural biomaterials such as **enamel**, bone etc. Also contains flouride, chloride or carbonate ions in place of the phosphate (PO_4^-) group.

hydroxyl –OH. A monovalent group consisting of a hydrogen atom and an oxygen atom linked together.

hygro- Prefix from Gk *hygros*, wet, moist.

hygroscopic Tending to absorb moisture; in the case of solids, without liquefaction.

Hypalon TN for chlorosulphonated poly-

ethylene rubber (US).

hyper- Prefix from Gk *hyper*, above.

hypereutectoid steel Steel containing more carbon than the eutectoid *pearlite*, ie in carbon steels, one containing more than 0.8% carbon. See **steels** p. 296.

hypo- Prefix from Gk *hypo*, under.

hypoeutectoid steel Steel with less carbon than is contained in **pearlite**, ie the iron-cementite eutectoid. In plain carbon steels, one containing less than 0.8% carbon. See **steels** p. 296.

hypothesis A prediction based on theory, an educated guess derived from various assumptions, which can be tested using a range of methods, but is most often associated with experimental procedure; a proposition put forward for proof or discussion.

hysteresis The retardation or lagging of an effect behind the cause of the effect, eg dielectric hysteresis, magnetic hysteresis etc. A property commonly seen in the stress-strain curves of polymeric materials where unloading gives a curve which does not coincide with the original. See **damping**.

hysteresis loss Energy loss in taking unit quantity of material once round a hysteresis loop. It can arise in a polymeric material subjected to a varying stress, in a dielectric material subjected to a varying electric field or in a magnetic material in a varying magnetic field. See **characterizing magnetic materials** p. 57, **dielectrics and ferroelectrics** p. 92.

hysteresis tester A device, invented by J A Ewing, for making a direct measurement of magnetic hysteresis in samples of iron or steel.

Hytrel TN for polyester rubber, a kind of thermoplastic elastomer (US).

Hz Symbol for hertz.

I

i- An abbrev. for: (1) *optically inactive*; (2) *iso-*, ie containing a branched hydrocarbon chain.

I Symbol for **second moment of area**.

I Symbol for **ionic strength**.

IACS Abbrev. for International Annealed Copper Standard, relating the electrical conductivity of a metal or alloy to that of copper in percentage terms.

ice colours Dyestuffs produced on the cotton fibre direct, by the interaction of a second component with a solution of a diazo-salt cooled with ice.

ideal gas Gas with molecules of negligible size and exerting no intermolecular forces. Such a gas is a theoretical abstraction which would obey the ideal gas law under all conditions:

$$pV = nRT,$$

where p = pressure, V = volume, n = number of moles, R = **gas constant** and T = absolute temperature. The behaviour of real gases becomes increasingly close to that of an ideal gas as their pressure is reduced. Also *perfect gas*.

idio- Prefix from Gk *idios*, peculiar, distinct.

IEC International Electrotechnical Commission. Main standards setting body at international level for electronic materials and devices etc.

IIR Abbrev. for **Isoprene-Isobutene Rubber**.

imaginary modulus Also *loss modulus*. See **complex modulus**.

imidazoles *Glyoxalines*. Heterocyclic, compounds produced by substitution in a five-membered ring containing two nitrogen atoms on either side of a carbon atom. Benzimidazoles are formed by the condensation of ortho-diamines with organic acids, and contain a condensed benzene nucleus.

imides Organic compounds containing the group –CO.NH.CO–, derived from acid anhydrides. See **polyimides**.

imitation parchment A wood-pulp paper to which some degree of transparency, and grease resistance have been imparted by prolonged beating of the pulp.

immiscibility The property of two or more liquids of not mixing and of forming more than one phase when brought together.

impact Elastic or inelastic collision between bodies during which the rate of change of momentum is high, so that large contact forces are generated. The duration of an impact is frequently of the same order or less than the time for elastic stress waves to prop-

agate through the body (*dynamic loading*) so that static analyses of the deformations of the body do not apply. See **impact tests** p. 165.

impacter forging hammer A horizontal forging machine in which two opposed cylinders propel the dies until they collide on the forging, which is worked equally on both sides.

impact extrusion A fast, cold-working process for producing tubular components by one blow with a punch on a slug of material placed on the bottom of a die, so that the material squirts up around the punch into the die clearance.

impact modifier Specific polymer added esp. to polyvinyl chloride to increase impact strength, eg ABS, MBS. Copolymers now preferred, eg HI-PVC.

impact strength A measure of the resistance of materials to impact loading applied in an impact test, it is not a true strength but the energy absorbed per unit area of fractured material. See **impact tests** p. 165, **strength measures** p. 301.

impact tests See panel on p. 165.

IMPATT diode *IMPact Avalanche Transit-Time diode*. A microwave diode which exhibits a negative resistance characteristic due to avalanche breakdown and charge carrier transit time in chips made of gallium arsenide or silicon. When linked to a wave guide or resonant cavity or similar structure it can be used as a microwave oscillator or amplifier.

impermeable Not permitting the passage of liquids, gases etc.

implant (1) A graft of an organ or tissue to an abnormal position. (2) Engineered devices constructed of man-made materials replacing and restoring living tissues or organs within the body and exposed to body fluids eg hip joint implant, corneal lens implant, heart valve implant.

impregnation (1) The partial or complete filling of the pores of a powder product with an organic material, glass, salt or metal, to make it impervious or impart to it secondary properties. Vacuum, pressure and capillary forces may be employed. It may include stoving to produce setting. See **infiltration**. (2) Strengthening of porous material such as wood, cement or plaster by exposure to low molecular mass monomer or oligomer. They diffuse rapidly into the porous solid and are polymerized catalytically.

impression One of several similar cavities in a single mould tool.

impact tests

Under very rapid loading, the response of a material may be very different to that under slower rates of loading. If the loading rate is comparable to the velocity of elastic waves in the material, interactions can occur which greatly increase the stress locally. Time-dependent materials can undergo a transition from ductile to brittle behaviour as the loading rate increases.

Testing the response of materials under impact loading, together with the tensile test and hardness tests (see **hardness measurements** p. 153), are among the most widely used mechanical tests on materials. Impact tests fall into two categories: product impact tests on whole products, and specimen impact tests on samples of material.

Specimen impact tests usually employ standard specimens whose shape and dimensions are prescribed by eg BSI, ASTM, ISO or DIN. The tests are classified by their deformation mode, as tensile, flexed beam or flexed plate (Fig. 1). In addition, the first two can use either notched or unnotched specimens.

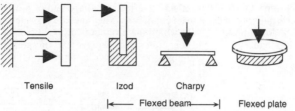

Fig. 1. **Impact test configurations**.

Tensile impact is essentially a high-speed variant of the *tensile test*. There are two versions of the flexed beam method. In the *Izod test*, the beam is a cantilever clamped at one end. The *Charpy test* uses a freely supported beam impacted at its centre (ie three-point bending). In the *flexed plate test*, the support is annular, and the specimen can either be clamped or freely supported. Loading in the tensile and flexed beam cases is conventionally by a falling pendulum (Fig. 2), hence *pendulum impact test*, whilst the flexed plate is usually loaded by a dropping weight. Impact speeds range from 1 to 6 m s^{-1}.

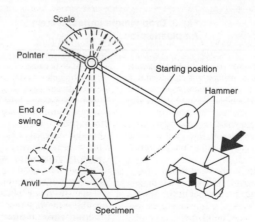

Fig. 2. **Pendulum impact tester**.

continued on next page

impact tests (contd)

Both types of loading provide a measure of the energy absorbed by the impacted specimen. Transducers on the specimen and the striker are increasingly being used to provide data on the impact event. The results of the pendulum tests are frequently expressed as energy absorbed per unit area of failed cross-section. When the specimen dimensions are fixed, the total energy absorbed may be used, as it is in the drop-weight tests.

Despite many of these tests being specified in standards, correlations between different tests are, at best, qualitative. Their utility in predicting the impact performance of products in service is, therefore, not high. They do, however, provide a quick and simple method of quality control, and are used to determine the **ductile-brittle transition temperature** in metals and plastics. Their limitations have led to impact tests on whole, or representative parts of, products.

Product impact tests better reflect the effects of shape and of processing on a product's integrity under impact conditions. They include crash tests on whole vehicles, drop impact tests (in which the product is dropped from a specified height) for battery cases and beer crates, drop weight tests on safety helmets or on plastic pipes (Fig. 3), and high-speed projectile impact tests on protective eye-wear.

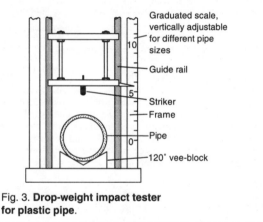

Fig. 3. **Drop-weight impact tester for plastic pipe**.

impurity Small proportion of *foreign matter*, eg arsenic, boron, phosphorus etc added to a pure semiconductor, eg silicon to obtain the required type of conduction and conductivity for solid-state devices. The impurity in the crystal lattice may add to, or subtract from the average densities of free electrons and holes in the semiconductor. See **acceptor, carrier, donor**.

impurity levels In the band theory of solids, localized energy levels in the band gap introduced by the presence of impurities in a crystal lattice.

in- Prefix from L. *in*, in(to), not.

in See **keruing**.

inactivation The destruction of the activity of a catalyst etc.

incandescence The emission of light by a substance because of its high temperature, eg a glowing electric-lamp filament. In the case of solids and liquids, there is a relation between the colour of the light and the temperature. Cf **luminescence**.

incineration One route to disposal of combustible materials following use and loss of function, particularly where costs of separation from domestic waste are prohibitive. Probably best route for tyre disposal, where the thermoset nature of several rubbers used in construction precludes solvent extraction, melt processing etc.

inclined shear mount Type of laminated bearing for controlling stiffness along several axes.

inclusion A particle of alien material retained in a solid material. In metals such inclusions are generally oxides, sulphides or silicates of one or other of the component metals of the alloy, but may also be particles of refractory materials picked up from the furnaces or ladle lining. In oxide glasses such particles are called *stones* and act as deleterious *stress concentrators*.

Incoloys Proprietary range of corrosion-resistant and high-temperature alloys containing 30% nickel, 20% chromium and 48% iron with small amounts of carbon, aluminium and titanium.

incompatibility Tendency of different polymers to form separate phases when mixed together. See **compatibility**.

incomplete reaction A reversible reaction which is allowed to reach equilibrium, a mixture of reactants and reaction products being obtained.

Inconels Nickel-based heat-resistant alloys containing some 13% of chromium, 6% iron, and a little manganese, silicon, or copper. See **nickel alloys.**

incremental permeability The gradient of the curve relating flux density to magnetizing force (the *B/H curve*). This represents the effective permeability for a small alternating field superimposed on a larger steady field. See **characterizing magnetic materials** p. 57.

indestructibility of matter See **law of conservation of matter**.

indeterminacy principle See **uncertainty principle**.

India paper A thin, strong, opaque rag paper, made for Bibles and other books where many pages are required in a small compass.

india-rubber See **rubber**.

indication A sign on inspection which indicates an imperfection of the material.

indirect-fired furnace One in which the combustion chamber is separate from the one in which the charge is heated.

indium A silvery metallic element in the third group of the periodic table. Symbol In. Electrical resistivity 9×10^{-8} Ω m. Found in traces in zinc ores. The metal is soft and marks paper like lead. Also used in manufacture of transistors and as bonding material for acoustic transducers. See Appendix 1, 2.

indium-doped tin oxide Used in thin film form as a transparent electrode material in electronic display devices. Also *ITO*.

induced dipole moment Induced moment of an atom or molecule which results from the application of an electric or magnetic field.

induced polarization That which is not permanent in a dielectric, but arises from applied fields.

induced reaction A chemical reaction which is accelerated by the simultaneous occurrence in the same system of a second, rapid reaction.

inductance (1) That property of a circuit element which, when carrying a current, is characterized by the formation of a magnetic field and the storage of magnetic energy. (2) The magnitude of such capability. Symbol *L*.

induction (1) Change in the electronic configuration and hence reactivity of one group in a molecule upon addition of a neighbouring polar group. (2) The driving of electric current by time-varying magnetic fields.

induction hardening Using high-frequency induction to heat a metal part for surface hardening. The heating is rapid and lends itself to control of the thermal gradient and hence the depth of hardening, since the penetration is inversely proportional to the frequency.

induction heating That arising from eddy currents in conducting material, eg solder, profiles of gear-wheels etc. Generated with a high-frequency source, usually oscillators of high power, operating at 10^6 to 10^7 Hz. Also *eddy current heating*.

induction period The interval of time between the initiation of a chemical reaction and its observable occurrence.

inductor (1) A substance which accelerates a slow reaction between two or more substances by reacting rapidly with one of the reactants. (2) A coil designed to exhibit **inductance**.

indurated Hardened, made hard. noun *induration*.

industrial design Methods used to plan and market products, with special emphasis on external shape and form (eg ergonomics). Complementary to engineering design. See **product design**.

inert Not readily changed by chemical means.

inert gases See **noble gases**.

infiltration Impregnation using capillary forces to soak up the impregnant.

infra- Prefix from L. *infra*, below.

infrared radiation Electromagnetic radiation in the wavelength range from 0.75 to 1000 μm approximately; ie between the visible and microwave regions of the spectrum. The *near* infrared is from 0.75 to 1.5 μm, the *intermediate* from 1.5 to 20 μm and the *far* from 20 to 1000 μm.

infrared spectroscopy Routine analytical tool for detection of functional groups by infrared absorption in molecules, esp. polymers, which can be easily examined in thin film form. Can also use fluid smear on NaCl discs. Absolute structure determination is more difficult than matching unknown spectrum with standard spectra, although additives present problems in obscuring spectrum of matrix polymer. Analysts then turn to **ultraviolet spectroscopy** or **nuclear magnetic resonance spectroscopy** if a polymer solution can be made.

infusible Not rendered liquid under specified conditions of pressure, temperature or chemical attack. Also *refractory*.

ingot A metal casting of a shape suitable for subsequent hot working, eg for rolling or forging.

ingot iron Iron of comparatively high purity, produced, in the same way as steel, but under conditions that keep down the carbon, manganese and silicon content.

ingot mould The mould or container in which molten metal is cast and allowed to solidify to form an ingot.

ingot stripper Mechanism for extracting ingots from ingot moulds.

inherent viscosity Natural logarithm of relative viscosity divided by polymer concentration in dilute solution. Also *log viscosity number*.

inhibitor Additive which retards or prevents an undesirable reaction, eg phosphates, which prevent corrosion by the glycols in antifreeze solutions; anti-oxidants in rubber.

inhibitory phase The protective colloid in a lyophobic sol.

in-house recycling Material reclamation within factory eg granulated thermoplastic runners and scrap mouldings fed back into injection moulding process.

initiation Start of polymerization, often catalytically induced. See **chain polymerization** p. 56.

initiator The substance or molecule which starts a chain reaction. See **chain polymerization** p. 56.

injection blow moulding Process for making plastic bottles, by injection moulding of **parison** followed by blow moulding of reheated parison. Widely used for polyethylene terephthalate (PET) beverage bottles.

injection efficiency The fraction of the current flowing across the emitter junction in a transistor which is due to the minority carriers.

injection moulding The principal method of shaping polymer products. See panel on p. 169.

inoculation (1) The introduction of a small crystal into a supersaturated solution or supercooled liquid in order to initiate crystallization. (2) Modification of crystallizing habit or of grain refinement in order to impart alloy qualities to molten metal in a furnace or ladle, by the addition of small quantities of other metals, deoxidants etc.

inorganic chemistry The study of the chemical elements and their compounds, other than the compounds of carbon; however, the oxides and sulphides of carbon and the metallic carbides are generally included in inorganic chemistry.

inorganic polymers Polymers whose chains are composed of atoms other than carbon. Many common materials contain such chains, eg silicate minerals, cement, where they perform an important reinforcement role. Chain bonds are Si–O (silicates), P–O (polyphosphates), P–N (phosphonitrilics), B–O (polyborates) etc. Silicone rubbers are of mixed lineage.

inquartation *Quartation*. Removal of silver from gold-silver bullion. The proportion of silver to gold must be raised by fusion to at least three to one, the silver being then dissolved in nitric acid. The silver can also be dissolved in concentrated sulphuric acid or converted to chloride by bubbling chlorine through the molten bullion. Also *parting of bullion*.

insert (1) In casting, a small metal part which is fitted into the mould or die in such a manner that the material flows around it to cast it in position. It may provide a hard metal wearing surface, sintered metal oil-retaining bearing surface etc. (2) A metal core inserted into a plastic article during the (injection) moulding process. (3) Special metal parts placed in a tool cavity which are designed to be incorporated in final product (eg screw thread) after injection moulding.

insoluble Incapable of being dissolved. Most 'insoluble' salts have a definite, though very limited, solubility.

instantaneous value A term used to indicate the value of a varying quantity at a particular instant. More correctly it is the average value of that quantity over an infinitesimally small time interval.

instantaneous velocity of reaction The rate of reaction, measured by the change in concentration of some key reagent. Abbrev. *ivr*. For instance, in the first order reaction, A $\rightarrow$AB, the *ivr* at any time t is given by $dC_A/dt = dC_B/dt = k\,C_A$, where C_A, C_B are the concentrations of reagent and product, respectively, and k is the 'velocity constant' of the reaction. Similarly, for the second order reaction, A+ B $\rightarrow$AC, the *ivr* is $dC_A/dt = dC_B/dt = -dC_C/dt = -k\,C_A\,C_B$.

Institute of Materials The principal UK professional institute for ceramic, composite, metal and polymer engineers.

insulating board **Fibre-board** of low density (300 kg m^{-3}) used for thermal insulation and acoustical control.

insulation (1) Any means for confining as far as possible a transmissible phenomenon (eg electricity, heat, sound, vibration) to a

injection moulding

The most effective way of shaping thermoplastics into complex products is compact injection moulding (CIM), a method where hot molten polymer is injected into a closed mould where it cools and solidifies. The mould tool comprises a cavity of the desired shape for the final product, which is split so that the product can be removed at the end of the moulding cycle (Fig. 1). The mould cavity connects via a gate, sprue and runner to the nozzle of the barrel of the melt preparation unit. The object of the latter is to prepare a shot of molten polymer using a travelling screw which fits inside a barrel or cylinder.

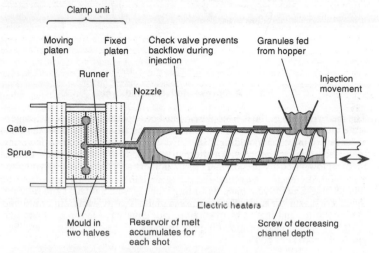

Fig. 1. **Mould tool and melt preparation unit**.

Like its operation in **extrusion**, the screw transports material by rotation to a reservoir immediately behind the nozzle. At this stage in the cycle, the nozzle is closed so that the screw is pushed back as the reservoir fills (unlike an extruder screw). Once the shot weight of material has accumulated, the nozzle valve opens, a check valve on the screw closes and the screw operates like a piston in the injection phase. Pressures within the reservoir are typically 120 MN m^{-2}, but are reduced to about 30 MN m^{-2} in the tool by the narrow runner, sprue and gate. The force needed to hold the male and female parts of the mould is known as the *locking force* and is the product of the moulding pressure and the projected area of the mould cavity. Locking forces range from less than 5 tonnes to 10 000 tonnes, the latter being required for car products like body panels and bumpers and even small boats. Products with simple re-entrant angles can be moulded using retractable cores. For more complex re-entrants, *fusible-core* technology is needed.

The production rate of injection moulding is determined by the mould cycle, the necessary cooling time forming the largest part (Fig. 2, over). This in turn is determined by the maximum section thickness, which becomes of the order of several minutes for sections in excess of about 4 mm in thickness. Structural **foam moulding**, **sandwich moulding** and **hollow moulding** are variants of the method designed to overcome the limit on wall thickness (and hence product weight) by creating partially filled or empty cavities within the mouldings.

continued on next page

injection moulding (contd)

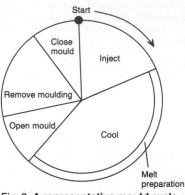

Fig. 2. **A representative mould cycle.**

The melt viscosity of the thermoplastic must be kept relatively low by temperature and molecular mass control. Shear thinning of the polymer at the high shear rate of injection moulding (10^3–10^5 s^{-1}) is important in further lowering melt viscosity in the process. An alternative strategy to reducing cycle time is to inject low-viscosity prepolymers or monomers together with a polymerization catalyst (*reaction injection moulding* or RIM). The method is used to mould large-area polyurethane mouldings (eg car wheel arches) and requires much less substantial tools. The increase in melt viscosity with fibre-reinforced thermoplastics and the problem of fibre-clumping have led to methods like **ZMC**, used in France for moulding car bonnets, wings, rear quarters etc. See **car body materials** p. 46.

Other thermoplastic shaping methods such as blow moulding have been applied to injection moulded parisons (rather than extruded parisons) to make polyethylene terephthalate (PET) bottles. Called *injection blow moulding*. Compact injection moulding can be applied to conventional thermosets and vulcanized rubber products by modification of the machines to allow for the greater control of the process necessary to prevent premature cross-linking. There are distant parallels between pressure die-casting methods for light alloys and compact injection moulding of polymers, but the latter offers a greater processing window. Compact injection moulding has been more directly applied to shaping ceramic components (in their green state) such as turbocharger blades.

particular channel or location in order to obviate or minimize loss, damage, or annoyance. (2) Any material (also *insulant*) or means suitable for such a purpose in given conditions, eg dry air suitably enclosed, polystyrene and polyurethane foam slab, glass fibre, rubber, porcelain, mica, asbestos, hydrated magnesium carbonate, cork, kapok, crumpled aluminium foil etc. See **lagging**, **shielding**.

insulator (1) A material which offers a high resistance to the passage of electric current. See **conductors and insulators** p. 70. (2) That part of an electrical or electronic device which is intended to block the passage of current. (3) A material which offers a high resistance to the flow of heat.

intaglio A form of decoration of glass in which the depth of cut is intermediate between deep cutting and engraving.

intarsia A weft-knitted fabric having designs in two or more colours.

integrated circuit An electronic sub-assembly in which several components are fabricated on the same semiconductor substrate. See **printed, hybrid and integrated circuits** p. 257.

intellectual property Legal term covering

patents, registered designs, design right,copyright, confidential information and moral rights.

intelligent materials See **smart materials**.

intensity of magnetization Vector of the magnetic moment of an element of a substance divided by the volume of that element.

interatomic forces Interactions between atoms in molecules or materials. Generally very strong covalent, ionic or metallic bonds. See **bonding** p. 33.

interatomic potential energy curve The diagram showing the relation between potential energy and interatomic distance for various bond types. It is a combination of a steep repulsion curve at close proximity and a broad, deep attraction curve as atomic electrons interact. See **bonding** p. 33.

intercrystalline failure Failures in fractures that follow the crystal boundaries instead of passing through the crystals, as in the usual transcrystalline fracture. It is frequently due to the combined effect of stress and chemical action, but may be produced by stress alone when the conditions permit a certain amount of recrystallization under working conditions.

interface The boundary between two different materials.

interference fit A negative fit, necessitating force sufficient to cause expansion in one mating part, or contraction in the other, during assembly.

interference microscopy Special optical microscopic method for examing polished material surfaces. Utilizes monochromatic light (eg Na D line) shone vertically onto surface through angled glass plate. Interference of reflected beam with incident beam gives fringe map of surface microtopography. Used to examine cracks, wear and scratches etc. See **Nomarski interference** which uses white light.

interferometer Instrument in which an acoustic, optical, or microwave interference pattern of fringes is formed and used to make precision measurements, mainly of wavelength.

intergranular corrosion Corrosion in a polycrystalline mass of metal, taking place preferentially at the boundaries between the crystal grains. This leads to disintegration of the metallic mass before the bulk of the metal has been attacked by the corrosive agent.

interlock A double-faced weft-knitted fabric made of two rib fabrics joined by interlocking loops. Although originally made from cotton for underwear the fabrics are now knitted from various fibres and also used for outerwear.

intermediate (1) Starting point for manufacture of materials or products, but usually excluding raw material. Normally a chemical compound or mixture of compounds. (2) A short-lived species in a chemical reaction.

intermediate constituent A constituent of alloys that is formed when atoms of two metals combine in certain proportions to form crystals with a different structure from that of either of the metals or the primary solid solutions based thereon. The proportions of the two kinds of atoms may be indicated by formulae, eg CuZn; hence these constituents are also known as *intermetallic compounds*.

intermediate phase A homogeneous phase in an alloy with a composition range different from the pure components of the system.

intermetallic compounds See **intermediate constituent**.

intermingled yarn A continuous filament yarn in which the constituent filaments are entangled by passing a turbulent air stream through the yarn.

Intermix TN for type of internal mixer for polymers using intermeshing rotors.

intermolecular forces Interactions between molecules generally involving van der Waals bonds and hydrogen bonds. Relatively weak compared to covalent, ionic or metallic bonds, but in the case of hydrogen bonds, critical for many biomaterials (eg **collagen**, **DNA**, **keratin**) and some synthetic polymers. See **bonding** p. 33.

internal energy The store of energy possessed by a material system. It is not usually possible to determine its absolute magnitude, but changes in its value can be measured. Changes in the internal energy of a system depend only upon the initial and final conditions, and are therefore independent of the paths of change. Symbol U.

internal friction That which gives rise to **hysteresis** and **damping** in elastic bodies. Also used for that which opposes flow in liquids and gases giving rise to viscosity.

internal mixer Type of mixer for polymers using rolls within a closed chamber, tending to replace two-roll mill. See **Banbury**, **Intermix**.

internal resistance That of any voltage source resulting in a drop in terminal voltage when direct current is drawn.

internal stress Residual stress in a material due to differential effects of heating, cooling or working operations, or to constitutional (eg phase) changes in a solid. To satisfy equilibrium, the net force on the body due to internal stresses must equal zero.

internal voltages Those, such as **contact potential** or **work function**, which add an effect to the external voltages applied to an active device.

International Annealed Copper Standard Standard reference for the conductivity of copper and its alloys. % IACS equals 1724.1/n, where n is the resistivity of the alloy per metre. Abbrev. *IACS*.

international paper sizes See ISO sizes.

International Rubber Hardness Degree A measure of the depth of penetration of an indenter into an elastomer, used to monitor the degree of cure. Abbrev. *IRHD*. See **hardness measurements** p. 153, **rubber hardness**.

international screw-thread A metric system on which the pitch of the thread is related to the diameter, the thread having a rounded root and flat crest.

International Standards Organisation Main international body setting standards by agreement with national standards bodies, eg BSI, DIN etc. Abbrev. *ISO*.

International System of Units See SI **units**.

International Union of Pure and Applied Chemistry A body responsible, among other things, for the standardization of chemical nomenclature, which it alters frequently. Abbrev. *IUPAC*.

interpolymerization Mixture of two or more individual homopolymers at a molecular scale. Made by polymerizing a monomer-swollen gel. Not to be confused with mixture of compatible polymers (eg Noryl).

interstice Space between atoms in a lattice where other atoms can be located, eg in close-packed metallic lattices.

interstitial compounds Metalloids in which small atoms of non-metallic elements (H,B,C,N,) occupy positions in the interstices of metal lattices. In general, the structure of the metal is preserved, though somewhat distorted. They are often (esp. those of group IV and V metals) characterized by exceptionally high melting points and hardness, by chemical inertness and by metallic lustre and conductivity.

interstitial solid solution Type of solid solution formed when there is a large difference in relative atomic sizes, usually the solute being less than 0.59 that of the solution. This enables the solute atoms to take up positions within the interstices of the crystal lattice of the solvent. The commonest example is that of carbon in iron.

intramolecular forces Interactions within a single molecule, generally involving van der Waals and hydrogen bonds. Although relatively weak compared to the covalent bonds of the backbone chain, they are important in polymers for determining chain flexibility, and hence a wide range of poly-

mer properties. See **rotational isomerism**.

intrinsic and extrinsic silicon See panel on p. 173.

intrinsic conduction That in a semiconductor when electrons are raised from a filled band into the conduction band by thermal energy, so producing hole-electron pairs. It increases rapidly with rising temperature.

intrinsic mobility The mobility of electrons in an intrinsic semiconductor.

intrinsic semiconductor See **intrinsic and extrinsic silicon** p. 173.

intrinsic viscosity Polymer viscosity as determined from dilute solutions, symbol η. Found by extrapolating curves of reduced viscosity and inherent viscosity to zero concentration. Gives measure of molecular mass (M_v) when inserted into the Mark–Houwink equation. Also *limiting viscosity number*.

intumescence The swelling of material on heating, often with the violent escape of moisture. Intumescent layers are used to enhance fire resistance by increasing the energy absorption before combustion.

Invar TN for iron-nickel alloy. Composed of 36% nickel, 63.8% iron and 0.2% carbon. **Coefficient of thermal expansion** is very small. Used for measuring tapes, tuning forks, pendulums and in instruments.

inverse segregation A type of segregation in which the content of impurities, inclusions and low melting point constituents in cast metals tends to be higher at the surface than in the axial regions. See **segregation, normal segregation.**

inverse spinel The crystal structure of **magnetite** and the other ferrites which are **ferrimagnetic**, similar to that of the mineral **spinel**, but with the divalent ions and half of the trivalent ions interchanged.

inversion In semiconductor devices, the local accumulation of nominally minority carriers to such an extent that they are actually in the majority. Such an effect occurs in the formation of a conducting channel between source and drain in **metal-oxide-silicon field effect transistors** (MOSFETs).

investment casting Forming a mould round a pattern whose shape can be destroyed to allow its removal. Patterns of complex shapes can be used which would otherwise be impossible to withdraw from the mould. Patterns may be of wax, plastic, frozen mercury etc. The ancient lost wax or *cire perdue* process is a good example. See **fusible core**. Cf **hollow moulding**.

iodine Non-metallic element in the seventh group of the periodic table, one of the halogens. Symbol I. It forms blackish scales with a violet lustre. See Appendix 1, 2.

ion Strictly, any atom or molecule which has

intrinsic and extrinsic silicon

Pure crystalline silicon is an *intrinsic semiconductor*. At absolute zero all four outer electrons of each atom in the bulk are involved in **covalent bonds** with surrounding atoms. The structure is like that of diamond. As the temperature is raised above absolute zero, an increasing fraction of bonding electrons is able to escape the covalent partnership and move almost freely from atom to atom. Such electrons are then termed *conduction electrons*. At the same time, the holes left behind in the network of bonds can be filled either by an electron from a neighbouring bond or else by recombination with one of the conduction electrons: in this way the hole apparently moves or vanishes. A hole is effectively a carrier of positive charge since it is drawn to regions of negative potential. Electrical conductivity arises from the presence and mobility of both conduction electrons and holes. See Fig. 1.

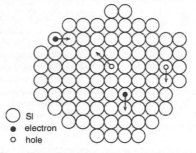

○ Si
● electron
○ hole

Fig. 1.**Two-dimensional schematic drawing of intrinsic silicon**. At room temperature there are only three electron-hole pairs for every 10^{13} silicon atoms.

At room temperature, on average about 3 in 10^{13} covalent bonds in silicon are incomplete, a tiny fraction but sufficient to endow the material with a conductivity of about 4×10^{-4} Ω m. An increase of 100 K in the ambient temperature raises the conductivity to about 0.3 Ω m.

In terms of the *band model* of electron energy states in a solid, the bonding levels (**valence band**) are separated from the higher energy states (**conduction band**) by a distinct energy gap in the otherwise near continuum of closely spaced energy levels found within the bands. Without thermal energy, all bonds are complete, the valence band is exactly full and the conduction band is empty. To leave a bond an electron must acquire enough energy to cross the gap into the conduction band, where the abundance of surrounding empty energy states provides ample opportunity for it to interact with an external electric field and so contribute to conductivity.

In practice, there are many factors which lead to departures from the model behaviour. In regions where the highly ordered crystal structure is disrupted, the energy gap is not so clearly defined. See Fig. 2, on next page. Crystal boundaries, defects and impurity atoms can all give rise to localized energy states within the gap. Charge carriers can be trapped by such localized states. If the localized levels are deep (separated from conduction or valence bands by amounts which are large compared with thermal energy) then the conductivity is adversely affected. On the other hand an electron held in a localized state close to the conduction band (or a hole close to the valence band) is easily dislodged by thermal energy, to join the stock of charge carriers free to contribute to conductivity.

continued on next page

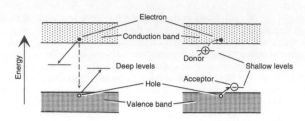

Fig. 2. **Trap (left) and dopant levels (right) in the band gap**.

Extrinsic silicon owes its conductivity to charge carriers provided from shallow levels deliberately introduced through the carefully controlled addition of dopant atoms. See Fig. 2. When a group V element is substituted into a silicon (group IV) lattice site only four of its five valence electrons are required for bonding. The fifth is weakly bound, efffectively occupying a shallow, localized level in the energy gap, from which it is easily dislodged thermally. Such a *dopant* is termed a *donor*. Equivalently, a group III element introduces a hole since it can only provide three of the four required bonding electrons. The lack of the fourth effectively forms a localized level in the gap, close to the valence band. A hole is launched into the valence band when the so-called acceptor dopant attracts an electron from a neighbouring silicon-silicon bond to make good its own deficit.

Doping with donors tends to make n-type semiconductors, introducing, almost *pro rata*, additional conduction electrons. See Fig. 3. The electrons are the majority charge carrier, being mostly balanced by an immobile positive charge associated with donor atoms which have been ionized. Holes form a group of minority carriers with a reduced life-time owing to the increased risk of recombination posed by the larger electron population. The net charge is zero. Doping with acceptors tends to make p-type semiconductor, with the holes (majority) balanced by negatively ionized acceptors and the minority species (electrons).

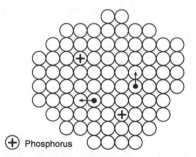

⊕ Phosphorus

Fig. 3. **Two-dimensional schematic drawing of n-type extrinsic silicon**. Doping typically substitutes 1 in 10^9 to 1 in 10^2 silicon atoms.

The majority carrier concentration in extrinsic semiconductor is almost independent of temperature. However, the material reverts to intrinsic behaviour when thermally generated carriers eventually outnumber those introduced by doping as the temperature is raised.

resultant electric charge due to loss or gain of valency electrons. Free electrons are sometimes loosely classified as *negative ions*. Ionic crystals are formed of ionized atoms and in solution exhibit ionic conduction. In gases, ions are normally molecular and cases of double or treble ionization may be encountered. When almost completely ionized, gases form a fourth state of matter, known as a **plasma**. Since matter is electrically neutral, ions are normally produced in pairs.

ion bombardment The impact on the cathode of a gas-filled electron tube of the positive ions created by ionization of the gas. The bombardment may cause electrons to be ejected from the cathode.

ion cluster Group of molecules loosely bound (by electrostatic forces) to a charged ion in a gas.

ion concentration That expressed in moles per unit volume for a particular ion. Also *ionic concentration*. See **pH**.

ion-exchange resins Term applied to a variety of materials, usually organic, which have the capacity of exchanging the ions in solutions passed through them. Different varieties of resin are used dependent on the nature (cationic or anionic) of the ions to be exchanged. Many of the resins in present-day use are based on polystyrene networks cross-linked with divinyl benzene. See **gel-permeation chromatography**.

ion flotation Removal of ions or gels from water by adding a surface active agent which forms complexes. These are floated as a scum by the use of air bubbles. See **Gibbs' adsorption theorem**.

ionic bond Coulomb force between ion-pairs in molecule or ionic crystal. These bonds usually dissociate in solution.

ionic concentration See **ion concentration**.

ionic conduction That which arises from the movement of ions in a gas or electrolytic solution. In solids, it refers to electrical conductivity of an ionic crystal, arising from movement of positive and negative ions under an applied electric field. It is a diffusion process and hence very temperature-dependent.

ionic conductor One in which conduction is predominantly by ions, rather than by electrons and holes.

ionic crystal Lattice held together by the electric forces between ions, as in a crystalline chemical compound.

ionic materials Solids with structure composed partly or wholly of charged species, anions and cations. Includes simple salts like magnesia (Mg^{2+}, O^{2-}) which has a crystal lattice of anions in which cations sit at inter-stitial sites, ceramics made of crystalline silicate sheets, chains or networks like mica, and glasses where the silicate chains form an amorphous network. See **silica and silicates** p. 285.

ionic migration Transport of ion-bearing particle to an electrode oppositely charged with electricity.

ionic product The product of the activities (see **activity** (2)) of the ions into which a pure liquid dissociates. For water, these ions are H_3O^+ and OH^-.

ionic strength Half the sum of the terms obtained by multiplying the **activity** (2) of each ion in a solution by the square of its valency; it is a measure of the intensity of the electrical field existing in a solution.

ionic theory The theory that substances whose solutions conduct an electric current undergo electrolytic dissociation on dissolution. This assumption explains both the laws of electrolysis and the abnormal **colligative properties**, such as osmotic pressure, of electolyte solutions.

ion implantation Technique by which impurities are introduced into semiconductors by firing high-energy ions at the substrate material.

ionization constant The ratio of the product of the activities of the ions produced from a given substance to the activity of the undissociated molecules of that substance. See **activity** (2), **dissociation constant**.

ionization potential Energy, in electron-volts (eV), required to detach an electron from a neutral atom. For hydrogen, the value is 13.6 eV. Atoms, other than hydrogen, may lose more than one electron and can be multiply ionized.

ionized (1) Electrolytically dissociated. (2) Converted into an ion by the loss or gain of an electron.

ionized atom One with a resultant charge arising from capture or loss of electrons; an **ion** in gas or liquid.

ion migration The movement of ions in an electrolyte or semiconductor due to applying a voltage across electrodes.

ion mobility Ion velocity in unit electric field (one volt per metre).

ionogenic Forming ions, eg electrolytes.

ionomer resins Ethylene copolymerized with small amount of acrylic acid and treated with zinc or sodium salt so that ion-acid groups act as physical cross-links, stabilizing the material. A type of **polyelectrolyte**.

ionotropy The reversible interconversion of certain organic isomers by migration of part of the molecules as an ion; eg hydrogen ion (prototropy).

ion source A device for producing ions for ion implantation and other applications.

Various configurations exist, eg deriving ions directly from ionized gases (**plasmas**), from liquid metals (by **field emission** from protruberances) or from solids (by surface ionization).

IR Abbrev. for: (1) **infrared** (spectroscopy); (2) *isoprene rubber* (see **elastomers** p. 106).

IRHD Abbrev. for **International Rubber Hardness Degree**.

iridium A brittle, steel-grey metallic element of the platinum family. Symbol Ir. Electrical resistivity 6×10^{-8} Ω m. Alloyed with platinum or osmium to form hard, corrosion-resistant alloys, used for pen points, watch and compass bearings, crucibles, standards of length. See Appendix 1, 2.

iroko A general utility timber from *Chlorophora excelsa*, a tropical African **hardwood** tree, it is golden-orange to brown, with interlocked grain and coarse but even texture. Used for structural work, shipbuilding, cabinet work, furniture and in plywood and veneers. Mean density 640 kg m^{-3}.

iron A metallic element in the eighth group of the periodic table. It exists in three forms; alpha-, gamma-, delta-. Symbol Fe. Electrical resistivity 9.8×10^{-8} Ω m. As basis metal in steel and cast-iron, it is the most widely used of all metals. See Appendix 1, 2.

iron dust core One used in a high-frequency transformer or inductor to minimize eddy-current losses. It consists of minute magnetic particles bonded in an insulating matrix.

iron loss The power loss due to **hysteresis** and **eddy currents** in the iron of magnetic material in transformers or electrical machinery.

iron (II) oxide See **ferrous oxide**.

iron (III) oxide See **ferric oxide**.

ironwood See **lignum vitae**.

irreversibility Physical systems have a tendency to change spontaneously from one state to another but not to change in the reverse direction. *Entropy* provides an indication of irreversibility.

irreversible colloid See **lyophobic colloid**.

irreversible reaction A reaction which takes place in one direction only, and therefore proceeds to completion.

isenthalpic Of a process carried out at constant enthalpy, or heat function H.

isentropic See **entropy**.

isinglass *Fish glue*. A white solid amorphous mass, prepared from fish bladders; chief constituent, **gelatin**. It has strong adhesive properties. Used in various food preparations, as an adhesive, and in the fining of beers, wines etc.

iso- A prefix indicating: (1) the presence of a branched carbon chain in the molecule; (2) an isomeric compound. From Gk *isos*, equal.

ISO Abbrev. for **International Standards Organisation**.

isobar A curve relating qualities measured at the same pressure.

isochore A curve relating quantities measured under conditions in which the volume remains constant.

isochromatic Interference fringe of uniform hue observed with white light source; esp. in **photoelastic analysis** where it joins points of equal phase retardation.

isochronous data Measurements taken at constant times, eg isochronous stress-strain curve for polymer is constructed by measuring constant time tensile creep strains at different stress levels, and replotting data in stress-strain form.

isoclinic Loci of points at which directions of principal axes of stress are parallel to the axes of the crossed plane polars in photoelasticity. Appear as black bands in white light, not observed in circularly polarized light. Cf **isochromatic**.

isocyanates Compounds with isocyanate group –NCO. Di-isocyanates such as TDI and MDI are used for making polyurethanes. They are highly reactive compounds, forming amines (with water), which can react with further isocyanate to give urea (–NH–CO–NH–) groups.

isodimorphous Existing in two isomorphous crystalline forms.

isodisperse Dispersible in solutions having the same **pH** value.

iso-electric point Hydrogen ion concentration in solutions, at which dipolar ions are at a maximum. The point also coincides with minimum viscosity and conductivity. At this **pH** value, the charge on a colloid is zero and the ionization of an ampholyte is at a minimum. It has a definite value for each amino acid and protein. See **gels** p. 143.

iso-electronic Said of similar electron patterns, as in valency electrons of atoms.

isohydric Having the same **pH** value, or concentration of hydrogen ions.

isomerism The existence of more than one substance having a given molecular composition and rel. mol. mass but differing in constitution or structure. See **optical isomerism**. The compounds themselves are called isomers or isomerides (Gk 'composed of equal parts'). Isobutane and butane have the same formula, C_4H_{10}, but their atoms are placed differently; one type of alkane molecule, $C_{40}H_{82}$, has over 5012 possible isomers. Isomerism is common in **polymers** p. 247.

isomerized rubber Rubber in which the molecules have been rearranged by

isothermal transformation diagrams

Also *TTT diagrams* (Time-Temperature-Transformation). These consist of a single C-shaped curve, or overlapping curves, displayed on logarithmic time and linear temperature axes which depict the beginning and end of a solid state transformation. They represent a balance between the nucleation and growth of a new phase during the transformation of a metastable phase to an equilibrium state.

The curves are determined by cooling specimens of the material from a temperature where the transforming phase is stable and holding it isothermally at a lower temperature where it is unstable, then measuring some physical property or characteristic (eg by microstructural examination, electrical resistivity measurements or dilatometry) which allows the time to be determined for transformation to begin and, later, to be completed. Conducting such experiments at different temperatures produces a *start* curve and an *end* curve displaced on the time axis (Fig. 1a).

A C-shaped curve is produced for each transformation which depends upon a nucleation and growth mechanism. However, in perhaps the most important solid state transformation, that of austenite in steels, three mechanisms occur. These are (1) austenite to ferrite below the upper critical temperature; (2) austenite to pearlite below the lower critical temperature and (3) austenite to bainite below approximately 820 K. These are followed by an *athermal transformation* at a still lower temperature. Hence in steels there are overlapping C-curves which usually appear and are described as an S-shape. The effect of alloying elements is to displace these curves to varying degrees, which gives rise to pronounced differences in the form of the diagrams and the times of transformation. A typical TTT curve for a medium-carbon low-alloy steel is shown in Fig. 1b.

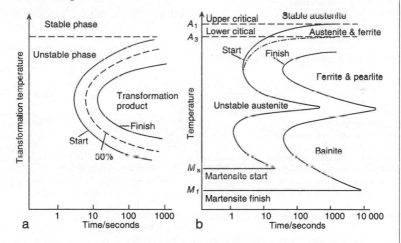

Fig. 1. **Typical C-curves.** (a) Single curve for a nucleation and growth-controlled transformation. (b) Overlapping curves for a medium-carbon low-alloy steel.

Because of their importance in the heat treatment of steels to achieve particular levels and combinations of engineering properties, TTT data are often redetermined in the form of **continuous cooling transformation** (CCT) diagrams. See p. 72. This is because in practice austenite transforms over a range of temperatures while it is being cooled whereas the data for a TTT diagram are determined isothermally under laboratory conditions.

continued on next page

177

isothermal transformation diagrams (contd)

The shape of the C-curve derives from the balance between nucleation of the new phase within the transforming material's crystal lattice and diffusion to enable the nucleus to grow. Just below the temperature where the transforming phase becomes unstable, diffusion is most rapid, but a large nucleus is necessary; so the delay is in awaiting the formation of a stable nucleus. At lower temperatures, the critical size of the nucleus is smaller and many will form in a given time, but the diffusion rate is low; so the delay is here caused by the slow growth process. There is always, therefore, an intermediate transformation temperature where a balance can be reached by optimizing the two factors. This is where the *nose* of the C-curve appears and overall transformation times are least.

In certain materials, and particularly steels, an *athermal transformation* of austenite may occur at temperatures too low for any nucleation and growth mechanisms. This is a crystallographic *shear transformation* to martensite. Such athermal transformations are usually plotted on the TTT diagram as a straight line at the temperature at which the shear begins and another at that where all the transforming phases have disappeared.

TTT diagrams usually record the product of the transformation close to the 'end of transformation' curve and include curves for intermediate stages in the transformation, eg 10%, 25%, 50%, 75% and 90%.

heating in solution in the presence of suitable catalysts.

isomers See **isomerism**.

isometric data Measurements taken with constant dimensions, eg isometric stress versus time creep curves for a polymer are obtained at constant strain, and can be derived from a creep curve or an isochronous stress-strain curve.

isopiestic Having equal pressure in the system or conditions described.

isopleth Line of constant composition on a **phase diagram** p. 236.

isoprene $CH_2 = C(CH_3).CH = CH_2$, a diene, colourless liquid, bp 37°C, obtained by dehydrogenation of 2-methylbutane, from propylene, or by several other methods. It is the monomer for synthesis of isoprene rubber and a co-monomer for isoprene-isobutene rubber.

isoprene-isobutene rubber A butyl rubber. See **elastomers** p. 106.

ISO sizes A series of trimmed, international, metric paper sizes based on a width to length ratio of 1:1.414 (ie 1:√2). The next smaller size in the series is produced by halving the longer dimension. The range comprises the A-, B- and C-series of sizes, based on basic sheets of 1 m^2 (= 2^0 m^2), $2^{\frac{1}{2}}$ m^2 and $2^{\frac{1}{4}}$ m^2 respectively. Thus AO is 841 mm × 1189 mm (= 1 m^2); A1, 594 mm × 841 mm; A2, 420mm × 594 mm etc: BO is 1000 mm × 1414 mm, and CO is 917mm × 1297 mm.

isostatic pressing See **ceramics pro-**
cessing p. 54.

isosteric Consisting of molecules of similar size and shape.

isotactic Term denoting linear-substituted hydrocarbon polymers in which the substituent groups all lie on the same side of the carbon chain. See **stereoregular polymers** p. 299.

isoteniscope An instrument for the static measurement of vapour pressure by observing the change of level of a liquid in a U-tube.

isothermal transformation diagram See panel on p. 177.

isothermal (1) Occurring at constant temperature. (2) A curve relating quantities measured at constant temperature.

isotopic symbols Numerals attached to the symbol for a chemical element, with the following meanings; upper left, mass number of atom; lower left, nuclear charge of atom; lower right, number of atoms in molecule, eg

$$_1^2H_2 , _{12}^{24}Mg.$$

isotropic Having properties which do not vary with direction. Cf **anisotropic**.

isotropic dielectric One in which the electrical properties are independent of the direction of the applied electric field.

isotropic etching Describes an etching process which proceeds equally in all directions such as in semiconductor processing when etching is accomplished with wet chemicals or by dry etching without substantial ion bombardment. Cf **anisotropic etching**.

ITO Abbrev. for **Indium-doped Tin Oxide**.

IUPAC Abbrev. for **International Union of Pure and Applied Chemistry**.

ivory The dentine of teeth especially the massive type occurring in elephants, mammoths etc. Formerly used for tools (eg harpoon tips) and still used for decorative products despite limitation attempts.

ivory board Genuine ivory board is a laminate formed from high-quality papers by starch-pasting two or more together.

IW An abbrev. for *Isotopic Weight*.

Izod test A flexed cantilever-beam, notched-specimen, impact test in which one end of a notched specimen is held in a vice while the other end is struck by a striker carried on a pendulum; the energy absorbed in fracture is then calculated from the height to which the pendulum rises as it continues its swing. See **impact tests** p. 165.

Izod value The energy absorbed in fracturing a standard specimen in an Izod pendulum impact-testing machine. See **impact tests** p. 165.

JK

J Symbol for: (1) joule; (2) yellow in the names of dyestuffs.

J Symbol for **polar second moment of area.**

jacaranda See **Brazilian rosewood**.

jacquard, Jacquard A device, frequently incorporating punched cards or punched continuous strip, used to produce patterned fabrics during weaving, warp-knitting, weft-knitting and lace making. Named after the French inventor, Joseph-Marie Jacquard, 1752–1834. Also applied to the fabrics so produced.

Jacquet's method Final polishing of metal surfaces by **electrolysis**.

Japanese paper Japanese hand-made paper prepared from mulberry bark. The surface is similar to that of Japanese vellum.

Japanese vellum An expensive handmade paper. Prepared from the inner bark of the mulberry tree, and thicker than Japanese paper.

jappe Lightweight, fine, plain-weave cloth usually made from silk, used for linings and dresses.

jarrah A dense **hardwood** (*Eucalyptus marginata*) from Australia, with a rich deep-red colour, usually straight-grained and even but medium-coarse-textured. The **heartwood** is very durable. It is used for construction, piles, ship building, heavy framing and railway sleepers and in veneers. Density 690–1040 kg m^{-3}, mean 800 kg m^{-3}.

jaspé (1) Plain woven fabric with a shaded appearance resulting from a warp-thread colour pattern. Used mainly for bedspreads and curtains. (2) Yarn made from two chemically different continuous filament yarns (eg nylon and polyester) textured together and then dyed in such a way that only one component is coloured.

jean Strong woven twilled fabric, used for overalls or casual wear. See **denim**.

jelutong Malayan **hardwood** (*Dyera costulata*) which is almost white in colour, straight-grained with a fine, even texture, but is non-durable. Used for carving, clogs, matches, interior joinery and in plywood. Mean density 460 kg m^{-3}. The tree also yields a **latex** which was once an important substitute for natural rubber, and is now used to make chewing gum.

jersey fabric The general name for knitted fabrics supplied in lengths.

jet (1) A fluid stream issuing from an orifice or nozzle. (2) A small nozzle, as the *jet* of a carburettor.

jetting An injection moulding defect where a thin stream of polymer is forced into the furthest part of the tool cavity.

JFET See **junction field effect transistor.**

jig A device used in the manufacture of (interchangeable) parts to locate and hold the work and to guide the cutting tool.

jig, jigger Machines with two rollers used for dyeing. The open-width fabric passes repeatedly from one roller to the other and back again while immersed in a bath of the appropriate solution. The machines are also used for scouring and bleaching fabrics.

JIT Abbrev. for **Just In Time**.

jointing Material used for making a pressure-tight joint between two surfaces; eg asbestos sheet, corrugated steel rings, vulcanized rubber etc.

Johnson noise Noise in resistors, thermally generated and having a flat power spectrum. Associated with the random motion of charge carriers within the material. Cf **flicker noise, shot noise.**

Jolly balance A spring balance used to measure density by weighing in air and water.

Jominy (end-quench) test A test for determining the relative **hardenability** of steels, in which one end of a heated cylindrical specimen is quenched from the austenitic region and the longitudinal hardness gradient along a ground flat measured. The hardness decreases from the martensitic level at the quenched end towards that at the air-cooled end; the greater the hardenability of the steel the farther the distance before a significant reduction appears.

Josephson effects Two effects which can occur when two superconductors are separated by a narrow insulating gap. By tunnelling through the gap a direct current can pass from one superconductor to another without an applied potential. Also, when a potential difference, V, is established between the superconductors there is an alternating current across the gap of frequency $\nu = 2Ve/h$, where e is the charge on the electron and h is **Planck's constant**. Applications include ultra-high-speed switching of logic circuits, memory cells and parametric amplifiers operating up to 300 GHz; the Josephson effect is being widely adopted as the basis of the standard volt.

joule SI unit of energy and work. 1 joule is the work done when a force of 1 newton moves its point of application 1 metre in the direction of the force. Symbol J. 1 erg = 10^{-7} J, 1 kW h = 3.6×10^6 J, 1 eV = 1.602×10^{-19} J, 1 calorie = 4.1868 J, 1 Btu = 1055 J.

Joule effect Slight increase in the length of

an iron core when longitudinally magnetized. See **magnetostriction**.

Joule magnetostriction That for which length increases with increasing longitudinal magnetic field. Also *positive magnetostriction*.

Joule's law (1) The internal energy of a given mass of gas is a function of temperature alone; it is independent of the pressure and volume of the gas. (2) The molar heat capacity of a solid compound is equal to the sum of the atomic heat capacities of its component elements in the solid state.

Joule–Thomson effect (1) When a gas is subjected to an adiabatic expansion through a porous plug or similar device, the temperature of the gas generally decreases. This effect is due to energy being used to overcome the cohesion of the molecules of the gas. The liquefaction of gases by the **Linde process** depends on this effect. (2) Thermodynamic heating when a rubber is stretched quickly under adiabatic conditions. Also *Joule-Kelvin effect*.

journal That part of a shaft which is in contact with, and supported by, a bearing.

jumping-up The operation of thickening the end of a metal rod by heating and then hammering it in an endwise direction. Also *upsetting*.

jump joint A butt joint made by **jumping-up** the ends of the two pieces before welding them together.

junction Area of contact between semiconductor material having different electrical properties.

junction diode One formed by the junction of n- and p-type semiconductors, which exhibits rectifying properties as a result of the potential barrier built up across the junction by the diffusion of electrons from the n-type material to the p-type. Applied voltages, in the sense that they neutralize this potential barrier, produce much larger currents than those that accentuate it.

junction capacitance Capacitance associated with the effective storage of charge (arising from the ionized, fixed dopant ions) in the depletion region of a p-n junction. Cf **diffusion capacitance**.

junction field effect transistor **Field effect transistor** in which the conducting channel is in effect actively controlled by a p-n junction which bounds it. Cf **metal-oxide-silicon field effect transistor**.

junction rectifier One formed by a p-n junction by **holes** being carried into the n-type semiconductor.

junction transistor See **bipolar transistor**.

juniper A conifer, therefore a **softwood**, *Juniperus virginiana*, yielding an essential oil used medicinally; its fruits are used to flavour gin, and its fragrant, aromatic wood is the standard material for pencils, and is also used for veneers, cigar boxes, etc. Density 530 kg m^{-3}. Also *Virginian pencil cedar*, though not a true cedar.

Just In Time. Manufacturing philosophy with the aim of reducing stock levels of parts needed for final product by minimizing delivery time to factory or production line. It may also involve tighter quality control and replacing external by internal suppliers (eg in-house injection moulding replacing trade moulding) as well as parts rationalization. Abbrev. *JIT*.

jute Strong, brownish, bast fibre from the Asian plants *Corchurus olitorius* and *C. capsularus*. Used in cordage, canvas, hessian and carpet backings.

k Symbol for: (1) the velocity constant of a chemical reaction; (2) **thermal conductivity**.

κ Symbol for: (1) electrolytic conductivity; (2) **thermal diffusivity**.

κ- Symbol for: (1) cata-, ie containing a condensed double aromatic nucleus substituted in the 1,7 positions; (2) substitution on the tenth carbon atom; (3) electrolytic conductivity.

K Symbol for: (1) **potassium**; (2) **thermal diffusivity**.

K A symbol for: (1) **equilibrium constant**; (2) **bulk modulus**; (3) **stress intensity factor**.

K_s Symbol for solubility product.

kanthals Alloys with high electrical resistivity, used as heating elements in furnaces. General composition iron 67%, chromium 25%, aluminium 5%, cobalt 3%.

kaolin *China clay*. The main constituents are *kaolinite*, a hydrated aluminium silicate, and *illite*, a mica-like aluminosilicate together with potassium, magnesium and iron. Basis of **porcelain** and **bone china** and used as a filler for paper, rubber and toothpaste.

kapok The seed hairs of the kapok tree, *Ceiba pentranda*. They are light and fluffy and in loose form are used as an insulating or flotation material eg in life jackets. They are not spun or converted into fabrics commercially.

Kapton TN for polyimide film (US).

kapur SE Asian **hardwood** (*Dryobalanops*), which also yields Borneo camphor (borneol). A uniform light- to deep red-brown, straight-grained with a coarse but even texture, and a camphor-like odour. **Heartwood** is very durable. When in contact with iron or steel, the wood develops a blue stain and eventually the metal corrodes. Used for many external structural purposes, and in plywood and veneers. Mean density 770 kg m^{-3}.

karri An Australian **hardwood** (*Eucalyptus*) with a dense deep-redwood similar to, but not so durable as **jarrah**, and also used as a structural timber. Density about 880 kg m^{-3}.

kata- Prefix from Gk *kata*, down. Also *cata-*.

kauri gum A gum found in New Zealand, used for varnishes and linoleum cements. It is the resinous exudation of the kauri pine (*Agathis australis*), a tree whose timber is of value for general joinery and decorative purposes.

kautschuk Ger. for natural rubber. See **caoutchouc**.

Kaye effect Phenomenon shown by concentrated polymer solutions when poured onto a liquid surface, where the impinging jet rebounds to form a second, rising jet. See **elastic liquids**.

K_c Critical **stress intensity factor** at which a sharp crack will propagate catastrophically in a strained material. See **fracture toughness**.

kd Abbrev. for kilodalton.

keel block A standard casting shaped like a ship's keel which is used to provide a test specimen for steel or other alloys subject to high shrinkage.

keeper Soft iron bar (or similar) used to close the magnetic circuit of a permanent magnet when not in use, thereby conserving its strength.

Kel-F TN for **polytrifluorochloroethylene** (US).

kelvin Symbol K. (1) The SI unit of temperature. It is defined by fixing the triple point of water (ie the temperature and pressure at which pure ice, water and water vapour can coexist at equilibrium) as exactly 273.16K above absolute zero. Note that K does not take the degree ° sign. (2) The SI unit of temperature interval. 1K is 1/273.16 of the interval between absolute zero and the triple point of water. The interval equals the interval 1°C.

Kelvin effect Same as **skin effect**.

Kematal TN for acetal resin (UK).

keratin Structural protein of hair and wool. See **biopolymers** p. 27.

kerf The part of the original material which was removed by cutting, pressing etc.

Kerr cell A light modulator consisting of a liquid cell between crossed polaroids. The light transmission is modulated by an applied electric field. See **Kerr effects**.

Kerr effects (1) Double refraction produced in certain transparent dielectrics by the application of an electric field. Also *electro-optical effect*. (2) Dispersion of the plane of polarization experienced by a beam of plane-polarized light on its passage through a transparent medium subjected to an electrostatic strain. Also *electrostatic Kerr effect*. (3) Modification of the state of polarization of light on reflection from the polished surface of a magnetized material. Also *magneto-optical effect*. See **Faraday effect**.

kersey A heavy woollen cloth, milled and raised giving a lustrous nap, similar to melton and used in eg overcoats.

keruing A moderately durable, dark reddish-brown wood from a number of related species of SE Asian evergreen, **hardwood** tree (*Dipterocarpus*), much used for house-building and for parquet flooring. Densities in the range 640–960 kg m^{-3}, mean 740 kg m^{-3}. Also *eng*, *in*.

ketones Compounds containing a carbonyl group, –CO–, in the molecule attached to two hydrocarbon radicals. The general formula is

$$R\text{---}\underset{\underset{O}{\|}}{C}\text{---}R'\,,$$

where R and R′ are alkyl or open organic groups. The simplest ketone is acetone (propanone).

kettle An open-top vessel used in carrying out metallurgical operations on low melting-point metals, eg in drossing and desilverizing lead.

Kevlar TN for aramid fibre (US).

key A piece inserted between a shaft and a hub to prevent relative rotation. It fits in a key-way, parallel with the shaft axis, in one or both members, the commonest form being the parallel key, of rectangular section.

key way A longitudinal slot cut in a shaft or hub to receive a key.

kieselguhr See **diatomite**.

killed steel Steel that has been killed, ie fully deoxidized before casting, by the addition of manganese, silicon and sometimes aluminium. There is practically no evolution of gas from the reaction between carbon and iron-oxide during solidification. Sound ingots are obtained. See **rimming**, **rimmed steel**.

kiln drying Accelerated seasoning of timber (typically 2–5 days as opposed to 2–10 years) in a temperature- and humidity-controlled chamber.

kilo- Prefix denoting 1000; used in the SI system, eg one kilogram = 1000 grams.

kilogram(me) Unit of mass in the SI and MKSA systems, being the mass of the *International Prototype Kilogram*, a cylinder of platinum-iridium alloy kept at Sèvres, France.

kilotex See **tex**.

kinematical theory of X-ray diffraction A treatment which does not take account of the attenuation of the incident beam as it passes through the crystal nor the interference between the incident beam and multiply diffracted beams; a theory which can be applied to very thin or very small crystals.

kinematic viscosity The coefficient of viscosity of a fluid divided by its density. Symbol ν. Thus $\nu = \gamma/\rho$. The unit is m^2s^{-1} in SI and the *stoke* (cm^2s^{-1}) in the CGS system.

kinetic energy Energy arising from motion. For a particle of mass m moving with a velocity ν it is

$$\tfrac{1}{2}M\nu^2,$$

and for a body of mass M, moment of inertia I_g, velocity of centre of gravity ν_g and angular velocity ω, it is

$$\tfrac{1}{2}M\nu_g^2 + \tfrac{1}{2}I_g\omega^2.$$

kinetic theory of elasticity Explanation of elastomeric properties in molecular terms using the **statistical chain model**. Chains are random coils in an elastomer at rest, but when strained, deform so that chains orient and elongate. The possible conformations are thereby reduced, so the entropy of the system decreases. When released, the elastomer snaps back to its original, equilibrium state, under the driving force to increase entropy. The theory accounts for the surprising increase in retractive force in elastomers as the temperature is increased. The retractive force in strained metals, ceramics and glasses is, by contrast, energetic in origin. Also *statistical theory of rubber elasticity*. See **elastomers** p. 106.

kingwood Dense Brazilian **hardwood** (*Dalbergia*) with a characteristic multi-coloured heartwood on a rich violet-brown background, a straight grain and lustrous, fine, even texture. Much used for high class furniture in the reigns of Louis XIV and Louis XV in France and in Georgian England. Now used by antique restorers, and for decorative ware and veneers. Density 1200 kg m^{-3}.

Kirchhoff's laws (1) The ratio of the coefficient of absorption of radiation to the coefficient of its emission is the same for all substances and depends only on the temperature. The law holds for the total radiative emission and also for the emission at any particular frequency. (2) Generalized extensions of Ohm's law employed in electrical network analysis. They may be summarized as: (1) $\sum i = 0$ at any junction, and (2) $\sum E = iZ$ round any closed path. $E =$ emf, $i =$ current, $Z =$ complex impedance.

Kirkendall effect If a piece of a pure metal is placed in contact with a piece of an alloy of that metal and the whole heated, the constituent elements from the alloy will diffuse into the pure metal, causing a shift of the original interface.

kish Solid graphite which has separated from, and floats on the top of, a molten bath of cast-iron or pig-iron which is high in carbon.

knee An elbow pipe ie short and rightangled.

knife coating Method of coating fabric with flexible layer (eg plasticized polyvinyl chloride) by passing fabric between roller and knife blade. Produces flat surface finish after curing, unlike reverse roll coating, which gives even finish over an often rough surface.

knife edge Support for a balance beam or similar instrument member, usually in the form of a hardened steel or an agate wedge, the apex of which gives line support.

Knight shift Shift in **nuclear magnetic resonance** frequency in metals from that of the same isotope in chemical compounds in the same magnetic field. It is due to the paramagnetism of the conduction electrons.

knitting The process of making a fabric from yarn by the formation of intermeshing loops. See **fibre assemblies** p. 130, **warp-knitting**, **weft-knitting**.

Knoop hardness Hardness measured with a Knoop diamond, an elongated rhombus pyramid with angles of 172.5° and 130° at the apex, giving a hardness value, $H_K = 14.233\ F/d^2\ \text{MN m}^{-2}$, where F is the load in N and d is the longer diagonal in mm. It is a shallower diamond than the **Vickers**, and so is better suited to measuring surface layers. See **hardness measurements** p. 153.

knot A hard and often resinous inclusion in timber, formed from the base of a branch which became buried as the trunk thickened.

Koch resistance The resistance of a vacuum photocell or phototube when its active surface is irradiated with light.

Kohlrausch's law The contribution from each ion of an electrolytic solution to the total electrical conductance is independent of the nature of the other ion.

Kovar TN for an alloy of Ni-Co-Fe for glass-to-metal seals over working ranges of temperature, when the temperature coefficients of expansion coincide.

krabak See **mersawa and krabak**.

kraft paper Strong paper made from a sulphate pulp (kraft pulp), and used in packaging, and in laminated plastics, eg melamine-formaldehyde laminates. Name derived from German for *strength*.

Kraton TN for styrene butadiene styrene thermoplastic elastomer.

Kroll's process Reduction to metal of tetrachloride of titanium or zirconium in vacuo or by reaction with magnesium in a neutral atmosphere.

Kronig–Penny model A relatively simple model for a one-dimensional crystal lattice from which the essential features of the behaviour of electrons in a periodic potential may be illustrated. See **band theory of solids**.

krypton One of the noble gases. Symbol Kr. It is a colourless and odourless monatomic gas, and constitutes about one-millionth by volume of the atmosphere, from which it is obtained by liquefaction. It is used in certain gas-filled electric lamps. It forms a few compounds, eg KrF_4. See Appendix 1, 2.

K-shell The innermost electron shell in an atom corresponding to a principal quantum number of one. The shell can contain two electrons. See **atomic structure**.

k-space Symbol for *momentum space* or *wave vector space*. This is an important concept in semiconductor energy band theory. See **Brillouin zone**.

K value Practical measure of molecular mass of polyvinyl chloride, obtained from intrinsic viscosity experiments. A K value of 69 corresponds to a M_w of 240 000.

L

l An abbrev. for **litre** (1 dm^3 is now used officially).

l- Abbrev. for **laevorotatory**.

l Symbol for: (1) specific latent heat per gramme; (2) mean free path of molecules; (3) (with subscript) equivalent ionic conductance, 'mobility'.

L- Abbrev. for **laevorotatory**.

L Symbol for **molar latent heat**.

labile Unstable, liable to change. Usually a kinetic, not a thermodynamic term.

lac See **shellac**.

lace (1) A fine open-work decorative fabric comprising an underlying net on which patterns are formed by looping, twisting, or knitting. The process may be done by hand using a needle or bobbin or by machine. (2) Long, thin extrudate which when cut by a rotating blade fitted after the die, gives polymer granules.

lacewood See **European plane**.

ladder In a knitted structure, esp. stockings and tights, a defect caused by the breaking of stitches resulting in the thread reverting in long runs in the wale direction to its original linear form.

ladder polymers Polymers consisting of two chains covalently bonded together, such as Black Orlon. See **high-performance fibres** p. 158.

ladle An open-topped vessel lined with refractory material; used for conveying molten metal from the furnace to the mould or from one furnace to another.

ladle addition Addition of alloying metal to molten metal in a ladle before casting.

laevo- Prefix from L *laevus*, left.

laevorotatory Said of an optically active substance which rotates the plane of polarization in an anti-clockwise direction when looking against the oncoming light.

lagging (1) The process of covering a vessel or pipe with an insulating material to prevent either loss or gain of heat. (2) The insulating material itself. See **insulation**.

laid paper Writing or printing paper watermarked with a pattern of spaced parallel lines (*chain lines*) generally disposed in the machine direction and usually accompanied by more closely spaced parallel lines at right angles (*laid lines*).

lakes Pigments formed by the interaction of dyestuffs and 'bases' or 'carriers', which are generally metallic salts, oxides or hydroxides. The formation of insoluble lakes in fibres, which are being dyed, is known as mordanting, the hydroxides of aluminium, chromium and iron generally being employed as **mordants**.

lamé Fabric with conspicuous decorative metallic threads.

Lamé constants **Hooke's law** written in its three-dimensional form for co-ordinate axes x, y, and z and with the stresses, σ expressed in terms of the strains, ε results in equations of the form:

$$\sigma_x = (\lambda + 2G)\,\varepsilon_x + \lambda\varepsilon_y + \lambda\varepsilon_z\,,$$

$$\text{where } \lambda = \frac{\nu E}{(1 + \nu)(1 - 2\nu)}$$

and the **shear modulus**, G, are the Lamé constants, ν is **Poisson's ratio** and E is **Young's modulus**.

Lamé formula A formula for calculating the stresses in thick (hydraulic) cylinders under elastic deformation.

lamellae Plate-like microscopic crystals found in partially crystalline polymers. They usually consist of chain-folded molecules, and may possess regular edges (eg truncated lozenges) as well as having hollow centres. See **crystallization of polymers** p. 82.

lamellar magnetization Magnetization of a sheet or plate distributed in such a way that the whole of the front of the sheet forms one pole and the whole of the back forms the other.

lamell-, lamelli- Prefixes from L. *lamella*, thin plate.

laminar flow A type of fluid flow in which adjacent layers do not mix except on the molecular scale. Also *streamline flow*.

laminate A structural sheet material made from two or more dissimilar layers (laminae) bonded together, eg laminated glass, paper, plastics. See **angle-ply laminate**, **balanced laminate**, **laminated plastics**.

laminated bearing Type of bridge and building support for absorption of movement and vibration. Comprises alternate layers of steel and rubber sheet (natural rubber or chloroprene rubber), laid perpendicular to main downward thrust of structure. See **shape factor**. Similar principle used in helicopter rotor bearings. See **bridge bearing**.

laminated bending The practice of bending several layers of material and at the same time joining them along their surfaces in contact to form a unit.

laminated core A core for a transformer or electrical machine made up from insulated laminations for the purpose of reducing losses associated with **eddy currents**.

laminated glass See **safety glass**.

laminated magnet (1) A permanent magnet built up from magnetized strips to obtain a high intensity of magnetization. (2) An electromagnet for a.c. circuits, having a laminated core to reduce **eddy currents**.

laminated paper Product formed by bonding the whole of the surface of a sheet of paper to another paper or sheet material such as metal foil, plastics film etc.

laminated plastics Superimposed layers of a synthetic resin-impregnated or coated filler (eg **kraft paper** or fibre reinforcement) which have been bonded together, usually by means of heat and pressure, to form a single piece. Also *laminate*.

laminated spring A flat or curved spring consisting of thin plates or leaves superimposed, acting independently, and forming a beam or cantilever of uniform strength.

lamination A sheet steel stamping shaped so that a number of them can be built up to form the magnetic circuit of an electric machine, transformer, or other piece of apparatus. Also *core plate*, *punching*, *stamping*.

lampblack The soot (and resulting pigment) obtained when substances rich in carbon (eg mineral oil, turpentine, tar etc) are burnt in a limited supply of air so as to burn with a smoky flame. The pigment is black with a blue undertone, containing 80–85% carbon and a small percentage of oily material. See **carbon black**.

lamp working Making glass articles, usually from tubing or rod, with the aid of an oxy-gas or air-gas flame.

lancewood Durable straight-grained wood chiefly from *Oxandra lanceolata*, a native of tropical America, it is used for archery bows, billiard cues, tool handles and in joinery, etc.

lancing A line cut made in a press which does not remove metal but only separates it.

Landau levels **Conduction electrons** of a solid in a magnetic field will describe complete orbits if $\omega\tau > 1$ where ω is the *cyclotron frequency* and τ is the time between scattering events. The electron density of states will be altered and the allowed energy levels, the *Landau levels*, will differ by $\hbar\omega$ with $\hbar = h/2\pi$, where h is **Planck's constant**.

Landau theory (1) Theory for calculating diamagnetic susceptibility produced by free conduction electrons. (2) That explaining the anomalous properties of liquid helium II in terms of a mixture of normal and superfluids. See **helium**, **superfluidity**.

Langevin equation (1) A classical expression for diamagnetic susceptibility produced by the orbital electrons of atoms. (The quantum mechanical equivalent of this was derived subsequently by Pauli.) (2) An expression for the resultant effect of atomic magnetic moments which enters into explanations of both **paramagnetism** and **ferromagnetism**.

lang lay A method of making wire ropes in which the wires composing the strands, and the strands themselves, are laid in the same direction of twist.

Langmuir adsorption isotherm The fraction of the adsorbent surface which is covered by molecules of adsorbed gas is given by $\theta = bp(1 + bp)$ where p is the gas pressure and b is a constant.

Langmuir–Blodgett film Monomolecular (organic) assemblies on a substrate. An integral part of **molecular electronics**.

Langmuir's theory (1) The assumption that the extranuclear electrons in an atom are arranged in shells corresponding to the periods of the periodic table. The chemical properties of the elements are explained by supposing that a complete shell is the most stable structure. (2) The theory that adsorbed atoms and molecules are held to a surface by residual forces of a chemical nature.

lanthanide contraction The peculiar characteristic of the **lanthanide series** that the ionic radius decreases as the atomic number increases, because of the increasing pull of the nuclear charge on the unchanging number of electrons in the two outer shells. Thus the elements after lanthanum, eg platinum, are very dense and have chemical properties very similar to their higher homologues, eg palladium.

lanthanides The rare earth elements at. nos 57–71, after lanthanum, the first of the series. Cf **actinides**. In both series an incomplete *f*-shell is filling. See Appendix 1.

lanthanum A metallic element in the third group of the periodic table, belonging to the rare earths group. Symbol La. See Appendix 1, 2.

lanthanum glass Optical glass used for high-quality photographic lenses etc with high **refractive index** and **low dispersion**.

lap (1) A rotating disk or other tool for grinding or polishing glass. (2) A similar tool for imparting a fine finish to metal. (3) A square piece of material, usually rubber, to protect the hands when handling glass. (4) A surface defect on rolled or forged steel. It is caused by folding a fin to the surface and squeezing it in; as welding does not occur, a seam appears on the surface. (5) Polishing cloth impregnated with diamond dust or other abrasive, used in polishing eg for metals and rock specimens.

lap joint A plate joint in which one member overlaps the other, the two being riveted or welded along the seam single, double or treble.

lapping (1) The finishing of spindles, bored holes etc to fine limits, by the use of laps of lead, brass etc impregnated with abrasive paste. (2) The final abrasive polishing of a quartz crystal to adjust its operating frequency. Also, smoothing the surfaces of crystalline semiconductors.

lap-shear test Method of testing adhesives using a **lap joint**, which is then tensioned, putting the joint into a state of shear.

large-scale integration Fabricating a very large number of electron devices on a single chip. Abbrev. *LSI*.

Larmor frequency The angular frequency of precession for the spin vector of an electron acted on by an external magnetic field.

Larmor precession The precessional motion of the orbit of a charged particle when subjected to a magnetic field. Precession occurs about the direction of the field.

Larmor radius That of the circular or helical path followed by a charged particle in a uniform magnetic field.

laser Abbrev. for *Light Amplification by Stimulated Emission of Radiation*. A source of intense monochromatic radiation in the ultraviolet, visible or infrared region of the spectrum. It operates by producing a large population of atoms with their electrons in a defined high energy level. By stimulated emission, transitions to a lower level are induced, the emitted photons travelling in the same direction as the stimulating photons. If the beam of inducing light is produced by reflection from mirrors or Brewster's windows at the ends of a resonant cavity, the emitted radiation from all stimulated atoms is in phase, and the output is a very narrow beam of coherent monochromatic radiation. Solids, liquids and gases have been used as the active medium. Also *masers* in the microwave region.

laser-beam cutting Using the intense narrow beam of radiation from a laser to cut often complex shapes in sheet or plate. Good finish and high precision can be achieved.

laser-beam machining The use of a focused beam of high-intensity radiation from a laser to vaporize and so machine material at the point of focus. Cf **laser beam cutting**.

laser diode See **optoelectronics** p. 226.

latent heat More correctly, **specific latent heat**. The heat which is required to change the state of unit mass of a substance from solid to liquid, or from liquid to gas, without change of temperature. Most substances have a latent heat of fusion and a latent heat of vaporization. The specific latent heat is the difference in *enthalpy* of the substance in its two states.

latent magnetization The property possessed by certain feebly magnetic metals (eg manganese and chromium) of forming strongly magnetic alloys or compounds.

lateral Situated on or at, or pertaining to, a side.

lateral contraction ratio Although not an **elastic constant**, it is the parameter corresponding to **Poisson's ratio** in anisotropic and/or non-linear elastic materials.

lateral load A force acting on a structure or a structural member in a transverse direction, eg wind forces on a bridge or building at right angles to its length, which trusses and girders are not primarily designed to withstand.

latex (1) A milky viscous fluid extruded when a rubber tree (eg *Hevea brasiliensis*) is tapped. It is a colloidal system of **caoutchouc** dispersed in an aqueous medium, density 990 kg m^{-3}, which forms rubber by coagulation. The coagulation of latex can be prevented by the addition of ammonia or formaldehyde. Latex may be vulcanized directly, the product being known as *vultex*. (2) In synthetic rubber manufacture, the process stream in which the polymerized product is produced. (3) An aqueous emulsion of synthetic rubber-like compounds used to increase the flexibility and durability of paper eg for bookbinding papers or base material for imitation leather. It may be added to the stock or used as an impregnant. Also extensively employed as the binder in mineral coating applications.

lathe A machine tool for producing cylindrical work, facing, boring and screw cutting. It consists generally of a bed carrying a head stock and tail stock, by which the work is driven and supported, and a saddle carrying the slide rest by which the tool is held and traversed.

lattice A regular spatial arrangement of points as for the sites of atoms in a crystal.

lattice dynamics The study of the excitations a crystal lattice can experience and their consequences for the thermal, optical and electrical properties of solids.

lattice energy Energy required to separate the ions of a crystal from each other to an infinite distance.

lattice girder A girder formed of upper and lower horizontal members connected by an open web of diagonal crossing members, used in structures such as bridges and large cranes.

lattice water **Water of crystallization**, which is present in stoichiometric proportions and occupies definite lattice positions, but is in excess of that with which the ions can be co-ordinated. This water apparently fills in holes in the crystal lattice, as with **clathrate** compounds.

Laue pattern Pattern of spots produced on photographic film when a heterogeneous X-ray beam is passed through a thin crystal, which acts like an optical grating. Used in the analysis of crystal structure.

laurel See **stinkwood**.

law A scientific law is a rule or generalization which describes specified natural phenomena within the limits of experimental observation. An apparent exception to a law tests the validity of the law under the specified conditions. A true scientific law admits of no exception. A law is of no scientific value unless it can be related to other laws comprehending relevant phenomena.

lawn Fine, lightweight, plain-weave cloth made of flax or cotton yarns. See **organdie**.

law of conservation of matter The law stating that matter is neither created nor destroyed during any physical or chemical change.

law of constant (definite) proportions The law that every pure substance always contains the same elements combined in the same proportions by weight.

law of Dulong and Petit The law that the atomic heat capacities of solid elements are constant and approximately equal to 25 (when the specific heat capacity is in J mole^{-1} K^{-1}). Certain elements of low atomic mass and high melting point have, however, much lower atomic heat capacities at ordinary temperatures.

law of equilibrium See **law of mass action**.

law of equivalent (reciprocal) proportions The law that the proportions in which two elements separately combine with the same mass of a third element are also the proportions in which the first two elements combine together.

law of Guldberg and Waage See **law of mass action**.

law of mass action Fundamental law applying to equilibrium chemical reactions of the general form

$$aA + bB + \ldots \rightarrow cC + dD + \ldots .$$

The law states that concentrations of reactants and products, shown thus [..], are related by the equation

$$K_c = \frac{[C]^c \cdot [D]^d \ldots}{[A]^a \cdot [B]^b \ldots} ,$$

where K_c is known as the **equilibrium constant**. When expressed in terms of activity, it is K_a, and in partial pressures, K_p. The concentration equilibrium constant is also the ratio of the forward and reverse reaction rate constants:

$$K_c = \frac{k_f}{k_r} .$$

It thus obeys **Arrhenius's rate equation**. Also *law of equilibrium*. See **order of reaction**.

law of mixtures See **rule of mixtures**.

law of multiple proportions The law that when two elements combine to form more than one compound, the amounts of one of them which combine with a fixed amount of the other exhibit a simple multiple relation.

law of octaves The relationship observed by Newlands (1863) which arranges the elements in order of atomic mass and in groups of eight (octaves) with recurring similarity of properties. See **periodic table**.

law of partial pressures See **Dalton's law of partial pressures**.

law of photochemical equivalence See **Einstein law of photochemical equivalence**.

law of reciprocal proportions See **law of equivalent proportions**.

LAXS Abbrev. for **Low Angle X-ray Scattering**.

lb Abbrev. for **pound**.

LCA Abbrev for **Life Cycle Analysis**.

LCD Abbrev. for **Liquid Crystal Displays**.

LCI Abbrev. for **Life Cycle Inventory**.

LDPE Abbrev. for **Low Density Poly-Ethylene**.

L/D ratio Property of an injection moulding screw or extrusion screw, defined by ratio of screw *length* to flight *diameter*. It is typically 20–30 to one for thermoplastics and 5–10 to one for rubbers. See **extrusion**.

lead A metallic element in the fourth group of the periodic system. Symbol Pb. Electrical resistivity 20.65×10^{-8} Ω m. Principal uses: in storage batteries, ammunition, foil and as a constituent of bearing metals, solder and type metal. Lead can be hardened by the addition of arsenic or antimony. See Appendix 1, 2.

lead-acid (lead) accumulator A *secondary cell* consisting of lead electrodes, the positive one covered with lead dioxide, dipping into sulphuric acid solution. Its emf is about 2 volts; very widely used. See **accumulator** (2).

lead disilicate Obtained by fusing lead (II) oxide and silica together. As lead **frit**, it is used as a ready means of incorporating lead oxide in the making of lead **glazes**.

leaded Term descriptive of copper, bronze, brass, steel, nickel and phosphor alloys to which from 1% to 4% of lead has been added, mainly to improve machinability.

lead frit See **lead disilicate**.

lead glass Glass containing lead oxide. The amount may vary from 3–4% to 50% or more in special cases. *English Lead Crystal*, used for tableware because of its high refractive index, has the composition (wt %): SiO_2 56, PbO 29, K_2O 13, Na_2O 2. A similar glass is used in the *pinch* of incandescent light bulbs. Lead glass is also extensively used as a transparent radiation shield, esp. for X-rays.

lead (II) chromate (IV) $PbCrO_4$. Precipitated when potassium chromate (VI) is added to the solution of a lead salt. Used as pigments, called chrome yellows. The colour may be varied by varying the conditions under which the precipitation is made.

lead (II) hydroxide $Pb(OH)_2$. Dissolves in excess of alkali hydroxides to form plumbites (plumbates (II)).

lead (II) oxide PbO. An oxide of lead, varying in colour from pale yellow to brown depending on the method of manufacture. An intermediate product in the manufacture of red lead.

lead (IV) oxide PbO_2. A strong oxidizing agent. Industrial application very limited. Present, in certain conditions, in accumulators or electrical storage batteries as a chocolate brown powder.

lead sulphate $PbSO_4$. Formed as a white precipitate when sulphuric acid is added to a solution of a lead salt.

lead tree Form, in shape of tree, which lead takes after electrodeposition from simple salts.

lead zirconate titanate A piezoelectric ceramic with a higher **Curie point** than barium titanate (IV). Abbrev. *PZT*. See **dielectric and ferroelectric materials** p. 94.

leaf spring A machine component comprising one or a group of relatively thin, flexible, or resilient strips, reacting as a spring to forces applied to a main surface.

leather A material made by **tanning** and other treatment of the hides or skins of a great variety of creatures (mostly domesticated animals, but including also the whale, seal, shark, crocodile, snake, kangaroo, camel, ostrich etc). Artificial or imitation leathers are based on plasticized polyvinyl chloride or polyurethane, suitably filled with pigment etc and surface embossed to simulate the natural product.

leatherboard Board made largely or wholly from leather scraps, generally on an intermittent board machine and containing **latex** as a binder and to impart flexibility.

leathercloth A woven or knitted fabric coated on one side with a polymer (eg rubber, cellulose derivative, polyvinyl chloride) which is embossed to simulate leather.

leathery region Glass-transition region of master curve of polymers, where material behaves like very stiff rubber. See **viscoelasticity** p. 332.

Le Chatelier's principle If any change of conditions is imposed on a system at equilibrium, then the system will alter in such a way as to counteract the imposed change. This principle is of extremely wide application. It is a statement of the **law of mass action**.

lecith-, lecitho- Prefixes from Gk *lekithos*, yolk of egg.

Leclanché cell A **primary cell**, good for intermittent use, has a positive electrode of carbon surrounded by a mixture of manganese dioxide and powdered carbon in a porous pot. The pot and the negative zinc electrode stand in a jar containing ammonium chloride solution. The emf is approximately 1.4 volt. The *dry cell* is a particular form of Leclanché.

LED Abbrev. for **Light Emitting Diode**. See **optoelectronics** p. 226.

Leduc effect A magnetic field applied at right angles to the direction of a temperature gradient in an electrical conductor, will produce a temperature difference at right angles to the direction of both the temperature gradient and the magnetic field.

LEED *Low-Energy Electron Diffraction*. Used to study the structure of surfaces using electrons of energy in the range 10–500 eV.

LEFM Abbrev. for **Linear Elastic-Fracture Mechanics**.

lehr An enclosed, tunnel-like oven or furnace used for annealing glass or other forms of heat treatment. Hot glass from the forming process is passed through it to cool slowly, so that strain is removed, and cooling takes place without additional strain being introduced. Lehrs may be of the open type (in which the flame comes in contact with the ware) or of the muffle type. Also *lear*, *leer*, *lier*.

Lennard–Jones potential Potential due to molecular interaction, also the *6–12 potential* because the attractive potential varies as the inverse 6th power of the intermolecular distance, and the repulsive potential as the inverse 12th.

leno fabric A woven fabric in which the warp threads are made to cross one another between picks. Lightweight fabrics of this kind are known as gauzes; heavier qualities of cotton are used as blankets.

lens grinding The process of grinding pieces of flat sheet-glass (or pressed blanks) to the correct form of the lens. Cast iron tools of the correct curvature, supplied with a slurry of abrasive and water, are used.

letting down The process of tempering hardened steel by heating until the desired temperature, as indicated by colour, is reached, and then quenching.

leuco-bases, leuco-compounds Colourless compounds formed by the reduction of dyes, which when oxidized are converted back into dyes.

leuco-, leuko- Prefixes from Gk *leukos*, white.

leuko- See **leuco-**.

levelling agent Agent added to a dye bath

to produce uniform precipitation of the dye on to textile fibres.

levelling solvent One which has a high dielectric constant and high polarity, so that most electrolytes appear strong in solution.

lever rule A method for estimating the proportions of phases present in a mixture. See **phase diagram** p. 236.

levitation melting A process in which the melt is suspended in an electromagnetic field while being heated by **induction** in vacuo. It is therefore not contaminated by the material of a container while molten. Only suitable for quantities of a few grams, depending on the density.

Lewis acids and bases The concept that defines any substance donating an electron pair as a base, and any substance accepting an electron pair as an acid. Conventional acids and bases fit this definition, as do complex-forming reactions. Thus in the reaction $AlCl_3 + Cl^- = AlCl_4^-$, $AlCl_3$ is classed as an acid and Cl^- as a base.

Lewis's theory The assumption that atoms can combine by sharing electrons, thus completing their shells without ionization.

Lewis structure A possible structure for a molecule in which all electrons are specifically associated with one or two atoms. Many structures can only be described by a mixture of two or more Lewis structures, the best known example being benzene. Also *resonance structure*.

Lexan TN for polycarbonate (US).

Liesegang rings The stratification, under certain conditions, of precipitates formed in gels by allowing one reactant to diffuse into the other.

life cycle analysis Flow of energy and materials through manufacturing system from raw material in ground, through processing to shape, assembly of finished product and disposal following use. Consists of three stages; an inventory of inputs and outputs, assessment of the impact and formulation of solution. Also *womb-to-tomb* (US), *cradle-to-grave* (Europe) analysis. Not to be confused with *product life cycle*. Abbrev. *LCA*.

life cycle inventory Measurement or calculation of all material and fuel inputs and ouputs from a process starting with raw materials in the ground and ending with waste disposal back into the earth or atmosphere. Abbrev. *LCI*.

life-time The mean period between the generation and recombination of a charge carrier in a semiconductor.

ligament A bundle of fibrous tissue joining two or more bones or cartilages. Replaced by fluoropolymer fibre ligaments in implants (eg **Gore-Tex**).

ligand field theory An essentially ionic in-

terpretation of bonding in transition metal compounds, in which the spectroscopic and magnetic properties are rationalized in terms of the distortion of the non-bonding d-electrons of the metal by the electric field of the ligands.

ligands In a complex ion, the ions, atoms or molecules, surrounding the central (nuclear) atom, eg (CN^-) in $Fe(CN)_6^{4-}$.

light Of optical glass, having a low refractive index.

light alloys See **aluminium alloys**.

light alloys Alloys based on the low-density metals aluminium and magnesium. See **aluminium alloys** p. 8.

light emitting diode See **optoelectronics** p. 226.

light resistance The resistance, when exposed to light, of a photocell of the photoconductive type. Cf **dark resistance**.

light scattering Method for determining molecular mass of molecules, esp. soluble polymers. Gives absolute measure of M_w. See **molecular mass distribution** p. 212.

light sensitive Said of thin surfaces of which the electrical resistance, emission of electrons or generation of a current, depends on incidence of light.

lightwood Any coniferous wood having an abnormally high resin content.

lignin A complex crosslinked polymer based on substituted p-hydroxy phenyl propane $(HO.C_3H_6-)$ units; found in wood as a matrix material for the cellulose fibres.

ligno-celluloses Compounds of lignin and cellulose found in wood and other fibrous materials.

lignum vitae One of the hardest and densest commercial timbers, some 25% of its dry weight is guaiac resin which makes it self-lubricating. A **hardwood** (*Guaiacum*) from the W Indies and tropical America, its **heartwood** is greenish-brown to black, with a heavily interlocked grain and fine, uniform texture. Used in bearings and bushes for rotating parts (eg ships' propeller shafts) and as the 'woods' in bowling. Mean density 1230 kg m^{-3}.

limba See **afara**.

limbric Plain-weave cotton cloth of light to medium weight in which the weft predominates.

lime (1) A substance produced by heating limestone to $825°C$ or more, as a result of which the carbon dioxide and water are driven off. Lime, which is much used in the building, chemical, metallurgical, agricultural and other industries, may be classified as high calcium, magnesian, or dolomitic depending on the origin and composition. Unslaked lime is commonly known as *caustic lime*, also *anhydrous lime, burnt lime, quick-*

lime. (2) W European hardwood (*Tilia*); creamy-yellow, straight-grained, fine, even texture. Typical uses include carving, inlaying, marquetry work and cabinet making. Density about 540 kg m^{-3}.

lime water Saturated calcium hydroxide solution.

limiting conductivity The molar conductivity of a substance at infinite dilution, ie when completely ionized.

limiting range of stress The greatest range of stress (mean stress zero) that a metal can withstand for an indefinite number of cycles without failure. If exceeded, the metal fractures after a certain number of cycles, which decreases as the range of stress increases. See **fatigue strength** p. 124.

limit of proportionality See **proportional limit**.

limit values A European Community term specifying environmental quality standards and emission standards, which defines the limits that member states must set, eg concentrations of smoke and sulphur dioxide at ground level.

Linde process A process for the liquefaction of air and for the manufacture of oxygen and nitrogen from liquid air.

Linde sieve See **molecular sieve**.

linear density Weight of a fixed length of textile yarn, eg **tex, denier**; the smaller the number, the finer the thread. Cf **yarn count**.

linear elastic-fracture mechanics See **fracture mechanics**. Abbrev. *LEFM*.

linear low-density polyethylene Grade of polymer made by low pressure catalytic route with controlled level of short chain branches. Abbrev. *LLDPE*.

linear superpolymer Polymer in which the molecules are essentially in the form of long chains with an average molecular mass greater than 10 000.

linen Yarns, fabrics and articles made from **flax** fibres.

linkage A chemical bond, particularly a covalent bond in an organic molecule.

linseed oil An oil obtained from the seeds of **flax** (*Linum usitatissimum*). It contains solid and liquid glycerides of oleic and other unsaturated acids. It oxidizes and polymerizes in air to form a brittle polymer. Once much used for the mixing of paints and varnishes and for the manufacture of linoleum. Still used in **putty** for glazing windows etc.

lint (1) The main seed hairs of the cotton plant. (2) A plain-woven cotton fabric raised on one side to make it highly absorbent; used, after sterilization, as a wound dressing.

linters, cotton linters Short fuzzy fibres remaining on cotton seeds after substantial removal of the longer fibres or lint. Removed before the seeds are crushed and re-

cycled chemically into cellulose nitrate, cellulose acetate, etc.

liquation Partial melting of an alloy due to heterogeneity of composition.

liquefied petroleum gases Gases such as propane and butane, used for fuels. Abbrev. *LPG*.

liquid A state of matter between a solid and a gas, in which the shape of a given mass depends on the containing vessel, the volume being independent. Liquids are almost as incompressible as solids.

liquid crystal displays See panel on p. 192.

liquid crystal phases See panel on p. 194.

liquid crystal polymers Chains or precursors aligned along the flow direction in processing or polymerization, similar to nematic liquid crystals. They exhibit high stiffness along this axis. Moulding materials include copolymers of polyethylene terephthalate and PHB (para-hydroxy benzoic acid) and aramid fibres, Kevlar and Twaron. See **high-performance fibres** p. 158.

liquid crystals Certain pure liquids which are turbid and, like crystals, anisotropic over a definite range of temperature above their freezing-points. See **liquid crystal displays** p. 192.

liquid helium II See **helium**.

liquid oxygen See **oxygen**.

liquid-penetrant inspection See **penetrant flaw detection**.

liquid-phase sintering Sintering in which a small proportion of the material becomes liquid. It may or may not speed up the sintering process by solution transfer of the phases forming the matrix. See **ceramics processing** p. 54.

liquidus A line in a phase diagram indicating the temperatures at which solidification of one phase or constituent begins or melting is completed. See **phase diagram** p. 236.

lisle Long-staple, highly-twisted, folded cotton hosiery yarn, gassed (ie passed through gas flame or over hot element to remove protruding fibre ends) and often mercerized to produce a lustrous effect. See **mercerization**.

litharge Lead (II) oxide, used in paint-mixing as a drier; used also in the rubber and electrical accumulator industries.

litharge cement See **glycerine litharge cement**.

lithia Lithium oxide, Li_2O.

lithium An element symbol Li. It is the least dense solid, chemically resembling sodium but less active. It is used in alloys and as a basis for lubricant grease with high resistance to moisture and extremes of temperature; also as an ingredient of high-energy fuel cells. See Appendix 1, 2.

liquid crystal displays

Liquid crystals are fluids containing stiff, rod-like organic molecules which have a tendency to form ordered structures. They exhibit some of the properties of crystalline solids. In particular, the elastic, optical and dielectric properties of liquid crystals are highly **anisotropic** and these are exploited in the construction of display devices. See **liquid crystal phases** p. 194.

Fig. 1 shows the principle of a seven segment 'twisted nematic' cell such as that used in watch and pocket calculator displays, in which the molecular order in a liquid crystal film is locally influenced by an external electric field to control the transparency of elements of the cell. The liquid crystal substance is sandwiched between transparent electrode structures (usually indium tin oxide) and two sheets of polarizer arranged in a crossed orientation.

Electrically a segment of the display looks like a parallel plate capacitor, with a liquid crystal **dielectric**. With no field across a segment, the molecular ordering is locally determined by a combination of molecule-molecule and molecule-surface interactions. The device illustrated uses a **nematic** phase. The inner surfaces of the cell are coated with a thin polymeric layer in which are fine parallel grooves; the rod-like molecules preferentially align themselves with these grooves. On opposite sides of the cell the grooves are at right angles so that the molecular orientation through the bulk twists smoothly through 90° (see left side of Fig. 2). Owing to the anisitropic refractive index of the nematic phase, the plane of polarization of light incident through one of the polarizers also twists smoothly through 90°, so that since the polarizers are in a crossed orientation, such light is also able to escape. The light then strikes a reflector and retraces its path through the cell and out again, so that in this region the cell appears relatively bright.

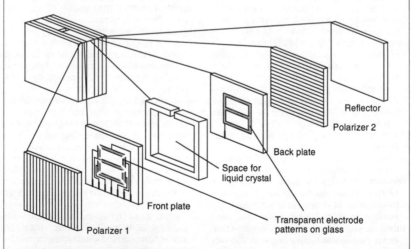

Fig. 1. **A simple type of liquid crystal display.**

A relatively strong electric field (typically around 300 kV m^{-1}) across a segment causes the molecules locally to re-orient along the field. Without the twisted molecular structure to rotate its plane of polarization, light entering the

continued on next page

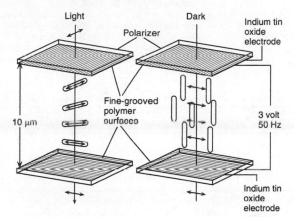

liquid crystal displays (contd)

Fig. 2. **Twisted nematic elements of a liquid crystal display.**

cell through one polarizer is therefore absorbed by the other. Where this occurs the segment appears dark (see Fig. 2). When the field is removed, thermal agitation allows the molecules to jostle each other back into the twisted orientation.

Liquid crystals for this application have to be engineered to give satisfactory operation over a useful temperature range and to give adequate switching speeds. A closely related type of liquid crystal display incorporates dye molecules which enables the construction of electrically switchable colour filters.

lithium 12-hydroxy stearate A lithium 'soap' widely use in high-performance greases as the main thickening agent. Helps to confer high water resistance and good low-temperature performance.

lithium hydroxide LiOH. A strong base, but not so hygroscopic as sodium hydroxide. Used in the form of a coarse, free-flowing powder as a carbon dioxide absorber, eg in submarines. Also to make lithium stearates etc in driers and lubricants.

lithium niobate See **dielectrics and ferroelectrics** p. 92. Abbrev. *LN.*

lithium tantalate See **dielectrics and ferroelectrics** p. 92. Abbrevs. *LT, LTO.*

lithographic paper High machine-finished or supercalendered paper, made so that any stretch occurs the narrow way of the sheet.

lithography Pattern transfer process used to define tracks on printed circuit boards and to create electronic device features. In semiconductor technology optical (ultraviolet), electron beam or X-ray irradiation through a mask transfers the pattern to a thin coating of appropriate resist on the wafer surface. Subsequent etching is able to proceed through open areas in the developed resist layer. See

semiconductor device processing p. 280.

litmus A material of organic origin used as an indicator; its colour changes to red for acids, and to blue for alkalis at **pH** > 7.

litre Unit of volume equal to one cubic decimetre or 10^{-3} m^3.

live load A moving load or a variable force on a structure; eg that imposed by traffic movement over a bridge, as distinct from a dead weight or load, such as that due to the weight of the bridge.

living polymers Anionic polymers which still have active chain ends and have not been terminated. Active ends are charged-pair complexes, which often produce highly coloured solutions. Useful for making block copolymers. See **chain polymerization** p. 56. Not to be confused with polymers from living organisms, eg biomaterials.

LLDPE Abbrev. for **Linear Low-Density PolyEthylene.**

LN Abbrev. for *Lithium Niobate.* See **dielectric and ferroelectric materials** p. 94.

load (1) Synonym for force applied to a body. (2) Measure for the equivalent area or volume of timber that will weigh approx. one ton, ie 600 ft^2 of 1-inch planks, 40 ft^3 of

---

liquid crystal phases

The first observation of liquid crystals is attributed to the Austrian biologist Reinitzer who in 1888 found two distinct melting temperatures in a derivative of cholesterol. Liquid crystals are mesophases arising between the normal solid and liquid phases of certain organic substances. There are two ways to describe these mesophases with reference to a phase diagram: either in terms of the composition range (lyotropic) or else in terms of their temperature range (thermotropic). For example, soap and water form a variety of lyotropic liquid crystal mesophases: as water molecules penetrate between the layers of soap molecules, the bonding between polar groups is weakened and new structural arrangements can form. A large number of biological systems exhibit this sort of behaviour.

In contrast with ordinary liquids, liquid crystal phases are **anisotropic** owing to the organic nature of the constituent molecules and the long range order. Thermotropic mesophases have found technological applications in various display devices. For this reason considerable effort has gone into extending and controlling the temperature range over which stable liquid crystal phases can be formed.

Figure 1 shows a schematic phase diagram of a liquid crystal system. There are several subclasses of liquid crystal phase according to the detail of the molecular arrangements. Three common ones are illustrated in Fig. 2 where individual molecules have been sketched as short rods to reflect the rod-like nature of thermotropic liquid crystals as typified by cyanobiphenyl compounds.

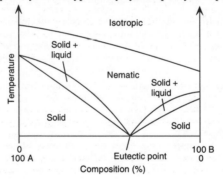

Fig. 1. **Schematic phase diagram of a liquid crystal system.**

continued on next page

---

unhewn timber, or 50 ft^3 of hewn timber.

load cell A load detecting and measuring element utilizing electrical or hydraulic effects which are remotely indicated or recorded.

load-extension curve A curve, plotted from the results obtained in a **tensile test**, showing the relations between the applied load and the extension produced.

loading (1) Adding filler to paper. (2) Increasing the weight of fabrics by use of starch, size, china clay etc. Also *filling*. (3) Addition of fillers to polymers, eg carbon black in rubbers. See **tyre technology** p.

325.

loadstone See **lodestone**.

loan High-quality bond paper, originally and occasionally made from a rag furnish and tub-sized, of durable character intended for documents which resist repeated handling and are required to last.

lock-fit Close-tolerance mechanical joint between two components. Used for joints not intended to be disassembled.

locking force See **injection moulding** p. 169.

locknit A fabric produced by **warpknitting**.

liquid crystal phases (contd)

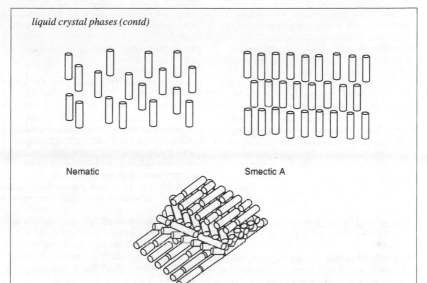

Nematic   Smectic A

Cholesteric

Fig. 2. **Three basic types of thermotropic liquid crystal structure.**

The nematic phase is least ordered. Molecules are lined up roughly parallel in a thread-like fashion. Thermal agitation causes some deviation from this idealized picture and beyond a critical (melting) temperature the isotropic, ordinary liquid phase is stable. The cholesteric (or chiral nematic) phase is locally similar to the nematic but with molecules arranged in layers. There is a progressive twist in the orientation between successive layers. The smectic phases are highly ordered layers of co-aligned molecules. Depending upon the extent of fixed and displaceable charge on the molecules these phases can give rise to anisotropic dielectric and optical properties which are the basis of liquid crystal displays.

Loctite TN for cyanoacrylate monomer adhesives.

loden A coarse-milled, woollen fabric used for jackets and coats because of its weather-resisting properties.

lodestone Iron oxide. A form of **magnetite** exhibiting polarity, behaving, when freely suspended, as a magnet. Also *loadstone*.

logarithmic decrement Logarithm to base e of the ratio of the amplitude of diminishing successive oscillations.

logarithmic strain See **strain**.

logwood The **heartwood** of *Haematoxylon campechianum*, having a distinctive sweet taste and a smell resembling that of violets. It yields a black dye, and also haematoxylin, a stain once much used in microscopy.

London forces Forces arising from the mutual perturbations of the electron clouds of two atoms or molecules, the forces varying as the inverse sixth power of the distance between the molecules. Also *van der Waals forces*.

London plane Same as **European plane**.

lone pair Pair of valency electrons in one orbital unshared with another atom. Such lone pairs are responsible for the formation of co-ordination compounds.

longitudinal heating Dielectric heating in which electrodes apply a high-frequency electric field parallel to lamination. See **glueline**.

longitudinal magnetization That of magnetic recording medium along an axis parallel to the direction of motion.

long-range elasticity Term used for elastomeric behaviour, often to distinguish it from the short-range Hookean elasticity shown by eg metals, ceramics and glasses.

long-range order A substitutional solid solution in which an ordered arrangement of solute extends across large numbers of lattice units or the entire crystal.

long ton A unit of mass, 2240 lb. See **ton**.

look-through Term for the appearance of paper by transmitted light.

loom A machine for weaving cloth, in which two sets of threads, *warp* and *weft*, are interlaced. Conventional models employ shuttles to carry the weft between the warp threads. Later types have small carriers or rapiers, or project the weft by jets of air or water (jet looms). Modern highly engineered models are called *weaving machines*.

looping mill A rolling mill in which the product from one stand is fed into another in the opposite direction, usually above the first.

loop pile Fabric made with loops protruding from the surface of a firm ground. See **moquette**.

loop-raised fabric A warp-knitted fabric in which some of the loops are plucked by wires in a subsequent raising process.

loose piece Part of a foundry pattern which has to be withdrawn from the mould after the main pattern, through the cavity formed by the latter.

Lorentz–Lorenz equation The equation by which *molecular refraction* is defined:

$$[R] = \frac{n^2 - 1}{n^2 + 2} \cdot \frac{M}{\rho},$$

where $[R]$ is the molecular refraction, n is the refractive index, M is the molecular mass, ρ is the density.

Loschmidt number (1) Number of molecules in unit volume of an **ideal gas** at standard temperature and pressure, ie 2.687×10^{25} m^{-3}. (2) Name sometimes given, esp. in continental Europe, to **Avogadro's number**.

Lossev effect The radiation due to the recombination of charge carriers injected into a pin or p-n junction biased in the forward direction.

loss factor Logarithmic decrement divided by π, a measure of the energy dissipation in a material, esp. polymeric. Equal to tan δ. Also *loss tangent*.

loss tangent Ratio of imaginary to real parts of dielectric constant. Also *tan δ*, *dissipation factor*. See **dielectric loss, dielectrics and ferroelectrics** p. 92.

lost wax casting See **investment casting**.

low angle X-ray scattering Method of structure determination for macromolecules esp. biomaterials and polymers. Where long-range order occurs, as in styrene-butadiene-styrene block copolymers, diffraction measurements allow calculation of domain

spacing in 3D array using the **Bragg equation**. Abbrev. *LAXD*. Also *SAXD, SAXS* (S= small).

low-carbon steel Arbitrarily, steel containing from 0.04% to 0.25% of carbon. See **steels** p. 296.

low-density polyethylene Original polymer from ethylene made at high pressure by a free radical mechanism. Has high ratio of both short and long chain branches, so low **degree of crystallinity** and hence low density (ca 920 kg m^{-3}). Mp ca 115°C. Used as insulant in cables etc, packaging, agricultural and building sheet. Abbrev. *LDPE*.

low hysteresis steel Steel with from 2.5 to 4.0% silicon, a high **permeability** and electric resistance, but low loss through **hysteresis**.

low melting point alloys See **fusible alloys, Wood's metal**.

low red heat A temperature between 550° and 700°C at which an object so heated emits radiation which makes it appear dull red.

lox Abbrev. for *Liquid OXygen*.

LPG Abbrev. for **Liquefied Petroleum Gases**.

L-shell The electron shell in an atom corresponding to a principal quantum number of two. The shell can contain up to 8 electrons. See **atomic structure**.

LSI Abbrev. for **Large-Scale Integration** of electronic devices on a single chip.

LT, LTO Abbrev. for *Lithium Tantalate*. See **dielectric and ferroelectric materials** p. 94.

lubricants Compounds (solid, plastic or liquid) entrained between two sliding surfaces. *Wetting* lubricants adhere strongly to one or both such surfaces while leaving the intervening film fairly non-viscous. Solids include graphite, molybdenite, talc. Plastics include fatty acids and soaps, sulphur-treated bitumen, and residues from petroleum distillation. Liquids include oils from animal, vegetable or mineral sources. See **boundary lubrication, fluid lubrication**.

Lucite TN for polymethyl methacrylate (US).

Lüders bands, lines Blemishes which appear as bands across the surface of a low carbon steel during the initial yielding stage of plastic deformation. The bands divide material which has yielded from that which has not and move across the surface while straining is continued, eventually disappearing when all the cross section has yielded and the bulk material is undergoing plastic deformation. See **yield point** p. 339.

lumber The term employed in N America for sawn wood of all descriptions. See **timber**.

luminescence Emission of light (other

than from thermal energy causes) such as *bioluminescence* etc. See **fluorescence**, **phosphorescence**.

luminescent centres Activator atoms, excited by free electrons in a crystal lattice and giving rise to **electroluminescence**.

luminophore (1) A substance which emits light at room temperature. (2) A group of atoms which can make a compound luminescent.

luminous flame One containing glowing particles of solid carbon due to incomplete combustion

luminous paints Paints which glow in the dark. Based on the salts, eg sulphide, silicate, phosphate of, eg zinc, cadmium, calcium, with traces of heavy metals, eg manganese. Phosphorescent paints glow for longer or shorter periods after exposure to light. Fluorescent paints glow continuously under the action of radioactive additives.

lutetium A metallic element, one of the **rare earth elements**. Symbol Lu. See Appendix 1, 2.

Lycra TN for first **spandex** fibre to be introduced into the market (US).

lye Strong solution of sodium or potassium hydroxide.

lyolysis The formation of an acid and a base from a salt by interaction with the solvent; ie the chemical reaction which opposes neutralization.

lyophilic colloid A colloid (lyophilic, *solvent-loving*) which is readily dispersed in a suitable medium, and may be redispersed after coagulation.

lyophobic colloid A colloid (lyophobic, *solvent-hating*) which is dispersed only with difficulty, yielding an unstable solution which cannot be reformed after coagulation.

lyosorption The adsorption of a liquid on a solid surface, esp. of solvent on suspended particles.

lyotropic series Ions, radicals, or salts arranged in order of magnitude of their influence on various colloidal, physiological and catalytic phenomena, an influence exerted by them as a result of the interaction of ions with the solvent. Cf **Hofmeister series**.

M

m- Abbrev. for: (1) meta-, ie containing a benzene nucleus substituted in the 1,3 positions; (2) **meso-** (1).

m Abbrev. for **milli-**.

μ Symbol for: (1) chemical potential; (2) dipole moment; (3) micron (obsolete); (4) **mobility**. Abbrev. for **micro-**, ie 10^{-6}.

μ- A symbol signifying, (1) **meso-** (2) a bridging ligand.

M A general symbol for a metal or an electropositive radical. Abbrev. for mega-, ie 10^6.

M Symbol for: (1) **relative molecular mass**; (2) **bending moment**.

machinability The ease with which a material can be machined.

machine direction The direction in the plane of a sheet or web of paper corresponding with the direction of travel of the paper machine. The direction at right angles to this is the cross direction.

machine finished Paper on which the requisite degree of surface finish has been attained by use of a **calender** or calenders forming part of the papermaking machine.

machine glazed Paper or board with a characteristic high finish on one side, produced by causing the damp web to adhere to the surface of a large diameter, highly polished, steam-heated drying cylinder (machine glazer or Yankee dryer). As water is evaporated the paper surface assumes the polish of the cylinder. Abbrev. *MG*.

machine wire In papermaking, originally a woven mesh of phosphor bronze for use on a **fourdrinier** or cylinder mould machine for forming the sheet of paper. Stainless steel is now used for machine wires and the term is retained for plastics cloths serving the same purpose.

machining allowance The material provided beyond the finished contours on a casting, forging, or roughly prepared component, which is subsequently removed in machining to size.

macro-axis The long axis in orthorhombic and triclinic crystals.

macr-, macro- Prefixes from Gk *makros*, large.

macro-defect-free cement Portland cement of much enhanced tensile strength, made by the addition of ca 5% polyvinyl alcohol followed by shear-mixing and compressing to eliminate pores over 15 μm in size. Abbrev. *MDF cement*.

macromolecule Term applied to a very large molecule such as haemoglobin (containing about 10 000 atoms) and often to a polymer of high molecular mass.

macroscopic Visible to the naked eye.

macroscopic state One described in terms of the overall statistical behaviour of the discrete elements from which it is formed. Cf **microscopic state**.

macrosection A section of material, mounted, cut, polished and etched as necessary to exhibit the **macrostructure**.

macrostructure Specifically, the structure of a metal or other material as seen by the naked eye or at low magnification on a ground or polished surface or on one which has been subsequently etched. Generally, the structure of any body visible to the naked eye. Cf **microstructure**.

Madelung's constant A constant representing the sum of the mutual potential coulombic attraction energy of all the ions in a lattice, in the equation for the lattice energy of an ionic crystal. Varies with lattice type.

magnesia See **magnesium oxide**.

magnesia alba Commercial basic magnesium carbonate.

magnesite Basic refractory used in open-hearth and other high-temperature furnaces; it is resistant to attack by basic slag. It is obtained from natural deposits (mostly magnesium carbonate, $MgCO_3$) which is **calcined** at high temperature to drive off moisture and carbon dioxide, before being used as a refractory.

magnesium A light metallic element in the second group of the periodic table. Symbol Mg. Electrical resistivity 42×10^{-8} Ω m. The metal is a brilliant white in colour, and magnesium ribbon burns in air, giving an intense white light, rich in ultraviolet rays. It is used as a deoxidizer for copper, brass and nickel alloys. A basis metal in strong light alloys which are used in aircraft and automobile construction and for reciprocating parts. Alloys with zirconium and thorium are used in aircraft construction. See Appendix 1,2.

magnesium carbonate $MgCO_3$. See **magnesite**.

magnesium oxide MgO. Obtained by igniting the metal in air. In the form of calcined **magnesite** and **dolomite**, it is used as a refractory material.

magnet A mass of iron or other material which possesses the property of attracting or repelling other masses of iron, and which also exerts a force on a current-carrying conductor placed in its vicinity. See **ferromagnetics and ferrimagnetics** p. 128, **characterizing magnetic materials** p. 57.

magnet core The iron core within the coil

of an electromagnet.

magnetic Said of all phenomena depending on magnetism, and also of materials which are **ferromagnetic** or **ferrimagnetic**. See **diamagnetism**, **paramagnetism**.

magnetic alloys Generally, any alloy exhibiting **ferromagnetism**, eg and most importantly, **silicon iron**, but also iron-nickel alloys, which may contain small amounts of any of a number of other elements (eg copper, chromium, molybdenum, vanadium etc), and iron-cobalt alloys.

magnetic annealing Heat treatment of magnetic alloy in a magnetic field, used to increase its **permeability**.

magnetic bias A steady magnetic field added to the signal field in magnetic recording, to improve linearity of relationship between applied field and magnetic *remanence* in recording medium.

magnetic bubble See **magnetic bubble memory**.

magnetic bubble memory Potentially cheap and reliable **chip** memory based on mixed ferrimagnetic **garnets**. Thin films of these materials can be prepared so that magnetization lies perpendicular to the plane of the film. Stable domains of reverse magnetization (bubbles) can be manipulated by a surface magnetic structure. The presence or absence of a magnetic bubble in a localized region of the film designates 0 or 1. Very compact, robust, high-capacity memories can be realized by this method.

magnetic circuit Complete path, perhaps divided, for magnetic flux, excited by a permanent magnet or electromagnet. The range of **reluctance** is not so great as resistance in conductors and insulators, so that leakage of the magnetic flux into adjacent non-magnetic material, esp. air, is significant.

magnetic domain Aggregated ferromagnetic group of atoms, usually well below one micrometre in diameter, forming part of a system of such groups. See **domain**, **ferromagnetics and ferrimagnetics** p. 128.

magnetic elongation The slight increase in length of a wire of magnetic material when it is magnetized. See **magnetostriction**.

magnetic energy See **characterizing magnetic materials** p. 57.

magnetic ferrites See **ferrite**.

magnetic field intensity Also *magnetizing field strength*, *magnetic intensity*, *magnetizing force*. See **characterizing magnetic materials** p. 57.

magnetic flux The surface integral of the product of the **permeability** of the medium and the magnetic field intensity normal to the surface. The magnetic flux is conceived, for theoretical purposes, as starting from a positive fictitious north pole and ending on a fictitious south pole, without loss. When associated with electric currents, a complete circuit, the magnetic circuit, is envisaged, the quantity of magnetic flux being sustained by a magneto-motive force, m.m.f. (coexistent with ampere-turns linked with the said circuit). Permanent magnetism is explained similarly in terms of molecular m.m.f.s (associated with orbiting electrons) acting in the medium. Measured in maxwells (CGS) or webers (SI).

magnetic flux density The basic magnetic quantity which accounts for the magnetization of a medium and the effects of any external magnetizing fields. See **characterizing magnetic materials** p. 57.

magnetic forming A fast, accurate production process for swaging, expanding, embossing, blanking etc, in which a permanent or expendable coil is moved by electromagnetism to act on the workpiece.

magnetic hysteresis See **hysteresis loop**.

magnetic hysteresis loss The energy expended in taking a piece of magnetic material through a complete cycle of magnetization. The magnitude of the loss per cycle is proportional to the area of the magnetic hysteresis loop. See **characterizing magnetic materials** p. 57.

magnetic induction Induced magnetization in a magnetic material, either by saturation, by coil-excitation in a magnetic circuit, or by the method of stroking with another magnet.

magnetic intensity See **magnetic field intensity**.

magnetic iron ore See **magnetite**.

magnetic moment Vector such that its product with the magnetic flux density gives the torque on a magnet in a homogeneous magnetic field. Also *moment of a magnet*.

magnetic oxides The ferromagnetic iron oxides. See **ferrites**.

magnetic particle inspection A rapid, non-destructive test for fatigue cracks and other surface and subsurface defects in steel and other magnetic materials, in which the workpiece is magnetized so that local flux-leakage fields are formed at cracks or other discontinuities. The position of these fields is shown by dusting the workpiece with a magnetic powder which may be coloured red or black with the workpiece painted white to facilitate detection. Abbrev. *MPI*.

magnetic polarization The production of optical activity by placing an inactive substance in a magnetic field and the extent of magnetization of the material. See **characterizing magnetic materials** p. 57.

magnetic rotation See **Faraday effect**.

magnetic saturation The limiting value of the magnetic induction in a medium when its **magnetization** is complete.

magnetic shield Skin of high permeability magnetic material (eg **Mumetal**) which screens one side from the effect of a magnetic field on the other side.

magnetic susceptibility See **characterizing magnetic materials** p. 57.

magnetite An oxide of iron, crystallizing in the cubic system. It has the power of being attracted by a magnet, but it has no power to attract particles of iron to itself, except in the form of lodestone. Also *magnetic iron ore*.

magnetization See **characterizing magnetic materials** p. 57.

magnetization curve See *M(H)* curve.

magnetizing roast Reduction of weakly magnetic iron ore or concentrate by heating in a reducing atmosphere to convert it to a more strongly ferromagnetic compound.

magneto-acoustic effect See **Barkhausen effect**.

magneto-caloric effect The reversible heating and cooling of a medium when the magnetization is changed. Also *thermomagnetic effect*.

magnetochemistry The study of the relation of magnetic properties to chemical structure. Particularly, the extent of paramagnetism in transition metal compounds can be related to the type of ligand bonded to the metal.

magnetocrystalline anisotropy The tendency for magnetic moments (and therefore **domains**) to align preferentially with certain crystal axes. This is quantified in terms of a set of anisotropy constants (in $J\,m^{-3}$) for a given material.

magnetoelectric Of certain materials, eg chromium oxide, the property of becoming magnetized when placed in an electric field. Conversely, they are electrically polarized when placed in a magnetic field. May be used for measuring pulse electric or magnetic fields.

magnetomotive force Line integral of the magnetic field intensity round a closed path. Abbrev. *m.m.f.*

magneton See **Bohr magneton**.

magneto-optical effects See **Kerr effects** (3), **Faraday effect**.

magneto-optic rotation See **magnetic polarization**.

magnetoresistance The resistivity of a magnetic material in a magnetic field which depends on the direction of the current with reference to the field. If parallel to one another, the resistivity increases, but if mutually perpendicular, it decreases.

magnetostriction Phenomenon of elastic deformation of certain ferromagnetic materials, eg nickel, on the application of magnetizing force. Used in ultrasonic transducers and, formerly, as basis of memory in computers.

magnetron (1) A two-electrode valve in which the flow of electrons from a large central cathode to a cylindrical anode is controlled by crossed electric and magnetic fields; the electrons gyrate in the axial magnetic field, their energy being collected in a series of slot resonators in the face of the anode. Magnetrons, used mainly as oscillators, can produce pulsed output power at microwave frequencies, with high peak power ratings. Used in microwave and radar transmitters and microwave cookers. (2) A low-pressure gas discharge in which magnetic fields (usually from permanent magnets) near one or more electrodes aid the confinement of electrons. Such discharges are often used for high rate sputtering.

magnet steel A steel from which permanent magnets are made. It must have a high remanence and coercive force. Steels for this purpose may contain considerable percentages of cobalt (up to 35%) as well as nickel, aluminium, copper etc. See **permanent magnet**.

magnet yoke Sometimes applied to the whole of the magnetic circuit of an electromagnet (or transformer etc) but strictly speaking, should refer only to the part which does not carry the windings.

magnolia metal A lead-base alloy, containing 78–84% lead; remainder is mainly antimony, but small amounts of iron and tin are present. Used for bearings.

magnon Quantum of **spin wave** energy in magnetic material.

Magnox Group of magnesium alloys for canning uranium reactor fuel elements. Best known are Magnox B and Magnox A 12. The latter is Mg with 0.8% Al and 0.01% Be.

mahogany Hardwoods from the family *Meliaceae*. See **African mahogany**, **American mahogany**. Also *Brazilian-*, *Cuban-*, *Uganda-mahogany*.

Maillifer screw Type of multiple extrusion screw to aid the mixing of polymers with additives.

majority carrier In a semiconductor, the electrons or *holes*, whichever carry most of the measured current. Cf **minority carrier**.

makoré W African hardwood (*Tieghemella*) whose **heartwood** is pinkish-red to red-brown, straight-grained, with a uniform, fine texture, and can have a **moiré** lustre. It is very durable, but liable to **blue stain**. Used in furniture and cabinetmaking, joinery, marine plywood and veneers. Density about $620\,kg\,m^{-3}$.

Makrolon TN for polycarbonate (Ger).

malachite green A triphenylmethane dyestuff of the rosaniline group. Obtained by the condensation of benzaldehyde with dimethylaminobenzene in the presence of $ZnCl_2$, HCl, or H_2SO_4, and by oxidation of the resulting **leuco-base** with PbO_2.

maleic acid *Cis-butenedioic acid.* HOOC.CH:CH.COOH, mp 130°C, with decomposition into its anhydride and water, readily soluble in water. It has the cis-configuration, whereas its isomer, fumaric acid, has the trans-configuration. Important in polyester resins.

male tool Part of a tool with a core which mates with female part to form cavity.

malleability Property of metals and alloys which affects their ease of plastic deformation on hammering, rolling, extrusion.

malleable cast iron A variety of cast-iron which is cast white, and then annealed at about 850°C to remove carbon (*white-heart process*) or to convert the cementite to rosettes of graphite (*black-heart process*). Distinguished from grey and white cast-iron by exhibiting some elongation and reduction in area in tensile test though not so much as in steels or spheroidal graphite cast irons. See **pearlitic iron**.

malleable iron Now usually means **malleable cast iron**, but the term is sometimes applied to **wrought iron**.

malleable nickel Nickel obtained by remelting and deoxidizing electrolytic nickel and casting into ingot moulds. Can be rolled into sheet and used in equipment for handling food, for coinage, condensers and other purposes where resistance to corrosion, particularly by organic acids, is required.

malthenes Such constituents of asphaltic **bitumen** as are soluble in carbon disulphide and petroleum spirit. See **asphaltenes, carbenes**.

mandrel (1) Accurately turned rod over which metal is forged, drawn or shaped during working so as to create or preserve desired axial cavity. A tapered mandrel is also used for holding and locating a bored component so that external diameters can be machined true to the bore. Also *arbor* (US), *mandril*. (2) Internal metal part of pipeforming die in polymer **extrusion**. Held in position by spider screws.

manganates See **permanganates**.

manganese A hard, brittle metallic element, in the seventh group of the periodic table, which exists in four polymorphic forms, α, β, γ and δ, the first two of which have complicated crystal structures. It is brilliant white in colour, with reddish tinge. Symbol Mn. Electrical resistivity 5.0×10^{-8} Ω m. Pure manganese is obtained electrolytically. Manganese is mainly used in steel manufacture, as a deoxidizing and desulphurizing agent. See Appendix 1, 2.

manganese alloys Manganese is not used as the basis of alloys, but is a common constituent in those based on other metals. It is present in all steel and cast-iron, and in larger amount in special varieties of these, eg **manganese steel, silico-manganese steel**; also in many varieties of brass, in aluminium-bronze, and aluminium and nickel base alloys.

manganese bronze Originally an alpha-beta brass containing about 1% of manganese; the term is now applied generally to high-strength brass, with up to 4% manganese. See **copper alloys** p. 76.

manganese dioxide *Manganese (IV) oxide*, MnO_2. Black solid, insoluble in water; occurs in pyrolusite. Basic and (mainly) acidic. Forms manganites. Oxidizing agent; uses: depolarizer in **Leclanché cells**, decolorizing oxidant for green ferrous ion in glass, laboratory reagent.

manganese steel A term sometimes applied to any steel containing more manganese than is usually present in carbon steel (ie 0.3–0.8%), but generally to austenitic manganese steel (known as *Hadfield's steel*) which contains 11–14%. This steel is resistant to shock and wear. Used for railway crossings, rock crusher parts etc. See **steels** p. 296.

manganin A copper-base alloy, containing 13–18% of manganese and 1.5–4% of nickel. The electrical resistivity is high (about 38×10^{-8} Ω m), and its temperature coefficient of resistivity low. It is therefore suitable for resistors. See **copper alloys** p. 76.

manifold bank A thin bank paper, generally around 30 g m^{-2}.

manila A rough bast fibre obtained from the leaves of *Musa textilis*, a banana like plant grown in the Philippines; used in sacking, ropemaking etc.

manilla paper Paper generally made from unbleached chemical woodpulp, with admixtures of waste paper and/or mechanical woodpulp and buff/brown in colour (unless dyed). Machine glazed varieties are generally intended for conversion into envelopes. Solid or pasted varieties are generally used for files or folders.

manio S American softwood (*Podocarpus chilensis*) similar and related to the **podo**.

'man-made' fibre A regenerated or synthetic fibre that has been manufactured, as distinct from a natural fibre. Man-made fibres are obtained as continuous filaments which may be cut or broken into staple fibres.

Mannesmann process A process for making seamless metal tubing from a solid bar of metal by the action of two eccentrically mounted rolls which simultaneously rotate the bar and force it over a mandrel. Also *Mannesmann piercing*.

maple Various species of the hardwood genus *Acer*, found in Europe, Asia and America, used for cabinet work, flooring, musical instruments, panelling etc and in veneers and plywood. They have in common a creamy-white to light tan colour, generally straight grain and a fine texture. They are non-durable. Density range 550–720 kg m^{-3}. Also *European-, Japanese-, rock-, soft-maple*.

maraging Heat treatment (*martensite ageing*) used to harden alloy steels (commonly those containing 18% nickel), involving precipitation of intermetallic compounds in a carbon-free martensite. These include nickel-iron martensites with high toughness and resistance to shock and saline corrosion.

Maranyl TN for nylon moulding material (UK).

marblewood A straight-grained, fine and evenly textured hardwood with figuring of dark black bands on a grey-brown background, from *Diaspyros kurzii*, a native of the Andaman and Nicobar Islands.

marcella Cloth, often cotton, used for the fronts of dress shirts, and with a **piqué** structure and superimposed fancy woven design.

marching modulus Term describing type of overcure of rubber where crosslinking continues at a slow rate after end of cure cycle.

Marform process A proprietary production process for forming and deep-drawing irregularly shaped sheet metal parts, in which a confined rubber pad is used on the movable platen of a press and a rigid punch on the fixed platen. The unformed portion of the blank is subjected to sufficient pressure during the process to prevent wrinkling.

Mark–Houwink equation Basis for calculation of viscosity-average molecular mass of polymers from intrinsic viscosity measurements. The equation is:

$$\eta = K M^a,$$

where K and a are experimentally determined constants.

marl yarn Folded yarn made from single yarns of two or more colours. May be continuous filament or woollen spun single yarn.

Marlex TN for high-density polyethylene (UK).

martempering Austenitizing a steel prior to hardening and then quenching to just above the M_s temperature (see **isothermal transformation diagram** p. 177) until the temperature has equalized throughout the section but not long enough for bainitic

transformation to begin, and then quenching to form martensite. Applied to steel components which are likely to crack as a result of thermal gradients if quenched directly from the austenitizing temperature. See **austempering**.

martensite Non-equilibrium microstructure formed in steel when the austenite phase is cooled too rapidly for carbon to diffuse out of solid solution to form Fe_3C and thus transform to pearlite. The entrapped carbon distorts the lattice and retards the shear transformation from **face-centred cubic** to **body-centred cubic**, causing the product of the transformation to be a tetragonal (**body-centred tetragonal**, BCT) lattice. This has a high dislocation density associated with it, and the BCT lattice lacks the five independent slip systems necessary for ductility. It becomes harder and less ductile the greater the carbon content. See **martensitic transformation**.

martensitic transformation Type of rapid transition from one crystal structure to another by shear (ie displacement) rather than the more common, and much slower, diffusion, nucleation and growth. Many materials show this transformation – it forms the basis of **shape-memory alloys** – but only in steels is the product much harder and less ductile than the untransformed material. Because of its importance, the name martensite usually refers to that in steel. See **steels** p. 296.

marver A flat cast-iron or stone (marble) block upon which glass is rolled during the hand method of working.

mash seam welding An electrical resistance welding process in which the slightly overlapping edges of the workpiece are forged together during welding by broad-faced, flat electrodes.

mask In semiconductor technology, a patterned metal film on a glass substrate containing selected open areas. During integrated circuit manufacture a series of different masks and lithographic processes is used to build up a complex circuit structure. See **semiconductor device processing** p. 280.

masking paper, tape A paper, generally semibleached and crêped, coated on one side with a self-adhesive compound. Usually sold in coil form and may be peeled from its support after use.

mass action See **law of mass action**.

mass effect The tendency for hardened steel to decrease in hardness from the surface to the centre, as a result of the variation in cooling throughout the section. Becomes less marked as the rate of cooling required for hardening decreases, ie as the

hardenability of the steel imparted by the content of alloying elements increases. See **continuous cooling transformation diagram** p. 72.

mass manufacture Method of producing discrete products in a continuous way using hand, automatic or robotic assembly methods.

mass resistivity The product of the volume resistivity and the density of a given material at a given temperature.

mass spectrograph A vacuum system in which positive rays of various charged atoms are deflected through electric and magnetic fields so as to indicate, in order, the charge-to-mass ratios on a photographic plate, thus measuring the atomic masses of isotopes with precision. System used for the first separation for analysis of the isotopes of uranium.

mass spectrometer A **mass spectograph** in which the charged particles are detected electrically instead of photographically.

master alloy An alloy enriched in the required components which can be added to a melt to bring it to the correct composition.

masterbatch Concentrate of polymer and additives supplied direct to moulders for addition to hopper. Eliminates cost of intermediate mixing.

master curve See **viscoelasticity** p. 332.

mastic (1) A pale-yellow resin from the bark of *Pistacia lentiscus*, used in the preparation of fine varnishes. (2) Now more widely applied to permanently waterproof but flexible seal in building applications. Usually a synthetic polymer eg silicone.

mastication Method used in rubber technology to increase processability of raw rubber stock, often on a calender mill. Natural rubber stock for example, has too high a molecular mass for mixing, so must be milled to break down chains mechanically.

masticator An apparatus consisting of two revolving and heated cylinders studded with teeth or knives; used for converting rubber into a homogeneous mass.

matchwood Wooden billets suitable for manufacturing into matches.

matelassé Woven fabric with a quilted effect made using two warps and two wefts, and used for formal dress wear.

materials database Collection of materials information (physical and chemical properties of available grades) usually kept in computerized form. The largest polymer manufacturers, for example, supply such databases for their own materials on request (eg *EPOS, CAMPUS*). Most materials selection systems have their own databases.

materials identification Various analytical methods applied to the problem of identifying unknown materials. Some methods are of general applicability, eg Energy Dispersive Analysis of X-rays (EDAX) combined with scanning electron microscopy, optical microscopy and thermal analysis, while others are material specific (eg gel permeation chromatography for polymers) or more useful when applied to a specific material (eg infrared and ultraviolet spectroscopy for polymers). In its widest sense, identification will include determining the elements and compounds present, their composition and phase, as well as such microstructural features as grain structure (for metals and ceramics) or spherulite structure (for polymers). See **trace element analysis**.

materials matching Method of materials analysis where unknown sample is identified by comparison with standard samples available to analyst. More comprehensive materials identification methods will be needed if standards not available.

materials science Study of chemical and physical properties of elements, compounds, mixtures (blends and alloys) and minerals for understanding of atomic and/or molecular causes.

materials selection Activity involving matching materials available to product specification, particularly in terms of mechanical properties (eg elastic modulus and strength), thermal properties (eg melting and glass transitions, thermal conductivity), chemical properties (eg corrosion resistance and environmental stress cracking) and electrical properties (eg electrical conductivity and tracking resistance). It also includes processing properties and ease of manufacture, as well as costing. Methods include merit index analysis and those based on value judgements. Some methods are computerized and cover metals, ceramics and polymers (eg PERITUS), others like PLASCAMS, only apply to polymers. Problems of application include neglect of difficult-to-quantify properties like appeal and marketability, the need to update data regularly, and the mismatch between standard material properties and actual properties of real products.

materials substitution Replacement of one material in a product by another, such as wood by glasss-reinforced plastic, or the latter by thermoplastic in small boats. Similarly, high-tensile steels for mild steel sheet in car bodies. See **car body materials** p. 57. Such replacement often demands substantial design changes to allow for different material properties. Motives include cost savings, rise in productivity, better properties or safety, eg polyethylene terephthalate for glass in beverage bottles.

materials technology Application of materials science to the development and practical use of conventional or new materials, esp. for manufactured products (eg metal-matrix composites, high-temperature polymers, mixed oxide superconductors). See **high-performance fibres** p. 158.

matrix (1) The component, usually continuous, of a composite material in which the fibres or filler materials are embedded, and which variously transfers stresses to them, prevents fibres from buckling and protects their surfaces. (2) The phase or phases which form the continuous skeleton of a powder body, thus forming the cells in which constituents imparting particular qualities may be held.

Matthiessen hypothesis, rule That the total electrical resistivity of a metal may be equated to the sum of the various resistivities due to the different sources of scattering of free electrons; this also applies to thermal resistivity.

matt, matte Smooth but dull; tending to diffuse light; said, eg of a surface painted or finished so as to be dull or flat.

mauveine Perkin's mauve, the first synthetic dyestuff.

maxillofacial implant Type of implant composed of silicone or polyurethane rubber and polytetrafluorethylene for soft facial tissue (such as gum, skin etc) or polymethyl methacrylate, vitallium, composite materials for harder tissue.

maximum tensile stress See **ultimate tensile stress**.

maxwell The CGS unit of magnetic flux, the MKSA (or SI) unit being the *weber*. One maxwell = 10^{-8} weber.

Maxwell–Boltzmann distribution law A law expressing the energy distribution among the molecules of a gas in thermal equilibrium.

Maxwell model Conceptual model for a viscoelastic material consisting of a Hookean spring and a dashpot which contains a Newtonian fluid arranged in series. Inadequate for modelling real materials since responses are linear rather than non-linear, as in real systems. See **Voigt model**.

Maxwell's demon Imaginary creature who, by opening and shutting a tiny door between two volumes of gases, could in principle concentrate slower (ie colder) molecules in one and faster (ie hotter) molecules in the other, thus reversing the normal tendency toward increased disorder or entropy and breaking the second law of thermodynamics.

Maxwell's field equations Mathematical formulations of the laws of Gauss, Faraday and Ampère from which the theory of electromagnetic waves can be conveniently derived.

MBE Abbrev. for *Molecular Beam Epitaxy*. See **epitaxial growth of semiconductors** p. 117.

MBS Abbrev. for **Methacrylate Butadiene Styrene**.

McLeod gauge Vacuum pressure gauge in which a sample of low pressure gas is compressed in a known ratio until its pressure can be measured reliably. Used for calibrating direct-reading gauges.

McQuaid–Ehn test A method of revealing grain size by carburizing a ferrous alloy in the austenitic temperature range. Grain sizes are classified from one (the finest) to eight.

MDI Abbrev. for **diphenyl-Methane DiIsocyanate**.

MDPE Abbrev. for **Medium-Density Poly-Ethylene**.

Me A symbol for the methyl radical $-CH_3$.

mean free path The mean distance travelled by a particle between collisions (eg electrons in a solid). It is dependent on the density and cross-section of scattering centres.

mean free time Average time between collisions eg of electrons with impurity atoms in semiconductors; also of intermolecular collision of gas molecules.

mean life See **life-time**.

mean stress The midpoint of a range of stress. When it is zero, the upper and lower limits of the range have the same value but are in tension and compression respectively.

mechanical alloying Producing an intimate dispersion of elements by milling together fine powders of the constituents in the desired proportions.

mechanical equivalent of heat Originally conceived as a conversion coefficient between mechanical work and heat (4.186 joules = 1 calorie) thereby denying the identity of the concepts. Now recognized simply as the **specific heat capacity** of water, 4.186 kJ kg^{-1}K^{-1}.

mechanical paper Paper composed substantially of **mechanical woodpulp**.

mechanical plating Forming a metallic layer on a surface by a hammering or tumbling process. See **sherardizing**.

mechanical woodpulp Pulp produced from wood entirely, or almost entirely, by mechanical means eg by grinding logs (*groundwood*) or by passing wood chips through a refiner (*refiner mechanical pulp*).

mechanical working Changing the shape of a metal or material by processes applying mechanical forces, eg rolling, forging, extrusion, spinning, pressing etc.

mechanics Study of the forces acting on bodies, whether moving (dynamics) or at rest (statics).

mechano-chemical engine Device designed to turn chemical energy directly into mechanical motion, as in muscle. Laboratory curiosities built from polyelectrolyte fibres can function in this way by cycling the **pH** of the solution enclosing them.

medi-, medio- Prefixes from L. *medius*, middle.

medium See **paper sizes**.

medium-density polyethylene Grade of polymer developed mainly for gas pipes, with enhanced resistance to **environmental stress cracking**. Intermediate density of ca 940 kg m^{-3} and T_m about 130°C. Abbrev. *MDPE*.

medulla Hollow cellular central portion of some animal fibres which is surrounded by the cortex.

Meehanite High-silicon cast-iron produced by **inoculation**.

mega- Prefix denoting one million, or 10^6, eg a frequency of one *megahertz* is equal to 10^6 Hz; *megawatt* = 10^6 watts; *megavolt* = 10^6 volts.

Meissner effect See **superconductors** p. 308.

MEK Abbrev. for **Methyl Ethyl Ketone** (butan-2-one).

melamine resins Synthetic resins derived from the reaction of melamine with formaldehyde or its polymers. See **thermosets** p. 318.

Melanex Thin plastic foil used as light shield over scintillation detectors for α-particles.

melded fabric Fabric made from bicomponent fibres in which the outer component melts at a lower temperature than the core. The cohesion is obtained by subjecting the pressed fibre mass to a temperature at which partial fusion occurs.

meld line Feature on moulded polymer surface where **flow lines** have visibly met, without forming a **weld line**.

Melinex TN for polyethylene terephthalate (PET) in film form. Very strong film with extremely good transparency and electric properties. Similar to *Mylar*.

melt elasticity See **melt memory**.

melt flow index Measure of melt viscosity of polyolefins, esp. polyethylene and polypropylene. Involves extrusion of polymer from standard orifice at temperature of 190°C (for polyethylenes). Melt flow index is the weight extruded in 10 mins, and is inversely related to molecular mass. (BS 2782). Abbrev. *MFI*.

melt fracture Break-up of extrudate at high shear rate. See **extrusion** p. 121.

melting point, temperature Symbol T_m, The temperature at which a solid begins to

liquefy. Pure crystalline materials, eutectics, and some intermediate constituents melt at a constant temperature; alloys generally melt over a range. Glasses, whether inorganic, metallic or polymeric, have no well-defined transition from solid to liquid. Abbrev. *mp*.

melt memory Tendency of a polymer to 'remember' its former state, esp. during processing. **Die swell**, for example, is caused by the extrudate reverting to its original state in the barrel prior to extrusion through the die. See **extrusion** p. 121. Also *melt elasticity*.

melton A strong heavily milled and cropped fabric used for making overcoats. Made from pure wool or cotton warp and woollen weft and having a felt-like appearance. Cf **kersey**.

melt spinning The formation of continuous filaments by extrusion of molten polymer. A subsequent drawing process is necessary to develop the desired mechanical properties, including strength, in the filaments.

melt transport Analysis of movement of molten or melting polymer along injection moulding or extruding machine. Involves analysis of drag, pressure and leak flows in barrel, and depends on screw characteristics, channel depth, melt viscosity, pressure gradient and screw speed. See **extrusion**.

melt viscosity Viscosity of molten polymer of specific importance for injection moulding. Strongly dependent on polymer grade, temperature and shear rate.

membrane filter A thin layer filter made by fusing cellulose ester fibres or by β-bombardment of thin plastic sheets, so that they are perforated by tiny uniform channels. Also *molecular filter*.

memory Term referring to tendency of polymers, esp. melts and solutions, to 'remember' their previous process history. Manifested in die swell etc. See **melt memory, shape memory alloys**.

Mendeleev's table See **periodic table**, Appendix 1.

meranti A SE Asian hardwood, from the genus *Shorea*. Four varieties are recognized, dark red-, light red-, yellow- and white-. They are moderately durable and used for joinery, furniture, flooring, etc and in veneers and plywood.

mercerization A process which greatly increases the lustre of cotton yarns and fabrics. It consists of treating the material, which can be under tension, with cold, aqueous sodium hydroxide (caustic soda) which causes swelling and results in increased strength and dye absorption properties. Invented by John Mercer in 1844.

merchant iron Term (now obsolete) for bar-iron made by repiling and re-rolling puddled bar. All wrought iron used to be treated in this way before being used for manufacture of chains, hooks etc.

mercurous (mercury (I)) chloride See **calomel**.

mercury A metallic element which is liquid at atmospheric temperature. Chemical symbol Hg. Electrical resistivity 95.8×10^{-8} Ω m. A solvent for most metals, the products being called *amalgams*. Its chief uses are in the manufacture of batteries, drugs, chemicals, fulminate and vermilion. Used as metal in mercury-vapour lamps, arc rectifiers, power-control switches, and in many scientific and electrical instruments. Also *quicksilver*. ^{198}Hg is a mercury isotope made from gold in a reactor, for use in a quartz mercury-arc tube, light from which has an exceptionally sharp green line, because of even mass number of a single isotope. This considerably improves comparisons of end-gauges in interferometers. See Appendix 1, 2.

mercury cell (1) Electrolytic cell with mercury cathode. (2) Dry cell employing mercury electrode, emf ca 1.3 volts; eg Mallory battery, Reuben–Mallory cell.

mercury intrusion method Finding the distribution of sizes of capillary pores in a body by forcing in mercury, the radius being found from the pressure, and the percentage from the volume of mercury absorbed at each pressure.

mercury seal A device which ensures that the place of entry of a stirrer into a piece of apparatus is gas-tight, while allowing the free rotation of the stirrer.

merino wool Wool of fine quality from merino sheep or similar. The name is also used in the woollen trade for wool fibre recovered from fine woollen and worsted clothing rags.

merit index Criterion for materials selection eg in terms of the best mechanical properties for least weight of material. Relevant for design of vehicles, aerospace products and devices fitted to or by hand (eg tools, building products). Products put in tension suggest a merit index of E/ρ, so for rope, a table of comparative indices can be made. The best materials with high merit indices can thus be filtered and further criteria applied (eg cost). See **high-performance fibres** p. 158.

mersawa and krabak SE Asian timber from a variety of related **hardwood** species (*Anisoptera*). **Heartwood** is a pink-tinged, yellowish brown, with a lightly interlocked grain and coarse but even texture. Moderately hard and durable; used in general construction, joinery, flooring, plywood, etc. Mean density 640 kg m^{-3}.

mes- Prefix from Gk *mesos*, middle.

mesa transistor A transistor using the *mesa method* of construction; the semiconductor wafer is etched down in steps so that the base and emitter regions appear as plateaux above the collector. Connections are terminated at the edges of the material, instead of coming to the surface, as in **planar process**.

mesh *Expanded metal* or *plastic* (eg **Tensar**) used as a reinforcement for concrete, asphalt, clay and many other building materials.

meso- (1) Optically inactive by intramolecular compensation. Abbrev. *m-*. (2) Substituted on a carbon atom situated between two hetero-atoms in a ring. (3) Substituted on a carbon atom forming part of an intramolecular bridge. From Gk *mesos*, middle.

mesocolloid A particle whose dimensions are 25–250 nm containing 100–1000 molecules.

mesomerism See **resonance**.

mesomorphous Existing in a state of aggregation midway between the true crystalline state and the completely irregular amorphous state. See **liquid crystals.**

meta- (1) Derived from an acid anhydride by combination with one molecule of water. (2) The 1,3 relationship of substituents on a benzene ring, eg *m*-cresol.

meta- Prefix from Gk *meta*, after.

metal Any material whose **Fermi surface** lies predominantly in its conduction band. More particularly an element which is held together by metallic bonds and shows characteristic properties, which include high reflectivities and electrical and thermal conductivities and relatively high density compared with non-metals. Most elements of the periodic table (see Appendix 1, 2) are metals. Further subdivided into alkali metals, alkaline earth metals, transition metals, actinides, lanthanides etc. See **bonding** p. 33.

metal-arc welding A type of electric welding in which the electrodes are of metal, and melt during the welding process to form filler metal for the weld.

Metalastik TN for type of **inclined shear mount**.

metaldehyde *Meta-aldehyde*, $(CH_3CHO)_4$. Long glistening needles which sublime at 115°C with partial decomposition into ethanal. Acetaldehyde is polymerized to metaldehyde by the action of acids at temperatures below 0°C. Sometimes used as a portable fuel, meta-fuel.

metal detector An instrument, widely used in industrial production, for detecting the presence of embedded stray metal parts in eg food products. It is usually incorporated in a conveyor line and gives visible or audible warning or automatically stops the line. Also used for detecting buried metal.

metal electrode A form of electrode used

in **metal-arc welding**.

metal-filament lamp A filament lamp in which the light is produced by raising a fine wire or filament of tungsten to white heat.

metal inert-gas welding Arc welding with a metal electrode shielded by an inert gas such as argon or carbon dioxide. Abbrev. *MIG*.

metallic bond Strong force of attraction between metal atoms. See **bonding** p. 33.

metallic-coated paper Wrapping or decorative paper coated with powdered metal eg tin, bronze, in a binder.

metallic conduction The transport of electrons which are freely moved by an electric field within a body of metal.

metallic-film resistor One formed by coating a high-temperature insulator, such as mica, ceramic, Pyrex glass or quartz, with a metallic film.

metallic glass One of a range of metal alloys which can be produced in the form of a glass by very rapid cooling. They tend to have good corrosion resistance and the absence of grain boundaries on which to pin **magnetic domains** means that they are well suited for transformer cores. See **glasses and glassmaking** p. 144.

metallization (1) The deposition of thin films of metal on to any surface for decorative or electrical purposes. Also the film itself. See **semiconductor device processing** p. 280. (2) The conversion of a substance, eg selenium, into a metallic form.

metallized yarn An effect yarn containing some metallic components which may be separate filaments or fibres made of metal or a thin metal strip (eg of anodized aluminium) which is protected by a transparent film (eg of cellulose acetate).

metallography The structural study of metals and their alloys by means of various procedures eg microscopy, X-ray diffraction etc.

metalloid An element having both metallic and non-metallic properties, eg arsenic.

metal matrix composite Class of **composite materials** that incorporates fibres (typically ceramics such as **alumina, silicon carbide**) in a metallic matrix. They offer much better high-temperature performance than composites based on polymeric matrices. Abbrev. *MMC*.

metallo-organic compounds Compounds in which the carbon atoms are linked directly with metal atoms (including alkali metals). For example, trimethyl gallium, which is used as a precursor in one of the growth processes for a GaAs semiconductor. See **epitaxial growth of semiconductors** p. 117, **organo-metallic compounds**.

metallurgy The science and technology of

metals and their alloys including methods of extraction and use.

metal-oxide semiconductor transistor An active semiconductor device in which a conducting channel is induced in the region between the electrodes by applying a voltage to an insulated electrode placed on the surface in this region. It is self-isolating by virtue of its construction, and so can be fabricated in a smaller area than a **bipolar transistor**.

metal-oxide-silicon An integrated circuit technology based on device structures comprising conductors, silicon dioxide insulator and semiconducting silicon (principally field effect devices). Abbrev. *MOS*.

metal-oxide-silicon field effect transistor Silicon field effect transistor in which the gate electrode is insulated from the conducting channel. Transistor action involves **inversion** of the semiconductor beneath the gate. Abbrev. *MOSFET*. Cf **junction field effect transistor**.

metal (powder) spraying A method of applying protective metal coatings or building up worn parts by spraying molten metal from a gun. The coating metal is supplied as wire or powder, melted by flame, and blown out of the gun as finely divided particles which form a mechanically adherent layer on the surface of the component. There is no diffusion or alloying.

metal spinning The shaping of thin sheet-metal disks into cup-shaped forms by the lateral pressure of a steel roller or a stick on the revolving disk, which is gradually pressed into contact with a former on the lathe faceplate.

metamerism A marked change in colour of material subjected to different lighting. Thus two such fabrics may match in daylight but appear different when examined in artificial light.

metastable state State which is apparently stable, often because of the slowness with which equilibrium is attained; said, for example, of a supersaturated solution.

meter US spelling of **metre**.

methacrylate butadiene styrene A co-**polymer** with tough mechanical properties. Abbrev. *MBS*. See **rubber toughening** p. 270.

methanal See **formaldehyde**.

methane CH_4. The simplest alkane, a gas, mp $-186°C$, bp $-164°C$; occurs naturally in oil-wells and as marsh gas. Fire-damp is a mixture of methane and air; coal-gas contains a large proportion of methane.

metre The Système Internationale (SI) fundamental unit of length. The metre is defined (1983) in terms of the velocity of light and is the length of path travelled by light in vacuo during a time interval of 1/299 792 458 of a second. Originally intended to represent 10^{-7} of the distance on the Earth's surface between the north pole and the equator, formerly it has been defined in terms of a line on a platinum bar and later (1960) in terms of a wavelength emitted from ^{86}Kr.

metric screw thread A standard screw thread in which the diameter and pitch are specified in millimetres with a 60° angle.

metric system A system of weights and measures based on the principle that each quantity should have one unit whose multiples and sub-multiples are all derived by multiplying or dividing by powers of 10. This simplifies conversion, and eliminates completely the complicated tables of weights and measures found in the traditional British system. Originally introduced in France, it is the basis for the *Système International* (*SI*) now universally adopted. See Appendix 3.

metrology The science of measuring.

MF Abbrev. for: (1) **Medium Frequency**, frequencies from 300 to 3000 kHz; (2) **Machine Finished**.

MFI Abbrev. for **Melt Flow Index**.

MG Abbrev. for **Machine Glazed**.

M(*H*) **curve** See **characterizing magnetic materials** p. 57. Also *M-H*, *M/H curve*.

M(*H*) **loop** See **hysteresis loop, characterizing magnetic materials** p. 57. Also *M-H*, *M/H loop*.

MIC Abbrev. for **Mineral Insulated Cable**.

mica A group of silicates which crystallize in the monoclinic system; they have similar chemical compositions and highly perfect basal cleavage. Mica is one of the best electrical insulators.

micafolium A composite insulating material consisting of a paper backing covered with mica flakes and varnish. Formerly much used for insulating wire and machine coils. Now widely replaced by polymers.

mica marks Silvery streaks found on visible surfaces of injection mouldings. A defect caused by impurities, such as water, in thermoplastic granules.

micelle (1) Colloidal aggregate of molecules formed in solution, esp. soaps in water. Particles are often spherical, with hydrophobic chains in the centre surrounded by hydrophilic groups. Only formed above a certain limit, known as the critical micelle concentration or cmc. (2) Supposed form of crystallite found in partly crystalline polymers. See **crystallization of polymers** p. 82, **fringed micelle model**.

micro- Prefix from Gk *mikros*, small. When used of units it indicates the basic unit × 10^{-6}, eg one micro-ampere (μA) = 10^{-6} ampere. Symbol μ. See Appendix 3.

microbeam analysis Use of fine collimated beam of radiation for identification of materials, eg energy dispersive analysis of X-rays (EDAX), infrared analysis in optical microscopes.

microcircuit isolation The electrical insulation of circuit elements from the electrically conducting silicon wafer. The two main techniques are **oxide isolation** and **diode isolation**.

microelectronics The technology of constructing and utilizing complex electronic circuits and devices in extremely small packages by using integrated-circuit manufacturing techniques.

microencapsulation A process whereby a substance in a finely divided state is enclosed in sealing capsules from which the materials is released by impact, solution, heat or other means. See **NCR**.

microengineering See **microfabrication**.

microfabrication Device or component manufacture (not necessarily electrical or electronic) on a small scale (typically of the order of micrometres), particularly using the techniques of **semiconductor device processing** p. 280.

microfibril Fine fibril, 5–30 nm wide, in a **cell wall**; of cellulose in vascular plants and some algae but of other polysaccharides (eg xylans) in other algae. Cf **matrix**.

microgram(me) Unit of mass equal to one millionth of a gram (10^{-9} kg). Sometimes known as *gamma*.

micrometer gauge A U-shaped length gauge in which the gap between the measuring faces is adjustable by an accurate screw whose end forms one face. The gap is read off a scale uncovered by a thimble carried by the screw, and by a circular scale which is engraved on the thimble. Commonly *micrometer*.

micrometre One-millionth of a metre. Symbol μm. Also *micron* in past.

micromicro- Prefix for 1-million-millionth, or 10^{-12}: replaced in SI by *pico-* (p).

micromodule Sometimes said of circuits or components formed from the same crystal of material, eg germanium. An *integrated circuit*.

micron Obsolete but still popular measure of length equal to 10^{-6} m, symbol μ. Replaced in SI by **micrometre**, symbol μm.

microporosity Minute cavities generally found in heavy engineering sections usually due to lack of efficient feeding during solidification or to release of dissolved gas.

microporous coatings Paints, stains or clear coatings which are permeable to water vapour. Primarily intended for use on timbers, these coatings do not trap moisture, unlike conventional paint systems.

microporous materials Polymers made permeable to gases or liquids by creation of very fine pores, used for filtration of liquids, battery separators or in clothing. Pores are made chemically with a blowing agent or physically, by microfibrillation, eg **Goretex**.

microprocessor A computer *central processing unit* (CPU) that is contained, usually, on one integrated-circuit chip, using **large-scale integration** (LSI) technology. Microprocessors with over a million transistors can be fabricated and they can be used as the main element of a *microcomputer* or as part of an automatic control system.

microscope An instrument used for obtaining magnified images of small objects. The *simple microscope* is a convex lens of short focal length, used to form a virtual image of an object placed just inside its principal focus. The *compound microscope* consists of two short-focus convex lenses, the objective and the eyepiece mounted at opposite ends of a tube. For most optical microscopes, the magnifying power is roughly equal to $450/f_o f_e$, where f_o and f_e are the focal lengths of objective and eyepiece in centimetres. See **electron microscope**, **scanning-electron microscope**, **ultraviolet microscope**.

microscope count method Technique for measuring the average particle diameter of a powder by microscope examination of a weighed drop of suspension.

microscopic stress Those set up at a level of the grain size of a material due to heat treatment etc

microsection A section of metal, ceramic etc, mounted, cut, polished and etched as necessary to exhibit the microstructure.

microstructure Units of microscopic size (about $1-100$ μm in diameter) which occur in materials. Such units include: spherulites, fibrils, lamellae in crystalline polymers (see **crystallization of polymers** p. 82); domains and rubber particles in amorphous polymers (see **rubber toughening** p. 270); grain structure in metals (see **creep** p. 79); ceramics, cell structure in natural materials (see **structure of wood** p. 305).

microwave heating Heating (induction or dielectric) of materials in which the current frequency is in the range 0.3×10^{12} to 10^9 Hz. Extensively used in domestic microwave ovens.

microwave resonance One between microwave signals and atoms or molecules of medium.

microwave spectroscopy The study of atomic and/or molecular resonances in the microwave spectrum.

MIG Abbrev. for **Metal Inert Gas welding**.

mil A unit of length equal to 10^{-3} in, used in measurement of small thicknesses, ie thin sheets. Also colloq. *thou*.

milanese fabric A warp-knitted fabric made with a warp containing twice as many threads as there are **wales** in the fabric with resultant reinforcement of the fabric.

mild steel Steel containing up to 0.15% by weight of carbon. See **low-carbon steel**, **steels** p. 298.

mile A unit of length commonly used for distance measurement in the British Commonwealth and the US. A *statute mile* = 1760 yd, = 1609.34 m.

mill Generally: (1) a machine for grinding or crushing, as a *flour mill*, *paint mill* etc; (2) a factory fitted with machinery for manufacturing, as a cotton mill, saw-mill etc.

millboards Paper boards, usually very dense, manufactured on an intermittent board machine from a variety of furnishes.

milled cloth Woven or knitted wool or woollen fabric in which the milling action causes felting. The cloth has a fibrous surface, the threads and structure being almost completely hidden. The process involves the application of pressure and friction to the cloth, while it is wet, eg with soapy water.

milled lead Sheet-lead formed from cast slabs by a rolling process.

millefiori glass Glassware in which a large number of sections of glass rods of various colours form a pattern and are fused together or set in a clear glass matrix. From Ital., *thousand flowers*.

Miller indices Integers which determine the orientation of a crystal plane in relation to the three crystallographic axes. The reciprocals of the intercepts of the plane on the axes (in terms of lattice constants) are reduced to the smallest integers in ratio. Also *crystal indices*.

milli- Prefix from L. *mille*, thousand. When attached to units, it denotes the basic unit $\times 10^{-3}$. Symbol m.

millimetre The thousandth part of a metre.

millimicron Obsolete term for nanometre, 10^{-9} m. Used in the measurement of wavelengths of eg light.

milling (1) A machine process in which metal is removed by a revolving multiple-tooth cutter, to produce flat or profiled surfaces, grooves and slots. (2) A felting process carried out on already woven woollen fabrics to make them thicker and hairier, carried out in a milling-machine by the agency of soap, alkali or acid (depending on the nature of the fabric and the dye), pressure and friction. Also *fulling*. See **milled cloth**.

milliradian 10^{-3} radian.

mill scale A thin flakey layer of blue/black iron oxide found on new hot rolled steel.

mineral flax A fibrized form of asbestos

formerly much used in the manufacture of asbestos-cement sheeting.

mineral-insulated cable One in which the conductor runs in an earthed copper sheath filled with magnesium oxide, which makes it fireproof and able to withstand excess loads. Also *copper-sheathed cable*. Abbrev. *MIC*.

mineral wool See **rock wool**.

Miner's rule, law Empirical but widely used method of estimating the fatigue lifetime. See **fatigue strength** p. 124.

minority carrier In a semiconductor, the electrons or **holes** which carry the lesser degree of measured current. See **majority carrier**.

minute (1) A 60th part of an hour of time. (2) A 60th part of an angular degree. (3) A 60th part of the lower diameter of a column.

mirror (1) A highly polished reflecting surface capable of reflecting light rays without appreciable scattering. The commonest forms are plane, spherical (convex and concave) and paraboloidal (usually concave). The materials used are glass silvered on the back or front, speculum metal or stainless steel. (2) Very smooth region of tensile fracture surface in brittle materials, esp. glass, associated with slow, but accelerating crack growth. Cf **mist, hackle**. See **fractography**.

mirror finish A very smooth, lustrous surface finish produced, for example, on stainless steels and other metals by electrolytic polishing or lapping, and on electroplated surfaces by mechanical polishing.

mist (1) A suspension, often colloidal, of a liquid in a gas. (2) Cloudy-looking region of tensile fracture surface in brittle materials, esp. glass, between **mirror** and **hackle**, associated with crack moving at up to its maximum velocity, but with insufficient energy to branch (or bifurcate). See **fractography**.

mixing (1) Mechanically blending staple fibres, esp. wools or cottons of similar staple and colour, to obtain the most suitable material for spinning uniform yarns economically. (2) For polymers additives must usually be homogeneously distributed through the material for full effect. It is a problem for high-viscosity melts with low **Reynolds number**, where fluid flow is laminar. Special methods may be needed (eg **cavity transfer mixing**). Pigments like carbon black form clumps, which must be broken up for optimizing properties. Hence **distributive mixing** and **dispersive mixing**. See **tyre technology** p. 325.

mixing index Measure of quality of mixing process, ratio of surface areas of particles after and before mixing, esp. when mixing fillers into polymers.

MKSA Abbrev. for **Metre-Kilogram(me)-Second-Ampere** system of units. See **SI units**.

ml Abbrev. for **millilitre**.

M-lines Characteristic X-ray frequencies from atoms due to the excitation of electrons from the M-shell. Only developed in atoms of high atomic number.

mm Abbrev. for **millimetre**.

MMC Abbrev. for **Metal Matrix Composite**.

mmf Obsolete abbreviation for *MicroMicroFarad*. Now replaced by *picofarad, pF*.

m.m.f. Abbrev. for **MagnetoMotive Force**.

mobility The average velocity of charge carriers per unit electric field in a given material. The mobility of electrons and holes differs greatly. In general mobility decreases with increasing temperature and with increasing defect and impurity densities.

mock leno Woven fabric with a cellular structure produced by allowing spaces to develop between groups of threads. The open structure makes it suited for resin **infiltration** when making a **composite** laminate by hand lay-up. Cf **leno fabric**.

MOCVD See **epitaxial growth of semiconductors** p. 117.

modacrylic fibre Fibres made from synthetic linear polymers containing 35–85% of acrylonitrile groups.

modal fibre See **polynosic fibres**.

modification Alteration of the structure or properties of a material by the addition of small quantities of another element, eg an addition of 0.05% sodium drastically changes the form of the aluminium-silicon eutectic structure and improves ductility.

modulus of elasticity See **elastic constant, elastic modulus**.

mohair The long fine hair from the angora goat, *Capra hircus aegagrus*.

Mohs scale Scale of hardness based on series of minerals, each of which scratches the one beneath it on the scale. See **hardness measurements** p. 153.

moil (1) Glass left on a *punty* or **blowing iron** after the gather has been cut off or after a piece of ware has been blown and severed. (2) Glass originally in contact with the blowing mechanism or head, which becomes **cullet** after the desired article is severed from it.

moiré fibre A ribbed fabric in which the yarns have been partially flattened by heat and pressure in calendering. This gives rise to the optical interference effect commonly known as *watered silk*.

moiré fringe A set of dark fringes produced when two ruled gratings or uniform patterns are superimposed. The separation D of the moiré fringes is equal to d/θ where d is the line spacing of the grating and θ is the angle of intersection of the gratings. Moiré fringes can be used to measure the displacement of one ruled pattern with respect to the other to a high degree of precision.

mol See **mole**.

molality The concentration of a solution expressed as the number of moles of dissolved substance per kilogram of solvent.

molal specific heat capacity The **specific heat capacity** of one mole of an element or compound. Also *volumetric heat* (for gases).

molar absorbance The **absorbance** of a solution with a concentration of 1 mol dm^{-3} measured in a cell of a thickness of 1 cm.

molar conductance The conductance which a solution would have if measured in a cell large enough to contain one mole of solute between electrodes 1 cm apart.

molar heat See **molar heat capacity**.

molar heat capacity The heat required to raise the temperature of one mole of a substance by 1 K. The symbol for that measured at constant volume is C_V and for that at constant pressure is C_P.

molarity The concentration of a solution expressed as the number of moles of dissolved substance per dm^3 of solution.

molar surface energy The surface energy of a sphere containing one mole of liquid; equal to $\gamma V^{2/3}$, where γ is the surface tension and V the molar volume. It is zero near the critical point and its temperature coefficient is often a colligative property.

molar volume The volume occupied by one mole of a substance under specified conditions. That of an **ideal gas** at s.t.p. is 2.2414×10^{-2} m^3 mol^{-1}.

Muldcool TN for CAD package which helps design of injection moulding tools, using for example, cooling analysis of mould cavity.

Moldflow TN for computer analysis of polymer flow within runner, gate and cavity of injection moulding tool.

mole The amount of substance that contains as many entities (atom, molecules, ions, electrons, photons etc) as there are atoms in 12 g of ^{12}C. It replaces in **SI** the older terms *gram-atom*, *gram-molecule* etc, and for any chemical compound will correspond to a mass equal to the relative molecular mass in grams. Abbrev. *mol*. See **Avogadro's number**.

molecular Pertaining to (1) a molecule or molecules or (2) 1 **mole**.

molecular association The relatively loose binding together of the molecules of a liquid or vapour in groups of two or more.

molecular beam epitaxy Abbrev. *MBE*. See **epitaxial growth of semiconductors** p. 117.

molecular biology The study of the structure and function of macromolecules in living cells. Notably successful in explaining the structure of proteins, and the role of

DNA as the genetic material, but with the ultimate aim of explaining the biology of cells and organisms in molecular terms. It is not primarily concerned with metabolic pathways or with the chemistry of natural products.

molecular electronics The technique of growing solid-state crystals so as to form transistors, diodes, resistors, in a single mass, for microelectronic devices.

molecular engineering Term applied to chemical methods of manipulating molecules to achieve a specific effect, esp. tailoring monomers and polymers with new and unusual properties. See **high-performance fibres** p. 158.

molecular formula A representation of the atomic composition of a molecule. When no structure is indicated, atoms are usually given in the order C, H, other elements alphabetically. Functional groups may be written separately, thus sulphanilamide may be written $C_6H_8N_2O_2^-S$ or $H_2N.C_6H_4.SO_2NH_2$. Cf **structural formula**.

molecular heat See **molar heat capacity**.

molecular mass distribution See panel on p. 212. Abbrev. *MMD*.

molecular models The three dimensional models of the structures of many molecules including complex molecules like proteins and DNA have been used as an aid both in determining their structure and understanding their function. *Space filling models* use truncated spheres of diameters corresponding to the atomic radius which can be fitted together to give the proper bond angles. *Stick models* show only the positions of a special repeating feature in the structure such as the *alpha carbon* of a peptide. These positions can be represented by a marker on a rod (*stick*) which can be set to the appropriate three-dimensional co-ordinates. These mechanical models are becoming redundant as computer modelling has become widely available.

molecular orbital A wave function defining the energy of an electron in a molecule. Molecular orbitals are often constructed from linear combinations of the atomic orbitals of the constituent atoms.

molecular refraction See **Lorentz–Lorenz equation**.

molecular sieve Framework compound, usually a synthetic zeolite, used to absorb or separate molecules. The molecules are trapped in 'cages', the sizes of which can be selected to suit solvent. See **clathrate**.

molecular structure The way in which atoms are linked together in a molecule.

molecular weight More formally *relative molecular mass*. (1) The mass of a molecule of a substance referred to that of an atom of

molecular mass distribution

One of the most important structural determinants of polymer properties is the *molecular mass distribution* (MMD), characterized by specific mass averages. They include the *number-average molecular mass* ($\overline{M}_n$) and the *weight-average molecular mass* ($\overline{M}_w$) as well as the simple ratio between the two, the *dispersion* (*D*), which characterizes the breadth of the distribution:

$$\overline{M}_n = \frac{\sum_i n_i M_i}{\sum_i n_i}\,;\quad \overline{M}_w = \frac{\sum_i w_i M_i}{\sum_i w_i} = \frac{\sum_i n_i M_i^2}{\sum_i n_i M_i}\,;\quad D = \frac{\overline{M}_w}{\overline{M}_n}\,;$$

where n_i is the number of molecules of molecular mass, M_i. The weight of molecules of mass M_i is w_i and is the product ($n_i\, M_i$). Absolute values of $\overline{M}_n$ and $\overline{M}_w$ can be determined by osmometry and light-scattering respectively, but most values are now computed from the MMD itself. The MMD is determined directly by gel-permeation chromatography (GPC). They are usually smooth monotonic curves with a simple peak (M_p), but if degradation to smaller molecules occurs eg by ultraviolet degradation, then a second peak or shoulder can frequently be detected using GPC.

Near-monodisperse polymers ($\overline{M}_w/\overline{M}_n \approx 1.0$) can now be made using *anionic polymerization* but most free-radical and co-ordination polymerizations yield very broad distributions eg high-density polyethylene has $D = 6–12$ and for low-density polyethylene, $D = 20–30$. Step-growth polymers like nylon 6,6 and polyethylene terephthalate possess sharper distributions with $D = 1.5–2.0$.

Several important physical properties are controlled by molecular mass, particularly the melt viscosity, the value of which determines the choice of processing method, eg injection moulding or extrusion. Ultimate properties (eg tensile strength) are highly dependent on molecular mass. At a critical point (known as the *critical entanglement molecular mass*, M_c), the strength rises steadily from a negligible value with increase in molecular mass until a plateau value is reached at $M \approx 10^6$. For high-density polyethylene the critical mass $M_c \approx 8000$. Injection moulding grades of this polymer lie typically in the range 10^4–10^5 while extrusion grades can reach 2×10^5. Ultra-high molecular mass polyethylene (UHMPE) possesses a molecular mass in excess of 10^6. Its high strength and low coefficient of friction make it important for artificial hip joint sockets. It is difficult to injection mould and is usually sintered to shape using powder processing methods.

^{12}C taken as 12.000. (2) The sum of the relative atomic masses of the constituent atoms of a molecule.

molecule An atom or a finite group of atoms which is capable of independent existence and has properties characteristic of the substance of which it is the unit. Molecular substances are those which have discrete molecules, such as water, benzene or haemoglobin. Diamond, sodium chloride and zeolites are examples of non-molecular substances.

mole fraction Fraction, of the total number of molecules in a phase, represented by a given component. Symbol x_i.

moleskin cloth A heavy **fustian** type of cotton fabric with smooth face and twill back; used for working clothes.

molybdenum A metallic element in the sixth group of the periodic table. Symbol Mo. Its physical properties are similar to those of iron, its chemical properties to those of a non-metal. Used in the form of wire for filament supports, hooks etc, in electric lamps and radio valves, for electrodes of mercury-vapour lamps, and for winding electric resistance furnaces. It is added to a number of types of alloy steels and certain types of Permalloy and Stellite. It seals well to Pyrex, spot-welds to iron and steel. See Appendix 1,2.

moment (1) Of a couple. See **couple**. (2) Of a force or vector about a point, the product of the force or vector and the perpendicular distance of the point from its line of action. In vector notation, $\mathbf{r} \times \mathbf{F}$, where $\mathbf{r}$ is the position vector of the point, and $\mathbf{F}$ is the force or vector. (3) Of a force or vector about a line, the product of the component of the force or vector parallel to the line and its perpendicular distance from the line.

moment of inertia Of a body about an axis: the sum $\sum mr^2$ taken over all particles of the body where m is the mass of a particle and r its perpendicular distance from the specified axis. When expressed in the form Mk^2, where M is the total mass of the body ($M = \sum m$), k is called the *radius of gyration* about the specified axis. Also used erroneously for **second moment of area**.

momentum A dynamical quantity, conserved within a closed system. A body of mass M and whose centre of gravity G has a velocity v has a *linear momentum* of Mv. It has an *angular momentum* about a point O defined as the moment of the linear momentum about O. About G this reduces to $I\omega$ where I is the **moment of inertia** about G and ω the angular velocity of the body.

Monel metal A nickel-base alloy containing nickel 68%, copper 29% and iron, manganese, silicon and carbon 3%. Has high strength (about 500 MN m^{-2}), good elongation (about 45%), and high resistance to corrosion. Used for condenser tubes, propellers, pump fittings, turbine blades, and for chemical and food-handling plant.

mon-, mono- Prefixes from Gk *monos*, alone, single.

monochromatic light Light containing radiation of a single wavelength only. No source emits truly monochromatic light, but a very narrow band of wavelengths can be obtained, eg the cadmium red spectral line, wavelength 643.8 nm with a *half-width* of 0.0013 nm. Light from some lasers can have an extremely narrow line width.

monochromatic radiation Electromagnetic radiation (originally visible) of one single frequency component. By extension, a beam of particulate radiation comprising particles all of the same type and energy. *Homogeneous* or *monoenergic* is preferable in this sense.

monochromator Device for converting heterogeneous radiation (electromagnetic or particulate) into a homogeneous beam by absorption, refraction or diffraction processes.

monocoque A structure in which all structural loads are carried by the skin. In a *semimonocoque*, loads are shared between skin and framework, which provides local reinforcement for openings, mountings etc.

monodisperse polymer Polymer in which all chains are of equal length. Can be made by anionic polymerization.

monofilament A single filament of indefinite length. Used for ropes, surgical sutures etc, and in finer gauge, for textiles.

monolithic integrated circuit Electronic circuit formed by diffusion or ion-implantation on a single crystal of semiconductor, usually silicon.

monomer Small molecule with high chemical reactivity, capable of linking up with itself to produce polymers, or with similar molecules to make copolymers. See **polymers** p. 247, **copolymer** p. 75.

monovalent Capable of combining with one atom of hydrogen or its equivalent, having an oxidation number or co-ordination number of one.

Mooney viscometer Type of viscometer used widely in the rubber industry to obtain cure curves by measurement of torque needed to shear vulcanizing system.

moquette A heavy warp pile fabric used for upholstery. The pile may be cut or left as uncut loops.

mordant A compound, frequently a metallic salt or oxide, applied to a fabric to form a stable complex with a dyestuff. The complex has superior fastness on the fabric compared with the dyestuff itself.

morphology General term for study of shapes of microstructural units in materials, such as spherulites in polymers and minerals, grain structure in metals and ceramics.

morphotropic Refers to the effect on crystal structure of atomic substitutions.

Morse equation An equation which relates the potential energy of a diatomic molecule to the internuclear distance.

mortar (1) A pasty substance formed normally by the mixing of cement, sand and water, or cement, lime, sand and water in varying proportions. Used normally for the binding of brickwork or masonry. It is **thixotropic** and its working life can be extended by addition of soap solutions. Hardens on setting and forms the bond between the bricks or stones. (2) A bowl, made of porcelain, glass or agate, in which solids are ground up with a pestle.

MOS Abbrev. for **Metal-Oxide-Silicon**.

mosaic structure Discontinuous structure of a compound or metal consisting of minute domains, each bounded by its discontinuity lattice at the interface with other domains of the like composition.

Moseley's law For one of the series of characteristic lines in the X-ray spectrum of atoms, the square root of the frequency of the lines is directly proportional to the *atomic number* of the element. This result stresses the importance of atomic number rather than atomic mass.

MOSFET Abbrev. for **Metal-Oxide-Silicon Field Effect Transistor**.

motor A machine used to transform power into mechanical form from some other form.

mottled iron Cast iron in which most of the

carbon is combined with iron in the form of cementite (Fe_3C) but in which there is also a small amount of graphite. The fracture has a white crystalline appearance with clusters of dark spots, indicating the presence of graphite.

mottle yarn Same as **marl yarn**.

mouldability index Term referring to the ease or difficulty of melt flow of a polymer. Thus polycarbonate has a high index and is thus difficult to mould, while high-impact polystyrene is easy to mould and has a low index. Temperature dependent.

mould breathing Injection moulding technique which allows moulding of products larger than the nominal locking force available, by opening the tool at the end of the cycle. See **vertical flash tool**.

mould defects Moulding faults which are minimized by careful tool design and tool setting. They include sink, mica and burn marks, weld and flow lines, and voids.

mould parting line A line on a moulding left by the junction of the male and female parts of the tool.

mould release agent Speciality material often applied in aerosol form to mould surfaces to aid release of polymer product at end of moulding cycle. Usually consists of silicone or fluoropolymer.

mould temperatures Temperatures at which injection moulding tool parts are kept, often below ambient for crystalline polyolefins, but well above ambient for many engineering thermoplastics. Polycarbonate, for example, usually demands mould temperatures above about 80°C to minimize frozen-in strains.

M-shell The electron shell in an atom corresponding to a principal quantum number of three. The shell can contain up to 18 electrons.

MSI Abbrev. for **Medium Scale Integration**. Refers to a silicon or other chip with no more than a few hundred components on it.

muffle furnace A furnace in which heat is applied to the outside of a refractory chamber containing the charge.

mule Machine, intermittent in action, which first drafts and twists the yarn on the outward run, then on the inward run winds it on a spindle. It can spin very fine counts but there are also woollen mules for heavy yarns and condenser mules for waste yarns. (Invented by Samuel Crompton, Bolton, ca 1770.) Now only used for wool spinning.

mull Liquid used to prepare specimen for infrared spectroscopy, by crushing and abrasion in small pestle or **mortar**. See **Nujol**.

multichip integrated circuits See **printed, hybrid and integrated circuits** p. 257.

multi-impression Term applied to injec-tion moulding tool possessing several identical cavities, so increasing productivity over the mould cycle time. Contrast family tools, where the cavities have different shapes.

multilive feed Type of injection moulding process where molten polymer (often with chopped fibres) can be fed at two or more channels into the mould recess. See **injection moulding** p. 169.

multilobal A 'man-made' fibre or filament which is extruded so that the cross-section has several rounded lobes. More precisely named according to the number of lobes eg *trilobal*.

multiple screws Design of extruder or injection moulding screw where two or more steel helices are used on the rotor to aid mixing of polymer and additives.

multiplets (1) Optical spectrum lines showing fine structure with several components, ie triplet or more complex structures. Due to spin-orbit interactions in the atom.

multiplex film Packaging film where different polymers are sandwiched together by co-extrusion, eg low-density polyethylene for strength and polyvinylidene chloride for barrier properties.

multipole moments These are magnetic and electric, and are measures of the charge, current and magnet (via intrinsic spin) distributions in a given state. These **static** multipole moments determine the interaction of the system with weak external fields. There are also *transition* multipole moments which determine radiative transitions between two states.

Mumetal TN of high-permeability, low-saturation magnetic alloy of about 80% nickel, requiring special heat treatment to achieve special low-loss properties. Useful in non-polarized transformer cores, a.c. instruments, small relays. Also for shielding devices from external field as in **cathode ray oscilloscopes**.

mungo A low grade of waste fibres recovered by pulling down old *hard*-woven wool rags, including tailors' cuttings and old felt. Cleaned and spun again as woollen yarns and used for weaving into lower grade fabrics. Cf **shoddy**.

Muntz metal Alpha-beta brass, 60% copper and 40% zinc. Stronger than alpha brass and used for castings and hot-worked (rolled, stamped or extruded) products. See **copper alloys** p. 76.

Munsell colour system A system of colour notation devised by Albert Munsell which breaks colour into three attributes: (1) *hue* colour, red, blue, green etc; (2) *value*, the lightness or darkness of a colour; (3) *chroma*, strength or saturation of a colour. Colour identification can be done rapidly with hand-held photometers.

Murray red gum See **Red River gum**.

muslin Lightweight, plain-woven cloths of open texture and soft finish, bleached, and dyed; used as dress fabrics. There are also unbleached butter, cheese or meat etc muslins once commonly used for wrapping purposes.

mutton cloth A plain circularly knitted cotton fabric often used in the unbleached state as a cleaning cloth.

Mycalex TN for mica bonded with glass. It is hard, and can be drilled, sawn and polished; has a low power factor at high frequencies, and is a very good insulating material at all frequencies.

Mylar TN for polyethylene terephthalate film (US).

myosin A fibrous protein which packs in a regular hexagonal array with actin to form muscle fibrils. The relative movement of the two sets of fibres provides the molecular basis for muscle contraction.

myrtle Commonly *acacia*. A tree of the Leguminosae family giving a coarse-textured **hardwood** that is reddish-brown in colour. Used for tool handles, vehicle parts, walking sticks and turned articles. Density 850 kg m^{-3}.

N O

n Symbol for **nano-**.

n Symbol for: (1) amount of substance; (2) number density expressed as particles per cubic metre (or per cubic centimetre), especially of negative carriers (electrons) in semiconductors. Cf *p*.

n- An abbrev. for *normal*, ie containing an unbranched carbon chain in the molecule.

N Symbol for **nitrogen**.

N- Symbol indicating substitution on the nitrogen atom.

NA Abbrev. for **neutral axis**.

N$_A$ Symbol for: (1) Avogadro's number; (2) number of molecules.

NaK Acronym for sodium (Na) and potassium (K) alloy which has been used as a coolant for liquid metal nuclear reactors. It is molten at room temperature and below.

nano- Prefix for 10^{-9}, ie equivalent to millimicro or one thousand millionth. Symbol n.

nanotechnology The engineering of matter at a a scale approaching that of individual atoms.

nap A fluffy surface on fabrics produced by the finishing process of raising. Napping is raising by means of a revolving cylinder covered with stiff wire brushes, or rollers covered by teazles.

naphtha A mixture of light hydrocarbons. Petroleum naphtha is a cut between gasoline and kerosine with a boiling range 120–180°C, but much wider naphtha cuts may be taken for special purposes, eg feedstock for high temperature cracking for chemical manufacture, monomer and polymer synthesis. The hydrocarbons in petroleum naphthas are predominantly aliphatic.

naphthalene di-isocyanate A type of aromatic isocyanate monomer used to make polyurethanes. Abbrev. *NDI*.

naphthenes *Cycloalkanes*, cyclic aliphatic hydrocarbons, like cyclohexane. Many occur in petroleum.

narra See **amboyna**.

National Bureau of Standards US federal department set up in 1901 to promulgate standards of weights and measures and generally investigate and establish data in all branches of physical and industrial sciences. Abbrev. *NBS*.

National Physical Laboratory The UK authority for establishing basic units of mass, length, time, resistance, frequency, radioactivity etc. Founded by the Royal Society in 1900; now government controlled and engaged in a wide range of research. Abbrev. *NPL*.

natural ageing Changes in physical or mechanical properties which take place in a material over a period of time at ambient temperature. It is usually a response to some previous heat treatment or operation which has left the material in a metastable condition. If such changes are brought about at above ambient temperatures, the processes are termed artificial ageing.

natural gas Mixture of simple gaseous alkanes, mainly methane, plus smaller quantities of ethane, propane and butane, usually found associated with petroleum or coal deposits. Important source of both energy and materials, eg carbon black, polymers by cracking of ethane etc.

natural rubber Although some 2000 plant species yield polyisoprenes (see **balata, gutta percha**) similar to that of *Hevea brasiliensis*, the rubber tree is the main and richest source of natural rubber. After tapping, the **latex** is concentrated from about 33% *dry rubber content* to 60% *dry rubber content*, and may then be coagulated direct to dry rubber or further processed in latex form. The main component is cis-polyisoprene of very high molecular mass (5×10^5 to 10^6) together with traces of resin, vegetable protein, minerals and plant sugars. Dry natural rubber baked to **standard Malaysian rubber** requirements must be masticated before compounding and shaping owing to its very high Mooney viscosity. For engineering applications (tyres, bearings etc), natural rubber must be crosslinked (vulcanized) to prevent excessive creep in use. Its low tensile modulus (ca 1 MN m^{-2}) is usually increased up to 10 fold by mixing with carbon black. As a crystallizing rubber (T_m = 15–50°C, depending on rate, temperature, degree of crosslinking, strain etc) its mechanical properties are better than non-crystallizing rubbers (eg styrene butadiene rubber). Once the only source of engineering elastomers, it has faced increasing competition with synthetic rubbers eg butadiene, isoprene, ethylene-propylene diene since World War II. For a most important use, see **tyre technology** p. 325.

natural strain See **strain**.

naval brass An alpha-beta brass (centred on 60% Cu/40% Zinc) containing an addition of 1% tin to improve corrosion resistance. See **Tobin bronze, copper alloys** p. 76.

NBR Abbrev. for **Nitrile-Butadiene Rubber**. Also *nitrile rubber*.

NBS Abbrev. for **National Bureau of**

Standards.

NBS smoke test Test method for grading burning polymers by the amount of smoke made, using optical density measure.

NCR Abbrev. for *No Carbon Required*. TN for stationery, esp. for office machine use, in which simultaneous duplicate copies are obtained without the use of carbon paper. Microcapsules containing dyes on the verso of the 'top' copy are ruptured by the impact of the writing medium (pen, typewriter etc), the image being then transferred to the under copy or copies, the receiving surface of which has a coating of attapulgite with a starch or latex binder. See **microencapsulation**.

NDI Abbrev. for **Naphthalene Di-Isocyanate**.

NDT Abbrev. for **Non-Destructive Testing**.

Ne The symbol for **neon**.

nearly-free electron model A model from which the band structure of simple metals can be calculated. The periodic part of the potential due to the crystal lattice is treated as a minor modification to the free electron gas model. See **band theory of solids**.

neck See **necking**.

necking The localized reduction in cross-section which occurs under uniaxial deformation of ductile materials, for example in a tensile test specimen shortly before failure. It reflects the reduction in rate of work hardening as degree of plastic strain increases, causing all the deformation to be confined to one region which progessively reduces in cross-section until fracture occurs.

needlecord A fine-ribbed **corduroy** used as a dresscloth.

needle paper An acid-free, black paper for wrapping needles, pins etc.

Néel temperature The temperature at which the susceptibility of an anti-ferromagnetic material has a maximum value.

negative catalysis The retardation of a chemical reaction by a substance which itself undergoes no permanent chemical change.

negative crystal Birefringent material for which the velocity of the extraordinary ray is greater than that of the ordinary ray.

negative electricity Phenomenon in a body when it gives rise to effects associated with excess of electrons. See **positive electricity**.

negative ion Radical, molecule or atom which has become negatively charged through the gain of one or more electrons. See **ion, anion**.

negative resistance A property of a device in which a decrease in the voltage drop across it causes the current to increase. Most electrical gas discharges have this property,

along with some valves and semiconductor devices, including the Gunn and tunnel diodes.

nematic Thread-like. See **liquid crystal phases** p. 194.

neodymium A metallic element, one of the **rare earth elements**. Symbol Nd. Nd^{3+} ions in a calcium tungstate glass form a solid-state laser medium. Compounds of neodymium, iron and boron can be used to manufacture extremely high energy (BH_{max}) permanent magnets. See Appendix 1, 2.

neon Light, gaseous, inert element, recovered from atmosphere, Symbol Ne. Used in many types of lamp particularly to start up sodium vapour discharge lamps. Pure neon was the first gas to be used for high voltage display lighting, being bright orange in colour. See Appendix 1, 2

Neoprene TN for polychloroprene; the first commercial synthetic rubber (US 1931).

nephelometric analysis A method of quantitative analysis in which the concentration or particle size of suspended matter in a liquid is determined by measurement of light absorption. Also *photoextinction method, turbidimetric analysis*.

neps Term applied in the cotton industry to small entanglements of fibres that cannot be unravelled; generally formed during the ginning process from dead or immature fibres.

Nernst heat theorem As the absolute temperature of a homogeneous system approaches zero, so does the specific heat and the temperature coefficient of the free energy.

Nernst effect A voltage which appears at opposite edges of a strip of metal that is conducting heat in the presence of a magnetic field which is perpendicular to the surface of the metal.

Nernst equation Version of the **Gibbs–Helmholtz equations** relating the **free energy** change in an electrochemical cell (ΔG) to the number of electrons transferred (z), Faraday constant (F) and electrochemical potential (E):

$$\Delta G = -zFE.$$

Nernst lamp One depending on the electric heating of a rod of zirconia in air, giving an infrared source for spectroscopy. The rod must be heated separately to start the lamp, as the material is insulating at room temperature.

Nernst's distribution law When a single solute distributes itself between two immiscible solutes, then for each molecular species (ie dissociated, single or associated molecules) at a given temperature, there exists a constant ratio of distribution between the two solvents, ie $(C_1/C_2)_i = K_i$ for each molecular species, *i*.

nerve Term used to describe defective state of vulcanized rubber where degree of cross-linking varies randomly through the thickness of the product.

nest A cushion upon which glass is placed to be cut with a diamond.

net A firm open-mesh fabric made by weaving, knitting, or knotting.

Netlon TN for polymer net used in packaging, agriculture and engineering. Made by extruding polymer melt through contra-rotating dies with matching slots in their mating faces.

network modifier Metal ions such as Na^+, Ca^{2+}, K^+, B^{3+} etc in silicate glasses, which open up the network of silica tetrahedra and thus modify the properties of the glass. See **glasses and glassmaking** p. 144.

network polymer A polymer in which crosslinking has occured.

network structure The type of structure formed in alloys when one constituent exists in the form of a continuous network round the boundaries of the grains of the other. Even if the grains included in the cells are themselves duplex, they are regarded as individual grains.

Neumann lamellae Straight, narrow bands appearing in the microstructure parallel to the crystallographic planes in the crystals of metals that have been deformed by sudden impact. They are actually crystallographic twins produced during rapid deformation processes, particularly in **body-centred cubic** crystals, and are often observed in meteorites.

neutral Glass of high chemical durability.

neutral axis The line of zero stress in a beam subjected to bending; more accurately, a transverse section of the longitudinal plane which passes through the centre of area of the section. Also *neutral surface*.

neutral flame In welding, flame produced by a mixture at the torch of acetylene and oxygen in equal volumes, having neither an oxidizing or a carburizing effect.

neutral flux One used to modify fusibility of furnace slags, but which exerts neither basic nor acidic influence, eg calcium fluoride.

neutralization The interaction of an acid and a base with the formation of a salt. In the case of strong acids and bases, the essential reaction is the combination of hydrogen ions with hydroxyl ions to form water molecules.

neutral solution An aqueous solution which is neither acidic nor alkaline. It therefore contains equal quantities of hydrogen and hydroxyl ions and has a **pH** value of 7.

neutral state Said of ferromagnetic material when completely demagnetized. Also

virgin state.

neutral surface See **neutral axis**.

neutron Uncharged subatomic particle, mass approximately equal to that of the proton, which enters into the structure of atomic nuclei. Interacts with matter primarily by collisions. The spin quantum number of a neutron = +1/2, the rest mass = 1.008 665 a.m.u., the charge is zero and the magnetic moment is −1.9125 nuclear Bohr magnetons. Although stable in nuclei, isolated neutrons decay by β-emission into protons, with a half-life of 11.6 minutes.

neutron diffraction The coherent elastic scattering of neutrons by the atoms in a crystal. If the scattering is by the nucleus then the atomic structure of the crystal can be deduced from the measurements of the diffraction pattern. If the scattering is by atoms with electron configurations that have a magnetic moment, then details of the magnetic structure of the crystal can be determined. See **neutron-magnetic scattering**.

neutron elastic scattering A beam of thermal neutrons, whose electrons are such that their associated de Broglie wavelength is of the same order of magnitude as interatomic distances, will be diffracted by a crystal. Most of the intensity of the beam will be diffracted with no loss of energy, ie no wavelength change, and this is said to be *neutron elastic scattering*. Cf **neutron inelastic scattering.**

neutron inelastic scattering A beam of thermal neutrons, whose energies are of the same order of magnitude as a quantum of lattice vibrational energy, a **phonon**, will be diffracted by a crystal with exchanges of energy with the excited travelling waves. A detailed examination of the change in direction and energy of the scattered neutrons gives valuable information about the *lattice dynamics* of the crystal. Cf **neutron elastic scattering**.

neutron-magnetic scattering The magnetic moment in a crystal has contributions from both electron spin and electron orbital angular momenta. Both of these will interact with the neutron spin magnetic moment and will give rise to the magnetic scattering of neutrons. This is a powerful method for the study of magnetic structures.

neutron radiography Beam of neutrons from a nuclear reactor traverse an object and impinge on an image-producing detector which reveals areas of absorption. Similar to X-ray radiography, but with the advantage that elements of low atomic number (eg H, Li, B) are strongly absorbed. Used as a **nondestructive testing** technique.

neutron spectroscopy Spectroscopic analysis that is based on neutron inelastic scattering.

newsprint Inexpensive printing paper intended for printing by letterpress or web offset for the production of newspapers, generally from 40–52 g m^{-2}. Generally made from mechanical woodpulp, but de-inked waste paper may also feature prominently.

newton Symbol N. The unit of force in the SI system, being the force required to impart, to a mass of 1 kg, an acceleration of 1 m s^{-2}. 1 newton = 0.2248 pounds force.

Newtonian viscosity Newton's law of viscosity states that the shear stress is directly proportional to the shear rate, the proportionality constant (η) being known as the *coefficient of shear viscosity*. It is temperature dependent but independent of shear rate. Fluids like water are strictly Newtonian, but many fluids are non-Newtonian. See **non-Newtonian flow**.

Newton's law of cooling The rate of cooling of a hot body which is losing heat both by radiation and by natural convection is proportional to the difference in temperature between it and its surroundings. The law does not hold for large temperature excesses.

Newton's laws of motion They are: (1) every body continues in a state of rest or uniform motion in a straight line unless acted upon by an external impressed force; (2) the rate of change of momentum is proportional to the impressed force and takes place in the direction of the force; (3) action and reaction are equal and opposite, ie when two bodies interact the force exerted by the first body on the second body is equal and opposite to the force exerted by the second body on the first. These laws were first stated by Newton in his *Principia*, 1687. Classical mechanics consists of the application of these laws.

niangon W African **hardwood** (*Tarrietia*) with pale pink to reddish-brown **heartwood**, with wavy, interlocked grain and coarse texture. Average density 560 kg m^{-3} (*T. densiflora* 760 kg m^{-3}). Durable and used for furniture, construction, joinery and in veneers. Also *nyankom*.

nibbling A sheet metal cutting process akin to sawing, in which a rapidly reciprocating cutter slots the sheet along any desired path.

Nichrome TN for a nickel-chromium alloy which is largely used for resistance heating elements because of its high resistivity ($\approx 110 \times 10^{-8}$ Ω m) and its ability to withstand high temperatures.

nickel Silver-white metallic element. Symbol Ni. Used for structural parts of valves. It is magnetostrictive, showing a decrease in length in an applied magnetic field and, in the form of wire, was much used in computers for small stores, the data circulating and

extracted when required. Used for electroplating, coinage and in chemical and food-handling plants. See **nickel alloys**, Appendix 1, 2.

nickel alloys Nickel is the main constituent in Monel metal, permalloy and nickel-chromium alloys. It is also used in cupro-nickel, nickel-silver, various types of steel and cast-iron, brass, bronze and light alloys.

nickel carbonyl Ni(CO)$_4$. A volatile compound of nickel, bp 43°C, formed by passing carbon monoxide over the heated metal. The compound is decomposed into nickel and carbon monoxide by further heating. Formerly used on a large scale in industry for the production of nickel from its ores by the Mond process.

nickel-chromium steel Steel containing nickel and chromium as alloying elements. 1.5% to 4% nickel and 0.5% to 2% chromium are added to impart **hardenability** which enables thick sections to be heat treated to levels of high tensile strength, hardness and toughness, eg highly stressed automobile and aero-engine parts, armour plate etc. See **steels** p. 296.

nickel-iron-alkaline accumulator, cell An **accumulator** in which the positive plate consists of nickel hydroxide enclosed in perforated steel tubes, and the negative plate consists of iron or cadmium also enclosed in perforated steel tubes. The electrolyte is potassium hydrate, the emf 1.2 volts per cell. It is lighter than the equivalent lead-acid accumulator. Also *Edison accumulator, Ni-Fe accumulator*.

nickel silver See **German silver**, **copper alloys** p. 76.

nickel steel Steel containing nickel as principal alloying element. Varying amounts, between 0.5% and 6.0%, are added to increase the strength in the normalized condition, to enable hardening to be performed in oil or air instead of water, or to increase the core strength of carburized parts Larger amounts are added to steels for special purposes. See **steels** p. 296.

Niclad Composite laminated sheets made by rolling together sheets of nickel and mild steel, to obtain the corrosion resistance of nickel with the strength of steel.

Nicol prism A device for obtaining plane polarized light, it consists of a crystal of Iceland spar which has been cut and cemented together in such a way that the ordinary ray is totally reflected out at the side of the crystal, while the extraordinary plane polarized ray is freely transmitted. Largely superseded by Polaroid.

niello A method of decorating metal. Sunk designs are filled with an alloy of silver, lead and copper, with sulphur and borax as fluxes, and fired.

Ni-Fe accumulator TN of a **nickel-iron-alkaline accumulator**.

Ni-hard Cast-iron to which a ladle addition of about 4% nickel has been made, to render the alloy martensitic and abrasion resistant.

Nimonic TN for alloy series used in such high temperature work as that called for in gas turbine blades. Based on the basic composition of 80% nickel and 20% chromium with closely controlled amounts and combinations of other elements, eg titanium, aluminium, manganese, iron, carbon, cobalt, silicon, copper, boron, zirconium and molybdenum.

niobium Rare metallic element, symbol Nb, used in high-temperature engineering products (eg gas turbines and nuclear reactors) owing to the strength of its alloys at temperatures above 1200°C. Combined with tin (Nb_3Sn) it has a high degree of superconductivity. See Appendix 1, 2.

nip The line of near contact between two rotating parallel rollers in a **calender** through which material passes while being compressed. Produces dispersive mixing in polymers.

Ni-resist A cast iron consisting of graphite in a matrix of austenite. Contains carbon 3%, nickel 14%, copper 6%, chromium 2% and silicon 1.5%. Has a high resistance to growth, oxidation and corrosion.

Nitralloy Steel specially developed for nitriding (which is less effective with ordinary steels). Contains carbon 0.2–0.3%, aluminium 0.9–1.5%, chromium 0.9–1.5% and molybdenum 0.15–0.25%.

nitric (V) acid HNO_3, a fuming unstable liquid, bp 83°C, mp 41.59°C, rel.d. 1.5, miscible with water. Old name *aqua fortis*. Prepared in small quantities by the action of conc. sulphuric acid on sodium nitrates and on large scale by the oxidation of nitrogen or ammonia. An important intermediate of fertilizers, explosives, organic synthesis, metal extraction and sulphuric acid manufacture.

nitrides Compounds of metals with nitrogen. Usually prepared by passing nitrogen or gaseous ammonia over the heated metal. Those of group I and II metals are ionic compounds which react with water to release ammonia. Those of group III to V of the periodic table exist as interstitial compounds, having great hardness and refractoriness, eg boron, titanium, iron.

nitriding Nitrogen case hardening. A process for producing hard surface on special types of steel by heating in gaseous ammonia. Components are finish-machined, hardened and tempered, and heated for about 600 h at 520°C. Case is about 0.50 mm deep and surface hardness is 1100 H_V. See Nitralloy.

nitrile rubber Copolymer made from acrylonitrile and butadiene, composition depending on use. Widely used in solvent-resistant applications, eg petrol hose, brake hose.

nitrocellulose See **cellulose nitrate**.

nitrogen Gaseous element, forming a diatomic, colourless, odourless gas. Symbol N. Molecular nitrogen occupies approx. 80% by volume of the Earth's atmosphere. Nitrogen occurs as a chemical component of minerals and all living matter. Gaseous nitrogen is used as an inert atmosphere in filament lamps, in sealed relays, in Van de Graaff generators, and in high-voltage cables as an insulant. Liquid nitrogen is used as a coolant. See Appendix 1, 2.

nitrogen case-hardening See **nitriding**.

nm Abbrev. for *nanometre* = 10 angstroms, = 10^{-9} m.

NMOS Metal-oxide-silicon technology based on n-channel devices in a p-type substrate.

NMR Abbrev. for **Nuclear Magnetic Resonance**.

NMR spectroscopy See **nuclear magnetic resonance spectroscopy**.

nn, n-n junction A junction formed within a n-type semiconductor where the donor dopant concentration changes abruptly.

nN, n-N heterojunction A junction in a **heterostructure** formed between a narrower band gap material (n-type) and a wider gap material (N-type).

noble gases Elements helium, neon, argon, krypton, xenon and radon, much used (except the last) in gas-discharge tubes. Their outer (valence) electron shells are complete, thus rendering them inert to all the usual chemical reactions; a property for which argon, the most abundant, finds increasing industrial use. The heavier ones, Rn, Xe, Kr, are known to form a few unstable compounds, eg XeF_4. Also *inert gases*, *rare gases*.

noble metals Metals, such as gold, silver, platinum etc, which have a relatively positive electrode potential, and which do not enter readily into chemical combination with non-metals. They have high resistance to corrosive attack by acids and corrosive agents, and resist atmospheric oxidation. Cf **base metal**.

nodular cast iron See **ductile cast iron**.

noil (1) Short or broken fibres from silk after opening and combing. (2) Short fibre removed from wool during combing; often added to woollen blends.

Nomarski interference Method of polarized optical microscopy of material surfaces producing colour contour map, esp. useful for semiconductor chip surfaces or others where it is important to observe slight changes in relief. See **polarized light microscopy**.

Nomex TN for aramid fibre based on meta-terephthalamide polymer, used in paper for a *stiff honeycomb* for aerospace applications and for fire-resistant clothing (US).

non-degenerate gas One formed of particles, the concentration of which is so low, that the **Maxwell–Boltzmann distribution law** applies. Particular examples are the electrons applied to a conduction band by donor levels in an n-type semiconductor, and those resulting from the passage of electrons from the normal band to an impurity band of acceptor levels in a p-type semiconductor.

non-destructive testing Methods of inspecting materials and products without affecting their subsequent properties and performance. These include acoustic emission, eddy current testing, neutron and X-ray radiography, optical and ultrasound holography, magnetic particle inspection, penetrant flaw detection, and ultrasonic imaging and testing. Abbrev. *NDT*.

non-ferrous alloy Any alloy based mainly on metals other than iron, ie usually on copper, aluminium, lead, zinc, tin, nickel or magnesium.

non-metal Element which is generally of low density and a poor conductor of heat and electricity, being held together by strong covalent bonds eg. carbon as in diamond. They occur towards the upper right-hand corner of the Periodic Table as drawn conventionally (see Appendix 1). Together with metals, they form a vast range of ionic materials. See **bonding** p. 33.

non-metallic inclusions See **inclusions**.

non-Newtonian flow Fluid flow which deviates from Newton's law of viscosity, especially pseudoplastic fluids, dilatant fluids and **Bingham solids**. See **dilatency**.

non-stoichiometric compounds Some solid compounds do not possess the exact compositions which are predicted from Daltonic or electronic considerations alone (eg iron (II) sulphide is $FeS_{1.1}$), a phenomenon which is associated with the so-called defect structure of crystal lattices. Often show semiconductivity, fluorescence and centres of colour. See **rusting** p. 271.

non-woven fabrics Cloth formed from a random arrangement or web of natural or synthetic fibres using adhesives, heat and pressure, needling techniques etc to confer adhesion. See **batt**, **felt**.

normal Containing an unbranched chain of carbon atoms; eg normal propyl alcohol is $CH_3.CH_3.CH_2.OH$, whereas the isomeric isopropyl alcohol, $(CH_3)_2 = CH.OH$, has a branched chain.

normal calomel electrode A calomel electrode containing molar potassium chloride solution.

normal electrode potential See **standard electrode potential**.

normality A concentration unit, abbreviated N. It is used mainly for acids or bases and for oxidizing or reducing agents. In the first case, it refers to the concentration of titratable H^+ or OH^- in a solution. Thus a solution of sulphuric acid with a concentration of 2 mol l^{-1} (2 M) will be a 4 N solution.

normalizing A heat treatment applied to steel. Involves heating above the critical range, followed by cooling in air, performed to refine the crystal structure and eliminate internal stress. See **annealing**.

normal magnetization Locus of the tips of the magnetic hysteresis loops obtained by varying the limits of the range of alternating magnetization.

normal pressure Standard atmospheric pressure, 101.325 kN m^{-2} = 760 torr, to which experimental data on gases are referred.

normal salts Salts formed by the replacement by metals of all the replaceable hydrogen of the acid.

normal segregation A type of segregation in which the concentration of low melting point constituents and inclusions tends to increase from the surface to the axial regions of cast metals. See **inverse segregation**.

normal solution See **normality**.

normal temperature and pressure The earlier term for **standard temperature and pressure**.

Noryl TN for blend of polystyrene and polyphenylene oxide (US). A rare example of compatible polymers, mixing at a molecular level.

not The unglazed surface of hand-made or mould-made drawing-papers. See **hotpressed**, **rough**.

notch brittleness Susceptibility to fracture as disclosed by **Izod** or **Charpy test**, due to weakening effect of a stress-concentrating flow at the surface.

notched-bar test Type of impact test using a notched specimen. See **Charpy test**, **impact tests** p. 165, **Izod test**.

notch sensitivity The extent to which the endurance of materials, as determined on smooth and polished specimens, is reduced by surface discontinuities, such as tool marks, notches and changes in section, which are common features of actual components. It tends to increase with the hardness and endurance limit.

notch toughness Misleading use of toughness for the energy required to break standard specimens under standard conditions in eg **Izod** or **Charpy tests**. It is also used to mean the opposite of **notch brittleness**.

novolak Type of phenolic resin prepolymer made by reacting excess phenol with formaldehyde under acidic conditions. Also *two-stage resin*, since more formaldehyde must be added for final crosslinking. Both novolaks and resols are A-stage resins.

nozzle Narrow orifice at front of injection moulding machine through which molten polymer passes to reach tool cavity. See **injection moulding** p. 169.

npin, n-p-i-n transistor Similar to n-p-n transistor but incorporating a layer of intrinsic semiconductor between the base and the collector to extend the high-frequency range.

NPL Abbrev. for **National Physical Laboratory**.

npn, n-p-n transistor A bipolar transistor having a p-type base between an n-type emitter and an n-type collector. In operation, the emitter should be negative and the collector positive with respect to the base.

NR Abbrev. for **Natural Rubber**.

n-type conduction Conductivity associated with charge transport by electrons. Also *n-type conductivity*.

n-type semiconductor One in which donor impurities predominate over acceptors. See **intrinsic and extrinsic semiconductors** p. 117.

N-type semiconductor A relatively wide band gap n-type semiconductor.

nuclear Bohr magneton See **Bohr magneton**.

nuclear magnetic resonance Certain atomic nuclei, eg ^{1}H, ^{19}F, ^{31}P, have a nuclear magnetic moment. When placed in a strong magnetic field, the nuclear moments can only take up certain discrete orientations, each orientation corresponding to a different energy state. Transitions between these energy levels can be induced by the application of radiofrequency radiation. This is known as nuclear magnetic resonance (NMR). For example, the protons in water experience this resonance effect in a field of 0.3 T at a radiation frequency of 12.6 MHz. The resonant frequency depends on the magnetic field at the nucleus which in turn depends on the environment of the particular nucleus. NMR gives invaluable information on the structure of molecules. It has also been developed to provide a non-invasive clinical imaging of the human body, magnetic resonance imaging (MRI); multiple projections are combined to form images of sections through the body, providing a powerful diagnostic aid. The nuclei of certain isotopes, eg ^{1}H, ^{19}F, ^{31}P, behave like small bar magnets and will line up when placed in a strong d.c. magnetic field. The direction of alignment of the nuclei in a sample may be altered by RF irradiation in a surrounding coil whose axis is at right angles to the magnetic field. The frequency at which energy is absorbed and the direction of alignment changes is known as the resonant frequency and varies with the type of nucleus; eg in a magnetic field of 14 000 gauss ^{1}H resonates at 60 MHz, ^{19}F at 56 MHz and ^{31}P at 24.3 MHz. Nuclei with even mass and atomic number, eg ^{12}C, ^{16}O, have no magnetic properties and do not exhibit this behaviour. Abbrev. *NMR*.

nuclear magnetic resonance spectroscopy Routine analytical tool for detecting atomic nuclei with spin (^{1}H, ^{14}C, ^{15}N, ^{19}F, ^{31}P etc) in molecules, by absorption at resonance. Nuclear magnetic field is modified by chemical environment (chemical shift), so protons etc in different parts of molecules can be differentiated. It is esp. valuable for polymers, enabling different stereoisomeric forms to be identified, and its sensitivity enables fingerprinting of unknown materials. Resolution depends on applied field strength, with many machines available now up to 500 MHz. Complementary to infrared and ultraviolet spectroscopy. Also *NMR spectroscopy*.

nuclear magneton See **Bohr magneton**.

nuclear paramagnetic resonance See **nuclear magnetic resonance**.

nucleating agents Substances added to molten materials and solutions to accelerate the onset and increase the rate of crystallization, eg stearates in polypropylene.

nucleation Initiation of processes, such as crystallization or fracture of materials. Hence homogeneous and heterogeneous nucleation of crystalline materials, depending on phase state of nuclei. Followed by growth processes, studied using eg the **Avrami equation**.

nuclei Centres or *seeds* from which crystals begin to grow during solidification or recrystallization. In general, they are minute crystal embryos formed spontaneously in the melt, but foreign particles and non-metallic inclusions may also act as nuclei.

nucleons Protons and neutrons in a **nucleus** of an atom.

nucleophilic reagents Term applied to reagents which react at positive centres, eg the hydroxyl ion (OH$^-$) in replacing Cl$^-$ from a C–Cl link in which the C atom can be considered as being the positive centre.

nucleus Composed of protons (positively charged) and neutrons (no charge), and constitutes practically all the mass of the atom. Its charge in units of the charge of the electron equals the atomic number; its diameter is from 10^{-15} to 10^{-14} m. With protons, equal to the atomic number, and neutrons to make up the atomic mass number, the positive charge of the protons is balanced by the same number of extra-nuclear electrons.

nugget (1) Term for a welding bead. (2) A mass of gold or silver found free in nature.

Nujol Trade name for a pure paraffin oil free of unsaturated compounds and used esp. for making mulls for infrared spectroscopy.

number-average molecular mass See **molecular mass distribution** p. 212.

nun's veiling Lightweight, plain-weave fabric made from worsted, silk or cotton yarns, usually dyed black, and used as the name suggests.

nyankom See **niangon**.

Nylatron TN for a range of filled nylon compounds reinforced for engineering purposes.

nylon Generic name for any long chain synthetic polymeric amide which has recurring amide groups as an integral part of the main polymer chain, and which is capable of being formed into a filament in which the structural elements are oriented in the direction of the axis. See **step polymerization** p. 298.

nylon-11 Polyamide plastic with eleven carbon atoms between each amide group in backbone chain. Also *nylon-12*. TN Rilsan.

nylon salt See **step polymerization** p. 298.

Nyrim TN for a *reaction injection moulding* process using block polymers formed from caprolactam and polyethers, so type of thermoplastic elastomer (Europe).

o- Abbrev. for **ortho-**, ie containing a benzene nucleus substituted in the 1.2 positions.

ω Symbol for: (1) specific magnetic rotation; (2) angular frequency.

ω- Symbol indicating: (1) substitution in the side chain of a benzene derivative; (2) substitution on the last carbon atom of a chain, farthest from a functional group.

O Symbol for **oxygen**.

O- Symbol indicating that the radical is attached to the oxygen atom.

oak A strong, tough and dense **hardwood**, from European, Asian and American varieties of *Quercus*. The European oak is hard and very durable with a straight-grained, light tan to biscuit coloured **heartwood**. Density about 720 kg m⁻³. Commonly used in constructional work for timber bridges, heavy framing and piles, as well as for high-class joinery, cooperage, furniture, flooring, and in veneers. Also *American red oak*, *American white oak*, *Japanese oak*.

Obach cell A dry **primary cell** having ingredients similar to the **Leclanché cell**, ie a carbon positive electrode surrounded by a depolarizer of manganese dioxide paste, and having the electrolyte in the form of a paste

of sal-ammoniac, plaster of Paris, and flour.

obeche Low-density, W African **hardwood** (*Triplochiton scleroxylon*), creamy-white to pale yellow, density about 380 kg m⁻³. Non-durable and used for light interior work and in plywood and veneers. Also *African white-wood*, *wawa*.

obscured glass Glass which is so treated as to render it translucent but not transparent, eg by sandblasting or etching.

octave See **law of octaves**.

octavo See **paper sizes**.

octet See **electron octet**.

ODR Abbrev. for **Oscillating Disk Rheometer**.

oersted CGS electromagnetic unit of magnetic field strength, such that 2π oersted is a field at the centre of a circular coil 1 cm in radius carrying a current of 1 abampere (10 amperes). Now replaced by the SI unit ampere per metre. $1 \text{ A m}^{-1} = 4\pi \times 10^{-3}$ oersted.

OFHC Abbrev. for *Oxygen-Free High-conductivity Copper*.

ohm SI unit of electrical resistance, such that 1 ampere through it produces a potential difference across it of 1 volt. Symbol Ω.

ohmic contact One in which the current flows with equal facility in both directions, with incremental junction voltage proportional to incremental current. Ohmic contacts in connections to semiconductors can usually be ensured by heavily doping the semiconductor so that electrons can tunnel through any barrier layers which may arise from the bulk and interface properties of the materials involved.

ohm-metre SI unit of **resistivity**. Symbol Ω m.

Ohm's law In an electrically conducting component, at constant temperature and zero magnetic field, the current I flowing through it is proportional to the potential difference V between its ends, the constant of proportionality being its conductance. So $I = V/R$ or $V = IR$, where R is the resistance of the component. The law is strictly applicable only to a direct current and for practical purposes to components of negligible reactance carrying alternating current. Can be extended by analogy to any physical situation where a pressure difference causes a flow through an impedance, eg heat through walls, liquid through pipes.

-oid Suffix after Gk *oides*, from *eidos*, form.

oil (1) Crude oil, see **petroleum**. Source of fuels and chemical intermediates, and materials such as polymers, carbon black, etc. (2) Viscous liquid of organic nature, having special properties, eg **bitumen of Judea**, aromatic oil etc. See **oils**.

oilcloth A waterproof material obtained by coating a cotton fabric with oxidized **linseed**

oil or other drying oil.

oiled paper Paper impregnated with non-drying oil, eg oiled wrapping, or with **linseed oil**, eg oiled manilla.

oil extension Absorption of aromatic oil by rubber to aid processing. Similar effect to plasticization of rigid polymers like polyvinyl chloride. Most widely used for styrene butadiene rubber in manufacture of tyres. See **tyre technology** p. 325.

oil hardening The hardening of steels of medium and high carbon content by heating to the austenitic condition, followed by quenching into a bath of oil, resulting in a cooling less sudden than is effected by water, and consequently a reduced risk of cracking.

oil-hardening steel Any alloy steel which will harden when cooling in oil instead of in water. The limiting diameter or cross-section which will harden fully in this manner must also be stated, since this depends on the alloy composition and transformation characteristics of the steel.

oil of Judea See **bitumen of Judea**.

oil quench The operation of quenching a heated component into a bath of oil.

oils A group of neutral liquids comprising three main classes: (1) fixed (fatty) oils, from animal, vegetable and marine sources, consisting chiefly of glycerides and esters of fatty acids; (2) mineral oils, derived from petroleum, coal, shale etc consisting of hydrocarbons; (3) essential oils, volatile products, mainly hydrocarbons, with characteristic odours, derived from certain plants. See **oil**.

olefins See **alkene, oligomer**. Several monomer units linked together.

olig-, oligo- Prefixes from Gk *oligos*, a few, small.

oligomer Low molecular mass polymer formed from two or more monomer units linked together. Used as **prepolymers** in eg epoxy adhesives.

oligosaccharides Carbohydrates containing a small number (two to 10) of monosaccharide units linked together, with elimination of water.

olive *Olea europaea*, from the Mediterranean regions of Europe, important for its fruit which produces edible oil when compressed and is itself edible after processing with dilute NaOH to neutralize the bitter glucoside content. E African olive (*O. hochstetteri*) is used for high-grade, abrasion-resistant flooring, decorative work and in veneers. It is a pale- to mid-brown, attractively marbled wood, moderately durable, and density about 900 kg m^{-3}.

Olsen ductility test Test in which a standard size ball is forced into a blank of sheet metal and the depth of cupping measured when the metal fractures.

one-electron bond Bond formed by the resonance of a single electron between two atoms or radicals of similar energy.

onion skin paper A glazed translucent paper having an undulating surface caused by the special beating and drying of the paper.

Onsager equation An equation based on the **Debye–Hückel theory**, relating the equivalent conductance Λ_c at concentration c to that at zero concentration, of the form $\Lambda_c = \Lambda_0 (A + B\Lambda_0) \sqrt{c}$, where A and B are constants depending only on the solvent and the temperature.

opalescence The milky, iridescent appearance of a solution or material, due to the reflection of light from very fine, suspended particles.

opal glass Glass which is opalescent or white; made by the addition of fluorides (eg fluorspar, cryolite) to the glass mixture.

open-hearth process Now obsolete process for steel making in a furnace having a dish shaped hearth, eg Siemens–Martin process (1866). Cold pig-iron and steel scrap are charged to open hearth and melted by preheated gases with flame temperatures up to 1750°C produced by regeneration as they enter through brick labyrinth, where heat is maintained by periodic reversal between these and exit gases. After melting (up to 1600°C) most impurities are in the covering slag. Remainder are removed by addition of iron ore, scale, or limestone before tapping into ladles, where ferrosilicon, ferromanganese, or aluminium may be added as deoxidants. See **acid process, basic process, steel-making**.

open loop recycling Material reclamation from one system to another eg collection of polyethylene terephthalate containers, granulation and melt spinning to fibre for low-grade filling. Also *secondary recovery*.

open pore A cavity within a particle of powder which communicates with the surface of the particle.

opepe W African **hardwood** (*Nauclea diderrichii*) whose **heartwood** is orange-brown, with an interlocked grain, fairly coarse texture, and has a copper lustre. It is very durable, and used in piling and marine work, flooring, furniture, joinery, and veneers. Density about 740 kg m^{-3}.

operating diagram Graph of volumetric melt flow in extruder barrel versus melt pressure. Intersection of screw and die operating curves defines the **operating point**. See **extrusion** p. 121.

operating point Optimum set of conditions eg temperature, pressure, shear rate for extrusion of polymers.

operational sphericity A shape factor defined as the cube root of the ratio of the volume of a particle of powder to the volume of the circumscribing sphere.

OPP film Abbrev. for **Oriented Poly-Propylene film.**

optical activity A property possessed by many substances whereby plane-polarized light, in passing through them, suffers a rotation of its plane of polarization, the angle of rotation being proportional to the thickness of substance traversed by the light. In the case of molten or dissolved substances it is due to the possession of an asymmetric molecular structure, eg no mirror plane of symmetry in the molecule. See **chirality.**

optical bleaches See **fluorescent whitening agents.**

optical crown Glass of low dispersion made for optical purposes. See **crown glass.**

optical fibre Fibres made from high-purity, low-loss, low-dispersion glass and used as medium for telecommunications by transmission of high-frequency pulses of light. Basis of operation is total internal reflection at interface between higher refractive index core and lower refractive index sheath or cladding. Core is usually vitreous silica doped with germania (GeO_2). Used in **fibre optics.**

optical flint Glass of high dispersion made for optical purposes. See **flint glass.**

optical glass Glass made expressly for its optical qualities. The composition varies widely both as to constituents and amounts, with very exacting requirements for freedom from streaks and bubbles. See **crown glass, flint glass.**

optical isomerism The existence of isomeric compounds, stereoisomers, which differ in their **chirality.** Important for polymers. See **stereoregular polymers** p. 299.

optical microscopy Method of analysing the microstructure of materials using visible light source, in transmission (transparent materials like many polymers, glasses etc) or reflection (opaque materials like crystalline metals and ceramics). Specimen preparation method is important for accurate interpretation of structure, and includes ultramicrotomy for polymers and biomaterials, polishing and etching for metals. Special methods use polarized light, ultraviolet etc.

optical rotary dispersion The change of optical rotation with wavelength. Rotatory dispersion curves may be used to study the configuration of molecules. Abbrev. *ORD.*

optical rotation The rotation of the plane of polarization of a beam of light when passing through certain materials.

optical whites See **fluorescent whitening agents.**

optic axis Direction(s) in a doubly refracting crystal for which both the ordinary and the extraordinary rays are propagated with the same velocity. Only one exists in uniaxial crystals, two in biaxial.

optic branch The lattice dynamics of a crystal containing n atoms per unit cell shows that the dispersion curve (frequency ω against wavenumber q) has $3n$ branches of which $3n - 3$ are *optic* branches. The branches are characterized by different patterns of movement of the atoms. For the optic branch ω is independent of q for small values of q. See **acoustic branch.**

optoelectronics See panel on p. 226.

orange lead Lead oxide. Pb_3O_4. Obtained by heating white lead (basic lead (II,II,V) carbonate) in air at approximately 450°C. Commercial varieties contain up to approximately 35% PbO_2.

orbital The properties of each electron in a many-electron atom may be reasonably described by its response to the potential due to the nucleus and to the other electrons. The **wave function**, which expresses the probability of finding the electron in a region, is specified by a set of four quantum numbers and defines the orbital of the electron. The state of the many-electron atom is given by defining the orbitals of all the electrons subject to the **Pauli exclusion principle.**

ORD Abbrev. for **Optical Rotatory Dispersion.**

order-disorder transformation A solid-state transformation in which the random arrangement of solute and solvent atoms in the crystal lattice takes up an ordered configuration or superlattice below a critical temperature.

ordered state A solid solution alloy existing as an ordered arrangement of atoms within the crystal lattice, eg CuAu and Cu_3Au.

order of reaction A classification of chemical reactions based on the index of the power to which concentration terms are raised in the expression for the instantaneous rate of the reaction, ie on the apparent number of molecules which interact. If $A + B \rightarrow$ products, then the rate of reaction is $k\,A^a\,B^b$; the order is a with respect to A and b with respect to B. k is the *rate constant.*

ordinary ray See **double refraction.**

Oregon pine See **Douglas fir.**

organdie Lightweight, cotton, plain-weave, dress fabric, stiffened and made transparent by treatment with conc. sulphuric acid. Cf **mercerization.**

organic chemistry The study of the compounds of carbon. Owing to the ability of carbon atoms to combine together in long chains (catenate), these compounds are far

opto-electronics

Opto-electronics concerns the interconversion of electrical and optical signals. Opto-electronic devices are used in visual displays and in optical communication systems. Cathode ray tube **phosphors**, indicators on hi-fi equipment, the lasers which are used to read compact discs and the light sensors in cameras are common examples of devices which exploit opto-electronic effects. The transmitters and receivers in remote controls for television/video equipment and optical fibre communication systems use opto-electronics in both the generation and the detection of signals.

The absorption and emission of light from materials is associated with electronic transitions between energy levels, differing in energy by an amount equivalent to that of the photons involved. Transitions between energy levels can also be accomplished by means of thermal energy (phonon) interactions and these processes are often in competition with radiative (optical) ones. Visible radiation has wavelengths in the range 400–800 nm but opto-electronic devices operate over a wider range from the infrared (3000 nm) to the near ultraviolet (200 nm), which correspond with photon energies from 0.4 to 6 eV.

Table 1. **Band gap and wavelength for various semiconductor materials**.

Semiconductor material	Direct or indirect gap	Band gap eV	Wavelength µm
Indium antimonide, InSb	D	0.17	7.29
Lead telluride, PbTe	I	0.31	4.00
Indium arsenide, InAs	D	0.36	3.44
Lead sulphide, PbS	I	0.41	3.02
Germanium, Ge	I	0.67	1.85
Gallium antimonide, GaSb	D	0.72	1.72
Silicon, Si	I	1.12	1.10
Indium phosphide, InP	D	1.35	0.92
Gallium arsenide, GaAs	D	1.42	0.87
Cadmium telluride, CdTe	D	1.56	0.79
Cadmium selenide, CdSe	D	1.70	0.73
Gallium phosphide, GaP	I	2.26	0.55
Cadmium sulphide, CdS	D	2.42	0.51
Silicon carbide, SiC	I	3.00	0.41
Gallium nitride, GaN	D	3.36	0.37
Zinc sulphide, ZnS	D	3.68	0.34

Many of the applications of opto-electronics use some form of light emitting diode (LED) to generate radiation and a photodiode to detect it. In these devices it is carriers in and around the **p-n junction** which are involved in transitions between **conduction** and **valence bands**, with associated photons having an energy comparable with the **band gap**. Table 1 lists the band gaps of a number of photoconductors. Diodes fabricated from those with so-called *indirect gaps* are extremely poor emitters; this is connected with the dynamics of recombination and with detail of the band structure. *Direct gap* materials achieve much higher efficiency, as high as 50%. Control over band gap energy (and so wavelength of emitted photons) can be achieved in practice by alloying: the GaAs/GaP system is used, for example, to make LEDs covering the range from red to green.

By combining different semiconductor materials in **heterostructures**, complex p-n devices can be built which confine the emitted radiation to certain regions within the material. Such structures are at the heart of semiconductor *diode lasers*.

continued on next page

opto-electronics (contd)

Photodiodes can be used to generate electric power as in solar cells or else simply to respond to changes in illumination. In either case, radiation incident on the junction region creates electron-hole pairs which are swiftly swept apart by the electric fields (inherent in p-n junctions) thereby constituting a current which is dependent on the intensity of the radiation.

The integration of electronic with opto-electronic devices is hindered by the dismal opto-electronic properties of silicon and the relative difficulty of electronic device fabrication in other semiconductors. However the development of **porous silicon** and of novel electronic devices in other materials may ultimately result in their easier integration with very important industrial advantages.

more numerous than those of other elements. They are the basis of all living matter.

organic electrical conductor An organic substance exhibiting electrical conductivity similar to that of metals. Eg TTF-TCNQ (tetrathiofulvalene-tetracyanoquinodimethane) has a room temperature conductivity of 5×10^4 S m^{-1} and its conductivity increases at low temperature, by a factor of at least 20 at 60 K. (TMTSF)$_2$ClO$_4$, where TMTSF is the tetra methyl selenium derivative of TTF, is one of a series of organic conductors which show superconductivity below 1.2 K.

organic phosphor Organic chemical used as solid or liquid scintillator in radiation detection.

organo-metallic compounds Wide range of compounds in which carbon atoms are linked directly with metal atoms, including the alkali metals, eg butyl lithium, C_4H_9.Li; lead, eg lead tetraethyl, $Pb(C_2H_5)_4$; zinc, eg zinc dimethyl, $Zn(CH_3)_2$. The compounds are useful reagents in polymerization reactions. Many are formed by transition elements. See **Ziegler–Natta catalyst**.

organosilicone Synthetic resin characterized by long thermal life and resistance to thermal ageing. See **silicones**.

organosol A coating composition based on a dispersed polyvinyl chloride resin mixed with a plasticizer and a diluent (see **plastisol**), a colloidal solution in any organic liquid.

organza Plain-weave fabric made from untreated, continuous filament silk, it is sheer and stiff (due to the silk gum); now also made from continuous 'man-made' fibre yarns.

orientation (1) The determination of the position of substituent atoms and groups in an organic molecule, esp. in a benzene nucleus. (2) The ordering of molecules, particles, or crystals so that they point in a definite direction. (3) The position of important sets of planes in a crystal in relation to any fixed

system of planes. (4) The direction of polymer molecules or crystallites in fibres relative to the fibre axis imposed either during growth (natural fibres) or while being extruded and drawn (man-made and synthetic fibres). (5) Way in which polymer chains are aligned in a moulded product. (6) Direction of fibres in **composite materials**. Control of orientation is critical to optimize product strength and stiffness.

oriented polypropylene film Film which has been biaxially stretched to improve its physical properties. Widely used for packaging because of its high gloss and clarity, high impact strength and low moisture permeability. Abbrev. *OPP film*.

O-ring A toroidal ring, usually of circular cross-section, made of neoprene or similar materials, used eg as an oil or air seal.

Orlon TN for synthetic fibre based almost wholly on **PAN** with a small amount of a different monomer to serve as a dye receptor. Widely used in knitted fabrics, as imitation fur and in carpets.

ormolu An alloy of copper, zinc and sometimes tin, used (esp. in the 18th century) for furniture mountings, decorated clocks etc.

oroide See **French gold**.

ortho- Prefix from Gk *orthos*, straight. Specifically with meaning: (1) derived from an acid anhydride by combination with the largest possible number of water molecules, eg orthophosphoric acid; (2) consisting of diatomic molecules with parallel nuclear spins and an odd rotational quantum number, eg orthohydrogen; (3) The 1,2 relationship of substituents on a benzene molecule. Cf **meta** (4), **para** (2).

orthogonal cutting A cutting process used to analyse the forces occurring in machining, in which a straight-edged cutting tool moves relative to the workpiece in a direction perpendicular to its cutting edge.

oscillating disk rheometer Device for measuring flow rate such as the Monsanto rheometer, which measures torque needed to shear rubber, typically during a **cure cycle**. Abbrev. *ODR*.

oscillator crystal A **piezoelectric** crystal used in an oscillator to control the frequency of oscillation.

osmiridium A very hard, white, naturally occurring alloy of osmium (17–48%) and iridium (49%) containing smaller amounts of platinum, ruthenium and rhodium. Used for the tips of pen-nibs.

osmium A metallic element, a member of the platinum group. Symbol Os. Osmium is the densest element, rel.d. at 20°C 22.48. Like platinum it is a powerful catalyst for gas reactions, and is soluble in *aqua regia*, but unlike platinum when heated in air it gives an oxide, volatile OsO_4. Alloyed with iridium it forms an extremely hard material, eg **osmiridium**. See Appendix 1, 2.

osmium tetroxide *Osmium* (VIII) *oxide*, OsO_4, yellow crystals which give off an ill-smelling, poisonous vapour. Its aqueous solution is used as a histological stain for fat, a catalyst in organic reactions, a fixative in electron microscopy and a stain for double-bonded polymers in blends and copolymers.

osmometry Method for determining molecular mass of molecules, esp. soluble polymers. Measurement of osmotic pressure across a semipermeable membrane allows calculation of the number-average molecular mass, $\overline{M}_n$. It is an absolute method for molecular mass, M, but relatively slow compared to **gel-permeation chromatography**, GPC, which also gives the whole distribution. See **molecular mass distribution** p. 212.

osmosis Diffusion of a solvent through a semipermeable membrane into a more concentrated solution, tending to equalize the concentrations on both sides of the membrane.

osmotic pressure Symbol Π. The pressure which must be applied to a solution, separated by a semipermeable membrane from pure solvent, to prevent the passage of solvent through the membrane. For substances which do not dissociate, it is related to the concentration (c) of the solution and the absolute temperature (T) by the relationship, $\Pi = cRT$, where R is the **gas constant**.

Ostwald's dilution law The application of the law of mass action to the ionization of a weak electrolyte, yielding the expression

$$\frac{\alpha^2}{(1 - \alpha)\,V} = K,$$

where α is the degree of ionization, V the dilution, and K the ionization constant, for the case in which two ions are formed.

Ostwald viscometer Type of capillary viscometer used for measuring viscosity of dilute polymer solutions, and hence, by a series of steps, viscosity-average molecular mass.

ottoman A woven cloth with a flat, prominent rib in the weft direction; originally with a silk warp and a worsted weft. Used in tailoring.

outgassing Removal of occluded, absorbed or dissolved gas from a solid or liquid. For metals and alloys, done by heating in vacuo.

oven-dry paper See **bone-dry paper**.

overblowing Prolonging the oxygen blast through the molten steel after the carbon has been removed in the **Bessemer process**, resulting in excessive oxidation of iron.

overcure Rubber technology term for effects occuring towards end of vulcanization, which may involve chain degradation or increased crosslinking.

overheated Said of metal which has been heated in preparation for hot-working, or during a heat-treating operation, to a temperature at which rapid grain growth occurs and large grains are produced. The structure and properties can be restored by treatment, and in this respect it differs from burning. See **burnt metal**.

overpoled copper See **poling**.

overpotential The extra voltage which must be applied to an electrode to initiate the electrode reaction in an electrochemical cell, over and above the equilibrium electrode potential. Symbol η. See **Tafel plot**.

overstrain The result of stressing an elastic material beyond its yield point; a new and higher yield point results, but the elastic limit is reduced.

Oxford shirting A plain-weave, warp-striped, cotton shirting material in which two ends are woven as one.

oxidation The addition of oxygen to a compound. More generally, any reaction involving the loss of electrons from an atom. It is always accompanied by reduction.

oxidation of polymers Effect of atmospheric oxygen on polymers, particularly at elevated temperature and during exposure to ultraviolet light. Starts at weak points in chain (eg tertiary carbon atoms) or chain defects (eg head-to-head junctions), with formation of carbonyl groups. If further activated, they break the backbone chain and the molecular mass falls.

oxidation-reduction indicators Those substances which exist in oxidized and reduced forms having different colours, used to give approximate values of oxidation-reduction potentials.

oxidation-reduction potential See **standard electrode potential**.

oxide-coated cathode One coated with

oxides of the alkali and alkaline-earth metals, to produce thermionic emission at relatively low temperatures.

oxide deposition In semiconductor processing, the deposition of silicon dioxide as a passivation layer, for example by **chemical vapour deposition** or as a **spin on glass**.

oxide growth In silicon semiconductor processing the growth of silicon dioxide by oxidation of substrate silicon at high temperatures ($\approx 1000°C$) in an oxygen ('dry') or steam ('wet') environment.

oxide isolation The isolation of the circuit elements in a microelectronic circuit by forming a layer of silicon oxide around each element.

oxides Compounds of oxygen with another element. Oxides are formed by the combination of oxygen with most other elements, particularly at elevated temperatures, with the exception of the noble gases and some of the noble metals. Many are used in ceramics, eg MgO and glasses, eg SiO_2, B_2O_3.

oxidizer A substance that removes electrons from another and produces heat.

oxidizing agent A substance which is capable of bringing about the chemical change known as oxidation in another substance. It is itself reduced.

oxidizing flame The outer cone of a nonluminous gas flame, which contains an excess of air over fuel.

oxyacetylene cutting Using the heat of an oxy-acetylene flame to cut through material. If the material is a metal which oxidizes readily, when once the temperature is high enough the acetylene may be turned off and the jet of oxygen alone continues the cutting action. See **flame cutting**.

oxyacetylene welding Welding with a flame resulting from the combustion of oxygen and acetylene. Many ferrous metals can also be cut by the same process. The cut is started by introducing a jet of oxygen on to the metal after it has been preheated.

oxygen A non-metallic element, symbol O. Molecular oxygen is a diatomic, colourless, odourless gas which supports combustion and is essential for the respiration of most forms of life. Oxygen is the most abundant

element on Earth, forming 21% by volume of the atmosphere, 89% by weight of the water, and nearly 50% by weight of the rocks of the Earth's crust. It is manufactured from liquid air, and is used in gas welding (usually with acetylene), in steel manufacture, in medical practice and in anaesthesia; liquid oxygen is much used in rocket fuels. Ultraviolet radiation forms ozone, O_3 from atmospheric oxygen. See Appendix 1, 2.

oxygen cracks Surface effect of oxidation of polymers, usually forming a randomly oriented pattern. In rubbers, forms a crazypaving pattern readily distinguishable from **ozone cracks**.

oxygen index Test for flammability of polymers, defined as the fraction of oxygen in a nitrogen mixture needed to maintain a candle-like flame on a burning specimen (held in a standard way). Polymers with an oxygen index (abbrev. *OI*) greater than the normal fraction of oxygen in air, are held to be flammable, those with an OI less than 0.21, inflammable. But see **flammability**. Also *limiting oxygen index* (LOI).

oxygen lancing A process used principally for cutting heavy sections of steel or cast iron, in which oxygen is fed to the cutting zone through a length of steel tubing which is consumed as the cutting action proceeds. The cutting zone or the end of the tube has to be preheated to commence cutting.

oxyhydrogen welding A method of welding in which the heat is produced by the combustion of a mixture of oxygen and hydrogen.

ozone O_3. Produced by the action of ultraviolet radiation or electrical corona discharge on oxygen or air. Powerful oxidizing agent.

ozone cracks Deep, sharp cracks oriented perpendicular to strain axis, commonly found on external surfaces of rubber products, esp. tyres made from natural rubber and butadiene polymers. Caused by chain scission at double bonds in polymers due to atmospheric attack by ozone (O_3), esp. found in air polluted from sunlight photolysis, and near electrical equipment. Inhibited by antiozonants. Cf **oxygen cracks**.

PQ

p Symbol for: (1) **pico-**; (2) **proton**.

p Symbol for the number density of positive carriers in semiconductors expressed as holes per cubic metre (or per cubic centimetre). Cf ***n***.

p- Symbol for **para-**.

φ Symbol for: (1) the phenyl radical C_6H_5; (2) amphi-, ie containing a condensed double aromatic nucleus substituted in the 2.6 positions; (3) **magnetic flux**; (4) **work function**.

π See under **pi**.

P Symbol for **phosphorus**.

P Symbol for: (1) electric polarization; (2) power; (3) pressure.

Π Symbol for pressure, esp. **osmotic pressure**.

Ψ Symbol for: (1) **pseudo-**; (2) **wave function**.

[*P*] Symbol for **parachor**.

Pa Abbrev. for **Pascal**.

PA Abbrev. for **PolyAmide**, often followed by numbers, eg PA 6,6, which represent the number of carbon atoms in repeat unit(s).

PAA Abbrev. for **PolyAcrylic Acid**.

pacemaker Bioengineered electronic/electrical implant designed to control heart beat. Powered by lightweight longlife cells, eg lithium battery.

packaging Material used to surround and thus protect products from their environment. Paper and board now giving way to thermoplastics, esp. polyolefins, SAN, for their design flexibility, as in blister packs and shrink-wrap film. Foamed polymers, eg foam polystyrene, provide additional protection with crush-proof properties and good thermal and electrical insulation. Some materials present recycling problems, esp. polyvinyl chloride, which can form HCl gas when burnt. Plasticizer migration into food etc can be a problem with plasticized polyvinyl chloride film.

pack-hardening Case-hardening, using a solid carburizing medium which is packed around the low carbon steel objects and the whole then heated for a time and at a temperature to allow carbon to diffuse into the surface to the desired depth. After the carburization process, the objects must be heat treated to develop hardness at the surface.

packing pressure Extra pressure applied towards end of injection moulding cycle to ensure complete filling of cavity.

padauk Three varieties of **hardwood** (*Pterocarpus*): African-, Andaman-, and Burma-padauk. Their **heartwood** has attractively shades of red-brown, with straight to interlocked grain and a moderately coarse texture. It is strong and very durable and is used in high class furniture and cabinetmaking, flooring, joinery etc and in veneers. Densities are about 720 kg m^{-3} (African-), 770 kg m^{-3} (Andaman-) and 850 kg m^{-3} (Burma-).

paddle mixer Type of machine used for mixing paints and polyvinyl chloride plastisols.

pad oxide In silicon semiconductor processing, a thin oxide growth used to protect active areas during early stages of fabrication.

Page effect A click heard when a bar of iron is magnetized or demagnetized. See **magnetostriction**.

paint A suspension of a solid or solids in a liquid which, when applied to a surface, dries to a more or less opaque, adhering solid film. The solids are colour- or opacity-imparting, finely ground pigments together with extenders; the liquid consists of suitable oils, solvents, polymers, aqueous colloidal solutions or dispersion etc, together with other lesser ingredients. Used for protective and decorative purposes.

palladinized asbestos Asbestos fibres saturated with a solution of a palladium compound, which is subsequently decomposed to give finely divided palladium dispersed throughout the asbestos.

palladium A metallic element, symbol Pd. Used as a catalyst in hydrogenation and increasingly used with platinum for catalytic converters in cars. Native palladium is mostly in grains and is frequently alloyed with platinum and iridium. See Appendix 1, 2.

PAN Abbrev. for **PolyAcryloNitrile**.

panama Plain-weave, worsted fabric used for tropical suiting and weighing about 200 g m^{-2}; may also be made from non-woollen fibres.

panel beating A sheet-metal working craft in which complex shapes are produced by stretching and gathering the sheet locally with hammers and dollys.

papaya A tree (*Carica papaya*), native to S America but common in the tropics. The trunk, leaves and fruit yield papain, a protease used commercially as a meat-softener.

paper Consists of a mat of cellulose fibres, freed from non-cellulose constituents and deposited from an aqueous suspension. Wood-pulp, **esparto** and rags are the chief raw materials, though other substances are used. During the process of manufacture, the fibres are reduced to the requisite

paper and papermaking

Paper is a mat of cellulose fibres held together by hydrogen bonds. In principle, any cellulose-based, natural material can be used as a source of the fibres; in the past, major sources have been hemp ropes, cotton rags and esparto grass. Now, over 90% of paper is made from wood pulp, containing 0.5 to 3 mm length fibres.

A suspension of about 2% of these fibres in water is subjected to a high shear rate, known as *beating* or *fibrillation*, to break down partially their cell wall structures. See **structure of wood** p. 305. A combination of hydrolysis (swelling the fibres and breaking the cellulose-cellulose hydrogen bonds) and the shearing action splits the cell walls into fibrils, which splay out from the parent fibres. The amount and nature of the fibrillation are critical to the properties of the resulting paper. The beaten suspension is then drained by spreading it over a fine mesh (the *web* or *felt*) which moves like a conveyor belt in continuous papermaking or onto a rotating, perforated cylinder. Drainage can be vacuum-assisted. The surface tension of the draining water, both in bulk and in capillaries, provides the consolidating forces on the fibrous mat which turn it into paper. These forces need to be sufficiently high to ensure that intersecting fibres and fibrils are close enough to form hydrogen bonds, and this is aided by the enhanced plasticity of the swollen fibres. About 1–2% of the total hydrogen bonds available in the cellulose are used in the bonding of paper.

The properties of paper are extensively controlled and modified by the use of additives and coatings. These include such physical properties as density, elastic modulus (it exhibits **viscoelasticity** p. 332) and strength (particularly when wet; see **wet strength paper**) as well as opacity, colour, gloss, liquid absorbance and printability. Typically, papers show the following range of properties: density, 200–1400 kg m^{-3}; tensile modulus, 1–5 GN m^{-2}; tensile strength, 30–80 MN m^{-2}; breaking strain, 4–12 %.

length, and their physical properties are modified by mechanical or chemical treatment. See **cellulose**.

paper and papermaking See panel on p. 231.

paper capacitor One which has thin paper as the dielectric separating aluminium foil electrodes, these being wound together and waxed.

paper sizes Now increasingly based on **ISO sizes**; previous UK sizes, shown in inches below, for esp. book printing were based on broadsides (ie unfolded sheets) of *crown* (15 × 20), *demy* (17½ × 22½), *medium* (18 × 23) and *royal* (20 × 25), which were then divided into quarter (*quarto*, or 4to) and eighth (*octavo* or 8vo) sizes. See **albert, elephant, foolscap, quad**.

papier mâché Substance made from paper mashed in water and built up in layers with glue, either over a preform or into a mould. Used for furniture in the 18th century, imitation interior wood-, stone-, and plasterwork in the 19th century, and aircraft fuel tanks in World War II.

papyrus An early form of writing material prepared from the water reed (*Cyperus papyrus* or *Papyrus antiquorum*) of the same name and used for records up to the 9th century AD.

para- (1) Prefix from Gk *para*, beside. (2) Containing a benzene nucleus substituted in the 1,4 positions. (3) Consisting of diatomic molecules with anti-parallel nuclear spins and an even rotational quantum number.

paraffins A term for the whole series of saturated aliphatic hydrocarbons of the general formula C_nH_{2n+2}. Also alkane hydrocarbons. They are indifferent to oxidizing agents, and not reactive, hence the name paraffin (L. *parum affinis*, little allied).

paraffin wax Higher homologues of alkanes (C_{22} and on), wax-like substances obtained as a residue from the distillation of petroleum; mp 45–65°C, rel.d. 0.9, resistivity 10^{13} to 10^{17} Ω m, permittivity 2–2.3.

Parafil TN for parallel-laid cable constructed from high-performance fibre (eg aramid) enclosed by thermoplastic protective sheath, and fitted with special end-connectors.

paramagnetic resonance See **electron spin resonance**.

paramagnetism Phenomenon in materials which have a positive susceptibility and a permeability slightly greater than unity. An applied magnetic field tends to align the magnetic moments of the atoms or molecules and the material acquires magnetization in the direction of the field; it disappears when the field is removed. Used to obtain very low temperatures by adiabatic demagnetization.

parana pine S American **softwood** (*Araucaria angustifolia*) which is not a true pine (*Pinus*). Its **heartwood** is soft, non-durable, pale brown, with straightish grain and uniform texture. Mostly used in internal joinery, plywood and veneers. Mean density 540 kg m^{-3}.

parchmentizing The process of passing paper through sulphuric acid, or zinc chloride, which causes the fibres to swell and the paper to become translucent, dense and greaseproof, possessing high wet-strength properties.

parenchyma See **structure of wood** p. 305.

parent metal A term used in welding to denote the metal of the parts to be welded.

parison Short length of moulded or extruded glass or polymer which is the precursor for subsequent **blow moulding**.

parison variator Device fitted to extrusion blow moulding machine which automatically controls wall thickness of **parison** to give even wall in final product.

Parkesine TN for first thermoplastic resin (1865), based on cellulose nitrate, camphor, vegetable oil and other ingredients.

partially oriented yarn A synthetic polymer in continuous filament form that already has a substantial degree of molecular orientation but which requires further orientation. This may be done by drawing during a subsequent process such as texturing. Abbrev. *POY*.

partial pressures The pressure exerted by each component in a gas mixture. See **Dalton's law of partial pressures**.

particle mean size The dimension of a hypothetical particle such that, if a powder were wholly composed of such particles, such a powder would have the same value as the actual powder in respect of some stated property. See **particle size**.

particle porosity The ratio of the volume of open pores to the total volume of a particle in a powder. See **porosity**.

particle size The magnitude of some physical dimension of a particle of powder. When the particle is a sphere, it is possible to define the size uniquely by a unit of length. For a non-spherical particle, it is not possible to define a size without specifying the method of measurement. See **particle mean size**.

parting of bullion See **inquartation**.

partition coefficient The ratio of the equilibrium concentrations of a substance dissolved in two immiscible solvents. If no chemical interaction occurs, it is independent of the actual values of the concentrations. Also *distribution coefficient*.

parts rationalization Design method which seeks to minimize the number of parts in a specific product, thereby saving assembly costs and material, eg in replacing many metal parts with a single injection moulding.

Parylene TN for **polyxylylene** film.

pascal The SI derived unit of pressure or stress, equals one newton per square metre. Abbrev. *Pa*.

passivation (1) To mask the normal electro-potential of a metal. A treatment to give greater resistance to corrosion in which the protection is afforded by surface coatings of films of oxides, phosphates etc. See **Pourbaix diagram** p. 252. (2) For non-metallic materials, eg oils, the introduction of substances which have a stabilizing action in preventing chemical reaction under service conditions.

passivation layer A thin coating for the purpose of passivating a surface In semiconductor processing, especially important in protecting high field surface regions (eg where p-n junctions cut the semiconductor surface) from the environment.

paste (1) Thick dispersion of powder in a fluid. Applied eg to fluid dispersion of polyvinyl chloride in a plasticizer used for a variety of dipping processes such as glove making, leather cloth and soft toys. (2) The combined ingredients of **porcelain**. (3) Alternative name for **strass**.

pasteboard A laminated product formed by pasting two or more webs or sheets of paper together. The middle layer(s) may, if required, be of cheaper material than the facings.

patent A document which gives an inventor monopoly rights over the manufacture or marketing of a new and non-obvious device, process, material or chemical for 20 years. It is divided into two parts, the specification and the claim. The claim part represents the kernel of the invention and its wording is critical to the validity of the invention.

patenting A process applied in the manufacture of steel wire to obtain high tensile strength. Consists of austenizing the wire at 900°, cooling quickly to 550–600° and allowing to transform to a very fine pearlitic microstructure. The wire may then be cold drawn to achieve tensile strength levels in excess of 1600 M N m^{-2} without becoming brittle.

patina The thin, often multicoloured, film of atmospheric corrosion products formed on the surface of a metal or mineral.

pattern Shape used to form cavity in a

mould in which metal is to be cast.

patterning See **semiconductor device processing** p. 280.

pattern maker's rule One with graduations lengthened so as to compensate for the cooling contraction which must be allowed for in the cast object.

Pauli exclusion principle Fundamental law of quantum mechanics, that no two *fermions* in a system can exist in identical quantum states, ie have the same set of quantum numbers. This constraint explains the electronic structure of atoms and also the general nature of the periodic table. As neutrons and protons are also fermions, nuclear structure is also strongly influenced by the exclusion principle,

Paxolin TN for a paper-reinforced laminated plastic usually of the phenolic class: used in the manufacture of sheets, tubes, cylinders and laminated mouldings and for low-grade printed circuit boards.

PBT Abbrev. for **PolyButylene Terephthalate**.

PC Abbrev. for **PolyCarbonate**.

PDS Abbrev. for **Product Design Specification**.

pE Negative logarithm of effective electron concentration in a redox system. Its use is analogous to that of **pH** in acid-base reactions.

PE Abbrev. for **PolyEthylene**.

peak molecular mass Symbol M_p. See **molecular mass distribution** p. 212.

pear Hardwood (*Pyrus communis*) from Europe and W Asia. Its **heartwood** is pinkish-brown, straight-grained and with a very fine and even texture. Used in decorative applications, musical instruments and veneers. Mean density 700 kg m^{-3}.

pearlite A microconstituent of steel and cast iron comprising an intimate mechanical mixture of ferrite and cementite (iron carbide), so called because in etched sections it appears iridescent and resembles mother of pearl. It is produced at the eutectoid by the simultaneous formation of ferrite and cementite from austenite, and normally consists of alternate plates (*lamellae*) of these two constituents (see, however, **granular pearlite**). A carbon steel containing 0.8% of carbon consists entirely of pearlite when cooled under equilibrium conditions. See **eutectoid steel, hypo-eutectoid steel, hypereutectoid steel.**

pearlitic iron A grey cast-iron consisting of graphite in a matrix which is predominantly pearlite, and consequently stronger than one with a ferritic matrix.

peau de soie Fine silk, satin-weave fabric, sometimes reversible, with a ribbed or grained appearance; from Fr *skin of silk*.

PEEK Abbrev. for **PolyEther Ether Ketones**.

peeling A layer of a coating becoming detached because of poor adhesion.

peel test Mechanical test for adhesive bonding between two surfaces.

peening Using hammer blows or shot blasting to cold-work the surface layers of a metal object.

pellet A compressed mass of plastic moulding material of prescribed form and weight.

Peltier coefficient Energy absorbed or given out per second, due to the **Peltier effect**, when unit current is passed through a junction of two dissimilar metals.

Peltier effect Phenomenon whereby heat is liberated or absorbed at a junction when current passes from one metal to another.

pencil hardness See **pencil hardness test**.

pencil hardness test Simple test for hardness by scratching surface with pencils of varying hardness (8H to 4B). See **hardness measurements** p. 153.

penetrant (1) Substance which increases the penetration of a liquid into porous material or between contiguous surfaces, eg alkyl-aryl sulphonate. (2) A wetting agent.

penetrant flaw detection The use of a dye, frequently fluorescent, which will penetrate any minute crack or flaw in a component. After immersion the component is wiped dry and any subsequent seepage from fissures is detected, eg by irradiation at the exciting wavelength or by drawing the liquid out into a white absorbent coating applied after drying off. Also *fluorescent penetrant inspection*.

penetration (1) In welding, the distance between the original surface of the metal and the position of the furthest fusion penetration of the weld. (2) In foundry work, metal penetration occurs when moulding sand allows the metal to enter the spaces between the grains and so mar the surface.

penetration depth (1) Depth to which an electromagnetic field is able to penetrate a conductor (also *skin depth*). (2) Depth to which a magnetic field is able to penetrate a superconducting material. See **superconductors** p. 308.

Penton TN for a chlorinated polyether widely used as a coating material for vessels etc, where very good chemical resistance is required. Relatively expensive but cheaper than the fluorinated plastics. The film form is called *Pentaphane*. See **polytetrafluoroethylene**.

PEO Abbrev for **PolyEthylene Oxide**.

peptides Substances resulting from the breakdown of proteins, characterized by the occurrence of the –CO–NH– or amide structure in the molecule which joins adjacent pairs of amino acids together. See **polyamides.**

peptization The production of a colloidal solution of a substance, esp. the formation of a sol from a gel. Also *deflocculation.*

perfluoroalkoxy polymers Melt processable fluoropolymer, a copolymer of polytetrafluorine ethlene and perfluoro (propyl vinyl ether). High T_m of ca 300°C with good mechanical properties. Used for high-performance insulation for cable and wire, and chemically resistant valves.

performance index Alternative term for merit index, applied to both materials and products. Also *performance indicator.*

peri- Prefix from Gk *peri*, around.

periodicity Location of element in periodic table.

periodic reverse Changing the polarity in some electrolytic procedures to eg clean the electrodes.

periodic system See **periodic table.**

periodic table Classification of chemical elements into periods (corresponding to the filling of successive electron shells) and groups (corresponding to the number of valence electrons). Original classification by relative atomic mass (Mendeleev, 1869). Also *periodic system.* See Appendix 1.

peripheral Situated or produced around the edge.

peritectic Physical reaction appearing in phase diagrams in which liquid reacts with a solid already separated to form a new solid phase.

peritectoid Similar reaction to **peritectic,** except that a liquid phase is not involved, two solid phases reacting together to form a new solid phase.

PERITUS TN for computer-aided materials selection procedure (Gk. for *expert*)

Perlon TN for polycaprolactam, nylon 6, a synthetic fibre.

permalloy An alloy with high **permeability** (from permeability alloy) and low **hysteresis loss.** Originally 78.5% Ni 21.5% Fe, but now includes compositions containing additions of Cu, Mo, Cr, Co, Mn etc.

permanent load The dead loading on a structure, consisting of the weight of the structure itself and the fixed loading carried by it, as distinct from any other loads.

permanent magnet Ferromagnetic body which retains an appreciable magnetization after excitation has ceased. Cobalt steel, sintered and ceramic materials and various ferritic alloys are used on a large scale in loudspeakers, relays, small motors, magnetrons etc. See **characterizing magnetic materials** p. 57.

permanent set (1) An extension remaining after load has been removed from a test piece, when the **elastic limit** has been exceeded. (2) Permanent deflection of any structure after being subjected to a load, which causes the elastic limit to be exceeded.

permanganates *Manganates (VII).* Oxidizing agents, the best known being *potassium permanganate* (manganate (VII)).

permeability (1) The facility with which a material permits the passage of a gas or liquid. (2) See **characterizing magnetic materials** p. 57.

permeameter Instrument for measuring static magnetic properties of ferromagnetic sample, in terms of magnetizing force and consequent magnetic flux.

permittivity See **dielectrics and ferroelectrics** p. 92.

perovskite Important crystal structure of some ferroelectics, such as those in the barium titanate family.

peroxides Strictly limited to compounds containing the ion O_2^{2-} or the organic function –O–O–, those containing the ion HO^{2-} or the group HO–O– being hydroperoxides. The term is loosely used for oxides of highvalent metals, eg lead peroxide is PbO_2, but is more properly called lead (IV) oxide.

persistence Continued visibility or detection of luminous radiation from a gas-discharge tube or luminous screen when the exciting agency has been removed. Long-persistence cathode-ray tubes are used for retaining a display which has been momentarily excited. Also *afterglow.*

persistence characteristic Graph showing decay with time of the luminous emission of a **phosphor,** after excitation is cut off.

Perspex TN for thermoplastic **polymethyl methacrylate.**

PET Abbrev. for **PolyEthylene Terephthalate.**

Petersburg standard See **standard.**

petrochemicals Those derived from crude oil or natural gas. They include light hydrocarbons such as butene, ethene and propene, obtained by fractional distillation or catalytic cracking.

petroleum Naturally occurring green to black coloured mixtures of crude hydrocarbon oils, found as earth seepages or obtained by drilling. Petroleum is widespread in the Earth's crust, notably in the North Sea, US, former USSR and the Middle East. In addition to hydrocarbons of every chemical type and boiling range, petroleum often contains compounds of sulphur, vanadium etc. Commercial petroleum products are obtained from crude petroleum by distillation, cracking, chemical treatment etc.

pewter An alloy containing 80–90% of tin

and 10–20% of lead. In modern pewter, antimony may replace tin, and from 1–3% copper is usual.

PF resins Abbrev. for *Phenol Formaldehyde resins*. See **phenolic resins, thermosets** p. 318.

PGC Abbrev. for **Pyrolysis Gas Chromatography**.

pH A logarithmic index for the hydrogen ion concentration in an aqueous solution. Used as a measure of acidity of a solution; given by pH = $\log_{10}(1/[H^+])$, where $[H^+]$ is the hydrogen-ion concentration. A pH below 7 indicates acidity, and one above 7 alkalinity, at 25°C.

phanero- Prefix from Gk *phaneros*, visible.

phase The sum of all those portions of a material system which are identical in chemical composition and physical state, and are separated from the rest of the system by a distinct interface called the *phase boundary*.

phase change See **phase transition**.

phase diagram See panel on p. 236.

phase reversal An interchange of the components of an emulsion; eg under certain conditions an emulsion of an oil in water may become an emulsion of water in the oil.

phase rule A generalization of great value in the study of equilibria between phases. In any system, $P+F=C+2$, where P is the number of phases, F the number of **degrees of freedom** (1), C the number of **components**.

phase transition Change of state in materials, whether first order (eg melting, boiling), second order (eg **Curie point** for ferromagnetic materials, **glass-transition temperature**) or higher order.

phenol C_6H_5OH, carbolic acid.

phenol A See **bisphenol A**.

phenolic resins A large group of plastics, made from a phenol (phenol, 3 cresol, 4 cresol, catechol, resorcinol, or quinol) and an aldehyde (methanal, ethanal, benzaldehyde, or furfuraldehyde). An acid catalyst produces a soluble and fusible resin (modified resin) used in varnishes and lacquers; while an alkaline catalyst results in the formation, after moulding, of an insoluble and infusible resin. Phenolics may be moulded, laminated or cast. See **thermosets** p. 318.

phenols *Hydroxybenzenes*. A group of aromatic compounds having the hydroxyl group directly attached to the benzene nucleus.

phenoxy resins Range of polymers now replaced by polycarbonates.

phenyl group The aromatic group C_6H_5-.

phonon A quantum of lattice vibrational energy in a crystal. The thermal vibrations of the atoms can be described in terms of *normal modes* of oscillation, each mode specifying a correlated displacement of the atoms.

The energy of a mode is quantized and can only exchange energy with other modes in units of hv, where v is the mode frequency and h is **Planck's constant**. This quantum is the *phonon*. Lattice vibrations can be described in terms of waves in the lattice and the phonon is the particle of the field of the mechanical energy of a crystal. (Cf **photon** as a particle of the electromagnetic field.) Phonons have been identified by neutron inelastic scattering experiments.

phonon dispersion curve A curve showing *frequency* as a function of *wave vector* for the modes of lattice vibrations in a crystal. Determined by neutron spectroscopy, their interpretation provides a powerful method of testing various models of interatomic forces in crystals.

phonon drag In the **Peltier effect**, due to the *electron-phonon* interactions the phonons gain momentum from the electrons and the electron current 'drags' the phonons with it.

phonon-phonon scattering The interaction of one mode of lattice vibration with another. One process determining the thermal conductivity of a solid.

-phore Suffix from Gk *pherein*, carry.

phosphating Treatment of metal surface with hot phosphoric acid in order to inhibit corrosion and also serving to act as a key prior to painting.

phosphor Generic name for materials which exhibit luminescence, especially the components of fluorescent coatings on cathode ray tubes, such as zinc doped zinc oxide.

phosphor-bronze A term sometimes applied to alpha (low tin) bronze deoxidized with phosphorus, but generally it means a bronze containing 10–14% of tin and 0.1–0.3% of phosphorus, with or without additions of lead or nickel. Used, in cast condition, where resistance to corrosion and wear is required, eg gears, bearings, boiler fittings, parts exposed to sea water etc. See **copper alloys** p. 76.

phosphorescence Luminescence which persists for more than 0.1 nanoseconds after excitation. See **fluorescence, persistence**.

phosphorized copper Copper deoxidized with phosphorus. Contains a small amount (about 0.02%) of residual phosphorus, which lowers the conductivity. See **copper alloys** p. 76.

phosphorus A non-metallic element in the fifth group of the periodic table. Symbol P. It is used in the manufacture of phosphoric acid for phosphate plasticizers and in phosphating steel as a protection against corrosion. See Appendix 1, 2.

photocatalysis The acceleration or retardation of rate of chemical reaction by light.

photocathode Electrode from which elec-

phase diagram

A mapping to show the stable states in which a system of chemical element(s) or molecular components can exist under particular physical conditions, usually of temperature and pressure but also of concentration of component(s) where appropriate. Otherwise known as *constitution diagrams*, *equilibrium diagrams* or *thermal equilibrium diagrams*, since they represent equilibrium states of a system. They are essential for understanding the behaviour of crystalline materials (metals, ceramics) and the effect of thermal treatments such as involved in melting and control of mechanical properties.

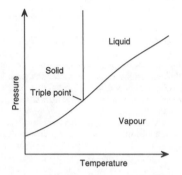

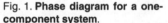

Fig. 1. **Phase diagram for a one-component system**.

A typical diagram for a single-component system is illustrated in Fig. 1. Because this is a unary (one component) system, the two physical variables temperature and pressure can be represented by the axes of a two-dimensional diagram and show the conditions under which each of the phases solid, liquid and vapour exist. In the areas, which are called phase fields, any of the three physical states may exist under conditions of temperatures and pressures within the natural limits of the phase field boundaries. Two of the phases may co-exist under conditions represented by the lines on the diagram. At one unique combination of pressure and temperature all three phases co-exist (the triple point).

The phase boundary between solid and liquid is virtually parallel to the pressure axis, ie the melting temperature is hardly influenced by pressure. As engineering materials are usually used and processed at constant pressure a composition axis may be used in place of pressure for systems containing two components (*binary systems*) as in Fig. 2 (next page). Here, composition of the components is plotted as the abscissa and temperature as the ordinate. The horizontal scale therefore ranges from pure component 'A' at one end through all ratios of A–B mixtures, to pure component 'B' at the opposite end. Such diagrams only apply to equilibria between solid and liquid phases.

Fig. 2 is the simplest possible binary system in which both components are mutually soluble in both the liquid and solid states. Few binary systems display complete mutual intersolubility. More usually the solubility in the solid state is limited and is sometimes so restricted that it cannot be shown on the same scale axis as the liquid solubility. Solid solutions are usually designated by letters of the Greek alphabet used in sequence across the diagram. Those based on the end member components are called *primary solid solutions* while phases within the diagram are either *intermediate solid solutions*, *electron* or *size factor compounds*, or strongly bonded *molecular compounds* which usually

continued on next page

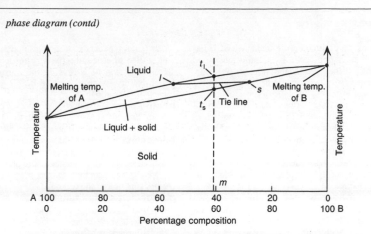

Fig. 2. **Phase diagram for a binary system exhibiting complete mutual solubility**.

exhibit very limited solubility range. Liquid solutions generally exhibit extensive intersolubility, except when there are strong chemical or ionic influences between the components.

The quantity of each phase in the two-phase region of a system such as Fig. 2 is predetermined at any selected temperature by the concentration represented by the solid and liquid compositions of the single phase boundaries intersected. If m is the concentration of the mixture which divides into liquid and solid at the temperature indicated by l and s, then the amounts are given by:

$$\text{solid fraction} = \frac{(l-m)}{(l-s)}, \qquad \text{liquid fraction} = \frac{(m-s)}{(l-s)}.$$

The above relationship is known as the *lever rule*, as it resembles a mechanical system of weights suspended on either side of a fulcrum at different distances along a balanced lever. The imaginary line linking the two compositions on the phase boundaries is referred to as a *tie-line*.

Binary diagrams are more complex than those for unary systems and new types of relationship are possible when three phases co-exist. Three phases of fixed compositions can exist in equilibrium at one unique temperature (the invariant). Two diagrams are shown in Fig. 3, (a) and (b) depicting the invariant conditions possible between one liquid and two solid phases (eutectic and peritectic, respectively).

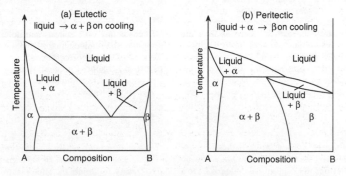

Fig. 3. **Two types of invariant three-phase equilibria in binary systems**.

trons are emitted on the incidence of radiation in a photocell. It is semi-transparent when there is photoemission on one side, arising from radiation on the other, as in the signal plate of a TV camera tube.

photochemistry The study of the chemical effects of radiation, chiefly visible and ultraviolet, and of the direct production of radiation by chemical change.

photochromic A photochromic material is one whose **transmittance** varies with the intensity of light incident upon it. Also *photosensitive*.

photoconductive cell Photoelectric cell using photoconductive element (often cadmium sulphide) between electrodes.

photoconductivity Property possessed by certain materials, such as selenium, of varying their electrical conductivity under the influence of light.

photodegradation Chemical degradation of a material such as a fibre caused by the absorption of light, esp. ultraviolet radiation. Usually leads to loss of strength of the material as the polymer breaks down.

photodiode A semiconductor diode fitted with a small lens which can focus light on the p-n junction. Certain types of junction exhibit marked variation of reverse current with illumination; this allows them to be used as compact and rugged light sensors.

photodissociation Dissociation produced by the absorption of radiant energy.

photoelastic analysis Method of examining transparent polymer models of structures etc. to isolate stress concentrations and other weak zones. Magnitude of the *stress concentration factor* can be found knowing the nominal applied stress and the stress-optical coefficient. Mapping isochromatic zones in transparent moulded polymers placed between crossed, circularly polarizing filters (eg **Polaroid** sheets) gives an idea of **frozen-in strain** due to chain orientation. It can be used for *non-destructive testing* analysis of eg injection moulded products subjected to severe stress in service. Analysis in white light gives coloured fringes, useful in finding fringe order.

photoelectric absorption That part of the absorption of a beam of radiation associated with the emission of photoelectrons. In general, it increases with increasing atomic number but decreases with increasing quantum energy when *Compton absorption* becomes more significant.

photoelectric effect Any phenomenon resulting from the absorption of photon energy by electrons, leading to their release from a surface, when the photon energy exceeds the **work function** (see **photoemission**), or otherwise allowing conduction when the inci-

dent energy exceeds an atomic binding energy. See **photoconductivity, photovoltaic cell**. Emission of X-rays on the impact of high-energy electrons on a surface is an *inverse photoelectric effect*.

photoelectricity Emission of electrons from the surface of certain materials by quanta exceeding a certain energy.

photoelectric threshold The limiting frequency for which the quantum energy is just sufficient to produce photoelectric emission. Given by equating the quantum energy to the work function of the cathode.

photoelectric work function The energy required to release a photoelectron from a cathode. It should correlate closely with the **thermionic work function**.

photoelectroluminescence The enhancement of luminescence from a fluorescent screen during excitation by ultraviolet light or X-rays when an electric field is applied, ie the field enhancement of luminescence.

photoelectron spectroscopy Using visible light, ultraviolet light or X-rays as the excitation source, the energy of photoelectrons emitted from the material can be analysed to give information on surfaces, interfaces and bulk materials. Also used to deduce binding energies for deep core levels in atoms with a high degree of precision. Also *electron spectroscopy, photoionization spectroscopy*.

photoemission Emission of electrons from the surface of a body (usually an electropositive metal) by incidence of light.

photolithography Process of pattern transfer using optics. See **lithography, semiconductor device processing** p. 280.

photolysis The decomposition or dissociation of a molecule as the result of the absorption of light.

photomagnetoelectric effect Generation of electric current by absorption of light on the surface of a semiconductor placed parallel to magnetic field. Due to transverse forces acting on electrons and holes diffusing into the semiconductor from the surface. Also termed *photoelectromagnetic effect*.

photomultiplier Photocell with series of dynodes used to amplify emission current by electron multiplication.

photon Quantum of light of **electromagnetic radiation** of energy $E = h\nu$ where h is **Planck's constant** and ν is the frequency. The photon has zero rest mass, but carries momentum $h\nu/c$, where c is the velocity of light. The introduction of this 'particle' is necessary to explain the photoelectric effect, the Compton effect, atomic line spectra, and other properties of electromagnetic radiation.

photonics Study and use of optical processing devices, analogous to electronics.

photopolymerization Polymerization initiated by visible light or ultraviolet photons, which react with monomer molecules to give free radicals. See **chain polymerization** p. 56.

photoresist Photosensitive material used in photolithography. See **resist, semiconductor device processing** p. 280.

phototransistor A three-electrode photosensitive semiconductor device. The emitter junction forms a photodiode and the signal current induced by the incident light amplified by transistor action.

phototropy (1) The property possessed by some substances of changing colour according to the wavelength of the incident light. (2) The reversible loss of colour in a dyestuff when illuminated at a definite wavelength.

photovoltaic cell A class of photoelectric cell which acts as a source of emf and so does not need a separate battery.

photovoltaic effect The production of an emf across the junction between dissimilar materials when it is exposed to light or ultraviolet radiation.

phthalic anhydride *Phthalic (benzene 1,2-dicarboxylic) anhydride.* Long prisms which can be sublimed, mp 128°C, bp 284°C. Used in the production of plasticizers, paints and polyester resins. Prepared by air oxidation of naphthalene in the presence of a catalyst such as vanadium (V) oxide.

phthalic glyceride resins See **alkyd resins**.

physical crosslink Type of crosslink which involves van der Waals or ionic bonds in copolymers rather than chemical covalent bonds.

physical electronics The branch of electronics which is concerned with the physical details of the behaviour of electrons in vacuum, in gases and in conductors, semiconductors, and insulators.

physical metallurgy The study of the properties of metals and their alloys at an atomic and microstructural level and the effect of deformation, composition, heat treatment, environment and other factors upon them.

physical vapour deposition A thin film deposition process involving physical mechanisms such as evaporation or **sputtering**, as distinct from one involving chemical reactions. Abbrev. *PVD.* Cf **chemical vapour deposition**.

pi The ratio of the circumference of a circle to its diameter. $\pi = 3.141\ 592\ 653$ to ten places.

PIB Abbrev. for **PolyIsoButylene**, or polybutene.

pick One traverse of the **weft**-carrying de-

vice through the **warp** array to lay a weft thread in weaving a cloth. A single weft thread in a fabric. Picks per inch (ppi) or per centimetre (ppcm) are used to denote the total number of weft threads per unit length of cloth.

pick-and-pick A woven fabric with alternate picks of yarns of two different colours or kinds.

pick-and-place Robotic device for automatic removal of injection mouldings etc from tool cavity at end of cycle.

pico- SI prefix for 1-million-millionth, or 10^{-12}. Formerly *micromicro-*. Abbrev. p.

picosecond The million-millionth part of a second, 10^{-12} s. Abbrev. ps.

piercing See **Mannesmann process**.

piezo- Prefix from Gk *piezein*, to press.

piezoelectric crystal One showing piezoelectric properties which may be shaped and used as a resonant circuit element or transducer (microphone, pick-up, loudspeaker, depth-finder etc). See **dielectrics and ferroelectrics** p. 92.

piezoelectric effect, piezoelectricity Electric polarization arising in some **aniso-tropic** (ie not possessing a centre of symmetry) crystals (**quartz, Rochelle salt, barium titanate**) when subject to mechanical strain. An applied electric field will likewise produce mechanical deformation. See **dielectrics and ferroelectrics** p. 92.

piezomagnetism An effect analogous to *piezoelectricity* in which mechanical strain in certain materials produces a magnetic field and *vice versa*.

piezoresistivity Strain-dependent resistivity exhibited by single-crystal semiconductors. Mechanical strain modifies carrier mobility in the crystal; the basis of semiconductor strain gauges.

pig (1) A handleable mass of metal (eg cast iron, copper or lead) cast in a simple shape for transportation or storage, and subsequently remelted for purification, alloying, casting into final shapes, or into ingots for rolling. (2) In glassmaking, an iron block laid against the pot mouth as a support for the **blowing iron**.

pig bed A series of moulds for iron pigs, made in a bed of sand. Connected to each other and to the tap hole of the blast-furnace by channels, along which the molten metal runs. The first subdivisions of the liquid run into larger sections called sows, and the pigs are the smaller sections which connect to these, resembling a row of piglets sucking the mother's teats when viewed from above the pig bed. Now replaced by pig casting machines using moulds on a moving belt.

pig iron The crude iron produced in the blast furnace and cast into pigs which are used for making steel, cast iron or wrought iron. Principal impurities are carbon, silicon, manganese, sulphur and phosphorus. Composition varies according to the ores used, the smelting practice and the intended usage.

pigment Insoluble, natural or synthetic, black, white or coloured materials reduced to powder form which, when dispersed in a suitable medium, are able to impart colour and/or opacity. Ideally, pigments should maintain their colour under the most unfavourable conditions; in practice, may tend to fade or discolour under the influence of acids, alkalis, other chemicals, sunlight, heat and other conditions. May also show bleeding. They are used in many industries, eg in paint, printing ink, plastics, rubber, paper etc. The two main types are the natural earth pigments which tend to be drab and limited in colour range and synthetic pigments with a comprehensive range of colours.

pile (1) A covering on the surface of a fabric, formed by threads that stand out from it. Pile in loop form is termed loop pile or terry; if the loops are cut, it is termed cut pile. The latter is produced by weaving two cloths face to face, and cutting them apart. Some carpets and moquettes are made this way. (2) A number of wrought iron bars arranged in an orderly pile which is to be heated to a welding heat and rolled into a single bar. (3) An early nuclear reactor.

pill Small unsightly balls of fibre that accumulate during wear on the surface of textiles, esp. knitted woollens.

Pilling–Bedworth ratio Volume ratio of metal oxide to that of its parent metal which provides a guide to the volume changes accompanying oxidation.

pilot Heavyweight, woollen cloth that has been highly milled and raised, used in naval jackets and overcoats.

pinacoid An open crystal form which includes two precisely parallel faces.

pinchbeck alloy Red brass, 6–12% zinc. See **copper alloys** p. 76.

pinch-off voltage In a **field-effect transistor** (FET), the voltage at which the current flow between source and drain ceases because the **channel** between these electrodes is completely depleted. For enhancement-mode FETs, using an n-type channel, the pinch-off voltage is positive; for depletion-mode FETs with a p-type channel, it is negative.

pine **Softwood** trees of the genus *Pinus*. See **pitch pine, Scots pine.**

pinhole porosity, pinholing Small rounded voids within the body of a casting caused by evolution of gas during solidification.

pin joint A joint between members in a structural framework in which moments are not transmitted from one member to another.

pinning The restriction of the passage of a **dislocation** or a **domain wall** by an impurity or other crystalline defect.

pint A unit of capacity or volume in the imperial system. Equal to 1/8 imperial gallon or 0.5682 dm^3. In US equal to 0.473 dm^3 (liquid) or 0.551 dm^3(dry). Contains 20 fluid ounces (US 16 fl. oz.).

PIN, p-i-n diode A semiconductor diode with a layer of intrinsic semiconductor material between the p-type and n-type material which would normally constitute the junction. Applications include high-frequency and microwave switching or attenuation, voltage-controlled resistors, photodiodes.

pin-twisting Formation of **false-twist** by passing continuous filament yarn through a rotating tube and around a hard-wearing ceramic pin.

pipe Tapering cavity formed in the upper parts of an ingot or casting due to the specific volume contraction when the liquid solidifies. May extend as far as half way down depending on the degree of contraction, the rate of solidification and whether a feeder head is used.

piqué Fabric with pronounced cord effects. In woven fabrics the woven cords run in the weft direction but in warp-knitted fabrics the cords are in the other direction.

PISE Abbrev. for *Plasma and Ion Surface Engineering*. See **surface engineering, dry etching, physical vapour deposition, chemical vapour deposition.**

pirn Small wooden bobbin which fits the shuttle of a weaving loom and carries the weft thread.

pitch (1) A dark-coloured, fusible, more or less solid material containing bituminous or resinous substances, insoluble in water, soluble in several organic solvents. Usually obtained as the distillation residue of tars. (2) Residual resinous material that may be present in unbleached sulphite or mechanical pulps. A form of **contrary** that can block the machine wires or felts of papermaking machines.

pitch pine *Pinus rigida* and *P. palustris* (American pitch pine) yielding very resinous, moderately durable **softwood** timber, yellow-orange to reddish-brown in colour. Used for heavy framing, piling, joinery, ship's masts and spars, crates and pallets, etc. Significant source of resins and turpentine. Density about 670 kg m^{-3}.

pit moulding Process by which extremely large castings are moulded in a pit, with brick-lined sides and cinder-lined bottom carrying connecting vent pipes to floor level.

pitting (1) Corrosion of metal surfaces due to highly localized chemical action. (2) A form of failure of gear teeth, due to imperfect lubrication under heavy tooth pressure.

pitting factor In assessment of metal corrosion, the depth of penetration of the deepest pit divided by average loss of thickness as calculated from loss of weight.

plain fabric (1) In a knitted plain fabric, all the loops are of the same type and are linked together in the same manner. (2) In a woven plain fabric, a weft thread passes over and under each succeeding warp thread whereas the next weft thread passes under and over the same warp threads. This is the simplest of all weaves.

planar diode Diode produced using the **planar process**, giving a diode with high forward conductance, low reverse leakage, fast recovery time and low junction capacitance.

planar process A silicon transistor and integrated circuit manufacturing technique in which an oxide layer, fractions of a micron thick, is grown on a silicon substrate. A series of etching, ion implantation, diffusion and deposition steps is then undertaken to produce the active regions and junctions inside the substrate. See **semiconductor device processing** p. 280.

planar transistor A transistor made using the **planar process**.

Planck's constant See **Planck's law**.

Planck's law Basis of quantum theory, that the energy of electromagnetic waves is confined in indivisible packets or quanta, each of which has to be radiated or absorbed as a whole, the magnitude being proportional to frequency. If E is the value of the quantum expressed in energy units and v is the frequency of the radiation, then $E = hv$, where h is known as *Planck's constant* and has dimensions of energy × time, ie action. Present accepted value is 6.626×10^{-34} J s. See **photon**.

Planck's radiation law An expression for the distribution of energy in the spectrum of a black-body radiator:

$$E_v \, dv = \frac{8\pi \, h \, v^3}{c^3 \, (e^{hv/kT} - 1)} \, dv \,,$$

where E_v is the energy density radiated at a temperature T within the narrow frequency range from v to $v + dv$. h is Planck's constant, c the velocity of light, e the base of the natural logarithms and k Boltzmann's constant.

plane See **European plane**.

plane polarized Term applied to particular type of polarized EM radiation, esp. visible light, where the plane of vibration is planar. Produced by plane polarizing filters, esp. **Polaroid** sheet. Used for photoelastic analysis of isoclinics and isochromatics. Commonly produced by reflection from polished surfaces, and at a maximum at a specific angle of polarization (about 60 degrees to the normal for sunlight). The polarization plane in sunglasses is so designed as to reduce horizontal reflections, as from still water.

plane strain Stress state in which the strains are biaxial, so that the associated stresses are triaxial. Important esp. in the interior of cracked materials under stress where through-the-thickness strains are constrained. See **fracture toughness, Griffith equation**.

plane stress Stress state in which the stresses are biaxial, so that the associated strain field is triaxial. Important esp. at the surfaces of cracked materials under stress where stresses normal to the material surface cannot exist. See **fracture toughness, Griffith equation**.

planishing Giving a finish to metal surfaces by hammering.

planon A log of hardwood timber roughly sawn or hewn to an octagonal shape, with a minimum of 10 in (250 mm) between opposite faces.

plant (1) The machines, tools and other appliances requisite for carrying on mechanical or processing operations; the term sometimes includes also the building and the site. (2) The permanent appliances needed for manufacture.

PLASCAMS TN for computer-aided polymer selection routine using input value judgements (on a scale of 1–10) of properties needed in the product being designed.

plasma A neutral, ionized medium, such as that in an electrical discharge through a gas.

plasma-arc cutting Cutting metal at temperatures approaching 3500°C by means of a gas stream heated by a tungsten arc to such a high temperature that it becomes ionized and acts as a conductor of electricity. Heat is not obtained by a chemical reaction; the process can therefore be used to cut any metal.

plasma deposition A coating process involving exposure to an ionized gas containing precursor material in the form of particles, radicals, ions or atoms.

plasma enhanced processing Material processing such as **chemical vapour deposition** assisted by the passage of electric current through the process gases.

plasma etching Etching by means of physical and/or chemical reactions at a surface exposed to a **plasma**. See **dry etching**.

plasma spraying Coating method in which material to be deposited is fed as a powder into a carrier gas flowing through an arc discharge, wherein it is melted. The method is useful for rapid deposition of a wide range of materials.

plaster mould casting Small, precision parts of non-ferrous alloys are cast in plaster moulds which are destroyed when the casting is removed.

plaster of Paris Partly dehydrated gypsum, $2CaSO_4.H_2O$ (hemihydrate). When mixed with water, it evolves heat and quickly solidifies, expanding slightly; used for making casts.

plastic bronze Bronze containing a high proportion of lead; used for bearings. Composition; 72–84% copper, 5–10% tin and 8–20% lead plus zinc, nickel and phosphorus.

plastic deformation Permanent change in the shape of a piece of material resulting from eg slip processes within the constituent crystals or by irreversible displacement of polymer chains, brought about by the application of mechanical force exceeding the elastic limit.

Plasticine TN for heavily filled polymer dough used for modelling etc.

plasticity A property of certain materials by which the deformation due to a stress is largely retained after removal of the stress.

plasticization Effect produced by addition of a plasticizer to a polymer; generally lowers T_g, increases elongation to break but lowers modulus of material.

plasticized PVC A polyvinyl chloride and plasticizer blend widely used as a substitute for leather in upholstery. Also used in stationery products, for conveyor belting, toys, cable covering, hose etc.

plasticizer cracking Loss of integrity of thermoplastics, esp. in presence of plasticized polyvinyl chloride. Bleeding of small molecules onto plastic surface causes cracking or crazing.

plasticizers High-boiling liquids used as ingredients in lacquers and certain plastics, eg polyvinyl chloride; they do not evaporate but preserve the flexibility and adhesive power of the cellulose lacquer films or the flexibility of plastic sheet and film. Well-known plasticizers are DOP, DAP, triphenyl phosphate, tricresyl phosphate, high-boiling glycol esters etc. DOP and DAP are commonly used in polyvinyl chloride for flexible film etc. Long-chain plasticizers have been developed where migration is a problem, eg blood and serum bags for medical use. Phosphate plasticizers are used where flame resistance is needed, eg coal mine conveyor belting.

plastic moulding Processes for manufacturing articles of plastic materials, using either injection moulding machines for the rapid production of small articles or hydraulic presses for compression moulding of rel-

atively large articles.

plastics A generic name for organic materials, mostly synthetic (see **polymers** p. 247) or semi-synthetic (casein and cellulose derivatives) polymers but also for certain natural substances (shellac, bitumen, but excluding natural rubber), which under heat and pressure become plastic, and can then be shaped or cast in moulds, extruded as rod, tube etc, or used in the formation of laminated products, paints, lacquers, glues etc. Plastics are **thermoplastic** or **thermosetting**. See **thermosets** p. 318.

plastic sulphur Allotropic form of elemental sulphur, created by polymerization of monomeric sulphur molecules (an eight-membered ring compound). It exhibits both ceiling and floor temperatures.

plastisol Paint-like mixture of plasticizer and powdered polymer, usually polyvinyl chloride, used in variety of processes to make flexible products. **Rotocasting** is used to make footballs, toys etc, while knife or **reverse roll coating** produces convey or belting.

plate A large flat body of steel, thicker than sheet, which is produced by the working of ingots, billets or slabs, in a rolling mill.

plate glass Form of flat glass of superior quality, originally cast on an iron bed and rolled into sheet form, and afterwards ground and polished. Modern methods have largely superseded this, save in special circumstances. Also *polished plate*. See **float glass**.

platens Steel plates to which fixed and moving parts of a tool are fitted in injection moulding. See **injection moulding** p. 169.

platinite Alloy containing iron 54–58% and nickel 42–46%, with a trace of carbon. Has the same coefficient of expansion as platinum, and is used to replace it in some light bulbs and measuring standards.

platinoid Alloy containing copper 62%, zinc 22%, nickel 15%. Has high electrical resistance and is used for resistances and thermocouples.

platinum A metallic element, symbol Pt. Platinum is the most important of a group of six closely related rare metals, the others being osmium, iridium, palladium, rhodium and ruthenium. It is heavy, soft and ductile, immune to attack by most chemical reagents and to oxidation at high temperatures. Used for making jewellery, special scientific apparatus, electrical contacts for high temperatures and for electrodes subjected to possible chemical attack. Also used as a basic metal for resistance thermometry over a wide temperature range. Native platinum is usually alloyed with iron, iridium, rhodium, palladium or osmium, and crystallizes in the cubic system. See Appendix 1, 2.

platinum metals A block of six transition metals with similar physical and chemical properties. Specifically ruthenium, osmium, rhodium, iridium, palladium and platinum.

pleated sheet Type of secondary structure formed in proteins by lateral hydrogen bonds, esp. beta-pleated sheet in silk. See **biomaterials** p. 57.

plissé See **seersucker**.

plug welding Method of welding in which apertures in one part allow penetration through to the other part to be joined, the apertures to be filled by welding in the process.

plunge-cut milling Milling without transverse movement.

plush (1) Woven fabric with cut pile, longer and less dense than velvet, on one side. Warp pile is generally made by weaving two cloths together, with a pile warp common to both, which is afterwards cut. (2) One type of weft-knitted fabric has the long loops on the back of the cloth, which is sometimes called knitted terry. In warp-knitted plush fabrics the loops may be cut or left uncut.

plywood A board consisting of a number of thin layers of wood glued together so that the grain of each layer is at right angles to that of its neighbour.

PM Abbrev. for **Powder Metallurgy**.

PMMA Abbrev. for **PolyMethylMeth-Acrylate**.

PMOS Metal-oxide-silicon technology based on p-channel devices in an n-type substrate.

pnip, p-n-i-p transistor Similar to p-n-p transistor with a layer of an intrinsic semiconductor between the base and the collector to extend the high-frequency range.

pn, p-n boundary The surface on the transition region between p-type and n-type semiconductor material, at which the donor and acceptor concentrations are equal.

pn, p-n junction Boundary between p- and n-type semiconductors, having marked rectifying characteristics; used in diodes, photocells, transistors etc.

pnp, p-n-p transistor A junction transistor having an n-type base region between a p-type emitter and a p-type collector. For normal operation, the emitter should be held positive and the collector negative with respect to the base.

Pockels' effect The **electro-optical effect** in a piezoelectric material.

podo Varieties of S and E African **softwood** (*Podocarpus*) whose **heartwood** is soft, non durable, light yellowish-brown, straight-grained and a uniform texture with no defined growth rings. Used in internal joinery, plywood and veneers. Mean density

510 kg m^{-3}.

poikil-, poikilo- Prefixes from Gk *poikilos*, many-coloured.

point-contact diode A semiconductor which uses a slender wire filament, touching a small piece of semiconductor material, to provide rectifying action. Junction capacitance is kept to a minimum and high-frequency operation is possible. Sometimes used as a detector in VHF and microwave devices.

Poiseuille's law An expression for the volume of liquid per second Q, which flows through a capillary tube of length L and radius R, under a pressure P, the viscosity of the liquid being η:

$$Q = \frac{\pi P R^4}{8 L \eta}.$$

Strictly, it is limited to laminar flow only. It is widely applied to non-Newtonian liquids which give an *apparent viscosity*.

Poisson's ratio One of the four elastic constants of an isotropic material, symbol ν. It is defined as the ratio of the lateral contraction per unit breadth, w, to the longitudinal extension per unit length, l, when a piece of the material is stretched, ie

$$\nu = -\frac{\Delta w / w}{\Delta l / l}.$$

Its value can range from -1 (no change in proportions) to $+\frac{1}{2}$ (no change in volume), but for most substances its value lies between 0.2 and 0.4. The relationships between Poisson's ratio, **Young's modulus** E, the **shear modulus**, G, and the **bulk modulus**, K, are given by:

$$\nu = \frac{E}{2G} - 1 = \frac{3K - 2G}{2(2K + G)}.$$

polarization (1) Non-random orientation of electric and magnetic fields of electromagnetic wave. (2) Change in a dielectric as a result of sustaining a steady electric field, with a similar vector character; measured by density of dipole moment induced.

polarized capacitor Electrolytic capacitor designed for operation only with fixed polarity. The dielectric film is formed only near one electrode and thus the impedance is not the same for both directions of current flow.

polarized light microscopy Method of examining thin, transparent materials using polarized light. Esp. useful for mineral sections, biomaterials and polymers, all of which may show birefringence. Can give data for identification of material, possibly improve contrast (eg spherulite structure in polymers) and aid microstructure determination. One variant of method used for surfaces is **Nomarski interference**.

polarizer Sheet of birefringent polymer (esp. Polaroid) used as light filter to produce either plane- or circularly-polarized light. Combined with analyser for photoelastic analysis etc.

polarizibility The degree to which a medium may be electrically (or magnetically) polarized. Electric polarization of a material arises from the displacement of dipoles, ions and electrons within the constituent matter. As a result it is a function of frequency, varying srongly when any component undergoes resonance.

polar molecule One with unbalanced electric charges, usually valency electrons, resulting in a dipole moment and orientation.

Polaroid TN for a range of photographic and optical products, including a transparent light-polarizing plastic sheet (*Polarscreen*) and methods of **instant photography** in black-and-white and colour.

Polaroid sheet TN for thermoplastic sheet containing optically active compound which acts as filter for visible light, giving either plane- or circularly-polarized light on transmission. Birefringent material which has replaced Nicol prisms in most analytical applications, eg photoelasticity, polarized light microscopy. Two sheets (polarizer and analyzer) are used to examine birefringent materials, such as polymer products, sandwiched between them.

polar second moment of area The polar second moment of area is a measure of resistance to twisting of a section. The polar second moment of area, J about an axis zz, perpendicular to the plane of the section is found from

$$J_{zz} = \int Az^2 \, dA$$

where dA is an element of the total area and z the distance from zz. Cf **second moment of area**.

pole The part of a magnet, usually near the end, towards which the lines of magnetic flux apparently converge or from which they diverge, the former being called a *south pole* and the latter a *north pole*.

pole piece Specially shaped magnetic material forming an extension to a magnet, eg the salient poles of a generator, motor or relay, for controlling flux.

poling (1) In the fire refining of copper, the impurities are eliminated by oxidation and the residual oxygen may in turn be removed by the reducing gases produced when green logs (poles) are burned in the molten metal. If the final oxygen content is too high the metal is said to be *underpoled*, if too low, *overpoled*; when the desired balance of small residual oxygen content is achieved it

is termed *tough pitch*. Poling with green tree trunks is now superseded by injection of reducing gas mixtures which has similar effect. (2) The application of an electric field to a ferroelectric material in order to establish a remanent polarization. *Depoling* is the removal of polarization.

polishing Sample preparation method for rigid materials when examined by reflection optical microscopy. Sample is abraded with succession of finer grade emery papers, then diamond polished (to 0.25 µm.) to give highly reflective surface. Art in method is to remove traces of all previous scratches. Usually followed by etching to expose microstructure (grains etc).

polishing stick A stick of wood, one end of which is charged with emery or rouge, used for finishing small surfaces. It is twisted in the hands or held in the chuck of a drilling machine.

polonium Radioactive element, symbol Po. Important as an α-ray source relatively free from γ-emission. See Appendix 1, 2.

poly- Prefix from Gk *polys*, many. In chemistry: (1) containing several atoms, groups etc; (2) a prefix denoting a polymer, eg polyethylene.

poly(4-methylpent-1-ene) Isotactic, transparent aliphatic polyolefin prone to brittle cracking.

polyacetals See **polyoxymethylenes**.

polyacetylene Electrically conductive polymer possessing conjugated double bonds along its backbone chain. Material still at developmental stage owing to poor mechanical properties, processing problems, cost etc.

polyacrylate A polymer of an ester of acrylic acid or its chemical derivatives.

polyacrylic acid Polyelectrolyte widely used as additive in range of different materials (eg building products) to increase viscosity. Unlike most synthetic polymers, it is water-soluble due to charged carboxylic groups on its backbone chain.

polyacrylimide Derivative of methacrylic acid and methacrylonitrile copolymer, where treatment with urea creates cyclic aliphatic units in the backbone chain (with imide functional units). Such partly cyclized material offers enhanced heat-resistance (to 160°C), and is used mainly in rigid foam products, eg composite helicopter blade cores. Urea also makes the product foam by gas formation, helping to press other components to shape in such uses.

polyacrylonitrile Fibre-forming polymer (under the name 'acrylic') which is also the starting material for carbon fibre. See **high-perfomance fibres** p. 158.

polyalkane Hydrocarbon polymers, essentially of long-chain molecules with only saturated carbon atoms in the main chain. More commonly called **polyolefins**.

polyamide Natural or synthetic fibres composed of polymers having the same amide group (–CO–NH–) repeated along the chain. Examples of the natural fibres are silk, wool and hair. For synthetic polyamides see **nylon**. Polyamides made with *aromatic* groups attached to the amide links are assigned to a different class, the **aramid fibres**.

polybenzimidazoles Family of heat resistant polymers with aromatic heterocyclic structures in the main chain. T_g generally above 400°C. See **thermosets** p. 318.

polybutadienes See **polymers** p. 247.

polybutene Isotactic, crystalline polyolefin made from 2-methyl propylene. Of very high molecular mass (up to three million), it shows high creep resistance, and is mainly used for extruded small-bore hot and cold water pipes.

polybutylene terephthalate Step-polymer made from 1,2- butanediol and terephthalic acid and used as a thermoplastic material. Abbrev. *PBT*.

polycarbonate Amorphous transparent and rigid polymer, made by step-growth mechanism from carbonate functional group (–O–CO–O–). Normally tough but notch sensitive, so care needed in design. Rigid backbone chain with bis-phenol A aromatic group in repeat unit, so high T_g of 145°C. Widely injection moulded for battery cases, containers and housings for consumer products. Abbrev. *PC*. See **CR39**.

polychloroprene Polymer made from chlorine substituted butadiene, with repeat unit structure same as polyisoprene (Cl replaces methyl group). Important rubber for its resistance to weathering, so used in bridge bearings, rubber dinghies etc.

polycrystalline materials Common state of crystalline metals and ceramics, formed by mass of interlocking single crystals. Their crystallographic axes are usually random, so giving isotropic properties. If preferential orientation of the axes occurs, properties will be **anisotropic** (eg rolled metal sheet).

polyelectrolyte Polymer in which some or all of the repeat units possess ionic groups. Includes many biomaterials (eg proteins and polypeptides) as well as synthetics (eg polyacrylic acid).

polyester amide A polymer in which the structural units are linked by ester and amide (and/or thio-amide) groupings.

polyester rubber Thermoplastic elastomer made from polytetramethylene ether glycol and polybutylene terephthalate by block copolymerization. By varying block size and polyester composition, a wide range of hardnesses can be made. It is used in moulded products for its high abrasion resistance (eg skimobile caterpillar tracks). Not to be confused with acrylic rubbers.

polyesters Polymers having ester functional group (–CO–O–) in repeat unit. Includes thermoplastics polyethylene terephthalate, polybutylene terephthalate, certain liquid crystal polymers, polyester rubbers, and thermosets acrylic rubbers, polyester resins etc.

polyester thermoset resins Range of polymers formed by the step-polymerization of polyhydric alcohols and polycarboxylic acids or anhydrides. Maleic and fumaric acids and ethylene and propylene alcohol are the usual starting materials, which are polymerized with eg styrene and reinforced with glass fibre for composite applications eg canoes, boat hulls, car body panels. Phthalic anhydride is frequently used to increase the stiffness of the final product. See **thermosets** p. 318.

polyether ether ketones Heat-resistant polymers with benzene rings linked by ether (–O–) and ketone (–CO–) groups in main chain. Tough thermoplastics with T_g ~144°C and T_m ~335°C for Victrex polyether ether ketone. Heat and flame resistance has led to aerospace applications.

polyether imides Type of melt-processable polyimide with ether links in main chain. TN Ultem.

polyethers Long-chain glycols made from alkene oxides such as ethylene oxide. Used as intermediates in the production of polyurethanes, anti-static agents and emulsifying agents. See **PTMEG**.

polyethylene oxide Water-soluble polymer with repeat unit –CH$_2$–CH$_2$–O–. Widely used in cosmetics, toothpaste etc as an inert matrix. Produces laminar flow in turbulent water streams at very low concentrations, so used by firefighters for increasing output from hose etc. Abbrev. *PEO*.

polyethylenes Polymers based on ethylene monomer. Include **low-density polyethylene**, linear low-density polyethylene, medium density polyethylene, **high-density polyethylene**, Sclair, **ultra-high molecular mass polyethylene**; copolymers: **ethylene vinylacetate copolymer**, **EPR**, polypropylene-copolyethylene; fibres: Dyneema, Spectra.

polyethylene terephthalate The chemical name for the polyester forming the basis of Terylene and Melinex. Abbrev. *PET*.

polyformaldehydes See **polyoxymethylene**

polyhydantoin Heat-resistant polymer having mixture of aromatic and aliphatic rings in its main chain. Used as wire enamel and insulating film for eg electric motors. Can also be used as solder-resistant base for flexible circuitry.

polyimides Polymers with the functional group

$$-CO.N.CO-,$$

therefore akin to polyamides. Commercial materials are aromatic, so are heat-resistant to over 300°C. Film used in electronic circuitry (eg flexible circuits), bulk in rod form. See **thermosets** p. 318.

polyisoprene Main polymer in natural rubber, and also available synthetically. See **elastomers** p. 106. Also *cis-polyisoprene*, *isoprene rubber*.

polymer crystallization P h e n o m e n o n where crystallizable polymers form crystallites, spherulites etc at a characteristic temperature not identical with polymer melting point, T_m. In general, crystallization temperature falls midway between T_g and T_m for most polymers. See **crystallization of polymers** p. 82.

polymer engineering Application of polymer science to practical problems involving properties and use of polymeric materials in demanding environments.

polymeric glass See **glasses and glassmaking** p. 144, **viscoelasticity** p. 332.

polymer identification Variety of methods used to determine components in polymeric material. Complete analysis specifies repeat unit(s) and molecular masses of chains (MMD) plus identity of fillers and other additives. methods include IR, ultraviolet and **nuclear magnetic resonance spectroscopy** for soluble thermoplastics and **pyrolysis gas chromatography** and swelling tests for thermosets. Controlled degradation to identifiable soluble products at ambient temperature may also be useful.

polymerization The combination of several molecules to form a more complex molecule, usually by a step- or chain-growth polymerization mechanism. See **polymers** p. 247.

polymer melting Phenomenon of crystalline polymers melting at a characteristic temperature, dependant on rate of heating, impurities present, polymer type and molecular mass. Not to be confused with polymer crystallization temperature. See **crystallization of polymers** p. 82, **viscoelasticity** p. 332.

polymers See panel on p. 247.

polymer selection See **PLASCAMS, EPOS, CAMPUS, CAPS, materials selection**.

polymethyl methacrylate The chemical name for Perspex and Lucite etc. Rigid glassy thermoplastic used for glazing, optical devices, baths and aircraft windows etc. Usually of very high molecular mass in sheet form (ca 1 million). Used as a positive resist for electron beam lithography in semiconductor processing. Abbrev. *PMMA*.

polymorphic transformation A c h a n g e in a pure metal from one form to another, eg the change from γ- to α-iron.

polynorbornene Speciality rubber with aliphatic ring system in main chain. When plasticized, gives very soft elastomer for rollers etc.

polynosic fibres Modified rayon fibres with improved properties, such as good dimensional stability on washing, arising from better uniformity of fibres.

polyolefins Polymers based on olefin monomers, such as polyethylenes, polypropylene and poly(4-methylpent-1-ene). Also *polyalkanes*.

polyoxymethylene Moulding polymer widely used in engineering products (eg gearwheels) for its strength and stiffness. Highly crystalline with T_m ~165°C. Homopolymer is easily depolymerized, so the commercial polymer is copolymerized with ethylene oxide for extra stability or the chain ends may be capped with a stable group. Commonly *acetal resin*, also *polyformaldehyde*. Abbrev. *POM*. TNs Kematal, Delrin.

polyoxymethylene resins A l t e r n a t i v e name for **acetal resins**.

polyparabanic acids Heat-resistant polymer similar in structure to polyhydantoin, and used in similar applications.

polyphenylenes Heat-resistant, intractable polymers consisting of benzene rings linked together in chains. Processability improved by inserting ether, ketone or sulphide links between rings, eg PPO, PPS, PEEK etc.

polyphenylene oxide Aromatic polyether, poly-(2,6-dimethyl-p-phenylene ether). Its rigid backbone chain gives a high T_g~208°C, but is mainly used blended with polystyrene (see **Noryl**). Abbrev. *PPO*.

polyphenylene sulphide Heat-resistant aromatic polymer with sulphide functional linking group (–S–) in main chain. Claimed heat stability to 500°C, also fire resistant (oxygen index to 0.53) but rather brittle. Used in under-the-bonnet moulded parts. Abbrev. *PPS*.

polyphosphates Glassy materials based on polymers of the phosphate group, $(PO_4)^{3-}$, susceptible to hydrolysis but widely used in protective coatings eg in car steel bodies. See **phosphating**.

polypropylene Polymers based on propylene (propene) monomer. The main parent

polymers

Polymers are long chain molecules built up by multiple repetition (*poly*) of groups of atoms known as repeat units (*-mer*). The structure of polymer chains is specified in two ways: by the shape of individual or groups of chains in space, *conformation*, and by the way each chain is constructed from its covalently bonded atoms, *configuration*. The chain configuration is determined mainly during polymerization, when the monomer units are linked together. Monomer configuration is often very similar to that of the repeat unit, but there are subtleties of chain structure which are not present in the monomer unit. The molecular size of the polymer helps to determine the mechanical properties, esp.strength.

The simplest repeat unit is that of polyethylene (HDPE) and comprises two carbon atoms covalently bonded together with four pendant hydrogen atoms ($-CH_2-CH_2-$). More complex linear polymers are created in several ways:

by substituting the hydrogen atoms eg vinyl polymers;

by substituting the carbon by other atoms eg heteropolymers;

by introducing double, triple or aromatic carbon bonds in the chain eg butadiene polymers;

by mixing different repeat units along the chain ie copolymers.

Replacing one hydrogen atom gives polymers like polyvinyl chloride, polystyrene and polyacrylonitrile. This creates an asymmetric carbon atom, where each of the groups attached is different. The grouping can exist in left-handed (L) or right-handed (R) forms, so that *stereoisomers* are formed: these are known as *tactic polymers* (see **stereoregular polymers** p. 299). Another form of isomerism is that found in doubly bonded repeat units like those in the polymer butadiene (Fig. 1).

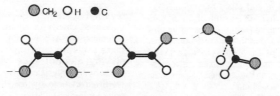

Cis-polybutadiene   *Trans*-polybutadiene   *Vinyl* or *1,2*-polybutadiene

Fig. 1. **Isomers of polybutadiene**.

The repeat units have identical elemental formulae (C_4H_6) and in the case of *cis* and *trans* forms, are known as *geometrical isomers*. The third (vinyl) isomer exists in stereoisomeric form, just like other vinyl polymers. Each of the isomeric forms is known and the properties of each are quite different from one another. Cis-polybutadiene is a highly resilient elastomer, but trans-polybutadiene is crystalline and rigid. Vinyl-polybutadiene is a rigid non-crystalline material when atactic. Commercial polybutadiene is usually a random mixture of all three different units, in effect a random copolymer.

material is isotactic and highly crystalline with repeat unit $-CH(CH_3)-CH_2-$. The T_m is high (ca 170°C) and the T_g is about –5°C, which makes it susceptible to cracking in a sharp frost. Copolymers with ethylene lower the T_g and improve toughness. Density of ca 905 kg m^{-3} (lowest of all solid polymers).

Used for small boat hulls, rope and string (fibrillated fibre), milk crates, battery cases etc. Atactic polypropylene is an important paper additive. Abbrev. *PP*.

polypyromellitimide Specific type of polyimide used as heat-resistant film. TN Kapton.

polysaccharides Polyoses; a group of

complex carbohydrates such as starch, cellulose etc. They may be regarded as derived from x monosaccharide molecules by the elimination of $x-1$ molecules of water. Polysaccharides can be hydrolyzed step to step, ultimately yielding monosaccharides.

polysilicates Minerals and materials based on polymers, sheets and networks of the silicate group $(SiO_4)^{4-}$. Widely exploited in ceramics, refractories and building materials. See **silica and silicates** p. 285.

polysilicon Polycrystalline silicon, deposited eg by **chemical vapour deposition**; used in semiconductor technology as a refractory gate material for **metal-oxide-silicon** devices.

polysiloxanes Oligomers and polymers based on the repeat unit:

Prepared by the hydrolysis of chlorosilanes R_3SiCl, R_2SiCl_2, ethers R_3SiOR, $R_2Si(OR)_2$, or mixtures. Frequently used as hydrophobic liquids on brickwork, cement etc. See **silicone rubbers**.

polystyrenes See **styrene resins, copolymer** p. 75.

polysulphide rubbers Polymers having linked sulphur atoms (eg –S–S–) in their backbone chain, and made by step-mechanism using chain-extension of thiol end groups. Vulcanized form used for lining aircraft fuel tanks, due to fluid resistance. Extensively used in prepolymer form as viscous fluid, which when mixed with crosslinking agent, can be applied to joints in buildings etc as sealant.

polysulphones Engineering polymers with aromatic back bone chains and T_gs from 190°C to 230°C. Step-polymers based on the sulphonyl functional group $(-SO_2-)$, and non-crystalline, so transparent. Mainly used in heat-resisting products, eg microwave oven parts.

polytetrafluoroethylene It has the repeat unit

$$-CF_2-CF_2- .$$

The principal fluoropolymer is noted for its very marked chemical inertness and heat resistance. It is difficult to shape by normal moulding methods and is often sintered. Widely used for a variety of engineering and chemical purposes, and as a coating for 'non-stick' kitchen equipment. Abbrev. *PTFE*. TNs Fluon, Teflon. Cf **Penton**.

polytetramethylene ether glycol A low molecular mass prepolymer (M=2000–3000), used in polyurethanes and polyester

rubber. Made by polymerization of tetrahydrofuran. Abbrev. *PTMEG*.

polythene See **polyethylene**.

polytrifluorochloroethylene A plastic material with many of the good properties of polytetrafluoroethylene (PTFE), but somewhat easier to fabricate as it can be injection moulded. Abbrev. *PTFCE*.

polyurethanes Versatile group of polymers forming tough material, either plastic or elastomeric. Step-polymer containing the urethane group (–NH–CO–) in its backbone chain. Formed by reaction of diols, dihydroxy polyesters, or hydroxyl-terminated polyethers (eg polytetramethylene ether glycol) with diisocyanates (eg tolylene-2,4-diisocyanate). Crosslinking can be achieved with polyols to give thermosets. Used as a casting polymer for shaped products, for foam rubber and as an extremely tough lining for eg ball mills. Also widely used for large car body parts when made by **reaction injection moulding**.

polyvinyl acetal (1) Group comprising polyvinyl acetal, polyvinyl formal, polyvinyl butyral and other compounds of similar structure. (2) A thermoplastic material derived from a polyvinyl ester in which some or all of the ester groups have been replaced by hydroxyl groups and some or all of these hydroxyl groups replaced by acetal groups. Generally a colourless solid of high tensile strength.

polyvinyl acetate *Polymerized vinyl acetate*, used chiefly in adhesive and coating compositions. Abbrev. *PVA*. See **vinyl polymers**.

polyvinyl alcohol Water-soluble polymer made by hydrolysis of polyvinylacetate, and modified chemically for fibre (vinal) and polyvinylbutyral.

polyvinyl butyral A polyvinyl resin made by reacting butanal with a hydrolyzed polyvinyl ester such as the acetate . Used chiefly as the interlayer in safety-glass windscreens to halt cracks and retain broken fragments.

polyvinyl carbazole Photoconductive polymer used in **xerography**. Made from polyvinyl alcohol and having large side-chain based on N-containing heterocyclic ring (carbazole group).

polyvinyl chloride See panel on p. 249.

polyvinyl chloride fibres Fibres of the linear polymer derived from vinyl chloride ie the repeat units are $[-CH_2-CH(Cl)-]$.

polyvinyl formal A polyvinyl resin made by reacting formaldehyde with a hydrolyzed polyvinyl ester. Used in lacquers and other coatings, varnishes and some moulding materials.

polyvinylidene chloride Based on vinylidene chloride (1,1-dichloroethylene). Usu-

polyvinyl chloride

Polyvinyl chloride (PVC) is one of the most versatile thermoplastics because of the way that its properties can be modified by additives. As a bulk commodity polymer it is one of the least costly, being made from chlorine (Cl_2) and ethylene (C_2H_4), both of which are among the most important chemical intermediates produced from inorganics (common salt, NaCl) and petrochemicals, such as naphtha from crude oil. See Fig. 1.

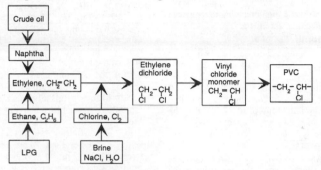

Fig. 1. **Synthetic pathway for polyvinyl chloride**.

A more efficient route for ethylene production, used in the US, UK and Norway, is from the ethane content of liquefied petroleum gases by catalytic cracking. The PVC is then made by addition, dehydrochlorination and polymerization.

Until about 40 years ago, PVC materials were of low strength due to thermal degradation during processing. Developments in polymerization (free radical, emulsion and suspension) and in additives such as the *stabilizers* have improved PVC beyond recognition. The applications of uPVC (u = unplasticized) include water pipes by extrusion, connections and supports for guttering, waste water etc by injection moulding. High-impact PVC is rubber toughened by graft copolymerization with methacrylate butadiene styrene (MBS) or **acrylate rubber** and is widely used for door and window frames. Plasticized PVC is formed by compounding PVC powder with esters like di-octyl phthalate (DOP) and di-alphanyl phthalate (DAP); they are absorbed and held within the PVC matrix by van der Waals bonding. Chain flexibility is increased because the glass transition temperature (T_g) is lowered from 80°C to less than ambient. Such materials are found as flexible film in stationery products (files, folders etc), imitation leather for car upholstery and serum bags etc for medical applications. One of the most widespread applications is for bottling detergents etc, where the material competes with polyolefins.

ally used as copolymer with small amounts of vinyl chloride or acrylonitrile. Basis of fibres, films for packaging, coatings for PET beer and cider bottles and chemically resistant piping. Highly crystalline and gas-impermeable polymer.

polyvinylidene chloride fibres Fibres from the polymer derived from vinylidene chloride ie the repeat units are

$$-CH_2-CCl_2-.$$

polyvinylidenefluoride Highly crystalline polymer with a repeat unit of:

$$-CH_2-CF_2-.$$

A *melt-processable* fluoropolymer of high toughness and piezoelectric properties in the drawn, oriented state. Used for large area flexible transducers, esp. in liquids (eg hydrophones, medical monitors).

polyvinyl polymers See **vinyl polymers**.

polyvinyl pyrrolidone Water-soluble polymer having a pyrrolidone cyclic unit as a pendant side group in the repeat unit.

Widely used in cosmetic products, and as a copolymer with methyl methacrylate etc in crosslinked soft contact lenses. The polyvinyl pyrrolidone part helps the lens swell in eye-fluid.

polyxylylene Polymer made by polymerizing paraxylene in film form. Heat stable, used as dielectric film.

POM See **PolyOxyMethylene**.

pongee Lightweight (75 g m^{-2}), plain weave cloth of wild silk, lighter and more regular in appearance than **shantung**.

pontie, pontil See **punty**.

poplar A common European **hardwood** tree (*Populus*) whose **heartwood** is creamy-white to grey, straight-grained, with a fine, even texture and mean density 450 kg m^{-3}. It is tough and non-splintering, but perishable and susceptible to insect attack. Used in the flooring of trucks and waggons, toys, interior joinery, crates etc and for veneers. Also *aspen, robusta*.

poplin A plain-woven fabric, with fine lines or cords running across the cloth (due to the ends per inch greatly exceeding the picks per inch); often of cotton and mercerized. Used for shirtings, pyjamas and dresses.

POR Abbrev. for **Propylene Oxide Rubber**.

porcelain Originally objects made of mother-of-pearl from the eponymous shell, but in the 16th century applied by the Portuguese to the ceramic whiteware developed in China. Now refers to any impermeable and translucent ware made from earthy raw materials, in particular **kaolin** (or China clay). 'Hard' porcelain is fired together with its glaze at up to 1400°C giving a much harder surface than 'soft' porcelain which is fired unglazed at 1250°, and then glazed and fired at about 1100°C. See **bone china**. Cf **earthenware, faience, stoneware, terracotta**.

porcelain insulator An insulator for supporting high-voltage electric conductors; made of a hard-quality porcelain.

pore distribution (1) The size and shape distribution of pores in a consolidated body such as a sintered metal compact. (2) In a loose powder compact, the space confined by adjacent particles can be considered as a system of interconnecting pores. The term pore distribution is used to describe the size variation in these so-called pores.

poromeric Permeable to water vapour; said of a polyurethane-base synthetic leather formerly used in the manufacture of shoe 'uppers'. See **Gore-tex**.

porosity (1) The fraction of voids in a material. If the porosity of a material is *p*, then its density ρ is related to *p* by

$$p = \frac{\rho_0 - \rho}{\rho_0},$$

where ρ_0 is the pore-free density. (2) In castings, unsoundness caused by shrinkage during cooling, or blow-holes. (3) In compaction of powders, the percentage of voids in a given volume under specified packing conditions. Also *powder porosity*.

porous bearing A bearing produced by powder metallurgy which, after sintering and sizing, is impregnated with oil by a vacuum treatment. Bearing porosity can be readily controlled and may be as high as 40% of the volume.

porous pot An unglazed earthenware pot serving as a diaphragm in a two-fluid cell.

porous silicon Bulk silicon which has been partially dissolved by electrochemical treatment in solutions containing hydrofluoric acid. The resulting porosity can be on a scale of greater than 50 nm (*macro*), between 50 and 2 nm (*meso*) and less then 2 nm (*micro*). Nanostructures within the material, which are made without recourse to lithography, form **quantum wires** having high luminescent efficiency which is not characteristic of the untreated bulk.

Portland cement The usual binder for concrete, named for its resemblance, when set, to Portland stone. Invented by Joseph Aspdin in 1824. See **cement and concrete** p. 52.

positive column Luminous plasma region in gas discharge, adjacent to positive electrode.

positive electricity Phenomenon in a body when it gives rise to effects associated with deficiency of electrons, eg positive electricity appears on glass rubbed with silk. See **negative electricity**.

positive ion An atom (or a group of atoms which are molecularly bound) which has lost one or more electrons, eg α-particle is a helium atom less its two electrons. In an electrolyte, the positive ions (*cations*) produced by dissolving ionic solids in a polar liquid like water, have an independent existence and are attracted to the anode. Negative ions likewise are those which have gained one or more electrons.

positive magnetostriction See **Joule magnetostriction**.

post- Prefix from L. *post*, after.

post (1) The formed glass gather prior to drawing by the hand process. (2) Term used for a pile of sheets or alternate felts and sheets in hand-made paper mills.

postal A non-preferred size of paper printing board 572× 725 mm (22× 28 in).

post-consumer recycling Reclamation of different materials in products after being discarded by consumer. Products such as lead-acid batteries, cars, paper and textile goods have well-established recycling routes but newer products containing large quantities of thermoplastics tend to be difficult to recycle easily. Incineration is one route where part of the feedstock energy of the polymer is recouped usefully (eg to raise steam). Also *secondary recycling*. See **reusability**.

post-forming sheet A grade of laminated sheet, of the thermosetting type, suitable for drawing and forming to shape when heated.

post-heating Annealing or tempering a weldment to remove strain or prevent local hardening.

post-tensioned concrete Form of **pre-stressed concrete**. See **cement and concrete** p. 52.

potassium A very reactive alkali metal, soft and silvery white, symbol K. In the form of the element, it has little practical use, although its salts are used extensively. In combination with other elements it is found widely in nature. It shows slight natural radioactivity due to ^{40}K (half-life 1.30×10^9 years). See Appendix 1, 2.

pot annealing See **close annealing**.

potential Scalar magnitude, negative integration of the electric (or magnetic) field intensity over a distance. Hence all potentials are relative, there being no absolute *zero potential*, other than a convention, eg earth, or at infinite distance from a charge.

potential barrier Maximum in the curve covering two regions of potential energy, eg at the surface of a metal, where there are no external nuclei to balance the effect of those just inside the surface. Passage of a charged particle across the boundary should be prevented unless it has energy greater than that corresponding to the barrier. Wave-mechanical considerations, however, indicate that there is a definite probability for a particle with less energy to pass through the barrier, a process known as *tunnelling*. Also *potential hill*. Cf **potential well**.

potential difference (1) The difference in potential between two points in a circuit when maintained by an emf or by a current flowing through a resistance. In an electrical field it is the work done or received per unit positive charge moving between the two points. Abbrev. *pd*. SI unit is the volt. (2) The line integral of magnetic field intensity between two points by any path.

potential energy Universal concept of *energy* stored by virtue of position in a field, without any observable change, eg after a mass has been raised against the pull of gravity. A body of mass m at a height h above the ground possesses potential energy mgh,

since this is the amount of work it would do in falling to the ground. In electricity, potential energy is stored in an electric charge when it is taken to a place of higher potential through any route. A body in a state of tension or compression (eg a coiled spring) also possesses potential energy.

potential well (1) Localized region in which the potential energy of a particle is appreciably lower than that outside. Such a well forms a trap for an incident particle which then becomes *bound*. Quantum mechanics shows that the energy of such a particle is quantized. Applied particularly to the variation of potential energy of a nucleon with distance from the nucleus. (2) Similarly, the region between atoms created by the balance between attractive and repulsive interatomic forces. See **bonding** p. 33.

potette A hood shaped like a pot, but with no bottom, which is placed in a tank furnace so that it reaches below the glass level. It protects workers gathering glass on their pipes or irons from furnace gases; also, the glass here is somewhat cooler than that in the main part of the furnace, where melting is taking place. Also *boot*, *hood*.

pot magnet One embracing a coil or similar space, excited by current in the coil or a permanent magnet in the central core. Main use is with a circular gap at an end of the core for a moving coil. Miniature split-sintered pot magnets are also used to contain high-frequency coils.

potty-putty TN for silicone polymer. See **viscoelasticity** p. 332. Also *bouncing putty*.

poultice corrosion That which occurs in crevices or on ledges which collect accretions of dirt and mud, esp. on the underside of motor vehicles.

pound The unit of mass in the old UK system of units established by the Weights and Measures Act, 1856, and until 1963 defined as the mass of the Imperial Standard Pound, a platinum cylinder kept at the Board of Trade. In 1963, it was redefined as 0.453 592 37 kg. The US pound is defined as 0.453 592 427 7 kg.

Pourbaix diagram See panel on p. 252.

pouring basin Part of the passage system for bringing molten metal to the mould cavity in metal casting. It is provided on large moulds, next to the sprue hole top, to simplify pouring and to prevent slag from entering the mould.

powder Discrete particles of dry material in the range 0.1 to 1000 µm.

powder core Core of powdered magnetic material with an electrically insulating binding material to minimize the effects of eddy currents, thus permitting use for high-frequency transformers and inductors with low

Pourbaix diagram

The plot of electrode potential against pH value which defines regions over which various ions and products are stable; used in electrochemistry and the prediction of metallic corrosion behaviour.

The example below represents the relationship for the system *iron-water-air* at 25°C.

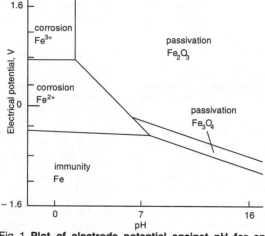

Fig. 1. **Plot of electrode potential against pH for an iron-water-air system**.

This exhibits three different types of zone.

First there are regions where metal ions are stable. If conditions are maintained corrosion occurs and metal ions form, so the regions are called *corrosion zones*.

In the second type, a solid film of oxide or hydroxide forms on the surface. If this is adherent and presents an impermeable barrier to air and water, corrosion will not continue. Such regions are called *passivation zones*. However, if the film is broken or not adherent, corrosion will continue; so such zones do not represent a stable state for the system.

The third type are *zones of immunity* and correspond to conditions under which the metal is stable with no tendency to ionize. Corrosion does not and cannot take place if such conditions are maintained.

loss in power.

powder metallurgy The working of metals and certain carbides in powder form by pressing and sintering. Used to produce self-lubricated bearings, tungsten filaments and shaped cutting inserts (carbide). See **cemented carbides**.

powder porosity The ratio of the volume of voids plus the volume of open pores to the total volume of the powder. Cf **porosity**.

powder-post beetle attack Attack on **sapwood** of timber for its starch content by Bostrychidae in the tropics and Lyctidae in temperate regions.

powder technology That covering the production and handling of powders, particle-size analysis and the properties of powder aggregates.

power Rate of doing work. Measured in joules per second and expressed in *watts* ($1 \text{ W} = 1 \text{ J s}^{-1}$). The foot-pound-second unit of power is the *horsepower*, which is a rate of working equal to 550 ft-lbf per second. one horsepower is equivalent to 745.7 watts. 1 watt is equal to 10^7 CGS units, that is 10^7 erg s^{-1}.

power law index Exponent of shear rate, symbol n, for modelling flow behaviour of non-Newtonian polymer melts etc. Power law equation is:

$$\tau = k\,(\dot{\gamma})^n,$$

where k is the viscosity coefficient and $\dot{\gamma}$ is the shear rate or consistency index. For Newtonian fluids, $n = 1$ and k is the shear viscosity (η). Dilatant behaviour occurs if $n > 1$, pseudoplasticity if $n < 1$. Also *flow behaviour index*. See **Newtonian viscosity**.

power transistor One capable of being used at power rating of greater than about 10 watts, and generally requiring some means of cooling.

POY Abbrev. for **Partially Oriented Yarn**.

PP Abbrev. for **PolyPropylene**.

ppi Abbrev. for *picks per inch*. See **pick**.

ppm Abbrev. for *Parts Per Million*.

PPO Abbrev. for **PolyPhenylene Oxide**.

pp, p-p junction A junction formed within in a p-type semiconductor where the acceptor dopant concentration changes abruptly.

pP, p-P heterojunction A junction in a **heterostructure** formed between a narrower band gap material (p-type) and a wider gap material (P-type).

PPS Abbrev. for **PolyPhenylene Sulphide**.

ppt, ppte Abbrevs. for *precipitate*.

praseodymium A metallic member of the rare earth elements, symbol Pr. It closely resembles neodymium and occurs in the same minerals. See Appendix 1, 2.

pre- Prefix from L *prae-*, in front of, before.

precipitation hardening See panel on p. 254.

precision grinding Grinding to a high finish and accurate dimensions on an appropriate machine.

preferred orientation During slip, metal crystals change their orientation; when a sufficient amount of deformation has been performed, the random orientation of the original crystals is converted into an arrangement in which a certain direction in all the crystals is parallel to the direction of deformation. Produces a texture in a rolled or wrought product.

pre-impregnation Reinforcing tapes etc which have been impregnated with resin and partially cured before assembly in a structure which is then finally cured in a mould. Abbrev. *prepreg*.

prepolymer Low molecular mass polymer, often a viscous liquid, used as an intermediate material prior to final shaping (eg polytetramethylene ether glycol).

prepreg Abbrev. for **pre-impregnation**.

press forging A forging process using a vertical press to apply a slow squeezing action as distinct from the sudden impact employed in drop forging. Press forging tends to work the material uniformly throughout the section whereas deformation in drop forging tends to concentrate near the surface.

pressing The application of pressure, often accompanied by heat and/or steam, to smooth fabrics or garments. Also the same process used to introduce chosen creases or pleats.

press rolls The heavy horizontal rolls situated in the press section of the paper- or board-making machine in various configurations to consolidate the web and remover water at the nips between them. The rolls may be constructed of granite, rubber covered metal or with a perforated metal shell over a fixed internal suction box.

press tool A tool for cutting, usually comprising at least a punch and a die, or for forming, or for assembling, used in a manual or in a power press.

pressure See **stress**.

pressure bag moulding See **bag moulding**.

pressure diecasting A process by means of which precision castings of various alloys are made by squirting liquid metal under pressure into a metal die. See **diecasting**.

pressure forging See **drop forging**.

pressure vessel A container for fluids stored under pressure above or below atmospheric. Classified as fired or unfired. Usually cylindrical with dished ends, for ease of construction and support, or spherical, for very high pressures.

pressure welding Welding parts which are pressed tightly together, as in forge welding or in various electrical resistance welding processes or in cold welding.

prestressed concrete Concrete beams and other structural members, in which the whole member is placed in compression by the tension in strained steel or other tendons within it. This enables the member to withstand higher bending and tensile loads than without prestressing. Developed by the French engineer, Freysinnet, in 1927. The two variants are pretensioned and post-tensioned. Cf **reinforced concrete**. See **cement and concrete** p. 52.

pretensioned concrete A form of **prestressed concrete**. See **cement and concrete** p. 52.

primary A substance which is obtained directly, by extraction and purification, from natural raw material; eg methane, ethane, propane and natural gas primaries. Cf **intermediate**.

primary acids Acids in which the carboxyl group is attached to the end carbon atom of a chain, ie to $-CH_2-$ group.

primary alcohols Alcohols containing the group $-CH_2.OH$. On oxidation, they form aldehydes and then acids containing the same number of carbon atoms as the alcohol.

primary bonds Term usually applied to the

precipitation hardening

Sometimes known as *age hardening*, it is a means of improving the strength of solid solution alloys by controlling the formation of precipitates on a crystal lattice scale, by ageing of the supersaturated solid solution. The essential requirement is for the phase diagram to display a change in solid solubility over a range of temperature, as illustrated by the sloping *solvus* in Fig. 1.

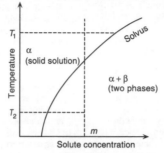

Fig. 1. **Solid solution solubility changing with temperature**.

A composition such as indicated by m will exist as a homogeneous solid solution if heated until equilibrium conditions are obtained at any temperature above the solvus, such as T_1. If this is then cooled slowly to temperatures below the solvus, eg T_2, so that equilibrium is still maintained, the second phase will separate and the solid solution composition will be that on the solvus curve. However, if the material is cooled rapidly from T_1 causing equilibrium to be suppressed, the single solid solution will be retained at T_2 in a metastable state and will be supersaturated with respect to the solvus at that particular temperature. Over a period of time this supersaturated solution may harden as the solute atoms diffuse to sites favourable to precipitate out the second phase. This is the phenomenon of precipitation hardening and the period over which the hardening occurs is known as the ageing time.

In some alloy systems, notably those based on **aluminium alloys,** the hardening takes place at ambient temperatures, reaching a maximum in about five days. This is referred to as *natural ageing*. Heating at moderately elevated temperatures (but below the solvus) gives similar, but slightly enhanced, hardening effects; this is called *artificial ageing*. These two effects are illustrated diagramatically in Fig. 2. In other alloy systems the ageing reponse is attained at considerably higher temperatures, for example nickel-chromium

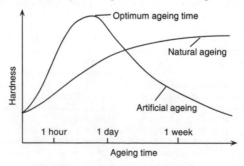

Fig. 2. **Natural and artificial ageing response**.

continued on next page

254

---

precipitation hardening (contd)

high-temperature alloys above 800°C, but the mechanism is essentially similar.

On the atomic scale the first stages in the hardening mechanism produces locally strained regions known as *Guinier–Preston* (GP) *zones*. The concentration in these zones may reach the nominal composition of the second phase without forming an interface with the solvent lattice, described as a *coherent precipitate*. In some alloy systems this condition represents the maximum of the ageing curve. After a further period of ageing, an interface forms which relaxes some of the local strains and the precipitate is then said to be *incoherent*, though it is still well below the resolution of an optical microscope. At this stage the hardening is due to the fine dispersion of the precipitate particles throughout the solid solution grains. Ageing for a very long time or at higher temperatures eventually causes the incoherent precipitate particles to grow to a microscopically visible size and the hardening effect is then lost. See **ageing**, **temper-hardening**.

The heat treatment process to achieve precipitation hardening must involve two essential steps: first the alloy has to be heated and brought into equilibrium in the single-phase region, referred to as the *solution heat treatment*. It must then be cooled rapidly to ambient temperature to retain the metastable state. In this condition components may be formed to finished shape, or parts such as rivets may be fixed, since the material is ductile. It is then either allowed to age naturally or subjected to an artificial ageing treatment to develop its optimum strength, when the ductility is greatly reduced.

Two examples of the magnitude of the response in a commercial aluminium alloy (4% copper) are set out below. The annealed data relate to the equilibrium condition of the alloy with all the $CuAl_2$ precipitated.

Condition	Annealed	Artificially aged
Hardness (H_v)	65	150
0.2% proof stress (MPa)	98	430
Tensile strength (MPa)	270	480
Elongation %	25	10

---

strong covalent, ionic or metallic bonds in materials. See **bonding** p. 33.

primary carbon atom Carbon atom linked to one carbon atom and three hydrogen atoms.

primary cell Voltaic cell in which chemical energy of the constituents is changed to electrical energy when current is permitted to flow. Such a cell cannot be recharged electrically because of the irreversibility of chemical reaction which occurs. It may be used as a sub-grade standard cell for calibration purposes.

primary energy Refers to energy such as crude oil, natural gas and coal, which is used in its natural form to provide *secondary energy* like electricity.

primary metal Ingot metal produced from newly smelted ore, with no addition of recirculated scrap.

primary solid solution A constituent of alloys that is formed when atoms of an element B are incorporated in the crystals of a metal A. In most cases solution involves the substitution of B atoms for some A atoms in the crystal structure of A (substitutional solid solution), but in a few instances the B atoms are situated in the interstices between the A atoms (interstitial solid solution).

primary standard A standard agreed upon as representing some unit (eg length, mass, emf) and carefully preserved at a national laboratory. Cf **secondary standard**.

primary stress An axial or direct tensile or compressive stress, as distinct from a bending stress resulting from deflection.

primary structure Covalent-bonded atoms and groups in biopolymers, determining properties of the substances. In natural homopolymers or copolymers, specified by repeat units and sequencing; most complex in proteins, where amino-acid sequence is usually unique. See **biomaterials** p. 57.

prima vera C American **hardwood** tree (*Tabebuia donnell-smithii*) whose **heartwood** is yellowish-rose streaked with orange, red and brown, straight- to wavy-grained, with a moderately coarse texture and mean density 450 kg m^{-3}. It is easy to work, but perishable and susceptible to insect attack. Used in furniture, panelling, high-class interior joinery and for decorative veneers.

primes Metal sheet and plate of the highest quality and free from visible imperfections.

Prince Rupert's drops Entertaining laboratory demonstration. Solid drops of silicate glass with long thin tails are formed by quenching gobs of the molten glass by dropping into water. The quenching puts the surface of the drop into compression and the interior into tension (see *thermally toughened glass* under **safety glass**). The drops shatter into a powder when their tails are broken off since the rupture of the compressive layer 'releases' the internal tensile stresses.

principal stress The component of a stress which acts at right angles to a surface, occurring at a point at which the shearing stress is zero.

printed circuit An electronic sub-assembly consisting of an insulating board or card with copper conductors laminated on. See **printed, hybrid and integrated circuits** p. 257.

printed, hybrid and integrated circuits See panel on p. 257.

proban Flameproof finish for textile fabrics based on a phosphorus compound, tetrakis (hydroxymethyl) phosphonium chloride.

process annealing Heating steel sheet or wire between cold-working operations to slightly below the critical temperature and then cooling it slowly. The process is similar to tempering but will not give as much softness and ductility as full annealing. See **annealing** p. 13.

process energy Energy needed to manufacture specific materials to shape and assemble them into finished products. The term usually extends to natural resources in the Earth's crust. See **ecobalance**.

process metallurgy The science and technology of extracting metals from their ores and purifying them.

product design All those methods used to plan, manufacture and evaluate industrial and consumer articles, includes both industrial and engineering design activities as well as selling, costing and legal constraints (eg product liability) etc.

product design specification Detailed outline of product function and the environment in which it will perform, with design geometry and materials of parts needed to fulfil that function. Abbrev. *PDS*.

product liability Risks associated with manufactured products when sold in the open market; specifically, risks to personal safety caused by faulty design, manufacturing defects, poor materials selection and/or marketing warnings. See **strict liability**.

product life cycle Life of industrial or consumer product between factory gate and scrapping. Real life cycles of a specific product may vary enormously, but specification life cycles are based on a notional formula based on market research, engineering design and materials performance etc.

products The elements and/or compounds formed in a chemical reaction. They are written on the right-hand side of a chemical equation.

projected area Area of moulding at right angles to injection direction. See **injection moulding** p. 169.

projection welding A process similar to spot welding but allowing a number of welds to be made simultaneously. One of the workpieces has projections at all points or lines to be welded, the other is flat. Both are held under pressure between suitable electrodes, the current flowing from one to the other at the contact points of the projections.

promethium A radioactive member of the **rare earth elements**, symbol Pm, having no known stable isotopes in nature. Its most stable isotope, ^{145}Pm, has a half-life of over 20 years. See Appendix 1, 2.

promoter A substance which increases the activity of a catalyst.

prompt gamma The γ-radiation emitted at the time of fission of a nucleus.

prompt neutrons Those released in a primary nuclear fission process with practically zero delay.

proof The impression, often in soft metal, of a die for inspection purposes.

proof stress The stress required to produce a certain amount of permanent set in metals which do not exhibit a sudden yield point. Usually it is the stress producing a strain of 0.1, 0.2 or 0.5%. See **strength measures** p. 301.

proof test A mechanical test carried out on a manufactured component to ensure that it is capable of meeting and exceeding foreseeable service requirements, eg gun barrels, pressure vessels, chains and lifting gear. The proof test usually stresses the component to double the **safe working load** to be used in the application. Not to be confused with **proof stress**.

---

printed, hybrid and integrated circuits

Fig. 1 shows a circuit board manufactured in the late 1980s. The main sub-assembly is a printed circuit board (PCB). Many of the components mounted on it are themselves complex circuits. These are hybrid and integrated circuits. The three types, printed, hybrid and integrated are distinguished by the type of the substrate material and the associated processing methods.

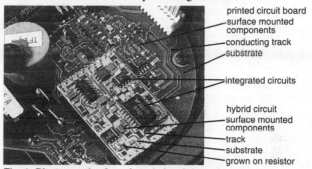

printed circuit board
surface mounted components
conducting track
substrate

integrated circuits

hybrid circuit
surface mounted components
track
substrate
grown on resistor

Fig. 1. **Photograph of a printed circuit board**.

PCBs are built from copper-clad polymeric boards on which conducting tracks are defined by photolithography. Components (resistors, capacitors, transistors as well as the hybrid and integrated sub-assemblies) are fixed by soldering, either directly onto the surface or else after passing leads through locating holes from one side of the board to the other. Multilayer boards can be made with eg buried earth planes or power lines and complicated assemblies use both sides of a board, allowing dense packing. Track and contact spacings less than 1 mm can be achieved. Most PCBs are now assembled by robot placement of components with soldering taking place in a single pass through a soldering machine.

Hybrid circuits use ceramic substrates on which some components are placed as with PCBs but the conducting tracks and other components are formed *in situ*, hence the technology is a hybrid of the printed and integrated circuit approaches. Tracks, resistors and insulating layers are built onto the substrate first by screen printing with special inks which are subsequently fired to bring them to their functional condition (conducting or insulating glassy enamel). Hence the use of a refractory substrate. Other components are then soldered in place as for PCBs. Hybrids are robust in hostile chemical, thermal and mechanical environments and are especially useful at microwave frequencies.

Integrated circuits (ICs) or *chips*, may contain thousands of interconnected components within a few square millimetres, capable of carrying out complex electronic operations. A silicon IC, for example, is built onto a wafer of single crystal silicon by a series of deposition, patterning and etching steps, performed simultaneously on hundreds of devices; the wafer is divided into individual chips only after all processing is complete. Each IC is then fitted with external connection leads and packaged for incorporation into one of the larger sub-assemblies. See **semiconductor device processing** p. 280.

---

propagation Repeated addition of monomer molecules to an activated polymer chain end, which may be anionic, cationic or free radical in nature. See **chain polymerization** p. 56.

propane C_3H_8, an alkane hydrocarbon, a colourless gas at atmospheric pressure and temperature. Bp $-45°C$; found in crude petroleum. In liquid form it constitutes liquefied petroleum gas (LPG), a clean-burning fuel, used by some as a petrol substitute. Cracking yields propylene.

Propathene TN for **polypropylene** (UK).

proportional limit The point on a stress-strain curve at which the strain ceases to be proportional to the stress. Its position varies with the sensitivity of the **extensometer** used in measuring the strain. Also *limit of proportionality*.

propylene C_3H_6, CH_2=$CH.CH_3$, an alkene hydrocarbon, a gas, bp $-48°C$. An important by-product of oil refining used for the manufacture of many organic chemicals including propanone, glycerine and polypropylene. Also *propene*.

propylene oxide rubber A speciality rubber based on the repeat unit

$$-O.CH(CH_3)CH_2-$$

made with allyl glycidyl ether as co-monomer. Similar to **epichlorhydrin rubber**. Abbrev. *POR*.

prosthesis Engineered device replacing or restoring tissue or organs but not normally exposed to body fluids such as blood etc eg artificial limbs, hearing aids etc.

protactinium A radioactive element. Symbol Pa, half-life of 2×10^4 years. One radioactive isotope is ^{233}Pa which is an intermediate in the preparation of the fissionable ^{233}U from thorium. See Appendix 1, 2.

protective coating A layer of a relatively inert substance, on the surface of another, which diminishes chemical attack of the latter, eg Al_2O_3 on metallic Al. Also includes paints and many polymers.

protective furnace atmosphere Inert gas produced from the products of combustion of gas and air in predetermined proportions, the atmosphere being first cooled, cleaned, dehydrated, and desulphurized before delivery to heating chambers of furnaces operated for bright-annealing of non-ferrous metals and bright treatment of steels.

protein Protein macromolecules consist of amino acids linked together by peptide (amide) bonds ($-NH-CO-$) in a sequence unique to each protein type. Unlike most synthetic homopolymers, where each repeat unit is chemically identical, the chain unit in proteins is one of 20 natural amino acids. They are thus heteropolymeric and more closely allied to multicomponent copolymers. The structural proteins (scleroproteins) all possess very high molecular masses (10^6) and are composed of hydrophobic amino acids, so are insoluble in water. Extra chemical stability is often provided by crosslinking through disulphide bonds etc especially crystine and lysine chain units. The properties of proteins are determined by its unique amino acid sequence (primary structure) which itself determines the correct folding of the chain in its native conformation. See **secondary structure**. Structural proteins are often highly crystalline and fibrous (eg silk fibroin, wool keratin, collagen in skin, bone and tendon) but some are non-crystalline and possess elastomeric properties (eg elastin, resilin). See **biomaterials** p. 57.

proton The nucleus of the hydrogen atom; of positive charge and atomic mass number of unity. With neutrons, protons form the nucleus of all other atoms, the number of protons being equal to the atomic number. It is the lightest *baryon*, 1.007276 a.m.u. and the most stable. Beams of high-energy protons produced in particle accelerators are used to study elementary particles.

proton resonance A special case of nuclear magnetic resonance. Since the nuclear magnetic moment of protons is now well known, those of other nuclei are found by comparing their resonant frequency with that of the proton.

prototype Generally, the first or original type or model from which anything is developed.

prunt An disc of glass bearing a monogram or badge, eg of the owner or vintner, and fused on to bottles or drinking vessels.

PS Abbrev. for **PolyStyrene**.

pseudo-elastic method See **quasi-elastic method**.

pseudo plasticity Type of **non-Newtonian flow**, where viscosity falls with increasing shear rate. Most polymer melts are pseudoplastic. Also known as shear thinning.

pseudo potential method A method of calculating the band structure of metals. See **band theory of solids**.

pseud-, pseudo- Prefixes from Gk *pseudes*, false.

PST Abbrev. for *lead scandium tantalate*. See **dielectric and ferroelectric materials** p. 94.

PSu Abbrev. for **PolySulphones**.

PTCR Abbrev. for *Positive Temperature Coefficient Resistor*. See **silicon resistor**.

PTFCE Abbrev. for **PolyTriFluoroChloroEthylene**.

PTFE Abbrev. for **PolyTetraFluoroEthylene**.

PTMEG Abbrev. for **PolyTetraMethylene EtherGlycol**.

p-type conduction, conductivity That associated with charge transport by holes.

p-type semiconductor That arising in a semiconductor containing **acceptor** impurities, with conduction by (positive) holes.

P-type semiconductor Relatively wide band gap p-type semiconductor.

pucella An implement for opening out the

top of a wine glass in the off-hand process.

puddled pig iron The agitation of a bath of molten pig iron, by hand or by mechanical means, in an oxidizing atmosphere to oxidize most of the carbon, silicon and manganese and thus produce wrought iron. Now obsolete. See **puddling**.

puddled bars Bars of wrought iron which have been rolled from the puddled ball after squeezing the ball to compact it and eliminate some of the slag. Now obsolete.

puddling The agitation of a bath of molten pig iron, by hand or by mechanical means, in an oxidizing atmosphere to oxidize most of the carbon, silicon and manganese and thus produce wrought iron. Now obsolete.

puddling furnace A small reverberatory furnace in which iron is puddled. Now obsolete.

pull A sample of textile fibres pulled by hand from a large quantity of raw material (eg a bale) and used for analysis; particularly for determining fibre length.

pulling (1) The conversion of rags and short lengths of yarn into fibres for re-use. (2) The removal of the wool fibres from skins taken from dead sheep. See **fellmongering**, **shearing**, **skin wool**.

pull-over mill A rolling-mill using a single pair of rolls. The metal, after passing through the rolls, is pulled back over the top roll to be fed through the mill a second time.

pulp Generic term for any fibrous raw material, often cellulose-based, after preparation in a pulp mill by grinding, refining, digesting with chemicals and/or bleaching. Pulp may be a suspension in water or in the form of fluff or dried or semi-dried sheets.

pulp board A printing board of the same **furnish** throughout.

pulping Any process, mechanical or chemical, that partially breaks down fibres from eg wood. Also *defibration*.

punch (1) A tool for making holes by shearing out a piece of material corresponding in outline to the shape of the punch; a machine incorporating such a tool. (2) A hand tool, struck with a hammer, for marking a surface or for displacing metal as in riveting.

punch-through Collector-emitter voltage breakdown in transistors.

punt The bottom of a glass container.

punty, puntee, pontie, pontil A short iron rod, at one end of which is either a button of hot glass or a suitably shaped piece of metal, which is applied hot to the end of a partially formed glass article in order that (1) it may be cracked off the blowpipe and manipulated on the punty, or (2) in the case of tube drawing, the mass of glass may be drawn out between punty and blowpipe.

purging Operation at beginning of batch

moulding, where previous polymer and accumulated debris are removed using a special polymer purging agent, often the polymer for the next run.

purl fabrics Knitted fabrics in which the reverse side stitches are brought to the surface for effect; used extensively for pullovers etc.

purpleheart Varieties of C American **hardwood** tree (*Peltogyne*) whose **heartwood** is a deep purple-violet, straight- to wavy-grained, with a moderate to fine texture and mean density 860 kg m^{-3}. Very durable and used in outdoor construction work, bridges, piling, flooring, furniture, and for decorative veneers. Also *amaranth*, *violetwood*.

pushed punt A concave bottom to a glass container. Also *push-up*.

push-fit Loose-tolerance mechanical joint between two parts (eg Lego bricks)

putty A mixture of whiting and linseed oil, sometimes including white lead, forming a plastic substance for sealing panes of glass into frames.

PVA Abbrevs. for **PolyVinyl Acetate**, **PolyVinyl Alcohol**. But see **PVAC** and **PVAL**.

PVAC Abbrev. for **PolyVinyl Acetate**.

PVAL Abbrev. for **PolyVinyl Alcohol**.

PVB Abbrev. for **PolyVinylButyral**.

PVC Abbrev. for **PolyVinyl Chloride** p. 249.

PVD Abbrev. for **Physical Vapour Deposition**.

PVDC Abbrev. for **PolyVinyliDeneChloride**.

PVDF Abbrev. for **PolyVinyliDeneFluoride**. See **dielectric and ferroelectric materials** p. 94.

PVK Abbrev. for **PolyVinyl Carbazole**.

PVP Abbrev. for **PolyVinyl Pyrrolidone**.

pycnometer (1) A small, graduated glass vessel, of accurately defined volume, used for determining the relative density of liquids. (2) Similar device for measuring the relative density of (soil) particles less than about 5 mm in diameter. Also *density bottle*.

pycn-, pycno-, pykn-, pykno- Prefixes from Gk *pyknos*, compact, dense.

pyknometer See **pycnometer**.

Pyrex TN for a borosilicate glass used for domestic ovenware and laboratory glassware. Its low thermal expansion coefficient makes it much less susceptible to thermal shock than conventional soda-lime-silica glass.

pyrolysis The decomposition of a substance by heat.

pyrolysis gas chromatography A method for analysing the products of thermal decomposition of substances using mass spectroscopy. Useful for intractable polymers (eg thermosets) which cannot be easily analysed using ordinary methods (eg infrared, ultraviolet and nuclear magnetic resonance spectroscopy). Abbrev. *PGC*.

pyrometallurgy Treatment of ores, concentrates or metals when dry and at high temperatures. Techniques include smelting, refining, roasting, distilling, alloying and heat treatment.

pyrometric cone equivalent A measure of the softening or melting point of a refractory, carried out by means of comparing with standard **Seger cones** placed alongside in a furnace. Abbrev. *PCE*.

pyrometric cones See **Seger cones**.

pyrophoric Materials and mixtures liable to spontaneous combustion, esp. finely divided metals. The nuclear fuels U, Th and Pu are all pyrophoric.

pyroxilins Nitrocelluloses with a low nitrogen content, containing from two to four nitrate groups in the molecule. Used in an ethanol-ethoxyethane solution to form collodion. Pyroxilin is a synonym for guncotton.

PZT See lead zirconate titanate.

Q (1) A symbol for charge. (2) Symbol of merit for an energy-storing device, resonant system or tuned circuit. Parameter of a tuned circuit such that $Q = \omega L/R$, or $1/\omega CR$, where L = inductance, C = capacitance and R = resistance, considered to be concentrated in either inductor or capacitor. Q is the ratio of shunt voltage to injected emf at the resonant frequency $\omega/2\pi$. $Q = f_r/(f_1 - f_2)$, where f_r is the resonant frequency and $(f_1 - f_2)$ is the bandwidth at the half-power points. For a single component forming part of a resonant system it equals 2π times the ratio of the peak energy to the energy dissipated per cycle. For a dielectric it is given by the ratio of displacement to conduction current. Also *magnification factor, Q-factor, quality factor, storage factor*.

Qiana TN for a speciality nylon with silken properties when spun into fibre (US).

quad Prefix to denote a paper size which is four times the area of that of the basic size (broadside), ie both dimensions of the basic size are doubled.

quadr-, quadri- Prefixes from L *quattuor*, four.

quality control A form of inspection involving sampling of parts in a mathematical manner to determine whether or not the entire production run is acceptable, a specified number of defective parts being permissible.

quality systems Ways of managing materials, components and products so as to ensure high quality control of manufactured products at all levels, as specified in, for example, BS 5750.

QuantiMet TN for machine which analyses material surfaces for microstructural variables such as grain size, diameter, orienta-

tion etc. Based originally on the optical microscope, extended to electron optical examination. Uses computer techniques to perform statistical analyses based on stereological methods.

quantization In quantum theory, the division of energy of a system into discrete units (*quanta*), so that continuous infinitesimal changes are excluded.

quantometer An instrument showing by spectrographical analysis the percentages of the various metals present in a metallic sample.

quantum (1) General term for the indivisible unit of any form of physical energy; in particular the *photon*, the discrete amount of electromagnetic radiation energy, its magnitude being $h\nu$ where ν is the frequency and h is **Planck's constant**. (2) An interval on a measuring scale, fractions of which are considered insignificant.

quantum efficiency Number of electrons released in a photocell per photon of incident radiation of specified wavelength.

quantum Hall effect The Hall resistivity changes by steps so that it is a fraction of h/e^2 where h is **Planck's constant** and e is the electronic charge. Observed in two-dimensional semiconductors (eg **metal-oxide-silicon**) at high magnetic fields and ultra-low temperatures. See **Hall effect**.

quantum mechanics A generally accepted theory replacing classical mechanics for microscopic phenomena. Quantum mechanics also gives results consistent with classical mechanics for macroscopic phenomena. Two equivalent formalisms have been developed, *matrix mechanics* (Heisenberg) and *wave mechanics* (Schrödinger). The theory accounts for a very wide range of physical phenomena.

quantum number One of a set, describing possible quantum states of a system, eg nuclear spin.

quantum wire A nano-structure proportioned like a wire so that electron behaviour is strongly constrained by quantum mechanical effects in two dimensions.

quartation See **inquartation**.

quarter girth The girth of a log or tree divided by four. A measure commonly used in countries where volumes are reckoned in **Hoppus feet**. Abbrev. *qg*.

quartering A piece of timber of square section between 2 and 6 in (50 and 150 mm) side.

quarter sawing A mode of converting timber, adopted when it is desired that the growth rings shall all be at a minimum of 45° to the cut faces. Also rift sawing.

quarter-wave plate A plate of quartz, cut parallel to the optic axis, of such thickness that a retardation of a quarter of a period is produced between ordinary and extraordinary rays travelling normally through the plate. By using a quarter-wave plate, with its axis at 45° to the principal plane of a Nicol prism, circularly polarized light is obtained.

quarto See **paper sizes**.

quartz Crystalline silica, SiO_2, occurs either in prisms capped by rhombohedra (low-temperature quartz, stable up to 573°C) or in hexagonal bipyramidal crystals (high-temperature quartz, stable above 573°C). Widely distributed in rocks of all kinds; igneous, metamorphic and sedimentary; usually colourless and transparent (rock crystal), but often coloured by minute quantities of impurities as in citrine, cairngorm etc; also finely crystalline in the several forms of chalcedony, jasper etc. See **silica and silicates** p. 285.

quartz crystal A disk or rod cut in the appropriate directions from a specimen of piezoelectric quartz, and accurately ground so that its natural resonance occurs at a particular frequency.

quartz glass See **vitreous silica**. Also *fused silica, silica glass*.

quartz-iodine lamp Compact high-intensity light source, consisting of a bulb with a tungsten filament, filled with an inert gas containing iodine (sometimes bromine) vapour. The bulb is of quartz, glass being unable to withstand the high operating temperature (600°C). Used for car-lamps, cine projectors etc. Also *quartz-halogen lamp, tungsten-halogen lamp*.

quartz oscillator One whose oscillation frequency is controlled by a piezoelectric quartz crystal.

quasi-elastic method Method of stress analysis for non-linear and/or time-dependent materials, especially polymers, in which elastic moduli in the elastic equations are replaced by the values of the corresponding secant modulus or creep modulus, at the required levels of strain or time, respectively.

quasi-Fermi levels Energy levels in a semiconductor from which the number of electrons or holes available for conduction under non-equilibrium conditions, esp. when light is falling on the semiconductor, can be calculated in the same way as from the true **Fermi level** which applies under equilibrium conditions.

quaternary diagram Phase diagram of four-component system.

Queensland blue gum See **Red River gum**.

Queensland walnut Not a true walnut (*Juglans*), but an Australian **hardwood** tree (*Endiandra palmerstonii*) whose **heartwood** is light to dark brown streaked with pink, grey, green and black, with irregular, interlocked grain, a lustrous, medium, even texture and a mean density of 680 kg m^{-3}. Non-durable and used in high-class furniture, panelling, joinery, and for decorative veneers.

quenching The process of cooling by plunging a heated object into a fluid, with the purpose of retaining the material in a metastable state. Quenching into water gives a more rapid cooling rate than into oil. The term also applies to cooling in salt and molten-metal baths or by means of an air blast. Applied to steels heated above their upper critical temperature in order to harden them prior to tempering and to other alloys for solution treatment prior to **precipitation hardening** p. 254. See **steels** p. 296.

quire Paper quantity; 25 sheets, or one twentieth of a ream.

R

R General symbol for an organic hydrocarbon radical, esp. an alkyl radical.

R Symbol for: (1) the **gas constant**; (2) the Rydberg constant.

R- Prefix denoting *right-handed*.

radial-ply Term applied to tyres which have a semirigid breaker strip under the tread and relatively little stiffening in the walls. Cf **cross-ply**. See **tyre technology** p. 325.

radiant heat Heat transmitted through space by **infrared radiation**.

radiant-tube furnace Modified form of muffle furnace, the heating chamber having a series of steel alloy tubes in which the fuel is burned, thus excluding products of combustion from the heating chamber.

radiation The dissemination of energy from a source. The energy falls off as the inverse square of the distance from the source in the absence of absorption. The term is applied to electromagnetic waves (radio waves, infrared, light, X-rays, γ-rays etc) and to acoustic waves. It is also applied to emitted particles (α, β, protons, neutrons etc). See types of radiation: **black-body, electromagnetic, infrared, ultraviolet, visible**. See **Planck's radiation law, Stefan–Boltzmann law.**

radiation chemistry That of radiation-induced chemical effects (eg decomposition, polymerization etc). Not to be confused with radiochemistry.

radiation crosslinking Crosslinking reaction induced by gamma-rays etc. Used on polyethylene to make shrink-on sleeves for cable connections etc.

radiation polymerization Chain reaction initiated by gamma-rays etc, which react with monomer to give free radicals. See **chain polymerization** p. 56.

radical (1) A molecule or atom which possesses an odd number of electrons, eg the methyl radical CH_3-. It is often very short-lived, reacting rapidly with other radicals or other molecules. See **free radical**. (2) A group of atoms which passes unchanged through a series of reactions, but is normally incapable of separate existence. This is now more usually called a group.

radioactivation analysis If a material undergoes bombardment by neutrons in a nuclear reactor, radioactive nuclei are frequently produced. This artificial radioactivity can be studied to give information about the isotopes present in the material.

radioactive atom One which decays into another species by emission of an α- or β-ray (or by electron capture). Activity may be natural or induced.

radioactive isotope Naturally occurring or artificially produced isotope exhibiting radioactivity; used as a source for medical or industrial purposes. Also *radioisotope*.

radioactivity Spontaneous disintegration of isotopes of certain elements accompanied by the emission of α-rays, which are positively charged helium nuclei; β-rays, which are fast electrons; and γ-rays, which are short X-rays.

radiofrequency heating See **dielectric heating, induction heating**.

radiofrequency plasma Plasma produced by either capacitively or inductively coupled radio frequency fields.

radioluminescence Luminous radiation arising from rays from radioactive elements, particularly in mineral form.

radiophotoluminescence Luminescence produced by irradiation followed by exposure to light.

radiothermoluminescence Luminescence produced by irradiation followed by exposure to heat. Now used for personal dosimetry.

radium Radioactive metallic element, one of the alkaline earth metals. Symbol Ra, half-life 1620 years. The metal is white and resembles barium in its chemical properties. It occurs in bröggerite, cleveite, carnotite, pitchblende, in certain mineral springs and in sea water. Pitchblende and carnotites are the chief sources of supply. See Appendix 1, 2.

radon A radioactive element, the heaviest of the noble gases. Symbol Rn. See Appendix 1, 2.

rag content papers Papers the fibrous furnish of which contains 25 wt% or more of rag (cotton etc) fibres.

raising Textile process in which wire-covered rollers or teazles on cylinders revolve over the cloth surface to produce a nap.

rake Angular relief, eg *top-rake, side-rake*, given to the faces of cutting tools to obtain the most efficient cutting angle. The face of a cutting tool with negative rake has no angular relief but, on the contrary, meets the workpiece with the cutting edge trailing, the tool being stronger in shear, impact and abrasion than one having positive rake.

r.a.m. Abbrev. for **Relative Atomic Mass**. Also *RAM*.

Raman scattering The scattering of light by molecules in which there is a change of frequency due to the molecules gaining or losing energy as a result of transitions between vibrational or rotational energy levels.

The scattering of light by the optic modes of vibration in a crystal, ie photon-phonon scattering. See **scattering**.

Raman spectroscopy A method making use of **Raman scattering** for chemical analysis. Like **infrared spectroscopy**, it investigates molecular vibrations and rotations.

ramie A **bast** fibre from the stems of *Boehmeria nivea*; can be bleached to give rather brittle white fibres that may be used in garments blended with other fibres. Formerly used for the manufacture of gas mantles.

ramin SE Asian **hardwood** (*Gonystylus macrophyllum*) whose **heartwood** is a uniform pale straw colour, straight-grained, with a fine, even texture and a mean density of 660 kg m^{-3}. Non-durable and used in furniture, panelling, interior joinery, flooring and for veneers and plywood.

random coil Conformation of single polymer chain in non-crystalline state and at equilibrium, a sphere of diameter determined by its molecular mass. Present in rubbers at rest, entangled with adjacent chains, but as isolated coils in dilute polymer solutions. See **elastomers** p. 106.

random copolymer See **copolymer** p. 75.

random sequence welding Adding welding beads at random along a seam to minimize distortion.

range of stress The range between the upper and lower limits of a cycle of stress in a **fatigue test**. The midpoint is the *mean stress*.

Raoult's law The vapour pressure of an ideal solution at any temperature is the sum of the vapour pressure of each component, P_i, multiplied by the mole fraction of that component, x_i, ie $P = \sum P_i x_i$.

RAPRA Abbrev. for **Rubber and Plastics Research Association**.

rare earth elements A group of metallic elements possessing closely similar chemical properties. They are mainly trivalent, but otherwise similar to the alkaline earth elements. The group consists of the lanthanide elements 57 to 71, plus scandium (21) and yttrium (39). Extracted from monazite, and separated by repeated fractional crystallization, liquid extraction or ion exchange.

rare gases See **noble gases**.

rate constant See **order of reaction**.

rate of crystallization Different materials crystallize at different rates when cooled from melt, most markedly in polymers. High-density polyethylene shows very rapid rate but inflexible chains such as polyethylene terephthalate crystallize very slowly. See **crystallization of polymers** p. 82.

rattle The crackling noise when paper is handled. It indicates the degree to which the fibre has been hydrated in the process of beating, and can be augmented by the addition of starch and other additives. Cf **scroop**.

raw material Starting point for manufacture of useful materials. Raw materials for polymers include oil, natural gas and liquid petroleum gas; for cement, coal, limestone and clay; for steel, iron ore and coking coal, oil or natural gas; for other metals, metal ore and reducing agent; for glass, silica sand and other metal oxides; for ceramics, metal oxides; for semiconductors, silicon plus dopants (arsenic, gallium etc). See **intermediate**.

ray The radial features in wood, made up of **parenchyma** or food storage cells. See **structure of wood** p. 305.

Rayleigh scattering See **scattering**.

rayon Man-made fibre produced by viscose process, a regenerated cellulose. Viscose solution is spun into an acid bath where the fibre coagulates to produce a characteristic three-lobed cross-section fibre. Finds application in tyre cord, textiles, etc.

reactance The imaginary part of the impedance. Reactances are characterized by the storage of energy rather than by its dissipation as in resistance.

reactants The elements and/or compounds which interact to form **products** and are written on the left-hand side of a chemical equation.

reaction injection moulding Injection moulding process where polymerization occurs in mould cavity. Since viscosities are much lower, locking forces are much less, so less substantial machinery is used. Used for polyurethanes and nylons, eg car bumpers and exterior trim. Abbrev. *RIM* or *RRIM* when reinforcing chopped fibres are added to polymer. See **injection moulding** p. 169.

reaction kinetics The study of the speed of chemical reactions or the rate of transformation of reactants to products. See **order of reaction**.

reaction sintering See **ceramics processing** p. 54.

reaction wood Abnormal types of wood tissue grown in response to bending loads due to eg steady winds or to non-vertical growth. In **softwoods**, *compression wood* is formed, in **hardwoods**, *tension wood*.

reactive ion etching Dry etching process which proceeds by the combined action of chemical species and ion bombardment.

reactive iron Iron inserted in the leakage-flux paths of a transformer to increase its leakage reactance.

real modulus See **complex modulus**. Also *storage modulus*.

ream (1) Twenty quires of paper; was 480, now usually 500 sheets. (2) A non-homogeneous layer in flat glass.

rebound resilience Ratio of rebound to drop height of a rubber ball, but temperature and rate sensitive owing to viscoelasticity. Rubbers of low hysteresis (eg polybutadiene 'superballs') are more resilient than high-hysteresis rubbers like butyl at ambient temperature.

recalescence Release of heat in a material as it cools through a temperature at which a change in crystal structure occurs,

reciprocal lattice The direct crystal lattice can be defined in terms of three vectors a_1, a_2, a_3. A reciprocal lattice whose vectors are b_1, b_2, b_3, defined by $a_i \cdot b_i = \gamma$ and $a_i \cdot b_j = 0$. It follows that

$$b_1 = \frac{\gamma}{V}(a_1 \times a_3) \text{ etc,}$$

where V is the volume of the unit cell of the direct lattice. In crystallographic work γ is chosen as 1, but in solid-state physics as 2π. The reciprocal lattice is extensively used to discuss diffraction and scattering effects by crystals and in the band theory of solids.

reciprocal theorem When a linear elastic body subject to a set of forces F_1 has a second set of forces F_2 applied to it, the work done by the displacements of F_1 in response to F_2 is equal to the work that would have been done by the displacements of F_2 in response to the application of F_1. This also applies to the case when moments as well as forces are applied to a body. Although often called *Maxwell's reciprocal theorem*, it was first stated in its general form by E Betti in 1872.

reciprocity The principle enunciated in the **reciprocal theorem**.

recombination (1) Chain termination reaction where two active free radical chain ends combine to form dead polymer. See **chain polymerization** p. 56. (2) Neutralization of free electron and hole in semiconductor, thus eliminating two current carriers. The energy released in this process must appear as a light photon or less probably as several phonons. See **exciton, phonon, photon**.

recombination coefficient Ratio of the rate of recombination per unit volume to the product of the densities of positive and negative current carriers.

recovery (1) First stage in the annealing process of cold-worked metals, in which some restoration of original properties (eg hardness, ductility, resistivity) is achieved by a reduction in the density of dislocations through their thermally stimulated mutual annihilation at temperatures around 0.4 T_m. Cf **recrystallization, grain growth**. (2) The extent of the return of a stressed material to its original state after distortion under load.

recrystallization Stage in the annealing process of cold-worked metals above about 0.4–0.5 T_m, in which deformed crystals are replaced by a new generation of crystals, which begin to grow at certain points in the deformed metal and eventually absorb the deformed crystals. The new crystals are more equal axes and contain far fewer dislocations than the deformed ones. See **annealing** p. 13. Cf **recovery, grain growth**.

recrystallization temperature That marking a change in crystal form at which new crystals nucleate and grow to consume the existing structure. The range of temperature through which strain-hardening disappears, approximately 0.6 T_m.

recti- Prefix from L. *rectus*, straight.

rectifier Component for converting a.c. into d.c.

recycling ratio In injection moulding, the ratio of waste thermoplastic (eg cold runners) to total shot weight; must be kept constant for quality control and low (eg 0.10) for critical, stressed products owing to polymer chain degradation during each moulding cycle.

recycling See **closed-loop recycling, in-house recycling, post-consumer recycling**.

red brass Copper-zinc alloy containing 15% zinc; used for plumbing pipe, hardware, condenser tubes etc. Red casting brass may contain up to 5% of lead and/or tin in place of the zinc.

red deal A light-yellow **softwood** obtained from the **Scots fir**; commonly used for the timbering of trenches, heavy framing, piles, joinery etc.

red gum See **Red River gum**.

red hardness The ability to retain hardness at temperatures at which a material becomes red hot. Applies to tool materials like high-speed steel and cemented carbides which become heated in use by friction. Plain carbon and low alloy steels would soften under such conditions whereas those which do not are said to possess red hardness.

red heat As judged visually, a temperature between 500°C and 1000°C.

redistilled zinc Zinc from which impurities have been removed by selective distillation. The process takes advantage of the different boiling points of zinc (907°C) and the impurities lead (1750°C) and cadmium (767°C). A purity of 99.9% zinc is obtainable.

redox Abbrev. for *reduction-oxidation*.

red pine See **Douglas fir**.

Red River gum A strong, durable, red-brown **hardwood** from Australia, with a close and even texture, but resinous with frequent gum pockets. Used for piles, heavy framing etc. Density about 825 kg m^{-3}. Also (*Queensland*) *blue gum*, (*Murray*) *red gum*.

red-short Lacking ductility or malleability when red hot. See **hot-short**.

reduced viscosity Specific viscosity of a dilute polymer solution divided by concentration. Also *viscosity number*.

reducing atmosphere One deficient in free oxygen, and perhaps containing such reactive gases as hydrogen and/or carbon monoxide; used in a reducing furnace to lower the oxygen content of mineral.

redundancy The provision of extra components or equipment, in parallel, so as to ensure continued operation after failure.

redwood (1) See **sequoia**. (2) Name given to **red deal** in N England.

Redwood second Unit used in viscometry. See **Redwood viscometer**.

Redwood viscometer One of several designs of standard viscometer in which viscosity is determined in terms of number of seconds required for the efflux of a certain quantity of liquid through an orifice under specified conditions.

re-entrant Angle or shape in moulding tool which would normally prevent product removal at end of cycle. If small, can often be jumped off, but if large, must be circumvented using a side core, or alternative process eg blow moulding. See **undercut**.

refined iron Wrought iron made by puddling pig iron. Now obsolete.

refinery Plant where impure metals or mixtures are treated by electrolysis, distillation, liquation, pyrometallurgy, chemical or other methods to produce metals of a higher purity or specified composition.

refining See **fining** (of glass).

refining of metals Operations performed after crude metals have been extracted from their ores, to obtain them in a condition of higher purity.

reflectivity Proportion of incident energy returned by a surface of discontinuity.

refraction Phenomenon which occurs when a wave crosses a boundary between two media in which its phase velocity differs. This leads to a change in the direction of propagation of the wavefront in accordance with **Snell's law**.

refractive index The absolute refractive index of a transparent medium is the ratio of the phase velocity of electromagnetic waves in free space to that in the medium. It is given by the square root of the product of the complex relative permittivity and complex relative permeability. See **refraction**, **Snell's law**. Symbol n.

refractories Materials used in lining furnaces etc. They must resist high temperatures, changes of temperature, the action of molten metals and slags and hot gases carrying solid particles. China clay, ball clay and fireclay are all highly refractory, the best qualities fusing at above 1700°C. Other materials are silica, magnesite, dolomite, alumina and chromite. See **silica**.

refractory alloy An alloy which is: (1) difficult to work at high temperatures; (2) heat resistant or having a very high melting temperature.

refractory clay See **refractories**.

refractory metals Term applied to transition group elements in periodic table which have high melting points. They include chromium, titanium, platinum, tantalum, tungsten and zirconium.

regain Weight of water present in a textile or textile fibre expressed as a percentage of the oven-dry weight. Dried textiles take up or regain moisture when left in any normal atmosphere.

regatta Twill fabric, usually cotton, containing alternate stripes of white and colour, used in eg nurses' uniforms.

regenerated cellulose Chemical dissolution of normally insoluble natural cellulose and reclamation from solution to produce fibre etc; techniques include cuprammonium method (now no longer used), hydrolysis of cellulose acetate and viscose process giving rayon.

register Exact correspondence of superimposed work such as when semiconductor processing masks are aligned with features defined in previous steps.

register marks Fine lines, cross marks or similar, added to artwork to provide reference points and thus aid fitting and positioning of images during film assembly, platemaking and printing.

regulus of antimony Commercially pure metallic antimony. Also, the impure metallic mixture which is produced during smelting. Regulus was the alchemical name for antimony, which readily combines with gold.

reheating furnace A furnace in which metal ingots, billets, blooms etc, are heated to temperature required for hot-working.

reinforced concrete Concrete reinforced with steel bars to increase its load-carrying capacity, especially in bending. Developed by Hennebique in France in 1892. See **cement and concrete** p. 52.

reinforced plastics General term for plastic composite materials, whether thermosetting or thermoplastic, in which the basic plastic has been reinforced by incorporating a fibrous material, eg paper, cloth, aramid, carbon or glass fibre, usually leading to enhanced stiffness, sometimes higher toughness but reduced impact strength.

Reiss microphone Device in which a large quantity of carbon granules between a cloth or mica diaphragm and a solid backing, such as a block of marble, is subjected to the applied sound wave. Characterized by high damping of the applied vibrational forces, and freedom from carbon noise by virtue of packing amongst the granules.

relative atomic mass Mass of atoms of an element in atomic mass units on the unified scale where the amu = 1.660×10^{-27} kg. For natural elements with more than one isotope, it is the average for the mixture of isotopes. Abbrevs. *RAM*, *r.a.m.* Also *atomic weight*.

relative density The ratio of the mass of a given volume of a substance to the mass of an equal volume of water at a temperature of 4°C. Originally *specific gravity*.

relative molecular mass Preferred term for **molecular weight**.

relative viscosity Ratio of time for solution to fall a standard distance in a capillary viscometer (eg. Ostwald or Ubbelohde) to time for solvent to fall the same distance in the same device. Also *viscosity ratio*.

relaxation Exponential return of system to equilibrium after sudden disturbance. Time constant of exponential function is *relaxation time*.

relaxation spectrum Curve describing relaxation times of polymers as a function of temperature etc Can be derived from the master curve. See **viscoelasticity** p. 332.

relaxation time The time for a perturbed system to return to equilibrium. More specifically: (1) the time constant of relaxation, eg from the higher to the lower energy level in nuclear magnetic resonance; (2) the time taken for all or part of a polymer chain to respond to stress or strain stimulus. Symbol τ. An important polymer property which determines the shape of the master curve. Temperature-sensitive since thermal motion of chains affects chain flexibility. See **viscoelasticity** p. 332.

rel. d. Abbrev. for **relative density**.

release paper Paper treated so that an adhesive surface will become easily detached from it without rupture of the paper surface. Used as protective backing of self-adhesive materials.

reluctance Ratio of magnetomotive force, *m.m.f.*, applied to a magnetic circuit or component to the flux in that circuit (it is the magnetic dual of resistance and is the reciprocal of *permeance*).

reliability Probability that an equipment or component will continue to function when required. Expressed as average percentage of failure per 1000 hours of availability. Also *fault rate*.

remanence See **characterizing magnetic materials** p. 57.

remanent flux density See **characterizing magnetic materials** p. 57.

remanent magnetization See **characterizing magnetic materials** p. 57.

repeat unit See **polymers** p. 247.

rep, repp A plain-weave fabric with weft ribs formed by using coarse and fine yarns in alternate order in both warp and weft.

reproducibility Precision with which a measured value can be repeated in a process, component or experiment.

residence time The term given by dividing the volume of any reservoir or pool or processing unit by the rate of flow through it. The expression gives the average time spent in the pool by the substance or organism under consideration.

residual magnetization Magnetization remaining in a ferromagnetic material after the exciting magnetic field has been removed. See **remanence**, **characterizing magnetic materials** p. 57.

residual flux density Same as **remanent flux density**.

residual resistance That persisting at temperatures near zero on the absolute scale, arising from crystal irregularities and impurities, and in alloys.

residual stress Stresses remaining within a material after processing. For equilibrium, residual tensile stresses must be balanced by compressive ones and vice versa. Also *internal stress*.

resilience Stored elastic energy of a strained material, or the work done per unit volume of an elastic material by a bending moment, force, torque or shear force, in producing strain.

resilin An elastomeric protein occurring in insects. Cf **elastin**.

resin (1) A constituent of varnishes or paints, classified as (a) natural resin, eg copal, rosin, amber, (b) synthetic resin, eg alkyd, phenolic, polyurethane. Natural resins are chiefly obtained as the exudation from plants or trees and have a glassy appearance with a slightly yellow-brown colouring. They consist of highly polymerized acids and neutral substances mixed with terpene derivatives. Synthetic resins are specially formulated to impart particular qualities to coatings such as hardness, flexibility, chemical resistance etc. (2) The term resin was widely but loosely used to describe any synthetic plastics material. Now, more precisely, it is applied to a thermosetting polymer prior to curing, eg epoxy resin, polyester resin. See **thermosets** p. 318.

resist Polymer material used in lithography. Positive resists become more soluble in a developer after exposure to radiation (eg ultraviolet) whereas negative resists become less so.

resistance In electrical and acoustical

fields, the real part of the impedance characterized by the dissipation of energy as opposed to its storage. Electrical resistance may vary with temperature, polarity, field illumination, purity of materials etc. SI unit is the **ohm**, symbol Ω. Reciprocal of *conductance*. See **impedance**, **reactance**, **Ohm's law**.

resistance noise Same as **thermal noise**.

resistance strain gauge Foil, wire, thin film or semiconductor resistor which has a value which varies with mechanical strain. Normally used in a bridge circuit.

resistance thermometer One using resistance changes for temperature measurement. Resistance element may be platinum wire for extreme precision or semiconductor (*thermistor*) for high sensitivity. Also *resistance pyrometer* for higher temperature.

resistivity Intrinsic property of a conductor, which gives the resistance in terms of its dimensions. If R is the resistance in ohms, of a wire l m long, of uniform cross-section a m^2, then $R = \rho l/a$, where the resistivity ρ is in ohm metres. Also, erroneously, *specific resistance*.

resistor Electric component designed to introduce known resistance into a circuit and to dissipate accompanying loss of power. Types are wirewound, composition, metal film etc.

resit C-stage phenolic resin made from a resol.

resitol B-stage phenolic resin made from a resol.

resol Type of phenolic resin prepolymer made by reacting phenol with excess formaldehyde under basic conditions. Also called *one-stage resin*, since no other agents need be added for final crosslinking. Both resols and novolaks are A-stage resins.

resonance (1) If a vibrating system, mechanical or acoustical, is set into forced vibrations by a periodic driving force and the applied frequency is at or near the natural frequency of the system, then vibrations of maximum velocity amplitude result. This is called resonance and it corresponds to minimum mechanical or acoustical impedance. The *sharpness of resonance* is measured by the ratio of the dissipation to the inertia of the system which also determines the rate of decay of the vibrating system when it is impulsed. (2) A description of a molecule whose structure cannot be represented by a single Lewis structure but only as a mixture of two or more of them. More appropriately called *mesomerism*, it is used in the sense that a mule may be said to be a resonance of a horse and a donkey.

resonant frequency See **nuclear magnetic resonance**.

resonator Any device exhibiting a sharply defined electric, mechanical or acoustic resonance effect, eg a stub, piezoelectric crystal or Helmholtz resonator.

resorcinol *1,3-dihydroxy-benzene*, C_6H_4 $(OH)_2$. A dihydric phenol, colourless crystal, mp 111°C, bp 276°C. With formaldehyde used in the preparation of cold-setting adhesives. Also *resorcin*.

retarder A 'negative' catalyst which is added to a reaction system to slow down the reaction rate.

retentivity Capacity to hold a response after the stimulus is removed. In *ferromagnetics*, equivalent to the remanent flux density.

reticulated foam A **foam** in which the cell structure is delineated by rod-like struts and ties of material rather than by cell walls.

reticul-, reticulo- Prefixes from L. *reticulum*, net.

retread Tyre carcass to which new, extruded tread is bonded and revulcanized. More frequently used with large truck tyres, where cost of carcass is higher than car tyre.

retro- Prefix from L. *retro*, backwards, behind.

retting Soaking flax straw in water to allow bacteria to soften the woody tissue so that the fibres can subsequently be separated by scutching (beating).

reusability Potential for a product to be used again following normal use; thus car and truck tyres with good carcasses can be retreaded at their end of their useful first life.

reverberatory furnace A furnace in which the charge is melted on a shallow hearth by flame passing above the charge and heating a low roof. Firing may be with coal, pulverized coal, oil or gas. Much of the heating is done by radiation from the roof. Has many applications.

reverse engineering Disassembly of a finished product for analysis of materials, design and manufacture. Usually performed on a competitor's product, so that innovations can be incorporated into one's own. Also used with computer software.

reverse roll coating Process for coating fabrics or textiles with flexible, elastomeric layer, often plasticized polyvinyl chloride. Involves passing fabric between **nip** of two **calender** rolls, onto one of which polyvinyl chloride plastisol is fed. Gives an even surface, which is cured by heating in oven or by hot air blast, when plasticizer diffuses into polyvinyl chloride particles to form a homogeneous coating.

reversible Said of a process whose effects can be reversed so as to bring a system to its original thermodynamic state.

reversing mill A type of rolling-mill in which the stock being rolled passes backwards and forwards between the same pair of rolls, which are reversed between passes.

reversion Rubber technology term for drop in modulus at end of cure cycle due to chain degradation or breakdown of crosslinks. One type of **overcure**.

Reynolds number The dimensionless number defined as

$$R_e = \frac{\rho \, v \, d}{\eta},$$

where ρ = density of a fluid travelling at velocity v in a pipe of diameter d. Above $R_e \approx 2000$, flow is *turbulent*. Below, it is *laminar*.

RF welding Method of joining similar thermoplastics using *radiofrequency* radiation to heat surfaces of parts.

rhenium A metallic element in the subgroup manganese, technetium, rhenium, symbol Re. A very rare element, occurring in molybdenum ores. A small percentage increases the electrical resistance of tungsten. Used in high-temperature thermocouples. See Appendix 1, 2.

rheo- Prefix from Gk *rheos*, current, flow.

rheology Scientific study of fluid flow, eg **Poiseuille's law, Reynolds number, Newtonian viscosity, non-Newtonian viscosity**. Of particular importance for polymer processing and solid/fluid mixtures.

rheopectic fluids See **thixotropy**.

rhodanizing The process of electroplating with rhodium, esp. on silver, to prevent tarnishing.

rhodium A metallic element of the platinum group, symbol Rh. A noble silvery white metal, it resembles platinum and is alloyed with the latter to form positive wire of the platinum-rhodium-platinum thermocouple. Used for plating silver and silverplate to prevent tarnishing, in catalysts and in alloys for high temperature thermocouples. See Appendix 1, 2.

Rhometal An alloy of permalloy type. Contains 64% iron and 36% nickel. Used in high-frequency electrical circuits.

rhometer One for the measurement of impurity content of molten metals by means of the variation in electrical conductivity.

rice paper The finely cut pith of *Fatsia* (or *Aralia papyrifera*), a tree originally from Taiwan. Not a true paper.

Richardson–Dushman equation The original Richardson formula, as modified by Dushman, for the emission of electrons from a heated surface, current density being

$$I = AT^2 \exp\left(-\phi/kT\right).$$

T is the absolute temperature, A is a material constant, k is Boltzmann's constant, with ϕ the work function of the surface.

riga last A unit of timber measure containing 80 ft^3 (2.265 m^3) of squared or 65 ft^3 (1.84 m^3) of round timber.

rigid expanded polyurethane Thermosetting **foam** used in thermal insulation and in providing light structural reinforcement. Reaction products of di-isocyanates with polyesters or polyethers in the presence of water. See **polyurethane**.

rigidity modulus Obsolete term for **shear modulus**.

rigid polyvinyl chloride (PVC) Polyvinyl chloride in its unplasticized form, widely used for applications such as chemical and building pipework. See **polyvinyl chloride** p. 249.

Rilsan TN for nylon-11, nylon-12 etc.

RIM Abbrev. for **Reaction Injection Moulding**.

rimming, rimmed steel Low carbon steel that has not been completely deoxidized before casting. Gas is evolved during solidification and, after initial solidification of a sound rim, small bubbles form within the body of the casting and the consequent volume increase counteracts solidification shrinkage. Rimmed ingots contain no **pipe** and the yield of useful metal is thereby increased. Impurities and inclusions are concentrated in the interior parts of the ingot. The internal voids close up during subsequent rolling or forging, but the rim of relatively pure and inclusion free metal is always present and enhances the surface quality. See **killed steel**.

ring-opening polymerization The polymerization of a cyclic monomer unit. See **step polymerization** p. 298.

ring spinning The most widely used fibre spinning method. Fibres in the form of drafted roving are twisted together to form yarn by a tiny traveller rotating in a ring around a collection package on a spindle which is driven at a speed greater than that of the traveller. Many of these spinning heads are mounted together on a spinning frame.

rivet A headed shank for making a permanent joint between two or more pieces. It is inserted in a hole which is made through the pieces, and *closed* by forming a head on the projecting part of the shank by hammering or other means. The head may be rounded flat, pan-shaped or countersunk.

riveted joint A joint between plates, sheets, or strips of materials (usually metallic) secured by rivets. See **butt joint, lap joint**.

r.m.m. Abbrev. for **relative molecular mass**.

r.m.s. value See **root mean square value**.

RNA Abbrev. for *RiboNucleicAcid*. Polynuclectide macromolecule similar in structure to DNA where the base group thymine is replaced by uracil. Like DNA, can hold genetic information (as in viruses) but is also the primary agent for transferring information from DNA to the protein synthesizing machinery of cells.

roak, roke A seam in mining.

robinia N American **hardwood** (*Robinia pseudoacacia*) whose **heartwood** is golden brown, straight-grained, with a coarse texture and a mean density of 720 kg m^{-3}. Tough and durable; used outdoors for fencing, gates, boat planking, wheels etc, and also for joinery and some veneers.

robotics The study of the design and use of robots, particularly for their use in manufacture and related processes. Distinguished from automatic handling devices (eg pick-and-place units) by their flexibility for multi-programming. Thus they can be used for part handling, assembly, painting, packing etc.

Rochelle salt Crystal of sodium potassium tartrate, having strong piezoelectric properties, but high damping. Used as a *biomorph* in microphones, loudspeakers and pick-ups. Its disadvantage is the limited range of piezoelectric property ($-18°C$ to $23°C$) and high temperature coefficient. See **piezoelectric effect**, **piezoelectricity**.

Rockwell hardness test A method of determining the hardness of metals by indenting them with a hard steel ball or a diamond cone, first applying a light load and then increasing to a specified higher load, and measuring the additional depth of penetration. See **hardness measurements** p. 153.

rock wool Fibrous insulating material made by blowing steam through molten **slag**. Also *mineral wool*, *slag wool*.

rolled gold Composite sheet made by soldering or welding a sheet of gold on to both sides of a thicker sheet of silver, or other suitable metal, and rolling the whole down to the thickness required.

rolled laminated tube Tube produced by winding, under heat, pressure and tension, a synthetic-resin impregnated or coated fabric or paper on to a former.

rolled steel joist Abbrev. *RSJ*. See **H-beam**.

rolled-steel sections Steel bars rolled into I- or T-shaped channel, angle, cruciform, or similar cross-sections for different applications in structural work, each section being made in graded standardized sizes. See **H-beam**.

roller-bearing A shaft-bearing consisting of inner and outer steel races between which a number of parallel or tapered steel rollers or elements are located by a cage. Suited to heavier loads than the ball-bearing.

roll feed A mechanism incorporating nipping rollers for feeding strip material to power presses. The rollers turn intermittently by a set distance and are timed in relation to the press so as to advance the required length of strip, correctly timed, for each press stroke.

roll forging A hot forging process for reducing or tapering short lengths of bar stock in the manufacture of axles, chisels, knife blades, tapered tubes etc. Parts are placed by tongues into profiled grooves in pairs of forging rollers.

roll forming A cold forming process in which a series of mating rolls progressively forms strip metal into tube or section, usually of complex shape, at high rolling speeds. Often associated with a continuous resistance welding machine.

rolling bearing A ball-bearing, roller-bearing, needle bearing or other type of bearing in which elements roll between load-bearing surfaces.

rolling mills Sets or stands of rolls, horizontally mounted, used in rolling metals either hot or cold into numerous intermediate and final shapes, eg blooms, billets, slabs, rails, bars, rods, sections, plates, sheets and strip. The roll in contact with the material is the work roll and others mounted behind to prevent it bowing are referred to as back-up rolls. For foils and very thin sheet, clusters of back up rolls may be required on account of the high pressures necessary.

rolling resistance When a wheel rolls on a surface, distortion of the contact surfaces due to the normal force between them destroys the ideal line contact. This introduces a force with a component, called the rolling resistance, opposing the motion.

room temperature vulcanizing Term usually applied to silicone and polysulphide prepolymers, which when mixed with cross-linking agent on site, can be applied to form final shape. Abbrev. *RTV*.

root mean square value Measure of any alternating waveform, the square root of the mean of the squares of continuous ordinates (eg voltage or current) through one complete cycle. For a simple sinusoid, it is $1/\sqrt{2}$ times the peak value. Used because the energy associated with any wave depends on its intensity, ie upon the square of the amplitude. Also *effective value*. Abbrev. *r.m.s. value*.

root of the joint In welding, the place where the original components were closest together.

root of the weld The place where the weld comes closest to the root of the joint.

rosewood See **Brazilian rosewood**.

rosin Natural resin extracted from long-leaved pine wood. Used in soldering flux, varnish, soap and size, as well as a drier in paints.

rubber toughening

Many homopolymers like polystyrene, polymethyl methacrylate and polyvinyl chloride are brittle, glassy solids at ambient temperatures and of restricted use for stressed, structural application. Such materials can be toughened by *co-polymerization* with elastomeric polymers, such as polybutadiene and poly-acrylate. The latter's polymer chains are highly flexible and non-crystalline (being atactic) at ambient temperatures but, like most polymers, they are incompatible with the host matrix and phase separate to form spherical domains up to about 2 μm in diameter. Polymerization is effected by dispersing a stable emulsion in the monomer followed by catalytic initiation in a chain-growth reaction. Since the rubber chains are covalently linked to the matrix chains, the material can be reprocessed (ie it is thermoplastic) with reformation of the domain structure.

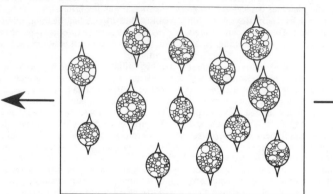

Fig. 1. **The effect of applying tension to the elastomeric particles (domains)**.

When a bulk sample is stressed in tension, each particle behaves as a stress concentration zone and two **crazes** form at right angles to the applied strain (Fig. 1).

Each craze absorbs energy and if particles are fine enough and well distributed through the matrix, a very large amount of energy will be absorbed, ie the material is toughened. When the craze size approaches the wavelength of light (ca 0.4 μm), they become visible owing to light scattering, an effect known as *stress* or *strain-whitening*. The method of rubber toughening has been applied to polystyrene in high-impact polystyrene and acrylonitrile-butadiene-styrene polymers, to polymethyl methacrylate in impact resistant acrylic (eg Plexidur sheet) using polybutyl acrylate and to polyvinyl chloride in high-impact PVC using methacrylate-butadiene-styrene terpolymer (MBS) or polyacrylate elastomers. It has also been applied to epoxy resins and nylon 6,6 in eg Zytel.

rotary lip-seal Special type of rubber seal for rotating shafts, eg crankshaft seal in a car engine.

rotating core Special core used to create internal screw thread in injection moulded product.

rotational isomerism Type of conformational isomerism, where steric hindrance between adjacent side groups in repeat units favours particular conformers, eg trans in polyethylene giving zig-zag conformation.

rotational moulding Process by which an object is shaped by rotating the mould to which the **powder** or **plastisol** material (eg polyvinyl chloride) has been added, in the heating oven and during cooling; used for some hollow articles. Also *rotocasting*, *rotomoulding*.

rotation of the plane of polarization A property possessed by optically active sub-

rusting

Rusting is the name given to the atmospheric corrosion of iron and steel. Unlike the other transition metals in their group, they corrode readily in the presence of oxygen and water, and this is enhanced by the presence of ions such as sulphides and chlorides common in industrial and/or marine atmospheres. The economic consequences of corrosion, in terms of prevention, repair and replacement, are enormous (well over $£10^9$ pa in the UK alone), and a major part of this is due to rusting.

Rust is the mixture of oxides of iron which forms on iron and steel surfaces in oxygenated aqueous environments. These oxides are:

green rust, hydrated ferrous oxide, FeO.OH, *goethite* or *lepidocrocite*

brown rust, hydrated ferric oxide, Fe_2O_3 with $Fe(OH)_3$, *haematite*

black rust, mixed ferrous and ferric oxides, Fe_3O_4 (=$FeO.Fe_2O_3$), *magnetite*.

Above a threshold relative humidity of about 60% (less in the presence of, for instance, hygroscopic dust), there is sufficient water around for some droplets to condense on the metal surface. These provide the electrolyte for electrochemical corrosion to start with

$$Fe \rightarrow Fe^{2+} + 2e^- \text{ at the } anode, \text{ and } \tfrac{1}{2}O_2 + H_2O + 2e^- \rightarrow 2OH^- \text{ at the } cathode.$$

In combination these give

$$Fe + \tfrac{1}{2}O_2 + H_2O \rightarrow Fe(OH)_2$$

and the ferrous hydroxide further oxidizes

$$Fe(OH)_2 + \tfrac{1}{2}O_2 + \tfrac{1}{2}H_2 \rightarrow Fe(OH)_3.$$

This ferric hydroxide changes rapidly to hydrated ferrous oxide

$$Fe(OH)_3 \rightarrow FeO.OH + H_2O$$

which can oxidize further

$$2FeO + \tfrac{1}{2}O_2 \rightarrow Fe_2O_3$$

and combine to give

$$FeO + Fe_2O_3 \rightarrow Fe_3O_4.$$

continued on next page

stances. See **optical activity**.

rotocasting See **rotational moulding**.

Rotocure TN for continuous vulcanization process where product of constant cross-section such as conveyor belting or flooring is made by reeling around a heated vulcanization drum. Care is needed to match speed of drum and crosslinking kinetics.

rotomoulding See **rotational moulding**.

roughing tool A lathe or planer tool, generally having a round-nosed or obtuse-angled cutting edge, used for roughing cuts.

roving A continuous strand of fibres sufficiently drawn to a diameter suitable for drafting and spinning into yarn, eg by ring spinning frames.

royal A former imperial paper size retained under the metric system for printing boards, 52 × 64 cm (20 × 25 in). Note also the following sizes recommended for book publishing papers. Metric small quad royal 96 × 127 cm, metric quad royal 102× 127 cm. See **paper sizes**.

RRIM Abbrev. for *Reinforced Reaction Injection Moulding*. See **reaction injection moulding**.

RTV Abbrev. for **Room Temperature Vulcanizing**.

rubber Generic term for any elastomer, but also used specifically for natural rubber. See **elastomers** p. 106.

Rubber and Plastics Research Association A leading research and development organization for the polymer industry.

rubber forming Pressing a rubber pad over a die to form a sheet placed between them.

rubber hardness Various empirical measures of shear modulus of rubbers, by resis-

rusting (contd)

These reactions therefore yield a mixture of nominally FeO.OH, Fe_2O_3 and Fe_3O_4. Each of these oxides tends to be non-stoichiometric due to the close structural relationship between them. They are all based on a cubic, close-packed array of oxide ions, with iron as Fe^{2+} and Fe^{3+} occupying the tetrahedral and octahedral interstices. If all the octahedral sites are occupied by Fe^{2+} ions then the ideal stoichiometric FeO results. If 14.1% of these are replaced by $\frac{2}{3} \times 14.1\%$ of Fe^{3+}, $Fe_{0.953}O$ is obtained. A 66.7% replacement, half of which migrate to tetrahedral sites, gives Fe_3O_4, while 100% replacement gives Fe_2O_3. Thus interconversion of the oxides is readily achieved merely by redistribution of ions between interstitial sites, and with no major structural changes.

Their lack of stoichiometry leads to defect structures which provide diffusion paths through their lattices. In Fe_2O_3 and Fe_3O_4, vacancy defects appear in the oxide anion sublattice, so that oxide ions can diffuse through the film towards the metal-oxide interface. At the same time, and also due to the non-stoichiometry, electrons can diffuse in the other direction (all three oxides are semiconductors), thus completing the electrochemical circuit.

The net result is that further oxidation takes place at the interface between the metal and the oxide. At first sight, the reaction should slow down as the oxide layer thickens and the diffusion paths lengthen. However, there is an approximate doubling in volume accompanying the reaction, and it is this expansion against the existing iron oxide layer that leads to *spalling*, exposing fresh metal to further direct attack, ie continued rusting.

In dry, unpolluted atmospheres, a thin (≈ 4 nm), protective oxide film can form on iron surfaces. This is anhydrous FeO, or *wüstite*. However, its composition is non-stoichiometric (ie there aren't an equal number of Fe^{2+} and O^{2-} ions). Its formula is more accurately written as $Fe_{0.953}O$, with the charge balance made up by Fe^{3+} ions. See **Pourbaix diagram**.

---

tance of a rubber to an impressed indentor needle. *International Rubber Hardness* and *Shore A* scales are roughly equal and proportional to shear modulus which in turn is proportional to degree of crosslinking. See **hardness measurements** p. 153.

rubber reclaim Rubber produced by chemical action (eg alkali digestion) on waste products such as worn tyres. Of limited use as recycled material for new tyre compounds.

rubber seal A common form of a device for preventing leakage of fluid from one part of a container to another, whether moving or static. Care needed in design, since poor tolerances can cause extrusion of seal under pressure. Material also critical to performance, depending on nature of fluids, so oil-resisting rubbers like nitrile, chloroprene rubber or Viton favoured for engine and fuel seals. Often an O-ring, but many other shapes exist for special uses.

rubber springs Devices designed to isolate a structure from its environment, ranging from bridge bearings and earthquake mounts to anti-vibration mountings for engines and torsion bushes.

rubber toughening See panel on p. 270.

rubidium A metallic element in the first group of the periodic system, one of the alkali metals, symbol Rb. The element is widely distributed in nature, but occurs only in small amounts; the chief source is carnallite. The metal is slightly radioactive. See Appendix 1, 2.

ruby The blood-red variety of the mineral corundum, being essentially alumina (Al_2O_3), with chromium impurity. Used as an optically pumped, solid-state laser medium in which the Cr^{3+} ions are the active component.

ruche A narrow woven or knitted fabric made with extended weft threads which may be bunched together for extra effect. The material is used as decorative trim round the edges of upholstery or applied to dresses.

rule of mixtures Method of determining properties of a mixture (eg a composite material) by summing for all constituents their value of the corresponding property multiplied by the volume fraction present (ie a volume-weighted average of the properties of the components). Applies generally to the density of any composite, and under certain conditions, to such properties as elastic modulus, tensile strength, thermal and electrical conductivity, and dielectric constant. Does not, in general, apply to **toughness**.

runner The channel through which molten metal flows to a mould or other receptacle. More specifically, the round channel through which molten plastic flows into a mould or cavity for shaping. Also applied to the solid material left in the channel at the end of the moulding cycle. Runners are often cold, in which case the solid product is recycled following granulation. Hot runners are now in-creasingly used in **injection moulding**, so reducing or eliminating the waste problem.

Rupert's drops See **Prince Rupert's drops**.

rust Product of oxidation of iron or its alloys, due either to atmospheric attack or electrolytic effect of cell action round impurities. See **rusting** p. 271.

rusting See panel on p. 271.

ruthenium A metallic element, symbol Ru. The metal is silvery-white, hard and brittle. It occurs with the platinum metals in osmiridium, and is used in certain platinum alloys. See Appendix 1, 2.

S

s Symbol for *second* (time).

σ Symbol for: (1) conductivity; (2) normal stress; (3) nuclear cross-section (barn); (4) Stefan–Boltzmann constant; (5) surface charge density; (6) surface tension; (7) wave number.

S Symbol for **siemens**.

S Symbol for **entropy**.

Σ Symbol for *sum of*.

saccharides Carbohydrates, which according to their complexity are usually divided into mono-, di-, tri- and polysaccharides.

sacking Coarse fabrics of jute, flax or polyolefin used for making sacks, etc.

sacrificial anode A metal electrode used to prevent the corrosion of a structure. The anode dissolves and requires regular replacement. Commonly made of magnesium alloy containing 6% aluminium and 3% zinc. The zinc coating on galvanized steel acts as a sacrificial anode. See **cathodic protection**.

sacrificial protection The prevention of electrolytic corrosion in a component by providing another electrochemically more active metal close by and electrically connected to it.

sacrificial-tape welding Joining method for similar thermoplastics in safety-critical products, like small boats and some types of battery cases. Involves placing a braid of metal and thermoplastic fibre in joint, passing an electric current through the braid while the two parts are loaded.

safety-critical Part or component, the loss or damage of which will endanger or destroy the parent product. See **failsafe**.

safety factor Provision of extra margin in stress calculations etc to allow for errors and uncertainties. Thus a safety factor of two allows for twice the allowable stress calculated for the product specification.

safety glass (1) *Laminated glass*, formed of a sandwich of a thin (0.4–0.750 mm) layer of polymeric material, usually polyvinyl butyral, between glass sheets. The functions of the layer are to hold the glass fragments in place following fracture and to act as a barrier to penetration. (2) So-called *toughened glass*, formed by putting the surfaces into compression, either by chilling them with air jets from close to the softening point (*thermal toughening*), or by exchanging sodium ions in the surfaces for larger ions (*chemical toughening*). Stronger than untreated glass, with blunter fracture fragments, but **toughness** is unaffected. (3) Glass incorporating wire mesh to hold fracture fragments in place, but weaker than plain glass. See wired glass.

safe working load Maximum load permitted in service for items used for lifting ie chains, slings, hoists etc, usually to 20% of their minimum breaking load. It is a statutary requirement that such items are **proof tested**, normally to 40% of the minimum breaking load before being allowed to enter service. All such items must be subjected to regular inspections for wear, corrosion and mechanical deterioration.

saggar A clay box in which pottery is packed for firing in the kiln.

sailcloth (1) Woven fabrics designed for use in sailing ships and yachts. Originally closely woven cotton or linen canvases were used but these have been largely displaced by nylon for spinnakers, and polyester or aramid for foresails and mainsails. The weaves are carefully designed to give just the right air porosity under a wide range of wind velocities. (2) A ribbed cotton fabric with a structure between those of poplin and repp, used in dresswear.

salt bath A bath of molten salts used for heating treatment, ie for hardening, tempering or solution treatments. Salt baths give rapid, uniform heating and protect against oxidation. Different salts are used for different temperatures, eg for tempering of steels, sodium and potassium nitrate are used. For hardening of steels, sodium cyanide, and sodium, potassium, barium and calcium chlorides. An *electric salt-bath furnace* is a conductor-type electric furnace in which the salt is melted by the passage of the current.

samarium A hard and brittle metallic element, symbol Sm. Found in allanite, cerite, gadolinite and samarskite. Feeble, naturally radioactive, can be produced by decay of fission fragments, and forms reactor poison. See Appendix 1, 2.

SAN Abbrev. for **Styrene AcryloNitrile**.

sandalwood A scented wood from several trees of the family *Santalaceae*, found in the E Indies. Important for its essential oil and used for joss-sticks and small ornaments. One species is found in Australia.

sand-blasting A method of cleaning metal or stone surfaces by sand, steel shot, or grit blown from a nozzle at high velocity; also used for forming a key on the surface of various materials requiring a finish, such as enamel.

sandpaper Stout paper or cloth with a thin coating of fine sand glued on to one side, for use as an abrading material. Cf **emery paper**, **glasspaper**.

sandwich beam Composite structural beam comprising outer 'skin' layers which are typically a fibre-reinforced **composite material** (eg **carbon** or **glass fibre reinforced plastic**) encasing a lightweight 'core' which may be a **foam** or **honeycomb** material. Its chief advantage is a high ratio of bending stiffness to weight because the main load-bearing elements (the skins) are situated some distance away from the **neutral axis**. Cf **structural foam**.

sandwich moulding Injection moulding process by which surface of product formed is solid but the interior is foamed. It normally involves two separate injection units.

sapele An African **hardwood** tree (*Entandrophragma cylindricum*), with a mahogany-like, silky-grained **heartwood** of mean density 620 kg m^{-3}. Moderately durable and used for high-class furniture and cabinet-making, musical instruments, sports goods etc, and as the veneer for plywood, as well as a decorative veneer in its own right.

sapphire Fine blue transparent variety of crystalline corundum (predominantly Al_2O_3). Also the single-crystal alumina used as a substrate for **silicon on insulator** technology.

sapwood The outer, lighter-coloured, younger part of the wood of a tree, surrounding the **heartwood**, and used for conduction of nutrients and storage of carbohydrates. Usually less than 1/3 of the total radius, it is enclosed by the **cambium**. See **structure of wood** p. 305.

Saran US name for synthetic fibre or film based on a copolymer of vinylidene chloride (at least 80% wt) and vinyl chloride – a *chlorofibre*.

Sardinian A heavy, woollen twill overcoating fabric whose face is raised to a dense nap and then rubbed into pills.

sassafras N American **hardwood** tree (*Sassafras officinale*), with a light to dark brown, straight-grained, coarse-textured **heartwood** of mean density 450 kg m^{-3}. Soft, flexible and moderately durable; used for furniture, interior fittings, containers, fencing, etc and as a decorative veneer. Its bark yields sassafras oil used in perfumery and cosmetics, and its root an extract used for flavouring.

sateen A smooth-surfaced fabric produced by a weft-faced, sateen weave, often treated to make it lustrous. Cf **satin**.

satin A lustrous, smooth-surfaced fabric produced by a warp-faced, satin weave. Cf **sateen**.

satin walnut Misleading UK name for the **heartwood** of American red gum.

satinwood See **East Indian satinwood**.

saturated compounds Organic compounds which do not contain any free valencies and to which no hydrogen atoms or their equivalent can be added, ie which contain neither a double nor a triple bond.

saturation For magnetic materials, the application of sufficient magnetizing field to achieve the maximum (saturation) magnetization. Likewise with analogous phenomena.

SAW Abbrev. for **Surface Acoustic Wave device**.

saxony A high-quality woollen cloth made from merino wool.

SBR Abbrev for *Styrene-Butadiene Rubber*. See **elastomers** p. 106.

SBS Abbrev. for *Styrene-Butadiene-Styrene*. See **copolymer** p. 75.

scalar A real number or element of a field. The term *scalar* is used in vector geometry and in vector and matrix algebra to contrast numbers with vectors and matrices.

scale The ratio between the linear dimensions of a representation or model and those of the object represented, as on a map or technical drawing.

scandium A metallic element, classed with the rare earth metals, symbol Sc. It has been found in cerite, orthite, thortveitite, wolframite and euxenite; discovered in the last named. Scandium is the least basic of the rare earth metals. See Appendix 1, 2.

scanning electron microscopy Routine method for examination of material surfaces using a focused electron beam raster accelerated at low voltages (~5 to 100 kV). Produces a three-dimensional image which allows detailed exploration of eg fracture surfaces. See **fractography**. Elemental analysis is possible with **energy dispersive analysis of X-rays** (EDAX) facility. Nonconductors such as ceramics, glasses and polymers usually need coating with a conductive layer (eg gold) to prevent static build-up. Abbrev. *SEM*.

scanning transmission electron microscope One which uses field emission from a very fine tungsten point as the source of electrons. The electrons transmitted through the sample are either unscattered, elastically scattered or inelastically scattered; they are collected, separated and analysed to produce an image. Abbrev. *STEM*.

scantling A piece of timber 50–100 mm (2–4 in) thick and 50–110 mm wide.

scarfing (1) Tapering the ends of materials for a lap joint, so that the thickness at the joint is substantially the same as that on either side of it. (2) Preparing metal edges for forge welding.

scattering General term for irregular reflection or dispersal of waves or particles, eg in acoustic waves in an enclosure leading to diffuse reverberant sound, *Compton effect* on electrons, light in passing through material, electrons, protons and neutrons in solids, radio waves by ionization. Particle scattering is termed *elastic* when no energy is surrendered during scattering process, otherwise it is *inelastic*. If due to electrostatic forces it is termed **coulomb scattering**; if short-range nuclear forces are involved it becomes *anomalous*. Long-wave electromagnetic wave scattering is *classical* or *Thomson*, while for higher frequencies *resonance* or *potential* scattering occurs according to whether the incident photon does or does not penetrate the scattering nucleus. Coulomb scattering of α-particles is *Rutherford scattering*. For light, scattering by fine dust or suspensions of particles is *Rayleigh scattering*, while that in which the photon energy has changed slightly, due to interaction with vibrational or rotational energy of the molecules, is **Raman scattering**. *Shadow scattering* results from interference between scattered and incident waves of the same frequency. See **acoustic scattering, atomic scattering**.

scavenging Addition made to molten metal to counteract an undesired substance.

schizo- Prefix from Gk *schizein*, to cleave.

Schottky diode A semiconductor diode formed by contact between a semiconductor layer and a metal coating. Hot carriers (electrons in n-type material and holes in p-type material) are emitted from the Schottky barrier of the semiconductor and move into the metal coating. Majority carriers predominate, so that injection or storage of minority carriers is not present to limit switching speeds.

Schottky effect (1) The removal of electrons from the surface of a semiconductor when a localized electric field is present at the surface; electrons can then surmount a semiconductor potential barrier and enter regions where allowed energy levels are available. (2) The increase in cathode current in a valve, beyond that available by normal thermionic emission, on account of the effective lowering of the work function of the cathode by the localized electric field.

Schrödinger equation F u n d a m e n t a l equation of *wave mechanics*. Solutions of this equation are *wave functions* for which the square of the amplitude expresses the probability density for a particle or a set of particles. If the system is isolated then a *time-independent* form of the equation is applicable. The solution for this version for bound particles shows that the energy for the system must be quantized.

scintillation Minute light flash caused when α-, β- or γ-rays strike certain **phosphors**, known as *scintillators*. The latter are classed as liquid, inorganic, organic or plas-

tic according to their chemical composition.

Sclair TN for polyethylene which crosslinks during **rotational moulding** using free radical catalyst. This gives lower creep rate for uses like kayak canoes.

scleroproteins Insoluble high molecular mass proteins forming the structural parts of living tissue eg keratin in horn, hoof, nail, hair etc, collagen and elastin in skin ligament etc. See **biomaterials** p. 57.

scleroscope hardness test The determination of the hardness of metals by measuring the rebound of a diamond-tipped hammer dropped from a given height. See **Shore scleroscope, hardness measurements** p. 153.

scorch Term used in rubber technology for premature vulcanization, before final shape of product has been achieved.

scorification Assay in which impurities are slagged while bullion metals dissolve in molten lead, from which they are later separated by **cupellation**.

scorifier A crucible of bone ash or fireclay used in assaying and in the metallurgical treatment of precious metals. See **scorification**.

scotophor Material which darkens under electron bombardment, used for the screens of cathode ray tubes in storage oscilloscopes. Recovers upon heating. Usually potassium chloride.

Scots fir Synonym for **Scots pine**.

Scots pine **Softwood** (*Pinus sylvestris*) native to many parts of Europe and the best known UK species for producing commercial timbers. **Heartwood** is a pale reddish-brown, resinous, with well-defined growth rings and a mean density of 510 kg m^{-3}. Non-durable and liable to insect attack. Best grades used for furniture, joinery etc, others in construction, telegraph poles, piling etc. Also used in plywood and as a decorative veneer. Also *Baltic redwood, red deal, Scots fir, Scots pine, yellow deal*.

scouring Processes for removing from textile materials fats, oils, waxes and other impurities which may be natural (eg wool grease) or which may have been added to aid processing (eg **size**). For cotton and flax, boiling aqueous sodium hydroxide solutions are used but other fibres require milder treatments eg with aqueous sodium carbonate or neutral solutions of detergent (for wool).

SCR Abbrev. for **Silicon-Controlled Rectifier**.

scrap (1) Materials reclaimed from products at the end of their specification life-cycle or product life-cycle. Scrap metal, glass and ceramic, paper, textile and some polymers are routinely reclaimed from used products. The extent of recycling depends on raw material or intermediate prices as well as the degree of mixing of different materials (and hence the costs of separation). See **tramp element**. (2) Defective products unfit for sale.

screening constant A number which when subtracted from the atomic number (Z) of an atom, gives the *effective* atomic number so that X-ray spectra may be described by 'hydrogen-like' formulae. It arises from the nuclear charge ($+ Ze$) being screened by the inner electron shells.

screen pack Gauze fitted behind polymer extrusion die to filter out impurities. See **extrusion** p. 121.

screw Steel cylinder with external flight(s) shaped in a helical form, so that when rotated in cylinder of injection moulding or extrusion machine, will transport molten polymer through the nozzle. Sometimes known as an *Archimedes screw*, it is also widely used for transporting powder and other materials or fluid (eg water).

screw characteristics Geometric properties of injection moulding or extrusion screws, which may vary depending on polymer type etc. Includes compression ratio, L/D ratio, flight angle and multiple screws. See **extrusion** p. 121.

screw dislocation Defect within a crystal in which the **Burgers vector** is parallel to the line of the dislocation. See **dislocation**.

screw thread A helical ridge of approximately triangular (or V), square or rounded section, formed on a cylindrical core, the pitch and core diameter being standardized under various systems.

scrim An open woven fabric used for reinforcement in bookbinding, upholstery, wall plastering and for the basecloths of certain non-woven fabrics.

scroop The crunching noise and the corresponding **handle** that is obtained when certain fabrics, particularly of silk, are crushed in the hand. The noise is rather like that obtained when dry snow is squeezed into balls. Cf **rattle**.

scuffing (1) A sign of inadequate lubrication in which adhesive wear produces scratches and tears in a mating surface. (2) Surface marks on a moulding running in the direction of mould opening, due to tool wear or damage. Type of **mould defect**.

scutching (1) Process of mechanically separating and cleaning closely packed cotton fibres before forming a yarn. (2) Process of separating flax fibres from the woody part of deseeded or retted flax straw.

sea cell Primary electrolyte cell which functions as a source of electric power when immersed in sea water. A battery of such cells is possible, even though cells are par-

tially short-circuited by the sea water. Fitted to life-belts etc, so that an indicating light is produced automatically in the event of use at night.

seam (1) A surface defect in worked metal, the result of a blowhole being closed but not welded; it remains as a fine discontinuity. (2) Ridge in casting, effect of enclosure of impurity or scale in working.

seaming machine A press for forming and closing longitudinal and circumferential interlocking joints used in the manufacture of containers from sheet metal.

seamless tube Tube other than that made by bending over and welding the edges of flat strip. May be made by extrusion (nonferrous metals), or by piercing a hole through a billet and then rolling down over a mandrel to form a tube of the required dimensions.

season cracking Form of **stress corrosion** cracking first experienced in the late 19th century in brass cartridge cases during the sub-tropical monsoon season, caused by a combination of high residual stress from the forming process and a mildly corrosive environment containing traces of ammonia. The cracking can be overcome by a simple stress relieving anneal immediately after manufacture.

seasoning The process in which the moisture content of timber is brought down to an amount suitable for the purpose for which the timber is to be used. See **kiln drying**.

secant modulus Elastic (or quasi-elastic) **modulus** derived from a non-linear **stress-strain curve** by taking the ratio of the stress to the strain at a particular point on the curve, which must be specified in terms of the level of stress or strain, eg '0.5% strain secant modulus'. The creep moduli of polymers are almost invariably secant moduli. Cf **tangent modulus**.

second (1) 1/60 of a minute of time, or 1/86 400 of the mean solar day; once defined as the fraction 1/31 556 925.974 7 of the tropical year for the epoch 1900 January 0 at 12 h ET. Since 1965 defined, in terms of the resonance vibration of the caesium-133 atom, as the interval occupied by 9 192 631 770 cycles. This was adopted in 1967 as the SI unit of time-interval. Symbol s. (2) Unit of angular measure, equal to 1/60 of a minute of arc; indicated by the symbol ". (3) In duodecimal notation 1/12 of an inch; indicated by ". (4) Unit for expressing flow times in capillary viscometers (Redwood, Saybolt Universal or Engler), eg an Engler second is that viscosity which allows 200 cm^3 of fluid through an Engler viscometer in one second.

secondary alcohols Alcohols containing

the group –CH(OH). When oxidized they yield ketones (alkanones).

secondary bonds Term usually applied to the weak van der Waals and hydrogen bonds in materials. See **bonding** p. 33.

secondary carbon atom Carbon atom linked to two other carbon atoms and two hydrogen atoms with stability between the more stable primary and less stable tertiary atoms.

secondary cell See **accumulator**.

secondary hardness Further increase in hardness produced in tempering high-speed steel after quenching.

secondary ion mass specrometry A surface analytical technique in which sputtering by primary ions leads to the ejection of some surface species as secondary ions. Subsequent mass analysis is used to identify the secondary ions. Abbrev. *SIMS*.

secondary metal Pure metal or alloy retrieved from engineering scrap and residues refined and worked up for return to manufacturing industry. Distinguishes it from primary metal which is that produced directly from ore and appearing in metallic form for the first time.

secondary standard A copy of a **primary standard** for general use in a standardizing laboratory.

secondary stress A bending stress, resulting from deflection, as distinct from a direct tensile or compressive stress. Now obsolete.

secondary structure The first level of three-dimensional folding of the backbone of a polymer. Thus a polypeptide chain can be folded into an α-helix or β-pleated sheet and the nucleotide chains of DNA into a double helix. See **biomaterials** p. 57.

secondary transitions Transitions in polymers secondary to T_g or T_m, such as those involving increase in side chain flexibility. Often detected in accurate damping experiments.

second moment of area The second moment of area is a measure of resistance to bending of a loaded section. A plank stood on edge is less easily bent when loaded than one which has the wide dimension horizontal. This is because the second moment of area on edge is very much greater than that when the plank is laid flat. The second moment of area, I about an axis xx in the same plane is defined as

$$I_{xx} = \int_A y^2 \, dA \, ,$$

where dA is an element of the total area and y the perpendicular distance from xx. I_{xx} can also be obtained from the second moment of area about a parallel axis through the centroid of the area, I_{gg} by the equation, $I_{xx} = I_{gg}$

$+ \, Ah^2$, where A is the total area and h the perpendicular distance between the two axes. Cf **polar second moment of area**.

section modulus Ratio of second moment of area to distance of the farthest stressed element from the **neutral axis**. It is an important property of structural members, for calculating bending stresses. Symbol Z. Also, loosely, *elastic modulus*. Units cm^3.

security paper Anti-counterfeit paper designed to show up the falsification or attempted falsification of documents. Achieved by special, intricate watermarking, the addition of selected chemicals to the stock or incorporation in the furnish of synthetic or animal fibres, threads, planchettes etc.

sedimentation Method of analysis of suspensions, by measuring the rate of settling of the particles under gravity or centrifugal force and calculating a particle parameter from the measured settling velocities. It is also used for measuring the molecular mass of soluble polymers. The same method can be applied to biological macromolecules using the ultracentrifuge. It gives an absolute measure of $\overline{M}_z$, a higher molecular mass than $\overline{M}_w$. See **molecular mass distribution** p. 212.

sedimentation potential The converse of **electrophoresis**; a difference of potential which occurs when particles suspended in a liquid migrate under the influence of mechanical forces, eg gravity.

sedimentation techniques Group of methods of particle-size analysis in which concentration changes within a suspension are measured, the changes being due to differential rates of settling of various sizes of particle.

Seebeck effect Phenomenon by which a thermoelectric emf is set up in a circuit in which there are junctions between different bodies, metals or alloys, the junctions being different temperatures. See **thermo-electric effect**.

seed Small bubbles in glass.

seerloup A plain-weave, lightweight cloth similar to gingham, containing some coloured, coarse warp threads which form loops on the cloth surface.

seersucker A woven fabric with stripes or checks of puckered material alternating with flat sections. The effect is called *plissé*, Fr. for *pleated*.

seg Dress or suiting dyed fabrics of simple twill weave, often made from wool but also from other fibres or wool-containing blends.

Seger cones Small cones of clay and oxide mixtures, calibrated within defined temperature ranges at which the cones soften and bend over. Used in furnaces to indicate, within fairly close limits, the temperature reached at the position where the cones are placed. Also *fusion cones, pyrometric cones*.

segregation Non-uniform distribution of impurities, inclusions and alloying constituents in metals. Arises from the process of freezing, and usually persists throughout subsequent heating and working operations. See **inverse segregation, normal segregation.**

seizure, seizing-up The locking or partial welding together of sliding metallic surfaces normally lubricated, eg a journal or bearing.

selective absorption Absorption of light, limited to certain definite wavelengths, which produces so-called absorption lines or bands in the spectrum of an incandescent source, seen through the absorbing medium. See **Kirchhoff 's laws.**

selective freezing A process involved in solidification, as a result of which the first crystals formed differ markedly in composition from the melt. Thus, in a **eutectic system** (except the eutectic alloy), crystals based on one component are formed from a melt containing two or more, and these early crystals may be separated from the remainder of the melt if desired.

selenium A non-metallic element, symbol Se. A number of allotropic forms are known. *Red selenium* is monoclinic. *Grey (metallic) selenium*, formed when the other varieties are heated at 200°C, is a conductor of electricity when illuminated. It is used as a decolorizer for glass, in red glass and enamels, and in photoelectric cells and rectifiers. Selenium is similar to tellurium and sulphur in chemical properties. See Appendix 1, 2.

selenium cell Early photoconductive cell which depends on the change in electrical resistance of selenium when illuminated.

self-aligned gate Descriptive of a design procedure and processing step in semiconductor technology in which the gate electrode of a **MOSFET** is fashioned early on and subsequently used as the mask for ion implantation of source and drain regions. This ensures accurate registration and minimizes stray capacitances or faults associated with misaligned gates.

self-annealing A term applied to metals such as lead, tin and zinc, which recrystallize at ambient temperature and consequently exhibit little **strain-hardening** when cold-worked.

self-discharge (1) Loss of capacity of **primary cell** or **accumulator** as a result of internal leakage. (2) Loss of charge from capacitor due to the finite insulation resistance between its plates.

self-hardening steel See **air-hardening steel**.

self-lubricating bearing Plain, sintered-

metal bearing made by powder metallurgy techniques, having a porous structure impregnated with lubricant or polymer, eg PTFE. No external lubricant need be supplied during normal service lifetime. Used extensively in small electric motors in eg domestic appliances. Also applied to all-plastic (eg acetal) bearings.

self-skinning foams The term applied to polymer foams which collapse to form a solid surface when they collide with the mould walls, eg in **reaction injection moulding**. Usually achieved by using volatile liquid blowing agent.

self-tapping screw Screw made of hard metal which cuts its own thread when driven into softer materials.

Sellotape TN for transparent polymer film with adhesive backing.

selvedge The strong edge at both sides of a woven cloth. It sometimes bears a woven trade-mark or name, but its function is to give strength to the fabric in the loom and in subsequent processes carried out on the open width.

SEM Abbrev. for **Scanning Electron Microscope** (or microscopy). See **microscope**.

semi- Prefix from L. *semi*, half.

semi-chemical pulp Pulp for papermaking produced from the raw material by a combination of chemical and mechanical means. The relatively light digestion is insufficient to resolve all the non-cellulose matter or to permit dispersion of the fibres without some mechanical treatment.

semiconductor An element or compound having higher resistivity than a conductor, but lower resistivity than an insulator. The limited conduction through the material is governed as follows. (1) In *intrinsic semiconductors*, charges cross forbidden energy gaps due to thermal energy. This effect is temperature-dependent and produces equal numbers of electron and hole carriers. (2) In *extrinsic semiconductors*, current carriers are introduced by **donor** and **acceptor** impurities with locked energy levels near the top or bottom of the forbidden gap. Important semiconductors include silicon and gallium arsenide. See **intrinsic and extrinsic silicon** p. 173. Semiconductor materials are the basis of diodes, transistors, thyristors, photodiodes and all integrated circuits.

semiconductor device processing See panel on p. 280.

semiconductor diode Two-electrode, point contact or junction, semiconducting device with assymmetrical conductivity.

semiconductor diode laser A laser in which the lasing medium is p- or n-type semiconductor diode, eg gallium arsenide. Capable of continuous output of a few milli-

semiconductor device processing

Semiconductor devices are manufactured by a sequence of processes during which various layers of material are grown and deposited and dopants are introduced in controlled quantities into selected areas of the surface of a semiconductor wafer. Operations are carried out on wafers of several centimetres diameter at a scale as small as one micrometre. The sequence of processes is specific to the devices being fabricated.

Boules (rods) of electronics-grade semiconductor material are produced by the **float zone** and **Czochralski processes**. Wafers are then cut and polished to produce the high-quality substrates on which devices are built. Often the next stage involves growing an epitaxial layer on the substrate when it is this material which is actually used as the active semiconductor. See **epitaxial growth of semiconductors** p. 117. All semiconductor processing takes place in extremely clean, dust free environments because of the small scale of the engineering, and uses high-purity chemicals and gases because of the sensitivity of electronic materials to impurities (below the parts per billion level).

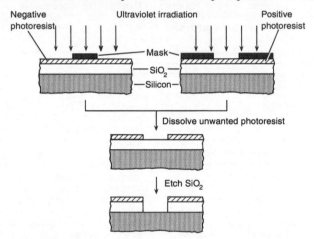

Fig. 1. **Opening a window in silicon dioxide by photolithography**.

To produce local **n-type** regions in a **p-type** layer it is necessary to introduce donor **dopant** atoms into restricted regions. In silicon technology this is done by first growing **silicon dioxide** over the entire surface. Next, **photolithography** is used to open apertures through the oxide and down to the underlying p-type silicon, according to the desired pattern, which is stored as a **mask**. Fig. 1 shows how one such aperture can be opened using either a negative or a positive photoresist. The role of the photoresist is to protect the silicon dioxide from exposure in those areas where access to the silicon is not required, the oxide being etched away from the unprotected areas in a subsequent step. The dopant is then introduced either by diffusing dopant atoms in from the surface or else by injecting ionized dopant atoms at high energy (ion implantation).

Diffusion processes enable dopant to penetrate several micrometres below the surface but require prolonged high-temperature ($\approx 1000°C$) exposure and have to be carefully planned to anticipate the effects of a series of diffusion cycles in complex device structures. Ion implantation penetrates only the first

continued on next page

semiconductor device processing (contd)

micrometre below the surface but allows a degree of control over the dopant profile and does not require prolonged high temperatures.

The fabrication of one device may require several cycles of oxidation, masking and doping and finishes with the opening up of contact apertures to the underlying semiconductor, the deposition of a metal interconnection layer and its subsequent patterning by further photolithography. Fig. 2 shows a schematic section through the surface of a wafer where **complementary metal-oxide-silicon, CMOS** devices have been fabricated. The changes in level of the silicon surface are a consequence of the various oxide growth cycles during which silicon from the wafer is consumed.

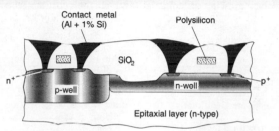

Fig. 2. **Cross-section through the surface of a wafer of a CMOS device.**

One wafer may contain several hundred separate integrated circuits, processed simultaneously. After testing and identifying defective circuits, the wafer is cut up (*diced*) into chips, the faulty ones discarded and the remainder then individually packaged, using fine wire to connect from the chip to the more robust legs of the finished device.

watts at wavelengths in the range 700–900 nm. See **optoelectonics** p. 226.

semiconductor junction One between the **donor** and **acceptor** impurity semiconducting regions in a continuous crystal; produced by one of several techniques, eg alloying, diffusing, doping, drifting, fusing, growing.

semiconductor radiation detector Semiconductor diodes, eg silicon junction, are sensitive under reverse voltage conditions to ionization in the junction depletion layer, and can be used as radiation counters or monitors.

semiconductor trap Lattice defects in a semiconductor crystal that produce potential wells in which electrons or holes can be captured.

semipermeable membrane A membrane which permits the passage of solvent but is impermeable to dissolved substances.

semisteel Cast-iron with low carbon content incorporating a proportion of melted down scrap steel.

semiworsted spun Textile yarn made from carded sliver or roving.

sensitized cheque paper   Paper for cheques and similar financial documents, containing chemicals intended to reveal attempts at falsification. Cf **security paper**.

sensitized paper Paper that has been coated to render it suitable for a reprographic process, generally light activated eg dyeline, blueprint, photographic. Also paper, made from stock to which selected chemicals have been added, or coated, so that a colour reaction is produced on exposure to other reagents.

sensor General name for detecting device used to locate (or detect) presence of matter (or energy, eg sound, light, radio or radar waves).

sequoia Fast-growing, N American **softwood** tree (*Sequoia sempervirens*), with a spongy bark and a dull reddish-brown, straight-grained, medium- to fine-textured **heartwood** of mean density 420 kg m^{-3}. Moderately durable, but liable to insect attack; used for exterior cladding, interior joinery, coffins, organ pipes, telegraph poles, etc in plywood and as a decorative veneer. Its bark is used in **chipboard**. Closely related to the giant redwood (*S. gigantea*), the world's largest known tree. Also *redwood*.

set (1) Percentage residual deformation left in a material, esp. viscoelastic polymers, after deformation for a given time (often after period of recovery following removal of load). Also *permanent set, compression set*. (2) A smith's tool similar to a short, stiff cold chisel; used for cutting bars etc without heating. Also *sett, cold sett*.

setting (1) Treatments usually by heating and cooling in dry or steamy atmospheres that confer stability on textile materials. Heat setting is particularly important for many fabrics. (2) Operation in **injection moulding** for optimizing production from new tools. It involves balancing mould temperatures, pressures, machine parameters so that mould defects are minimized.

sett, set The number of threads per unit length in the weft and/or warp of woven fabrics. In a square sett fabric, the two values are equal and the yarns in both directions are of the same count. In an unbalanced sett fabric these values are significantly different.

SF Abbrev. for **structural foam**.

S-glass A high tensile strength, high modulus, glass fibre, composition (wt %): SiO_2 65, Al_2O_3 25, MgO 10, developed for reinforcing composite materials in aerospace applications.

shadow-mark A defect of paper showing in the look-through as a faint reproduction of the holes of the suction couch or press rolls.

shaft furnace One in which ore and fuel are charged into the top and gravitate vertically, reacting as they proceed to bottom discharge.

shake A partial or complete separation between adjacent layers of fibres in wood. See **cup shake**.

shank A ladle for molten metal.

shantung Plain-weave silk cloth, with a randomly irregular surface; made from *tussah*, the silk produced by the wild silkworm.

shape factor (1) Ratio of single loaded to total unloaded surface area of rubber in laminated bearing, the higher it is, the greater the resistance to downward thrust from the structure. (2) A parameter describing the shape of a particle of a powder or the ratio of two average particle sizes determined by methods in which the shape of the particles influences the measured parameter in different ways.

shape memory alloys Alloys which undergo mechanical twinning, or a martensitic or similar reversible solid-state transformation which involves a dimensional change, occurring usually over a narrow temperature range. This enables a shape produced in one state to be recovered if the temperature is altered back again despite the shape having changed in the interim. Useful for thermostats and on-off switches and for clamping devices such as pipe couplings where a loose sleeve can be caused to contract and clamp tightly by heating or cooling to induce the shape recall dimensions.

sharkskin (1) Woven or warp-knitted fabric with a characteristically firm construction and stiff handle, often with a dull appearance, used as a dress and suiting material. (2) Defective surface on an extrudate, shown by lines at right angles to polymer extrusion direction. See **extrusion**.

shear Type of deformation in which parallel planes in a body (solid or fluid) remain parallel but are relatively displaced in a direction parallel to themselves. A rectangle, if subjected to a shearing force parallel to one side, becomes a parallelogram. Pure shear is a distortion of the body, without deformation, ie the dimensions of the body remain unchanged. See **shear strain, torsion**.

shear force A force which tends to cause sliding of adjacent layers, relative to each other.

shear heating See **adiabatic heating**.

shearing (1) Cutting the wool from a living sheep (cf **fellmongering, pulling, skin wool**). (2) Trimming a pile on a fabric to a uniform height. Also *cropping*.

shear lip Small, often linear zone on fracture surface where brittle crack meets edge of specimen, and yielding occurs. See **brittle fracture**.

shear modulus One of the four basic elastic constants of an isotropic, linear elastic material. Usual symbol G, sometimes μ. Defined by $G = \tau/\gamma$, where τ is the shear stress and γ is the shear strain. Related to the other constants by

$$G = \frac{E}{2\,(1+\nu)} = \frac{3\,EK}{9\,K - E} \ ,$$

where E is **Young's modulus**, K is the **bulk modulus**, and ν is **Poisson's ratio**. Also *rigidity modulus*.

shear rate Rate at which the velocity of a fluid under shear changes through its thickness ($\dot{\gamma}$). Units are reciprocal seconds. Basic parameter in Newtonian viscosity. See **extrusion** p. 121, **injection moulding** p. 169.

shear strain The angular displacement of adjacent parallel planes in a body subject to shear. In a body of length l twisted through an angle θ when loaded in torsion, the shear strain γ at a distance r from the torsional axis is $\gamma = r\theta/l$.

shear stress Shearing force divided by the area over which it is acting. Shear stress occurs across the section of a beam loaded transversely and also across shaft sections subject to torque. The magnitude of the stress varies across the section depending on its geometric shape.

sheath Excess of positive or negative charge in a plasma, giving a shielding or **space-charge** effect.

sheath voltage Electrostatic potential across a space charge **sheath** region. In low-pressure plasma processing (**sputtering**, **etching** etc) use is often made of the fact that positive ions are accelerated by the sheath voltage onto adjacent surfaces.

shed The opening created by separating the warp threads during weaving so that the shuttle or other device can take the weft thread through.

sheet glass Form of **flat glass**, now largely superseded by **float glass**, it was used largely for glazing purposes; produced by drawing a continuous film of glass upwards from a molten bath and, after a suitable time interval for cooling, cutting-up the product into sheets. It is not of such good quality, nor so flat, as **plate glass** which is ground and polished.

sheeting A medium-weight, closely woven, plain or twill weave fabric used for bed coverings.

sheet materials Materials in which structure is formed of very strong covalently bonded sheets of atoms, held together by relatively weak bonds. Thus mica and talc minerals contain sheets of silicon-oxygen, held together electrostatically, so are easily cleaved. Graphite has a similar structure of carbon-carbon sheets.

shellac The refined form of lac, the secretion of the insect, *Lacifer lacca*, parasitic on trees in SE Asia. Consists of a mixture of a red dye, wax and natural polymer etc. The polymer fraction is a low molecular mass polyester with numerous hydroxyl (–OH) and carboxylic acid (–COOH) side and end groups on the polymer chains. Thermal polymerization at 150–160°C gives a hard, brittle solid used for electrical insulation when reinforced with mica particles. Also used for French polishing.

sheradizing A zinc-zinc oxide corrosion resistant coating applied to iron and steel articles which also acts as a paint key. Objects are packed in a finely divided mixture of zinc powder and zinc oxide. Then heated at 350–450°C for a a period depending on the coating thickness required.

shielded metal-arc welding The use of covered consumable electrodes to provide protection for the weld, as the covering vaporizes.

shift (1) Change in wavelength of spectrum line due, eg to Doppler or Zeeman effects or to Raman scattering. (2) Change in value of energy level (*level shift*) for quantum mechanical system arising from interaction or perturbation.

shift factor Term described by **Williams–Landel–Ferry equation** which is a measure of the distance a stress relaxation curve needs to be moved in order to create master curve. Symbol a. See **viscoelasticity** p. 332.

shiners Particles, usually from mica in the china clay, or undissolved alum, showing on the surface of paper.

shish-kebabs Polymer microstructure where lamellae form at right angles (ie epitaxially) to an oriented fibril, strung out along it at regular intervals like the eponymous product of the Middle East.

shives (1) Undigested particles of wood showing as pale yellow to brown splinters in wood pulp or paper. (2) Vegetable matter found in wool fleece. (3) Small particles of woody tissue removed from flax fibres in **scutching** and **hackling**.

shoddy (1) Short, fibrous material obtained from old, cleaned, loosely woven or knitted wool cloth after treatment, in a rag-tearing machine. Sprayed with oil, it is re-used in blends for cheap suits and coatings. (2) The short dirty fibres which drop from woollen cards.

Shore hardness Two fundamentally different ways of assessing the hardness of materials. One, *Shore A*, for softer materials such as elastomers and plastics, measures the depth of penetration of an indenter under specified conditions; the other (see **Shore scleroscope**) is a rebound test for harder materials such as metals and rocks. See **hardness measurements** p. 153.

Shore scleroscope Instrument for determining a hardness value for materials by measuring the rebound of a diamond-tipped hammer dropped from a given height. See **Shore hardness**.

short Term used for a glass that is fast-setting, ie has a short freezing range.

short column A column the diameter of which is so large that bending (ie **buckling**) under compressive load may be neglected, and in which failure would occur by crushing; commonly assumed as a column of height less than 20 diameters.

short range order **Substitutional solid solution** in which the strictly ordered arrangement exists over only a few lattice units within the crystal.

short shot Defect in **injection moulding** where molten polymer fails to reach end of tool cavity.

shot blasting Blasting with metal shot to remove scale and other deposits from castings etc.

shot effect The variable colour seen when certain fabrics are viewed at different angles.

The effect is obtained by having the warp threads of one colour and the weft threads of a contrasting colour.

shot noise Broadband noise associated with the quantized nature of electric charge.

shot peening Bombarding with rounded shot to locally harden surface layers. Results in significant improvement in fatigue resistance.

shot weight Mass of polymer required for each moulding cycle Similarly, shot volume. See **injection moulding** p. 169.

shrinkage (1) The reduction in dimensions of a fibre, yarn, or fabric that takes place in processing or in wear. Particularly effective shrinking treatments are wetting, including any laundering cycle. (2) Dimensional misfit between cool plastic product and tool cavity used to make it. Useful for ease of removal at end of moulding cycle, but needs careful control for final performance of product. Closely related to orientation and polymer crystallization.

shrinkage porosity Cavities produced within a solidified mass of eg ingot or casting, due to specific volume contraction on solidification. See **pipe**. Usually appears as ragged edged cavities associated with the last regions to solidify in contrast to the smooth rounded voids resulting from **gas porosity**.

shrinking-on The process of fastening together two parts by heating the outer member so that it expands sufficiently to pass over the inner and on cooling grips it tightly, eg in the attachment of steel tyres to locomotive wheels.

shrink-wrap Thermoplastic film used to cover products, then heat treated so that film shrinks onto product to provide seal against atmosphere.

shuttle (1) Loom accessory which carries the weft (in the form of a cop or pirn) across a loom, through the upper and lower warp threads. Often made of hardwood (eg persimmon), it is boat-shaped, with a metal tip at each end. One end has a porcelain eye through which the weft passes. More modern weaving machines do not have a shuttle. (2) In a sewing machine, a sliding or rotating device that carries the lower thread to form a lock-stitch.

SI Abbrev. for *Système International* (d'Unités). See **SI units**.

sialons Hard, tough ceramic based on β-silicon nitride, Si_6N_8, in which a proportion of the Si is replaced by Al and a proportion of the N by O, to give a range of compositions, $Si_{(6-z)}Al_zO_zN_{(8-z)}$. Some also contain yttrium to form yttrium aluminium garnet (TAG) which further improves high-temperature properties. Used

for high-performance cutting tools, bearings and parts for the combustion zones of internal combustion engines.

side chains Alkyl groups or longer chains, which replace hydrogen in polymer materials.

side-wall See **tyre technology** p. 325.

siemens SI unit of electrical conductance; reciprocal of **ohm**. Symbol S.

sieve analysis The measurement of the size distribution of a powder by using a series of sieves of decreasing mesh aperture.

sieve mesh number The number of apertures occurring in the surface of a sieve per linear inch. Unless the size of the wire used in weaving the mesh is specified, the mesh number does not uniquely specify the aperture size.

sigma phase A microconstituent causing embrittlement sometimes found in stainless and heat-resistant steels. Based on FeCr but can take other elements into solid solution, it tends to form after long periods at temperatures in the range 450–1000°C.

silanes A term given to the silicon hydrides: silane, SiH_4, disilane $H_3Si–SiH_3$, trisilane, $H_3Si(SiH_2)$ SiH_4 etc. Cf **methane**. Silane and chlorosilanes are used in chemical vapour deposition processes for silicon and in silicon dioxide deposition in semiconductor technology.

Silastic TN for a range of silicone rubbers. Noted for very good heat resistance and a wide temperature range of application. Excellent chemical resistance and electrical properties.

silal High (6%) silicon cast iron suitable for chemical plant and heat resistant applications.

silesia A smooth-faced, cotton lining fabric for garments, originally a plain weave but now a twill.

silica Dioxide ((IV) oxide) of silicon, SiO_2. See **silica and silicates** p. 285. Dioxide ((IV) oxide) of silicon, SiO_2, occurring in crystalline forms as quartz, cristobalite, tridymite; as cryptocrystalline chalcedony; as amorphous opal; and as an essential constituent of the silicate groups of minerals. Used in the manufacture of glass and refractory materials. Refractory materials containing a high proportion of silica (over 90%) are known as *acid refractories* (eg gannister), and are used in open-hearth and other metallurgical furnaces to resist high temperatures and attack by acid slags.

silica and silicates See panel on p. 285.

silica gel Hard amorphous granular form of hydrated silica, chemically inert but very hygroscopic. Used for absorbing water and vapours of solvents in eg dessicators, electronic equipment. When saturated, it may be regenerated by heat.

silica and silicates

Silica, or silicon (IV) oxide, SiO_2, forms the basis of silicates which, in combination with other elements, chiefly aluminium, make up the bulk of the Earth's crust. It occurs in crystalline form as quartz, cristobalite, tridymite; as cryptocrystalline chalcedony; as amorphous opal; and as an essential constituent of the silicate groups of minerals, over a thousand of which are known. It is the main ingredient of the silicate glasses (see **glasses and glassmaking** p. 144) and, in mineral form, is important in Portland cement (see **cement and cement making** p. 52), as well as in most ceramic and refractory products. Refractory materials containing a high proportion of silica (over 90%) are known as *acid refractories*, eg *gannister*. They are used as a lining for metallurgical and other furnaces to resist the high temperatures and attack by acid slags.

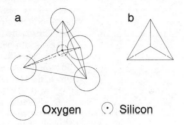

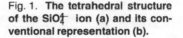

Fig. 1. **The tetrahedral structure of the SiO_4^{-} ion (a) and its conventional representation (b).**

The basis of these materials is the SiO_4^{-} tetrahedron (Fig. 1) in which four oxygens surround a silicon to share most of the silicon's four outer electrons, the bonding being about 65% covalent and 35% ionic. The smaller silicon is located in the interstice between the four, much larger, oxygens. Charge neutrality is maintained by each oxygen atom being shared by two neighbouring tetrahedra. The tetrahedra can combine together in many different ways producing chains and extended two- and three-dimensional networks leading to a range of crystalline forms (eg Fig. 2). The low mobility in the melt of these tetrahedral structures means that it is often relatively easy to form glasses when cooling, since there is insufficient time for the rearrangements necessary for crystal formation to take place.

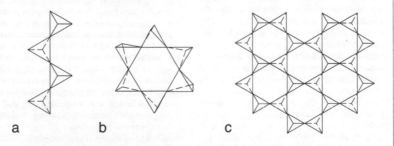

Fig. 2. **Different networks of the silicate tetrahedra giving (a) spodumene, (b) emerald and (c) talc.**

silica glass See **vitreous silica.**

silico-manganese steel See **manganese alloys.**

silicon A non-metallic element, symbol Si. Amorphous silicon is a brown powder; rel.d. 2.42. Crystalline silicon is grey; rel.d. 2.42. This element is the second most abundant, silicates being the chief constituents of many rocks, clays and soils. Silicon is manufactured by reducing silica with carbon in an electric furnace, and is used in glass and in making certain alloys, eg *ferro-silicon.* It has semiconducting properties, being used for a large range of electronic components. See **intrinsic and extrinsic silicon** p. 173, **semiconductor device processing** p. 280, Appendix 1, 2.

silicon brass A corrosion-resistant alloy based on copper and zinc but with a small addition of silicon. See **copper alloys** p. 76.

silicon carbide SiC. Formed by fusing a mixture of carbon and sand or silica in an electric arc furnace. Used as an abrasive and refractory. Also a wide band gap semiconductor with potential for high-temperature applications.

silicon-controlled rectifier A three-terminal semiconductor switching device consisting of a sandwich of p-n-p-n-type materials. It is normally open-circuit, but application of an appropriate control signal to the *gate* allows it to conduct, in one direction only, like a conventional rectifier. It continues to conduct even with the gate signal removed, until it is reverse biased by the voltage it is intended to switch. Used in voltage-control of power circuits.

silicon copper An alloy (20–30% Si), a *getter* used to remove oxygen from molten copper alloys.

silicon dioxide The oxide of silicon (SiO_2), also *silica*. In semiconductor technology, silicon dioxide grown on silicon wafers is a key material in device manufacturing processes and also is vital as a dielectric within device structures. See **semiconductor device processing** p. 280.

silicone rubbers *Polysiloxanes*. They have a main backbone chain which is inorganic and the repeat unit is

Different types are created by varying the substituent organic groups (R and R′) on the silicon atom. The commonest type has two methyl groups (–CH$_3$) in this position (MQ systems). Copolymers where different substituents occur (V = vinyl, P = phenyl, F = fluorine-containing groups) modify properties, eg PVMQ is a low-temperature resistant rubber, crosslinked via vinyl side groups. Generally used for gaskets where resistance to temperature extremes is needed, despite low strength. Room temperature vulcanizing materials are also used for water-resistant seals eg aqarium seals, building seals.

silicones Open-chain and cyclic organosilicon compounds containing –SiR$_2$O– groups. The simpler substances are oligomeric oils of very low melting point, the viscosity of which changes little with temperature, used as lubricants, shock-absorber fluids, constituents of polishes etc. Polymers, stable to heat and cold, and chemically inert, are exceptionally good electrical insulators which are highly hydrophobic and used for electric motors etc. Also found in gaskets and a wide variety of special applications. See **silicone rubbers.**

silicon iron Iron or low carbon steel to which 0.75–4.0% silicon has been added. Has low magnetic hysteresis, and is resistant to mild acids. Used for sheets for transformer cores. Typical composition: silicon 4%, manganese under 0.1%, phosphorus 0.02%, sulphur 0.02%, carbon 0.05%.

silicon nitride Hard engineering ceramic used for cutting tools and in wear-resistant and high-temperature applications. Composition Si$_3$N$_4$ but exists in two crystalline forms, α and β; formed from powders by **hot isostatic pressing.** See **sialons.**

silicon on insulator A semiconductor technology in which silicon devices are defined in a thin layer of silicon on an insulating substrate; devices are isolated by etching through to the substrate between active regions. Abbrev. *SOI.* See **silicon on sapphire.**

silicon on sapphire A **silicon on insulator** technology in which silicon is grown on a single crystal alumina substrate, with which there is a close enough match of lattice constant to allow epitaxial growth. Abbrev. *SOS.* See **epitaxial growth of semiconductors** p. 117.

silicon rectifier A semiconductor diode rectifier usually based on p-n junction in silicon crystal.

silicon resistor A resistor of doped silicon material which has a fairly constant positive temperature coefficient, making it suitable as a temperature-sensing element.

silk The protein fibre obtained in fine strands, in lengths of the order of kilometres, from the cocoon of silkworms, esp. of the moth *Bombyx mori*. The fibre is composed of **fibroin** surrounded by another protein, *sericin*, which is a gum removed during wet processing. Wild silk (eg *tussah*) is a similar fibre, but of more elongated cross-section, from the cocoon of other moths.

Silsbee rule A wire of radius r cannot carry a superconducting current greater than $1/2rH_c$, where H_c is the critical field. The self-magnetic field would then destroy superconductivity.

Silurian A mottled effect in special papers produced by adding deeply dyed fibres to the main furnish. Under certain circumstances this can be a paper defect.

silver A pure white metallic element, symbol Ag. The metal is not oxidized in air. Occurs massive, or assumes arborescent or filiform shapes. The best electrical conductor and the main constituent of photographic emulsions. Native silver often has variable admixture of other metals; gold, copper or sometimes platinum. Used for ornaments, mirrors, cutlery, jewellery etc and for certain components in food and chemical industry where cheaper metals fail to withstand corrosion. See Appendix 1, 2.

silver grain The light greyish, shining flecking seen in oak timber, caused by vascular rays exposed in preparing the timber when it is cut radially through the centre of a log. Also seen in beech.

silvering Traditional process of forming a reflective film eg for mirrors on a clean surface of glass by pouring onto it an ammoniacal silver solution, mixed with Rochelle salt or with a nitric-acid/cane-sugar/alcohol mixture. The silver film so formed is then washed, backed with varnish and painted. Alternative nowadays is to use sputtering.

silver solder A range of hard soldering alloys with excellent flow properties, based on the silver copper eutectic with other alloying additions, eg zinc, tin, cadmium, to further reduce melting temperature to as low as 600° C. See **brazing solders, copper alloys** p. 76.

silver steel Workshop description for bright-drawn carbon steel containing up to 0.3% silicon, 0.45% manganese, 0.5% chromium and 1.25% carbon; low in phosphorus and sulphur. See **steels** p. 296.

SIMS Abbrev. for **Secondary Ion Mass Spectrometry**.

Sindanyo TN designating materials, composed principally of asbestos, for the mounting of switchgear of all types, for electrical insulation work generally and for arc shields, barriers, furnace linings and other purposes.

single crystal Crystal formed by propagation of unit cell in three dimensions, usually from a single nucleus. Common in many minerals but absent in polycrystalline metals and ceramics. Massive single crystals of silicon grown for semiconductors. Of micro-

scopic size (typically 0.1 μm) in most crystalline polymers, where spherulites are normal. See **crystallization of polymers** p. 82.

single-domain particle Part of a magnetic material of sufficiently small size that it is energetically unfavourable for it to be further subdivided into **domains**.

single jersey Weft-knitted fabrics of a simple plain nature although patterned effects may be included eg with a **jacquard**.

single yarn The thread obtained from one unit of a spinning machine.

sink mark Depression on outer surface of moulding. Often an inevitable defect created at thick sections.

sintered carbides See **carbide tools, sintering**.

sintering The coalescing or fusing together of small particles to form larger masses. Used for ceramics (see **ceramics processing** p. 54), in powder metallurgy, and for ores and concentrates. High temperatures are frequently required because sintering occurs by solid state diffusion, either along the external surfaces or through the crystalline interior. In the latter case, crystal defects such as grain boundaries are preferred paths. Sometimes impurities in the batch or deliberate additions (fluxes) can speed up the sintering processes or allow lower temperatures to be used. See **liquid phase sintering, hot pressing, hot isostatic pressing**. Also used as a shaping process for intractable polymers, esp. **polytetrafluoroethylene** as in hip joint implant sockets.

sisal The **bast** fibre from the sisal plant (*Agave sisalana*). Grown in Mexico and used to make string and cord but being displaced for many uses by polyethylene or polypropylene.

sitka spruce N American softwood tree (*Picea sitchensis*), with a light pinkish-brown, straight-grained, medium-textured heartwood of mean density 430 kg m^{-3}. Non-durable and susceptible to insect attack. used for boat-building, interior fittings, soundboards of pianos, cooperage, etc. Veneers are used in **laminates** for aircraft, gliders, rowing shells and oars.

SI units System of coherent metric units (*Système International d'Unités*) adopted internationally in 1959. It was developed from the Giorgi/MKSA system (see **MKSA**) by the addition of the **kelvin** and **candela**, and later the **mole**, as base units. There are numerous *derived units* (**newton, joule** etc) and scales of decimal multiples and submultiples, all with agreed symbols. See Appendix 3.

size Film-forming substances (eg starch), sometimes containing lubricants, that are applied to warp threads to protect them during weaving, to paper to reduce its rate of water absorption and to glass fibres to protect their surfaces and reduce the incidence of strength-impairing flaws. Also *sizing*.

size factor compound An intermetallic compound or phase in which the composition is determined by the best fit of the different sizes of atom within a crystal lattice.

size fraction A portion of a powder composed of particles between two given size limits, expressed in terms of the weight, volume, surface area or number of particles.

sizing See **size**.

skelp Mild steel strip from which tubes are made by drawing through a bell at welding temperature, to produce lap-welded or butt-welded tubes.

skillet Mould for casting bullion.

skin (1) The protective tissue layers of the body wall of animals. Formed of collagen fibres and elastin. Animal skin is tanned to make leather. (2) The hard surface layer found in castings due to the rapid cooling effect of the mould, or on metal plates, strip and sheet, due to rolling, or on other materials or products due to the surface-hardening effect of the finishing process.

skin depth The distance into a conductor of conductivity σ and absolute permeability μ, at which the amplitude of an electromagnetic wave (frequency $\omega/2\pi$) falls to $1/e = 0.3679$ of its value at the surface, given by

$$\delta = \sqrt{2/\sigma\omega\mu}.$$

In copper at 50 Hz, δ is about 10^{-2} m. See **skin effect**.

skin effect The phenomenon by which high-frequency currents tend to be confined to a thin outer layer (skin) of conductors. It follows that the resistance of a given conductor will be higher at high frequencies, because of the effective reduction in cross-sectional area available for the current flow. See **skin depth**.

skin wool Wool removed from the fleeces of slaughtered sheep by chemical or biochemical processes. Cf **fellmongering, pulling, shearing**.

slab Metal, twice as wide as thick, and thus intermediate between an ingot and a plate in a rolling mill.

slag The top layer of the two-layer melt formed during smelting and refining operations. In smelting it contains the gangue minerals and the flux; in refining, the oxidized impurities.

slag wool. Fibrous material very similar to fibre glass but made from slag produced by an iron blast furnace as distinct from silicate glass.

slaked lime See under **caustic lime**.

slip The process involved in the plastic de-

formation of metal crystals in which the change in shape is produced by parts of the crystals sliding with respect to each other along certain crystallographic planes. The force required to cause block slip in perfect crystals would be very high and in practice it is the gliding of **dislocations** along the slip planes which causes the movement to occur at the much lower levels of stress found in practice.

slip bands Steps or terraces produced on the surfaces of crystals as a result of the parts having moved with respect to each other during slip.

slip casting See **ceramics processing** p. 54.

slipper satin A heavy, smooth, high-quality satin fabric made from continuous filament yarns (silk or man-made) used particularly for wedding dresses and evening shoes.

sliver Continuous untwisted strand of textile fibres as formed by carding or combing.

slow neutron See **neutron**.

slub A fault in drafted, twisted, textile yarns, which appears as a thicker part, with little twist. Slubs can also be deliberately made in yarn at regular or random intervals for ornamentation and special weave effects. They occur naturally in **tussah silk**.

slubbings Relatively thick, open but coherent strands of textile fibres; converted into yarn by subsequent spinning processes.

slug Any non-fibrous glass in a glass-fibre product.

slush moulding Method based on (1) injecting metal into a die in the pasty stage between **liquidus** and **solidus** and (2) using certain plastics, particularly polyvinyl chloride, in **plastisol** form. This is placed in a hollow heated mould which is rotated until the paste forms the solid replica of the mould configuration. Used, eg for dolls' heads.

slush pulp Pulp which is pumped direct from the pulp mill to the paper mill for use without passing through the pulp drying stage.

smart materials Metals, ceramics or polymers which respond to an external stimulus in a specific, controlled way. They include **photochromic** lenses which darken in bright sunlight, plastic film containing **liquid crystals** which change colour with changing ambient temperature, **piezoelectric crystals** which generate a voltage when stressed. Widely used in consumer products etc for controlling function by sensing changes in the environment. Also *intelligent materials*.

smear head Type of solid head of extruder screw which aids mixing of polymers and additives. But see **cavity transfer mixing**.

smectic Said of a mesomorphous substance whose atoms or molecules are oriented in parallel planes. Cf **nematic**.

smelting Fusion of an ore or concentrate with suitable fluxes, to produce a melt consisting of two layers – on top a slag of the flux and gangue minerals, molten impure metal below.

smoke Visible cloud of airborne particles derived from combustion, or from chemical reaction; the particles are mainly smaller than 10μm.

SMR Abbrev. for **Standard Malaysian Rubber**.

snakewood (1) Strong, tough and very durable **hardwood** from a C and tropical S American tree (*Piratinera*). **Heartwood** is red-brown with darker markings resembling a snake skin. Density about 1300 kg m^{-3}. Used for fancy goods, drum sticks, cutlery and umbrella handles, violin bows, archery bows, etc, and also for decorative veneers. Also *letterwood, leopard wood, tortoiseshell wood*. (2) The *Strychnus nux vomica*, a tree species yielding strychnine.

snap-fit Joint where closure is achieved over a small lip, usually in plastics products; reversible allowing disassembly.

snarl Small extra thickness of textile yarn where the twist has run back on itself; caused by excessive twist and inadequate tension on the yarn.

S/N curve Abbrev. for *Stress-Number of cycles to failure curve*. See **fatigue** p. 123.

Snell's law A wave refracted at a surface makes angles relative to the normal to the surface, which are related by the law

$$n_1 \sin \theta_1 = n_2 \sin \theta_2.$$

n_1 and n_2 are the refractive indices on each side of the surface, and θ_1 and θ_2 are the corresponding angles. The two rays and the normal at the point of incidence on the boundary lie in the same plane.

Snook effect When an interstitial solid solution is subjected to small elastic strains, solute atoms tend to move to interstices which align with the strain direction in order to minimize the total strain energy in the lattice. If the direction of strain is altered the atoms jump to adjacent sites in order to realign with the new strain axis. This effect can be used to study diffusion rates at ambient temperature by measuring the increase in internal friction associated with repeated realignments caused by eg a torsional pendulum with a period of roughly one second.

soaking A stage of a heating operation during which metal or glass is maintained at the requisite temperature until uniformly heated, and/or until any required phase transformation has occurred.

soda-lime-silica glass The commonest type of glass, used (with small variations in composition) for windows, containers and electric lamp bulbs. Contains silica, soda and lime in approx. proportions (wt%): SiO_2 70–74, Na_2O 12–15, CaO 8–12, plus MgO 0–4, Al_2O_3 1–2.

sodium One of the alkali metals, symbol Na. Sodium does not occur in nature in the free state, owing to its reactivity, but it is widely distributed combined as the chloride, nitrate etc. Metallic sodium is made by electrolysis of fused caustic soda. It is a soft, silver-white metal, very active, with a moderate thermal neutron cross-section. Used in reactors as a liquid metal heat-transfer fluid. See Appendix 1, 2.

sodium hydroxide NaOH, a deliquescent substance, with a soapy feel, whose solution in water is strongly alkaline. It is manufactured by treating quicklime with hot sodium carbonate solution; and its main industrial use is the manufacture of soap. Also *caustic soda*.

soft Applied to glass with a relatively low softening point.

softening (1) The end result of annealing or tempering, ie a reduction in hardness and strength. See **annealing** p. 13. (2) A process, for removing arsenic, antimony and tin from lead, after drossing.

softening point One of the reference temperatures in **glassmaking** p. 144.

soft iron Iron low in carbon with low hysteresis and unable to retain magnetism; useful as solenoid cores etc.

softness Tendency to deform easily. It is indicated in a **tensile test** by low ultimate tensile stress and large reduction in cross-section before fracture. Usually the elongation is also high. In a notched bar test, specimens bend instead of fracturing, and the energy absorbed is relatively small. See **brittleness, toughness**.

soft segment Term applied to elastomeric part of block copolymers, esp. polyurethanes and block polyesters, usually made of oligomeric polyglycols or polyesters.

soft solder Alloys of lead and tin used in soldering. Tin content varies from 63% to 31%. The remainder is mainly lead, but some types contain about 2% antimony and others contain cadmium. The best-known types are *tinman's solder* and *plumber's solder*. Owing to the toxicity of lead in plumbing systems, food cans etc, solders based on tin-copper and tin-silver are preferred.

SOG Abbrev. for **Spin On Glass**.

SOI Abbrev. for **Silicon On Insulator**.

solar cell Photoelectric cell using silicon, which collects photons from the Sun's radiation and converts the radiant energy into electrical energy with reasonable efficiency. Used in satellites and for remote locations lacking power supplies, eg for radiotelephony in the desert.

solarization Changes of the light transmission properties of a glass or other transparent material as a result of exposure to sunlight or other radiation in or near the visible spectrum.

solar panel Arrays of solar cells fitted to spacecraft and satellites in order to gather solar energy for conversion into electrical power for the equipment on board. In some satellites these may be in the form of extensible *paddles* which are stowed away during launching.

solder A general term for alloys, frequently of eutectic composition, used for joining metals together. The principal types are soft solder (lead-tin alloys) and brazing solders (alloys of copper, silver and zinc). Cf **hot melt adhesive**.

soldering Hot joining of metals by adhesion using, as a thin film between the parts to be joined, a metallic bonding alloy having a relatively low melting point. A **flux** is usually necessary to aid wetting of the surfaces to be joined.

solder paint Mixture of powdered solder and flux, applied by brush to surfaces to be joined by soldering to form the bonding film.

solenoidal magnetization The distribution of the magnetization on a piece of magnetic material when the poles are at the ends. Cf **lamellar magnetization**. Also *circuital magnetization*.

sol-gel process Chemical route to glass formation, using reactions in solution to produce a gel precurser, which yields the glass on drying-out, thus avoiding the melting stage.

solid diffusion Movement of atoms through a solid material, as when carbon diffuses into or out of steel during carburizing or decarburizing respectively, or when dopants are diffused into semiconductors. See **Fick's laws of diffusion**.

solid solubility The extent to which one material is capable of forming solid solutions with another.

solid solution An arrangement of atoms or molecules of different species within the same crystal lattice. See **interstitial solid solution, substitutional solid solution**.

solid state Pertaining to a circuit, device or system, which depends on some combination of electrical, magnetic or optical phenomena within a material which is usually a crystalline semiconductor. Loosely applied to all active devices or circuits which do not rely on valves or tubes.

solid-state physics Branch of physics which covers all properties of solid materials, including electrical conduction in crystals of semiconductors and metals, superconductivity and photoconductivity.

solid-state polymerization Polymerization carried out in a crystalline monomer.

solid-state welding Welding which does not involve melting, brazing or soldering, but may involve pressure.

solidus Line on a phase diagram representing temperatures above which mixtures begin to melt and below which mixtures are completely solid under equilibrium conditions. See **phase diagram** p. 236.

solubility of polymers Extent to which polymers pass into solution. Unlike small molecule substances or materials, solubilization may be very slow owing to the time needed for large chain molecules to diffuse into the fluid. The tendency of a polymer to be soluble in a fluid can be assessed by matching their **solubility parameters**. Good solubility occurs if they match exactly, poor solubility if they differ greatly, although hydrogen bonding and degree of crystallinity can affect the outcome. Uncrosslinked rubbers and amorphous thermoplastics tend to be soluble in organic solvents rather than water, although slight absorption can occur, esp. if the polymer is hydrogen-bonded (eg nylons). Polyelectrolytes tend to behave in the opposite way. Polyolefins are usually difficult to dissolve in any solvent at ambient temperature, high temperatures (ca 160°C) being needed.

solubility parameter Square root of cohesive energy density, symbol δ. Used to assess polymer solubility, by matching δ (solvent) and δ (polymer). See **solubility of polymers**.

solution heat treatment Heating suitable alloys (eg **Duralumin**) in order to take the hardening constituent into solution. This is followed by quenching, to retain the solid solution, and the alloy is then age hardened at atmospheric or elevated temperature. See **precipitation hardening**.

solution polymerization Polymerization conducted in an inert solvent where components are homogeneously dissolved. Contrast with emulsion and suspension polymerizations. See **chain polymerization** p. 56.

solvent bonding Textile process in which an organic liquid is used to soften fibres so that they adhere to each other and form a non-woven fabric. Cf **solvent welding**.

solvent welding Use of a good solvent or mixture of solvents to create joint between similar or identical thermoplastics. Formation of good bond may take some time, since solvent must diffuse away through solid material before solid joint is formed. A similar polymer is often added to the solvent to give a solvent cement to aid bond formation.

solvus See **phase diagram** p. 236.

Sommerfeld atom, model Atomic model developed from **Bohr atom**, but allowing for elliptic orbits with radial, azimuthal, magnetic and spin quantum numbers. Modern theories modify this by regarding the electrons as forming a cloud, the density of which is described in terms of their wave function.

sorbite Product formed by tempering of **martensite** to produce rounded particles of **cementite** in a ferrite matrix, thus distinguishing it from pearlite which forms directly from austenite by eutectoid decomposition.

SOS Abbrev. for **Silicon On Sapphire**.

source The electrode from which majority carriers flow into the **channel** of a field-effect transistor. Comparable with the emitter in a conventional bipolar transistor.

space charge As distinct from surface charge, the term applied to the local net charge when it is distributed through a finite volume, such as in the depletion region in a semiconductor device structure (p-n junction etc). Gradients in electrostatic fields are associated with regions of space charge.

space frame Collection of connected struts, rods or wires which are stressed when whole assembly is stressed, eg trestle bridge, half-timbered cottage. Cf **monocoque**.

spallation Any nuclear reaction in which several particles result from a collision, eg cosmic rays with atoms of the atmosphere; chain reaction in a nuclear reactor or weapon.

spallation neutron source A powerful pulsed neutron source for research. Protons, accelerated in a synchrotron, are focused on to a target of ^{238}U; 25 to 30 neutrons per proton are released, moderated and collimated into beams.

spalling The flaking-off of material from a surface due to eg impact (as in flint knapping), expansion of sub-surface material (see **rusting** p. 271), ingress of water and subsequent freezing (common in stonework), etc.

spandex Term for elastic fibre used in foundation garments etc, based on thermoplastic polyurethane rubber.

spar A round timber more than 6 in (150 mm) in diameter in the middle.

spark erosion Electrochemical metal machining process in which an electrode of male form is maintained very close to the workpiece, both being submerged in a dielectric liquid. High local temperatures are produced by passing a current through the gap, detaching and repelling particles from the workpiece.

spark machining See **spark erosion**.

speciality materials Commercial materials made in very low quantities compared to general purpose materials, generally of high price but with specific effect or properties not shared with others.

specific Term often used to indicate that the property described relates to unit mass of the substance involved, eg specific entropy is entropy per kilogram.

specification (1) A detailed description, including dimensions and other quantities of the function, construction, materials quality of a manufactured article. Also applied to an engineering project. (2) A description by an applicant for a patent of the operation and purpose of the invention.

specific charge The charge/mass ratio of an elementary particle, eg the ratio e/m_e of the electronic charge to the rest mass of the electron $= 1.759 \times 10^{11}$ coulomb per kilogram.

specific gravity See **relative density**.

specific heat capacity The quantity of heat which unit mass of a substance requires to raise its temperature by one degree. This definition is true for any system of units, including SI. In all earlier systems a unit of heat was defined by putting the specific heat capacity (shc) of water equal to unity, but SI employs a single unit, the joule, for all forms of energy including heat, which makes the shc of water equal to 4.1868 kJ kg^{-1}K^{-1}. See **mechanical equivalent of heat**.

specific latent heat The heat which is required to change the state of unit mass of a substance from solid to liquid, or from liquid to gas, without change of temperature. Most substances have a latent heat of fusion and a latent heat of vaporization. The specific latent heat is the difference in enthalpies of the substance in its two states. Unit J kg^{-1}.

specific modulus A merit index, usually the ratio of the Young's modulus of a material to its density. Also *specific stiffness*, *stiffness-to-weight ratio*.

specific permeability Same as **relative permeability**. See **permeability**.

specific resistance See **resistivity**.

specific strength A merit index, usually the ratio of the tensile strength of a material to its density. Also *strength-to-weight ratio*.

specific surface The surface area of the particles in a unit mass of the powder determined under stated conditions. Also sometimes used to describe the surface area of the particles in unit volume of the powder.

specific viscosity Relative viscosity minus unity.

specific volume The reciprocal of the **density** of a material.

spectacle crown White crown glass of refractive index 1.523, used for ophthalmic purposes.

spectacle flint A glass of higher refractive index which is fused to spectacle crown in the manufacture of bifocal spectacle lenses. See **flint glass**.

Spectra TN for gel-spun polyethylene fibre (US). See **high-performance fibres** p. 158.

spectroscopy The practical side of the study of spectra, including the excitation of the spectrum, its visual or photographic observation, and the precise determination of wavelengths.

speculum Alloy of one part tin to two of copper, providing wide spectral reflection from a **grating** ruled on its highly polished surface.

spelter Zinc of about 97% purity, containing lead and other impurities.

spew pip Small knob extruded from holes in tyre mould during final step in tyre manufacture. Removed before sale. See **tyre technology** p. 325.

sp. gr. Abbrev. for **specific gravity**.

sphalerite Zinc sulphide which crystallizes in the cubic system as black or brown crystals with resinous to adamantine lustre. The commonest zinc mineral and ore, deposits with fluorite, galena etc. Also *blende, zinc blende*.

sphen-, spheno- Prefixes from Gk *sphen*, wedge.

sphericity The ratio of the surface area of a particle eg in a powder to the surface area of the sphere having the same volume as the particle.

spheroidizing Process of producing, by heat treatment, a structure in which the cementite in steel or cast iron is in a spheroidal form, giving improved ductility and machinability.

spherulite Common morphology of crystalline polymers. See **crystallization of polymers** p. 82.

spherulitic, spheroidal graphite cast iron Cast iron in which the graphite formed during solidification is induced to form as isolated spheroidal particles, as distinct from interconnected flakes. This greatly improves the **fracture toughness** and imparts a degree of ductility. See **ductile cast iron**.

spider lines Linear defects on plastic pipe analogous to weld lines in injection moulding. See **extrusion** p. 121.

spiders Special fixtures which hold mandrel in extruder barrel for manufacture of polymer pipe or tubing. See **extrusion** p. 121.

spiegeleisen Pig-iron containing 15–30% manganese and 4–5% carbon. Added to steel as a deoxidizing agent and to raise the manganese content of the steel. Also *spiegel*.

spin The intrinsic angular momentum of an electron, nucleus or elementary particle.

Spin is quantized in integral multiples of half the *Dirac unit* ie $\hbar/2h$, where $\hbar = h/2\pi$ is **Planck's constant**. Those particles with spin of odd multiples are *fermions* (eg electrons, proton, neutron) and those with even multiples are *bosons* (eg photon, phonon). It is the quantized electron spin angular momentum combined with the orbital angular momentum that gives rise to the fine structure in atomic line spectra.

spin-draw A process for producing manmade fibres in which drawing takes place as extrusion proceeds.

spin-draw texturing In the manufacture of synthetic fibres, the combination on one machine of the extrusion, drawing and texturing processes.

spinel A group of closely related oxide minerals, crystallizing in the cubic system.

spinneret, spinnerette A disk with fine holes through which the molten polymer or a solution of it is forced to form the continuous filaments of a man-made fibre in the spinning process.

spinning (1) Two distinct textile processes: (a) the drafting and twisting together of staple fibres to form a yarn; (b) the extrusion of continuous filaments of man-made fibres (or of silk) through spinnerets. (2) Forming a metal disk into a hollow shape by rotating it in a lathe over a former or chuck, and pressing a round-ended tool against it.

spinning jenny First spinning machine employing vertical spindles, invented by James Hargreaves, Blackburn, in Lancashire, UK in 1767.

spin on glass An organic silicate used to deposit planarizing layers over semiconductor devices during fabrication; it is subsequently transformed to silicon dioxide by heating. Abbrev. *SOG*.

spin wave The spins of the magnetic ions in ferro- and ferrimagnetic and in anti-ferromagnetic materials are oriented along preferred directions. Coherent deviations of the spins from these directions are propagated in space and time like waves. See **magnon**.

spin welding Method of joining similar thermoplastics by rotating parts against one another so that frictional heat causes fusion bond.

spiral flow test Measurement of polymer melt flow in a specially designed tool with a narrow, spiral cavity. The distance along the spiral the melt flows at a specific temperature gives some idea of molecular mass for moulding purposes. See **melt flow index**.

spiro polymers Chain molecules comprising cyclic structures linked together at their apices. High-temperature stable materials but still mainly laboratory curiosities.

splash marks Defective marks on surface

of injection moulded products caused by contamination of melt by volatiles, eg water. Recognized by silvery streaks radiating from gate. Also *silvering, mica marks*.

splay marks Marks on surface of moulded products, consisting of fine grooves concentric to gate, and caused by low pressurization towards end of cycle.

splicing Joining two pieces of yarn or rope together by intertwining the constituent strands manually or by machine so that the joint is nearly the same diameter as the original and certainly smaller than a knot.

sponge Porous metal formed by chemical reduction or decomposition process without fusion.

spontaneous process A process which occurs of its own accord, such as dissolution of a soluble salt or the melting of ice when warmed. The **free energy** for such processes is zero or negative in sign.

spot welding A process of welding in which pieces of sheet are placed together between two electrodes and a pulse of heavy current passed while pressure is applied. Produces a nugget of fused metal which acts rather like a rivet, pinning the sheets together. Usually made in seams or rows, but spacing cannot be too close as the welding current would shunt through the lower resistance path of adjacent welds, thus preventing the interface between the electrodes reaching its fusion temperature.

spray bonding A method of treating a web of fibres with a spray of adhesive so that a non-woven fabric is obtained.

spraying The process of coating the surface of an article by projecting on to it a spray of molten metal or other substance. The adhesion is mechanical and does not involve welding or diffusion. Metal spraying is often used for building up worn parts of old machinery. See **plasma spraying**.

spring A device capable of deflecting so as to store energy, used to absorb shock or as a source of power or to maintain pressure between contacting surfaces or to measure force.

spring-back Elastic recovery of a workpiece after completion of a bending operation.

spring constant, rate Ratio of load to deflection of a spring, measured eg in newtons per millimetre. See **stiffness**.

spruce Trees of the genus *Picea*, giving a valuable structural softwood. See **whitewood, Canadian spruce, sitka spruce**.

sprue Waste plastic formed during injection moulding processes, being the material setting in the main inlet passages of the mould. See **gate**.

spun silk Yarn made from silk waste which is spun in a manner very similar to the woollen systems.

spun yarn A yarn made from staple fibres twisted together.

sputtering The non-thermal removal of atoms from a surface under energetic ion bombardment. Sputtering yields are strongly dependent on the ion and its energy but rise steadily from thresholds at a few eV to several atoms per ion in the keV range. The ejected atoms deposit on any adjacent surfaces. In plasma sputtering the ions come directly from a low-pressure gas discharge. Sputtering is used for thin film deposition (eg metallization layers in semiconductor technology), for surface (discharge) cleaning and for surface analysis (eg **secondary ion mass spectrometry**). Different types of plasma source can be used, hence *DC-, RF-, magnetron-sputtering*. Sputtering is particularly useful for making films from refractory materials.

square A piece of square-section timber of side up to 6 in (150 mm).

SQUID Abbrev. for **Superconducting QUantum Interference Device**.

stabilizer (1) Important additive for polyvinyl chloride, which was formerly very unstable due to chain defects etc. There is a variety of chemicals which are thought to react with such defects, so inhibiting their action. Includes lead sulphate, carbonate; cadmium-barium salts; organo-tin compounds and, lately, thiol systems. (2) A negative catalyst (cf **retarder**). (3) A substance which makes a solution stable.

stable Systems not exhibiting sudden changes, esp. chemical or physical transformations.

stacking fault A fault which occurs in close-packed crystal systems appearing as a mismatch of one of the close-packed planes as the crystal grows or is deformed. It amounts to a half **dislocation**. For example, FCC crystals are built up in close packed layers with centres of each succeeding layer in different positions which can be represented by A, B and C, so the perfect stacking would be ABC/ABC/ABC/A ... and so on; occasionally a layer appears in a misplaced position and the order becomes ABC/AB/ABC/A ... . The plane where the C is missing is termed a stacking fault. Stacking faults are significant in the deformation and creep of metals with close-packed crystal structures, since they affect the ease with which a dislocation may cross slip from one slip plane to another. See **creep, deformation map** p. 87.

staggered In **conformational analysis**, this represents a conformation in which the substituents of one atom in a bond are situated as far as possible from those of the other.

staining A general method of increasing contrast in thin polymer and biomaterial films for transmission electron microscopy examination. OsO_4 is a common stain for styrene-butadiene-styrene copolymers since it reacts preferentially with double bonds in the butadiene parts of the structure, making the domains visible.

stainless steel Corrosion-resistant steels of a wide variety of compositions, but always containing a high percentage of chromium (12–25%) since the stainless property derives from chromium oxide film on the surface. Exhibit passivity and therefore highly resistant to corrosive attack by organic acids, weak mineral acids, atmospheric oxidation etc. Used for cutlery and domestic appliances, furnace parts, chemical plant equipment, stills, valves, turbine blades, ball-bearings etc. See **steels** p. 296.

standard (1) Established unit of measurement, or reference instrument or component, suitable for use in calibration of other instruments. Basic standards are those possessed or laid down by national laboratories or institutes (NPL, BSI etc). (2) A detailed description, including dimensions and other quantities, of the function, construction, materials, and quality of a manufactured article or an engineering project. (3) Any specification for a material or process or test method etc as developed by an official body (eg BS, DIN, ASTM) or by private treaty. (4) Volume measure for timber, particularly softwoods, equal to 165 ft^3. Also *Petersburg standard*. In the US, 1 standard equals 16⅔ ft^3. Cf **board foot**.

standard electrode potential The potential of a chemical element dipping into a solution of its ions at unit activity, referred to that of hydrogen under a pressure of one atmosphere as zero.

standard heat of formation Enthalpy change when a compound is formed from its constituent elements under standard conditions (temperature 298.1 K, pressure one atmosphere or 101.325 kN m^{-2}). By convention, the enthalpies of formation of the elements are defined as zero. Standard free energy of formation is defined in a similar way. See **Ellingham series**.

standard knot A knot in timber which is 1.5 in (40 mm) or less in diameter.

Standard Malaysian Rubber A strict specification for various grades of **natural rubber**. Abbrev. *SMR*.

standard specification A specification incorporating widely accepted quantities or features, themselves standards, to ensure interchangeability, quality and reliability for least cost. Usually drawn up by national Standards Institutions.

standard state The condition of elements and compounds under standardized conditions of temperature (298.1 K) and external pressure (one atmosphere, 101.325 kN m^{-2}).

standard temperature and pressure See **s.t.p.**

staple fibre Natural or man-made fibres that are comparatively short in length eg 10–500 mm.

starch *Amylum*, $(C_6H_{10}O_5)_x$; polysaccharide found in all assimilating (green) plants. A white hygroscopic powder which can be hydrolyzed to dextrin and finally to D-glucose. Diastase converts starch into maltose.

stark rubber Natural rubber which has crystallized over several years by storage at low temperature. It is very stiff due to a high **degree of crystallinity**, and is difficult to process.

star polymer Macromolecule created by multifunctional chain extending agent reacting with growing and active chain ends. Usually formed by anionic polymerization.

starting sheet A sheet of pure metal used as the initial cathode on which the metal being refined is deposited during electrolytic refining.

static electrification Tendency of insulators to accumulate an electric charge, esp. when rubbed by another material. High voltages can be generated, leading to spark discharge, which may be hazardous where flammable gases are present such as ether, methane, petrol etc. See **triboelectric series**.

static fatigue The phenomenon of a material failing at a smaller load than that required to cause short term failure, after a period of constant loading by the smaller load; the load necessary to produce static fatigue decreases with increasing time under load. In brittle materials, static fatigue is due to the slow growth of sub-critical cracks to a length at which they will propagate catastrophically; in ductile and/or viscoelastic materials it is due to the progress of plastic or viscoelastic deformation (ie creep) to the point where catastrophic yielding can occur. Also *creep rupture*, esp. in relation to polymeric materials. Assisted by **environmental stress cracking**.

static friction See **friction**.

statics That branch of applied mathematics which studies the way in which forces combine with each other usually so as to produce equilibrium. Until the early part of the 20th century the term also embraced the study of gravitational attractions, but this is now normally regarded as a separate subject.

statistical chain Model of polymer chain used as basis for **kinetic theory of elasticity**.

The links in the chain are freely rotating, unlike a real chain, where rotational isomers occur due to steric hindrance. The chains are free to pass through one another, while real chains experience the **excluded volume** effect. Nevertheless, the model accounts for many elastomeric properties. See **elastomers** p. 106.

statistical mechanics Theoretical predictions of the behaviour of a macroscopic system by applying statistical laws to the behaviour of component particles. *Quantum mechanics* is an extension of classical statistical mechanics introducing the concepts of the quantum theory, esp. the **Pauli exclusion principle**. *Wave mechanics* is a further extension based on the **Schrödinger equation**, and the concept of particle waves.

steatite, soapstone A coarse, massive or granular variety of talc, greasy to the touch. On account of its softness it is readily carved into ornamental objects.

steel-making The process of making steel from pig-iron, with or without admixture with steel scrap

steels See panel on p. 296.

Stefan–Boltzmann law Total radiated energy from a black body per unit area per unit time is proportional to the fourth power of its absolute temperature, ie $E = \sigma T^4$ where σ (Stefan–Boltzmann constant) is equal to 5.6696×10^{-8} W m^{-2} K^{-4}.

Stellite TN for a series of alloys with cobalt, chromium, tungsten and molybdenum in various proportions. The range is chromium 10–40%, cobalt 35–80%, tungsten 0–25% and molybdenum 0–10%. Very hard. Used for cutting-tools and as hard facings protecting surfaces subjected to heavy wear.

Stelvetite TN for polyvinyl chloride-coated steel sheet used for a variety of purposes including cabinets, wall sections, domestic equipment.

STEM Abbrev. for **Scanning-Transmission Electron Microscope** (or microscopy)

step polymerization See panel on p. 298.

stere A stacked cubic metre of timber, which equals 35.3 stacked cubic feet.

stereoblock polymer Homopolymer possessing regular sequences of different stereoisomers.

stereochemistry Branch of chemistry dealing with arrangement in space of the atoms within a molecule. See **stereoregular polymers** p. 299.

stereo-isomerism The existence of different substances whose molecules possess an identical connectivity but different arrangements of their atoms in space. Also *allo-isomerism*. See **geometrical isomerism**, **optical isomerism**.

stereology Quantitative method of relating two-dimensional analysis of structure of materials (eg, as seen in an optical microscope) to their structure as it exists in three dimensions. Based on methods of mathematical statistics and often used in the **QuantiMet** analysis.

stereoregular polymers See panel on p. 299.

stereotype alloys Lead-based alloys with 5–10% tin and 10–15% antimony, used in certain printing processes.

steric hindrance The retarding influence by virtue of the size of neighbouring groups on reactions in organic molecules. In polymers, interference between adjacent atoms or side groups in repeat units or short chain lengths, giving rise to different energy states. See **chain flexibility**.

Sterling board TN for processed wood board and used for flooring, panelling etc.

sterling silver A silver alloy with not more than 7.5% base metal. Legal requirement that articles described as silver shall contain minimum 92.5% silver content.

sthene Unit of force in the metre-tonne-second system, equivalent to 10^3 newtons.

sticking probability Probability of an incident particle (atom, molecule, radical etc) being adsorbed on reaching a surface.

stiction *Static frICTION*. See **friction**.

stiffness The ability of a material to resist elastic or viscoelastic deformation. Formally, it is the reciprocal of **compliance** and equals force/deflection.

stiffness-to-weight ratio See **merit index**.

stinkwood A strong, hard, but non-durable timber from the S African hardwood tree *Ocotea bullata* (Cape walnut, olive), known for its beautiful and distinctive figuring. Its density is about 800 kg m^{-3}. It has an unpleasant smell when green. Used in high class furniture and cabinetmaking, joinery and as a decorative veneer.

stitch (1) In sewing: fastening with thread carried through the fabric by a needle. A variety of stitches (eg buttonhole, chain, lock) are used for different purposes. (2) In knitting: a single loop.

stitch-bonded fibre A web of fibres stitched together to form a non-woven fabric.

stitch density The number of stitches per unit area in a knitted fabric.

stock chest A vessel, usually of upright cylindrical form, generally tile lined, to contain stock (ie fibres in suspension) from which the paper machine draws its supply. Fitted with a propeller or agitator to ensure a uniform suspension and even mixing of additives such as filler, size, colour etc, which are usually added at this point.

steels

A very important versatile group of engineering materials based on the iron carbon system, which may contain up to 2% carbon although usually below 1% and accompanied by other elements in small amounts. When carbon is the principal alloying component they are referred to as *plain carbon steels* but, even so, these normally contain up to 0.8% manganese and 0.3% silicon together with 0.05% maximum of sulphur and phosphorus as impurities. *Low alloy steels* contain in addition to carbon, up to 5% in total of manganese, nickel, chromium, vanadium and molybdenum, at least two and often three or four of these elements being present in combination. Larger quantities of some of these elements can be added to impart specific properties in amounts significantly above 5%; such materials are referred to as *high alloy steels*.

Carbon forms an *interstitial solid solution* with iron and dramatic differences in properties can be achieved by controlling the crystal structure by heat treatment. This arises from the allotropy of iron as the equilibrium solubility of carbon is only 0.03% in the *body-centred cubic* alpha form, stable up to 803°C, whereas it is a maximum of 2% in the *face-centred cubic* gamma form. With more than 2% carbon, some **eutectic** appears during solidification and materials which undergo this reaction constitute the family of *cast irons*, the useful range of which extends to 4.3% carbon. Steels may be *cast* or *wrought*, although some in the latter group are even nowadays mistakenly described as 'cast steel', a description resulting from the long obsolete method of making high-quality steels by the crucible process.

Steels are conveniently divided into three broad categories, depending on their carbon content which determines their potential response to heat treatment:

Mild steels contain up to 0.25% carbon and essentially can only be strengthened by *cold work*, ie by **work hardening**, although some of the modern compositions, eg *microalloyed* or *high strength low alloy* (HSLA), develop higher yield and tensile strengths by **precipitation hardening**.

Medium carbon steels, 0.3–0.7% carbon, are heat treated to produce a wide range of engineering properties. Heat treatment usually consists of two stages, the first a *hardening* operation which involves heating to a temperature around 860°C at which the steel is wholly *austenitic* and then quenching into oil or water to achieve a *martensitic* condition which is hard and extremely brittle. (These characteristics derive from the carbon atoms being trapped in interstitial lattice positions during the shear transformation.) After hardening, the steel is reheated to a lower temperature which allows some diffusion of carbon and relief of the lattice strain, a process called *tempering* or *drawing*. Mechanical properties change during this process as indicated schematically in Fig. 1, and the tempering temperature is chosen to produce the desired combination for that particular carbon content.

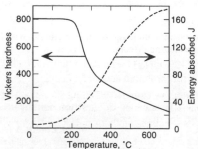

Fig. 1. **The effect of increasing tempering temperature on the hardness and impact energy absorbed (toughness) of a medium (0.4%) carbon steel**.

continued on next page

Both the above categories of steel with less than 0.8% carbon are termed *hypoeutectoid*, distinguishing them from the *hypereutectoid* or **high-carbon steels**. High carbon steels respond to heat treatment like medium carbon steels but they also contain free *cementite* (Fe_3C) which tends to reduce their ductility and shock resistance. However, the cementite present in their microstructures makes them particularly suitable for retaining the sharp edge required for metal forming, machining and cutting tools.

In addition to a multiplicity of trade names, many types of steel have traditionally been given names based on their application or form:

Ball bearing steel, 1% chromium, 1% carbon capable of high hardness and wear resistance.

Black bar, steel in the form of hot rolled bar which still retains its mill scale coating.

Bright bar, steel with the scale removed by acid pickling and subsequent cold drawing.

Chrome (and *chrome manganese* etc), colloquial terms for chromium containing steels.

Cryogenic steels, suitable for service at very low temperatures, such as those containing about 9% nickel.

Die (or *hot die*) *steels*, those with medium or high carbon together with strong carbide forming elements (chromium, tungsten, molybdenum, vanadium) to impart hardness and wear resistance at moderately elevated temperatures.

Free machining steels, containing manganese sulphide particles or lead additions to act as chip breakers during machining.

Heat resistant steels, containing large amounts (30%) of chromium to resist oxidation at high temperatures in furnace parts etc.

High speed steels, highly alloyed, medium or high carbon steels containing 5–20% tungsten, cobalt, chromium and vanadium and sometimes other elements, which retain their hardness at red heat and are therefore suitable for machining at the high cutting speeds at which frictional heating occurs.

Magnet steels, high carbon with tungsten, cobalt, chromium and other elements in amounts to impart high hysteresis and coercivity.

Manganese (or *Hadfield's*) *steel*, 13% manganese and 1.2% carbon which has a very high resistance to dry, abrasive wear.

Maraging steels, 18% nickel plus other elements which promote intense precipitation hardening effects.

Nickel steels, 1–9% nickel for hardenability and low-temperature toughness, usually with chromium and molybdenum.

Silicon iron, very low carbon containing 4% silicon with a low magnetic hysteresis, making it suitable for transformer cores etc.

Silver steel, heat-treatable medium carbon steel with a ground or bright surface.

Stainless steels, non-tarnishing and corrosion resistant family, which may be ferritic, austenitic or heat treatable, containing a minimum of 12% chromium, along with nickel, molybdenum and other elements to promote specific mechanical properties.

step polymerization

Materials like nylon-66, polyethylene terephthalate and polyurethane rubber are manufactured by the step-growth mechanism, which is quite different to chain-growth where only activated chain ends grow into monomer. Well-known reactions like amidation and esterification are exploited in step-growth poly-merization, and to form linear chains, monomers must possess two functional groups. Thus the monomer of polyethylene terephthalate is the compound:

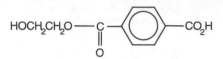

Fig. 1. **Polyethylene terephthalate monomer**.

which possesses an hydroxyl and carboxyl group at either end. When catalyzed, they will react together to produce longer and longer chains:

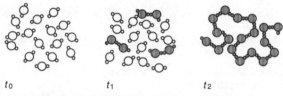

t_0 t_1 t_2

Fig. 2. **Polymerization of polyethylene terephthalate**.

The reaction is usually carried out in the melt without solvent and the water produced must be removed from the system to achieve high molecular masses and prevent hydrolysis, which results in depolymerization. Compared to chain growth, reaction rates are relatively low and high mass polymer is not produced until the very final stages.

With nylon 6,6 *two* different monomers are used, one possessing two amine groups (hexamethylene diamine) and the other two carboxyl groups (adipic acid). The concentration of each must be identical to make high molecular mass, and is achieved by using nylon salt, where the two components co-crys-tallize and must therefore be present in equimolar proportions.

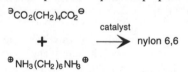

Fig. 3. **Polymerization of nylon 6,6**.

An alternative material is nylon 6, which is formed in a ring-opening polymer-ization from caprolactam:

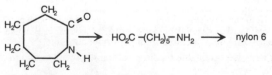

Fig. 4. **Polymerization of nylon 6**.

Step growth polymerization is also exploited in thermosetting reactions by employing molecules with more than two reactive functional groups. Also *condensation polymerization* since water is often produced.

stereoregular polymers

In vinyl polymers like polypropylene (PP) every other carbon atom along the main chain is *asymmetric*. Since such *chiral* atoms can exist in L- or D- forms (the one the mirror image or *enantiomer* of the other), three different types of polypropylene can exist.

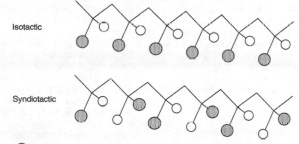

Isotactic

Syndiotactic

 CH$_3$ ◯ H

In *isotactic* polypropylene, all methyl pendant side groups possess the L- (or D-) configuration while the groups regularly alternate in *syndiotactic* polypropylene. Where no control of the chiral centre is possible, the placements are entirely random, producing the *atactic* polymer:

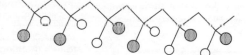

Atactic

Stereoregular polymers (iso- and syndiotactic) are produced by solid-state catalysts (eg **Ziegler–Natta catalysts**), the chain regularity created by stereospecific sites on the catalyst surface. Such polymers are usually highly crystalline. By contrast atactic polymers are totally non-crystalline and amorphous. Although not possessing a chiral centre, polyethylene can also be prepared using similar catalysts to give high-density polyethylene. Ultra-high molecular mass polymer (UHMPE) can be produced and, unlike low-density, medium-density or linear low-density polyethylenes, it possesses no chain branches. Head-to-head irregularity in vinyl polymers is also strictly limited in co-ordination polymerization. Other stereospecific polymers made commercially include cis-polyisoprene or isoprene rubber (the synthetic equivalent of natural rubber) and cis-polybutadiene, a highly resilient rubber widely used in car tyre sidewalls.

stoichiometry Determination of exact proportion of elements to make pure chemical compounds. Non-stoichiometric compouds or salts have non-integral numbers of atoms in their formulae. See **vacancies**.

stokes The CGS unit of kinematic viscosity (10^{-4} m^2 s^{-1}). Abbrev. *St*.

Stokes' law (1) The resisting force offered by a fluid of dynamic viscosity γ to a sphere of radius r, moving through it at steady velocity v is given by

$$F = 6\pi\eta\, r\, v$$

whence it can be shown that the **terminal velocity** of a sphere of density ρ, falling under gravitational acceleration g through fluid of density ρ_0 is given by

$$v = \frac{2gr^2}{9\eta}\,(\rho - \rho_0).$$

Applies only for viscous flow with **Reynolds number** less than 0.2. (2) Incident radiation is at a higher frequency and shorter wavelength than the re-radiation emitted by an absorber of that incident radiation.

stone See **inclusion**.

Stoner–Wohlfarth model Model describing the magnetization curves of an aggregation of single domain particles.

stoneware A form of ceramic ware which is dense, impermeable and hard enough to resist scratching with a steel point, but differs from **porcelain** in being more opaque and having a smaller proportion of glassy phase. Cf **bone china**, **earthenware**, **faience**, **terra-cotta**.

storage battery See **accumulator**.

storage modulus See **complex modulus**.

s.t.p., STP *Standard Temperature and Pressure*; a temperature of 0°C and a pressure of 101 325 N m^{-2}.

strain When a material is distorted by forces acting on it, it is said to be in a state of strain, or strained. Strain is the ratio

$$\frac{\text{change in dimension}}{\text{original dimension}}$$

and thus has no units. The main types of strain are **direct** (*tensile* or *compressive*) **strain**:

$$\frac{\text{elongation } or \text{ contraction}}{\text{original length}} \;;$$

shear strain:

$$\frac{\text{deflection in direction of shear force}}{\text{distance between shear forces}} \;;$$

volumetric (or *bulk*) **strain**:

$$\frac{\text{change of volume}}{\text{original volume}} \;.$$

These definitions, known as *engineering strains* are strictly only applicable to infinitesimal strains, and cannot be added at large deformations, where *true* (*logarithmic* or *natural*) *strain* is more convenient. For direct strains, true strain is:

$$\log_e \frac{\text{deformed length}}{\text{original length}} =$$

$$\log_e (1 + \text{engineering strain})$$

which is additive.
For the large elastic deformations of elastomers, the *extension* (or *elongation) ratio*, or

$$\frac{\text{deformed length}}{\text{original length}}$$

is used.

strain age embrittlement Loss of ductility and fracture toughness resulting from strain ageing processes in cold-worked metals.

strain ageing An increase in metal strength and hardness that proceeds with time, after cold-working. It takes place slowly at ambient temperature and is accelerated by heating. It is most pronounced in iron and steel, but also occurs in other metals. May result in strain age embrittlement.

strain crystallization Crystallization of elastomers when strained to high extension ratios ($\lambda > 5$). The crystallites are aligned along the strain axis and increase the breaking strength of the rubber. On retraction, the crystallites melt so the effect is reversible. The heat given out in stretching is partly due to the **Joule–Thomson effect**, and partly to the heat of crystallization. The **hot-stretching** of fibres exploits strain-crystallization.

strain disk A glass disk of calibrated birefringence, used as a comparative measure of the degree of annealing of glass.

strain energy The energy stored in an elastically deformed body equal to the work done on the body in deforming it (ie the area under the elastic portion of the force-deflection curve).

strain energy release rate The elastic strain energy required for unit area of crack propagation. At its critical value, it equals the **toughness**, G_c of the material. Also **crack driving force**.

strain gauge Metal or semiconductor filament on a backing sheet by which it can be attached to a body to be subjected to strain, so that the filament is correspondingly strained. The strain alters the electrical properties of the filament, this alteration forming the basis of measurement.

strain-hardening The increase in resistance to plastic deformation (ie in hardness) produced by deformation. See **cold-working**, **work-hardening**.

strain point One of the reference temperatures in **glassmaking** p. 144.

strain-whitening Yielding effect noticed in tough polymers. Also *stress-whitening*. See **rubber toughening** p. 270.

strass A very dense glass of high refractive power; used largely in making artificial jewellery. Also *paste*.

strength measures See panel on p. 301.

strength-to-weight ratio See **merit index**.

stress (1) The force per unit area acting on a material and tending to change its dimensions, ie cause a **strain**. The stress in the material is the ratio of force applied, to the area of material resisting the force, ie

$$\frac{\text{force}}{\text{area}} \;.$$

True stress is evaluated in terms of the actual area, *engineering stress* in terms of the original area before the force was applied. The two main types of stress are *direct* or *normal* (ie tensile or compressive) stress (usual symbol σ) and *shear* stress (usual symbol τ). The **SI units** are N m^{-2} (= Pa) and its multiples, kN m^{-2}, MN m^{-2} (= N mm^{-2}), GN m^{-2},

strength measures

There is a large number of quantities used to characterize a material's resistance to failure. Few have as well-defined properties as eg **Young's modulus** or electrical resistivity; many depend on details of how they are measured, specimen geometry and, especially for brittle materials, on the flaw size distribution in the surface.

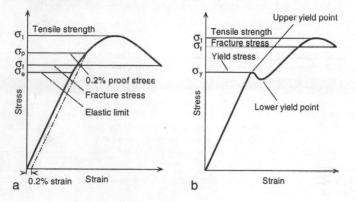

Fig. 1. **Schematic tensile stress-strain curves**. Without a yield point in (a) and with one in (b).

One group of these quantities is based around the *tensile test*, schematic results from which are sketched in Fig. 1. This group includes the following.

Tensile strength, the maximum tensile force in the test divided by the original cross-sectional area (also *ultimate tensile stress* (uts), *breaking stress* and *modulus of rupture* (now obsolete), *disruptive strength*, *tenacity*, but see below). For brittle materials, eg ceramics, whose strengths are inherently variable, *Weibull's modulus* provides a measure of the scatter in strength values.

Fracture stress, the tensile load at failure divided by the final cross-sectional area (also fracture strength); in brittle materials, eg glass, equal to tensile strength.

Elastic limit, the maximum stress (or strain) for fully recoverable, linear elastic behaviour (not applicable to viscoelastic or elastomeric materials).

Yield stress, the stress at the onset of plastic deformation (also *yield strength*) determined from the *yield point* (Fig. 1b) if it occurs, otherwise from the stress required for a defined amount of plastic strain known as the *proof stress* (Fig. 1a shows the 0.2% proof stress).

Work of fracture , the total area under the force-deflection curve.

Tenacity, tensile strength of fibres, expressed in newtons per **tex** (also *specific tensile strength*), $1 \text{ N tex}^{-1} = 10^{-6}/\rho \text{ N m}^{-2}$, where ρ is the density in kg m^{-3}. A variant, used for ropes and paper, is *breaking length*, the maximum length that will just support its own weight, hanging vertically from one end.

Compressive strength and ***shear strength***, are the equivalents of tensile strength in their respective loading modes. ***Flexural strength*** (also *bending strength*, *cross-rupture strength*) is the maximum tensile stress in a bend test specimen before failure. Other terms apply to specific types of material: ***cube strength*** is the stress required to crush a 100 mm cube of concrete; ***green strength*** and ***crushing strength*** are the compressive strengths of *green compacts* and *powders* or *agglomerates*; ***tear strength*** is the resistance to tearing

continued on next page

(eg in a *trouser-leg specimen*, Fig. 2a) of elastomers, textiles and paper, the last also characterized by *wet strength* and *folding strength*: **peel strength** is determined from the force required to peel a bonded layer away from its substrate (Fig. 2b). See **fatigue** and **impact testing** for other related terms.

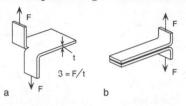

Fig. 2. **Specimens for testing tear strength (a) and peel strength (b)**.

Thermal shock resistance provides an index of whether the tensile strength of the material will be exceeded due to temperature change under prescribed heat transfer conditions.

Crack propagation

There is a group of measures of a material's resistance to the propagation of a crack; strictly, all materials that fail by crack propagation rather than by gross yielding are brittle, and are differentiated by having higher or lower values of *toughness* or of *fracture toughness*. The first is based on a fracture criterion that requires a critical quantity of *elastic strain energy* to be released from the material per increment of crack growth, as enshrined in the **Griffith equation**, originally known as *fracture energy*, it is now called the *toughness*, *crack driving force* or *critical strain energy release rate*, (G_C). When applied to elastomers, it is sometimes called the *tearing energy* (T) and has units $J\ m^{-2}$. Fracture toughness is based on a critical level of stress being necessary at the crack tip for crack growth, achieved through the crack concentrating the applied stress (see **stress concentration** p. 303). The mechanics leads to *fracture toughness* or *critical stress intensity factor* (K_C), which has units $N\ m^{-3/2}$. Depending on which of the three fundamental *fracture modes* apply as shown in Fig. 3, G_C and K_C are designated with subscripts I, II or III, eg G_{IC} or K_{IC} for the crack-opening mode.

Note that brittleness and toughness are sometimes used qualitatively to describe the response of esp. metals to an impact test: low impact energy = brittle, high impact energy = tough.

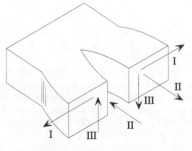

Fig. 3. **The three fundamental fracture modes**.
I, opening; II, sliding; III, tearing.

stress concentration

An abrupt, local increase of otherwise uniform stress created by some geometric configuration at the surface or within the body of a component. It may vary in size from a fine score mark or pit, through inclusions or voids within the microstructure to a major change in section of a large engineering structure. The sharper the change the greater the stress concentration, and the more brittle the material the greater its damaging effect; a fine score by a glass cutter guides the fracture of a large sheet of glass, whereas the same mark on a steel plate would have virtually no effect.

Fig. 1, below, shows diagrammatically the stress concentrating effect of a notch at the surface of a section subject to uniform tensile stress. Each line represents a given force and their density (ie lines per unit area) is greatest at the tip of the notch. This is where stress induced failures are most likely to start.

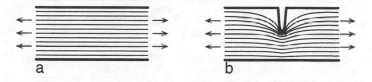

Fig. 1. (a) **The imaginary lines of force within a bar of uniform section.** (b) **The redirection of lines of force caused by a change in section in the bar.**

When deformation is elastic, the stress at the tip of the notch (σ_m compared with the initial or normal stress σ_o is given by

$$\sigma_m = \sigma_o (1 + 2\sqrt{D\,r^{-1}})$$

where D is the depth of the notch and r its tip radius, as illustrated in Fig. 2. Thus a circular hole has a stress concentration of 3× corresponding to two semicircular notches arranged side by side.

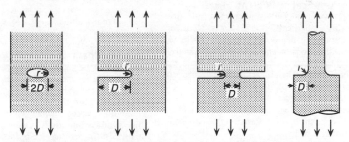

Fig. 2. **The depth of notch (D) and the radius (r) for various changes in section on a bar.**

Brittle materials are *notch sensitive* because they can exhibit very high stress concentrations but *ductile* materials blunt the tip of the notch by plastically deforming, thereby increasing the radius r and rendering them *notch insensitive*. It is this which imparts *toughness*. See **strength measures** p. 301.

Stress concentration can also occur independently of geometric features if materials of dissimilar elastic moduli are in contact, as in **composites**. When elastic strains are the same in both materials stresses must be of different magnitudes in each. This leads to localized stress concentration at the interface.

but lbf in^{-2}, tonf in^{-2} and bar were frequently used. (2) To stress may also mean *to calculate stresses*; thus a structure has been stressed when the distribution and levels of stress in it have been determined by calculation.

stress concentration See panel on p. 303.

stress concentration factor The factor by which the stress is increased at a stress concentrating feature in a loaded body. Abbrev. *scf*.

stress corrosion Postulated enhanced rate of chemical attack on a material when it is subjected to an applied stress. Advanced as a mechanism for **environmental stress cracking** and **static fatigue** by which crack growth is accelerated due to the increased reaction rate in the highly stressed neighbourhood of a crack tip.

stress fracture Medical term for **fatigue** fracture of bone.

stress intensity factor A measure of the increase in stress which occurs at the tip of a loaded crack in a material. Symbol K, with various subscripts to show loading conditions etc. Units N m$^{-3/2}$.The critical value for crack propagation, K_C, is a materials property, the **fracture toughness**. See **fracture mode**, **strength measures** p. 301.

stress-number curve *S/N* curve. A curve obtained in fatigue tests by subjecting a series of specimens of a given material to different ranges of stress and plotting the range of stress (S) against the number of cycles required to produce failure (N). See **fatigue strength** p. 124.

stress-optical coefficient Material constant, symbol C, which relates birefringence Δn to applied stress, σ: $\Delta n = C\sigma$. Used for determining degree of frozen-in strain in polymer products and in photoelastic analysis. Highest values of C are possessed by some polyurethanes, polycarbonate etc, so most useful for **photoelasticity**.

stress relaxation The decrease in stress with time in a viscoelastic material held at constant strain (or deformation). See **viscoelasticity** p. 332.

stress-relaxation modulus Modulus obtained from a stress-relaxation experiment, where sample is held at constant strain and the decaying stress recorded. Given by the equation:

$$E_r(t) = \sigma(t)\,\varepsilon^{-1}\,.$$

Commonly used for viscoelastic materials such as polymers, where moduli can be used in stress calculations provided time scale (t) is taken into account.

stress relief Annealing metals by heating to a temperature below that liable to alter the crystalline structure and with the object of reducing or eliminating any harmful residual stresses arising from other processes. See **annealing** p. 13.

stress-strain curve A curve similar to a load-extension curve, except that the load is divided by the original cross-sectional area of the test piece and expressed in units of stress, while the extension is divided by the length over which it is measured and expressed as a ratio. Depending on the material, the curve can reflect elastic or viscoelastic behaviour, brittle fracture, high elasticity, plastic flow, work-hardening and residual strain (permanent set). See **strength measures** p. 301.

stretcher strains See **Lüders lines**.

stretch forming Process for forming large sheets of thin metal into symmetrical shapes by gripping the sheet edges in horizontally sliding stretcher jaws and moving a forming punch, without a die, vertically between them against the sheet.

stria, striation A faint ridge or furrow; a streak; a linear mark.

strict liability Direct responsibility to user or consumer by manufacturer for defective product. See **product liability**.

striking In electroplating, the flash deposition of a very thin layer of one metal to facilitate subsequent plating with another.

strontium Metallic element, symbol Sr. Silvery-white in colour, it is found naturally in *celestine* and in *strontianite*; it also occurs in mineral springs. Similar chemical qualities to calcium. Compounds give crimson colour to flame and are used in fireworks. The radioactive isotope, ^{90}Sr, is produced in the fission of uranium and has a long life, hence its presence in 'fall-out' after a nuclear explosion. See Appendix 1, 2.

structural foam Products whose interior is foamed to decrease overall weight, and hence improve bending stiffness-to-weight ratio. Abbrev. *SF*.

structural foam moulding Injection moulding process where a blowing agent like **azodicarbonamide** is added to polymer so that a foamed interior is formed in final product.

structural formula A representation of the chemical structure of a substance which shows not only its composition but also its **connectivity** and the order of the bonds connecting the atoms.

structure (1) Assembly of atoms and molecules in a material. See **primary structure**. Also for larger assemblies; see **secondary structure**, **tertiary structure**, **microstructure**, **macrostructure**. (2) Assembly of struts in a space frame, monocoque etc, any collection of inter-connected parts forming a

structure of wood

There is little in the external appearance of a tree to suggest that wood is not only a *cellular composite material*, but is also highly **anisotropic**. Essentially, it's a foam with anisotropic cells, whose walls are themselves an anisotropic composite. Wood is itself used to make a range of composites as well as paper. Being based on cellulose, the cell wall structure of wood shares common features with cotton, but there are significant differences.

Since it is made up of a very large number of cells, wood has a more complex structure than single-celled cotton, and at different levels of scale. Fig. 1 shows some of the main structural features for a typical softwood, which has a simpler structure than a hardwood. This is apparent at scales between a complete tree (ca 50 m) and the unit cell in crystalline cellulose ($\approx$ 1 nm).

The bulk of a tree is made up of dead cells, the live and growing region being confined to the cambium immediately beneath the bark. Wood, of course, displays the well-known, annual pattern of growth rings (in cotton they're daily ones). The cell wall structure contains about 40–50% by weight of cellulose and 20–25% of hemicelluloses. The remaining material is largely composed of lignin, a complex, tarry, aromatic compound. The hemicelluloses mainly act as the cement for the cellulose, but the bulk of the **lignin** is found deposited as a final layer around the outside of the cell wall before the cell dies. The high stiffness of wood relative to other cellulosics such as cotton is due to the higher content of hemicelluloses and the presence of lignin.

A horizontal slice through the trunk reveals the annual growth rings, whilst removing a pie-slice from this and examining it at higher magnification shows that the main structural elements are an array of hollow, vertical cells, known as *tracheids*, and these are connected by a much smaller number of horizontal, radial cells known as *parenchyma*, located in the *rays*. Tracheids have **aspect ratios** of the order of 100, and a closer look at their cell walls shows that these have a layered structure.

The outermost primary (P) layer, is the first one to form as the cell is growing. In this layer the microfibrils of cellulose are arranged in an irregular network. When a cell has stopped growing, wall thickening starts and the three secondary layers are formed. In the outer (S_1) layer the microfibrils lie almost perpendicularly to the long axis of the cell. The middle layer (S_2) is very much thicker and has the microfibrils running at about 20° to the long axis. Finally, the inner layer (S_3) has the microfibrils oriented almost at right-angles to the long axis again. In each of these layers the cellulose microfibrils are embedded in a mainly hemicellulose matrix.

Finally, the individual cells are joined together with an amorphous, lignin-rich material which contains practically no cellulose. The relative thicknesses of the different components of the cell wall are approximately:

primary	P	$\approx 1\%$
secondary	S_1	10–20%
	S_2	70–90%
	S_3	$\approx 5\%$

The microfibrils themselves, in a similar way to those in cotton, are composd of more or less parallel cellulose molecules, which lie alternately in crystalline and amorphous regions of the microfibrils. In the crystalline regions the cellulose molecules form a regular array held together by **hydrogen bonds**, with the **unit cell** dimensions some 2×10^{-11} times the size of the tree from which they came.

The parenchyma in softwoods constitute some 5–10% of the cells and are used for food storage. Hardwoods have a similar proportion of parenchyma, but

continued on next page

structure of wood (contd)

very few tracheids. Instead, the tracheids' dual role in softwoods of conduction and support is performed by two different types of cell, called *vessels* and *fibres*, respectively. This is the main difference in structure between hardwoods and softwoods. Although some hardwoods, such as oak, are undoubtedly hard, others, such as balsa are softer and less dense than any softwood.

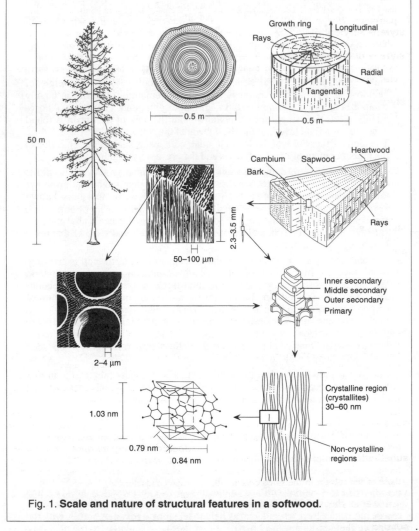

Fig. 1. **Scale and nature of structural features in a softwood**.

whole.

structure of wood See panel on p. 305.

structure-property relation Any clear dependence of the physical or chemical properties of a material on atomic, molecular stucture or microstructure. Thus allowing control of mechanical properties of metals and ceramics by grain size (see **Hall–Petch equation**) or the strength of polymers by the control of molecular mass. See **molecular mass distribution** p. 212.

strut Any light structural member or long column which sustains an axial compressive load. Failure occurs by bending or buckling before the material reaches its ultimate compressive stress. See **column**.

stuff Paper stock or pulp ready for making into paper.

S-twist See **twist direction**.

stylo- Prefix from Gk *stylos*, pillar.

styrene Phenylethene, $C_6H_5.CH:CH_2$. Intermediate for PS, ABS, HIPS etc. It is a colourless aromatic liquid, bp 145°C, widely used in polyester thermoset resins. See **copolymer** p. 75.

styrene acrylonitrile A copolymer. Abbrev. *SAN*. See **copolymer** p. 75.

styrene-butadiene rubber The main synthetic rubber used in tyre treads. See **elastomers** p. 106, **tyre technology** p. 325. Formerly GRS.

styrene-butadiene-styrene Abbrev. *SBS*. See **copolymer** p. 75.

styrene resins Polymers and copolymers based on styrene monomer. The parent homopolymer is a transparent, brittle and rigid material with a T_g of nearly 100°C. The backbone chain is atactic so the solid is amorphous. Copolymers like ABS, high-impact polystyrene and SAN are now more important for their improved properties. See **copolymer** p. 75.

styrol resins Obsolescent name for **styrene resins**.

sub- Prefix from L. *sub*, under, used in the following senses: (1) deviating slightly from, eg *subtypical*, not quite typical; (2) below, eg *subvertebral*, below the vertebral column; (3) somewhat, eg *subspatulate*, somewhat spatulate, (4) almost, eg *subthoracic*, almost thoracic in position.

substance (1) Kind of matter, with characteristic properties, and generally with a definite composition independent of its origin. (2) The thickness of flat glass or paper, usually expressed as mass per unit area. Largely superseded by millimetres for glass. For paper, the preferred term is **grammage** or **basis weight**.

substitution The replacement of one group, esp. hydrogen by another, eg halogen, alkyl, hydroxyl etc.

substitutional solid solution Type of solution in which atoms of solute(s) replace those of the solvent in the same crystal lattice, eg nickel and copper or the alpha solid solutions of aluminium, tin and/or zinc in copper.

substrate The surface onto which thin or thick films are deposited. Any object on which layers are to be deposited or fashioned, such as the wafer of semiconductor onto and into which electronic circuitry is to be integrated.

succinic acid *Butandioic acid*. $HOOC.CH_2$. $CH_2.COOH$, dibasic acid; mp

185°C, bp 235°C, with partial decomposition into its anhydride. It occurs in the juice of sugar cane, in the castor-oil plant and in various animal tissues, where it plays an important part in metabolism. Succinic anhydride is used in the manufacture of alkyd thermoset resins.

suction box A shallow compartment extending across a papermaking machine connected to a vacuum pump for the removal of water from the web or felt. Generally positioned under the forming web on the wire table or in a perforated roll, eg couch or press.

Suhl effect The reduction in lifetime of holes injected into an n-type semiconducting filament, by deflecting them to the surface using a powerful transverse magnetic field. Reverse of **Hall effect**.

sulphite process One of the acid pulp digestion processes in papermaking, using a bisulphite liquor with some free sulphur dioxide.

sulphones Compounds having the formula $R–SO_2–R'$. The sulphur is hexavalent.

sulphoxides Compounds having the formula $R–SO–R'$. The sulphur is in the (IV) oxidation state.

sulphur A non-metallic element occurring in many allotropic forms, eg plastic sulphur. Symbol S. Chemically, sulphur resembles oxygen, and can replace the latter in many compounds, organic and inorganic. Sulphur is used in the manufacture of sulphuric acid and carbon disulphide, and for **vulcanizing** rubber. See Appendix 1, 2.

sunburner An excessive local thickness of material in a mouth-blown glass article.

superalloy Alloy capable of service at high temperatures, usually above 1000° C, eg for turbine blades and components. See **heat-resisting alloy**.

supercalendered paper Paper to which the surface finish has been applied by passing it through a supercalender. The gloss is usually greater than that attained by machine calenders. See **calendered paper**.

superconducting magnet Very powerful electromagnet, made possible by the exceptionally large current-carrying capacity of superconductors.

superconducting quantum interference device A family of devices capable of measuring extremely small currents, voltages and magnetic fields. Based on two quantum effects in superconductors; (1) *flux quantization* and (2) the *Josephson effect*. They can detect changes in magnetic flux densities of ≈1 nT (10^{-9} T). Examples are the study of fields generated by the action of the human brain and in other magneto-biological research. Abbrev. *SQUID*.

superconductors

Superconductors are a class of materials which lose their resistance to the flow of direct current when cooled below a characteristic transition temperature, the critical temperature, T_c. It is a property exhibited by tens of metallic elements and thousands of compounds. Superconductors usually behave as poor metals at room temperature with d.c. resistivities typically around 10^{-3} Ω m. This resistivity becomes immeasurably small on transition to the superconducting state (ie less than 10^{-25} Ω m). Superconductivity is a quantum mechanical effect resulting from a cooperative interaction between conduction electrons close to the **Fermi surface**. It is usually interpreted within the framework proposed by Bardeen, Cooper and Schrieffer (the *BCS* theory) which has been applied successfully to most known superconducting materials.

In a normal metal the lattice vibrations (phonons) scatter the conduction electrons and hence contribute significantly to its electrical resistance. In a superconductor, however, the same lattice vibrations bind electrons together in so-called *Cooper pairs*. These electron pairs are then able to move unimpeded through the lattice when an electric field is applied. If induced in a sample of toroidal geometry, therefore, a d.c. supercurrent will flow indefinitely, ie without any dissipation of energy.

A second manifestation of superconductivity is that bulk superconductors behave as perfect diamagnets when exposed to a weak magnetic field. A state of zero magnetic field is maintained in the interior of the superconductor by eddy currents induced in a thin layer at the sample surface, known as the *penetration depth*, ≈ 100 nm thick. This flux expulsion is known as the *Meissner effect* and its presence is the hallmark of superconductivity.

A superconductor will revert to its normal conducting state when the applied magnetic field exceeds a critical value. At this point the sum of the stored magnetic energy and the energy of the free superconducting state exceeds that of the normal state. Superconductors are classified as either type I or type II according to their behaviour in an applied magnetic field. Type I materials lose their properties abruptly whereas type II superconductors undergo a gradual transition as the magnetic field is increased. The diamagnetic response of type I and II superconductors to an applied magnetic field is shown in Fig. 1.

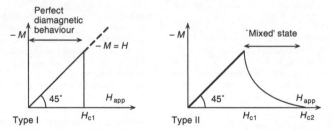

Fig. 1. **The response of superconductors to an applied magnetic field.**

A type II superconductor is said to exist in a *mixed state* in applied fields between H_{c1} and H_{c2}. Under these conditions filaments of magnetic flux are able to thread the bulk material in regions that become non-superconducting to accommodate the magnetic flux. The density of these filaments increases with increasing field and, hence, decreasing superconducting character. Type II superconducting materials are of more practical value than type I since H_{c2} is typically much greater than H_{c1}, ie 0.05 T compared with 20 T equivalent flux density.

continued on next page

superconductors (contd)

Superconductivity is also destroyed by current densities in excess of a critical value. The normal state is separated from the superconducting state by an energy gap. Properties such as critical current density, critical field and energy gap increase sharply as the temperature of the material is reduced below T_c before levelling off as absolute zero is approached. This is illustrated in Fig. 2. For most practical applications superconductors have to be stabilised at a temperature less than 2/3 T_c to yield close to optimum properties.

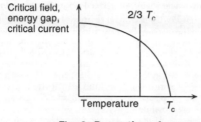

Fig. 2. **Properties of super-conductors below** T_c.

The discovery in 1986 of high-temperature superconducting ceramics which exhibit T_c values up to 125 K is significant in that liquid nitrogen can be used as a coolant (boiling point ca 77 K) rather than the more expensive and difficult to contain liquid helium (boiling point ca 4 K). Yttrium barium copper oxide ($YBa_2Cu_3O_{7-x}$, $T_c \approx 92$ K) and bismuth calcium strontium copper oxide ($Bi_2Ca_2Sr_2Cu_3O_z$, $T_c \approx 105$ K) , in particular, offer great potential for a variety of commercial ceramic and thin film applications. It is unlikely that these materials will replace liquid helium cooled Nb_3Sn ($T_c \approx 18$ K) in *superwire* applications in the short term, however, owing to their refractory nature.

Superconductors find application in zero-resistance devices such as high-field electromagnets, Josephson devices such as SQUIDS (**superconducting quantum interference devices**) which are used to detect minute magnetic fields ($< 10^{-10}$ T) and magnetic devices such as nuclear magnetic resonance based medical imaging systems.

---

superconductivity See **superconductors** p. 308.

superconductors See panel on p. 308.

super exchange energy The **exchange energy** in **ferrimagnetics** which is analogous to that in ferromagnetic materials. It is associated with the coupling of neighbouring dipole moments of ions through an interaction with the electron spins on intermediate oxygen ions.

superfinishing An abrasive process, resembling honing and lapping, for removing smear metal and scratches and ridges produced by grinding and machining operations, from bearing surfaces.

superfluidity At temperatures below 2.19 K, the lamda point, the helium isotope ^{4}He shows a striking change in its liquid properties; in particular its viscosity effectively vanishes and it is said to show *super-*

fluidity. At temperatures above the lamda point, ^{4}He is known as helium I and below helium II. Superfluid helium II behaves as if it consists of two parts, a super fluid and a normal fluid. The lower the temperature, the greater the fraction of superfluid in the mixture. The superfluid part consists of helium atoms in their lowest quantum state.

super-glue Popular name for cyanoacrylate adhesive, supplied as fluid monomer or prepolymer, which polymerizes when in contact with surfaces.

superlattice An ordering of solute atoms in a **substitutional solid solution** which results in a regular pattern of solute atoms superimposed upon the solvent lattice structure, like Cu_3Au and Ni_3Al (face-centred cubic) and CuZn (body-centred cubic). The ordering may extend over a few lattice units (short range) or over the whole crystal (long range) when it may significantly effect mechanical properties.

superparamagnetism The apparent paramagnetic behaviour exhibited by small single-domain particles when their thermal energy is comparable with the magnetization energy penalty exacted by their departure from an easy axis.

superplasticity The extremely high ductility shown by certain alloys when deformed at low strain rates at moderately high temperatures. Zinc-aluminium alloys near their eutectoid composition behave this way around 275°C. Normally ductile metals work harden as they deform and neck down locally which rapidly leads to fracture. Superplastic materials do not neck down and can undergo considerable plastic deformation provided the rate is kept below a critical level, thus resembling thermoplastic materials. Similar forming processes are possible but at lower production rates because of the slow deformation rate needed.

super-, supra Prefixes from L. *super*, over, above.

surface acoustic wave See **surface acoustic wave device**.

surface acoustic wave device Device which manipulates electronic signals by coupling them with (Rayleigh) surface acoustic waves on a piezoelectric crysal; used in high-frequency filter circuits and delay lines. Surface acoustic waves travel at a speed close to that of sound waves in the bulk material. See **dielectric and ferroelectric materials** p. 94.

surface active agent A substance which has the effect of reducing the surface tension of water and other liquids or solids, eg a detergent or soap. Also *surfactant*.

surface barrier Potential barrier across surface of semiconductor junction due to diffusion of charge carriers.

surface charge See **bound charge**.

surface conductivity See **surface resistivity**.

surface engineering The treatment of surfaces to confer on them properties distinct from those of the bulk; eg surface heat treatments, carburizing and nitriding, laser annealing, thin film coating and the techniques associated with semiconductor fabrication.

surface hardening Processes for locally increasing the hardness of surfaces of finished components to improve wear resistance or other performance related properties, eg fatigue resistance. They include those in which elements are diffused into the surface layers, eg **carburizing, nitriding, flame** and **induction hardening**.

surface lifetime The lifetime of current carriers in the surface layer of a semiconductor (where recombination occurs most readily). Cf **volume lifetime**.

surface measure A method of measuring timber in quantity, by the area of one face, irrespective of thickness. Cf **board measure**.

surface recombination velocity Electron-hole recombination on the surface of a semiconductor occurs more readily than in the interior; hence the carriers in the interior drift towards the surface with a mean speed termed the *surface recombination velocity*. It is defined as the ratio of the normal component of the impurity current to the volume charge density near the surface.

surface resistivity That between opposite sides of a unit square inscribed on the surface. Its reciprocal is *surface conductivity*.

surface strength The resistance of a paper to an adhesive force acting normally to the surface.

surface tension A property possessed by liquid surfaces whereby they appear to be covered by a thin elastic membrane in a state of tension, the surface tension being measured by the force acting normally across unit length in the surface. The phenomenon is due to unbalanced molecular cohesive forces near the surface. Units of measurement are $N\,m^{-1}$.

surfactant Abbrev. for **surface active agent**.

Surlyn TN for ionomer resin (US).

susceptibility See **electric susceptibility**, **magnetic susceptibility**.

suspension A system in which denser particles, which are at least microscopically visible, are distributed throughout a less dense liquid or gas, settling being hindered either by the viscosity of the fluid or the impacts of its molecules on the particles.

suspension polymerization See **chain polymerization** p. 56.

swaging Reducing of cross-section of metal rod or tube by forcing it through a tapered aperture between two grooved dies.

swarf (1) The cuttings from a machining operation. (2) US for the mixture of abrasive, bond and metal particles formed during grinding.

sweated joint See **sweating**.

sweating The operation of soft soldering pieces together by *tinning* the surfaces and heating them while pressed into contact. See **soft solder**.

Swedish iron Wrought iron of high purity, with some 300 $MN\,m^{-2}$ ultimate tensile strength, and 33% elongation before fracture.

sweeps Dust and debris in jeweller's workshops, gold refineries, bullion assay offices etc collected and treated periodically to recover valuable contents.

sweet Glass that is easily workable.

sweet crude oil One containing no hydrogen sulphide, H_2S.

swelling Process where polymers or biomaterials absorb fluids and expand. Negligible in rigid thermosets, it is widespread in thermoplastics prior to solubilization (depending on the **solubility parameter** of the solvent). In crosslinked rubbers, the chains cannot pass into solution, but the degree of swelling is a guide to the type of rubber and the degree of crosslinking.

SWG Abbrev. for **Standard Wire Gauge**.

switchboard model Suggested morphology of bulk crystallized polymers, with lamellar single crystals connected to one another by tie molecules, which form the amorphous regions in most crystalline polymers. An alternative model involves chain-folded crystals. See **crystallization of polymers** p. 82.

SWL Abbrev. for **Safe Working Load**.

sycamore Timber from the largest **hardwood** tree (*Acer pseudoplatanus*) native to Europe and W Asia; creamy-white and lustrous, straight-grained but often curly, with a fine, even texture. Density about $610 \, kg \, m^{-3}$. Perishable, so used in internal applications, eg furniture, bobbins, flooring, cooperage, musical instruments (the curly grained wood is used for the backs of violins, hence known as *fiddleback*), and for decorative veneers. In N America, 'sycamore' is the American plane or buttonwood (*Platanus occidentalis*). See **European plane**.

sym- See **syn-**.

symmetry The quality possessed by crystalline substances by virtue of which they exhibit a repetitive arrangement of similar faces. This is a result of their peculiar internal atomic structure, and the feature is used as a basis of crystal classification.

symmetry class Crystal lattice structures can show 32 combinations of symmetry elements, each combination forming a possible symmetry class.

syndiotactic Term describing very long-chain hydrocarbon polymers in which the substituent groups lie alternately on each side of the main chain. See **stereoregular polymers** p. 299.

syndiotaxy Polymerization exhibiting regular alternation of differences in stereochemical structure. Adj. *syndiotactic*.

syneresis Spontaneous expulsion of liquid from a gel.

synergist Substance which increases the effect of another.

syn-, sym- Prefixes from Gk *syn*, with, generally signifying fusion or combination.

syntactic foam Foam in which the voids are hollow fillers such as **hollow glass microspheres**.

synthetic fibre A fibre that has been manufactured by synthesis of a suitable linear polymer followed by extrusion of the molten material through a spinneret.

synthetic paper Paper made by conventional means but utilizing man-made fibre as the whole or a substantial part of the furnish. Generally a paper of considerable durability, permanence and chemical resistance. The term is (incorrectly) applied to other sheet materials of similar appearance to paper and used for similar purposes.

synthetic-resin adhesive One made from a thermosetting resin, eg urea or phenol formaldehyde, and used with an accelerator to regulate setting conditions, or from a thermoplastic resin, eg polymethyl methacrylate or polyvinyl acetate.

synthetic resins Resinous compounds made from synthetic materials, as by the condensation or polymerization of phenol and formaldehyde, formaldehyde and urea, glycerol and phthalic anhydride, polyamides, vinyl derivatives etc. See **phenolic resins, thermosets, vinyl polymers**.

synthetic rubber Any of numerous synthetic elastomeric materials, eg polymers of isoprene or its derivatives. See **butyl rubber, Neoprene, Thiokol**.

system A portion of matter, or a group or set of things that forms a complex or connected whole.

Système International d'Unités See **SI Units**.

T

t Symbol for tonne (metric **ton**).

t Abbrev. for: (1) trans-, ie containing the two radicals on opposite sides of the plane of a double bond or alicyclic ring; (2) tertiary, ie substituted on a carbon atom which is linked to three other carbon atoms.

τ Symbol for a time interval, esp. half-life or mean life.

T Symbol for: (1) **tera**-; (2) **tesla**.

T Symbol for **temperature**.

tack Ability of a material to bond with another when contact is brief and pressure is light. Important in rubber industry when bringing parts together to make products like tyres.

tack welding Making short, provisional welds along a joint to hold it in position and prevent distortion during subsequent continuous welding.

tacticity The stereochemical arrangement of units in the main chain of a polymer. See **atactic, isotactic, syndiotactic, stereoregular polymers** p. 299.

Tafel plot Graph of **overpotential** against logarithm of the current in an electrochemical cell. Useful in evaluating the kinetics of cell reactions.

taffeta A lightweight, plain-weave, crisp fabric with a faint weft rib produced from filament yarns and used for blouses etc.

tagger Thin sheet-iron or tin-plate.

talc A monoclinic hydrated magnesium silicate, $MgSi_8O_{20}(OH)_4$. It is usually massive and foliated and is a common mineral of secondary origin associated with serpentine and schistose rocks; also found in metamorphosed siliceous dolomites. Purified talc is used medically, in toilet preparations and in polymers, as fillers. See **steatite**.

tan delta Equivalent to **loss factor, loss tangent**.

tandem mill A rolling mill with two or more stands operating continuously and synchronized so that successive stands can accommodate the increasing velocity of the rolled material as its thickness is reduced.

tangent modulus Elastic (or *quasi-elastic*) modulus derived from a non-linear **stress-strain curve** by taking the slope of curve at a specified level of stress or strain, eg '0.5% strain tangent modulus'. This is usually less than the **secant modulus** at the same point.

tank furnace Essentially a large *box* of refractory material holding from 6 to 2000 tonnes of glass, through the sides of which are cut *ports* fed with a combustible mixture (gas and air, or oil spray and air), so that flame sweeps over the glass surface. With the furnace is associated a regenerative or recuperative system for the purpose of recovering part of the heat from the waste gases. See glasses and glassmaking p. 144.

tannin A mixture of derivatives of polyhydroxy-benzoic acids. When pure it forms a colourless amorphous mass, easily soluble in water, of bitter taste and astringent properties. Occurs in many trees, eg *Quebracho colorado*.

tanning Process of converting animal hide into **leather** by treating with eg **tannin**.

tantalum A metallic element, symbol Ta. It occurs in crystals and grains (usually containing, in addition, small amounts of niobium) in the Ural and Altai Mts. It is used as a substitute for platinum for corrosion-resisting laboratory apparatus, as acid-resisting metal in chemical industry, and in the form of carbide in cemented carbides. Used in surgical insertions because of its lack of reaction to body fluids. See Appendix 1, 2.

taper line gratings Transparent plastic plates engraved with lines that are more widely spaced at one end than the other. By placing the appropriate grating over a fabric the number of threads per cm (for woven fabrics) or courses per cm (in knitted fabrics) may be determined from the resulting diffraction pattern.

tapestry Originally a handwoven furnishing fabric in which a design was produced by hand stitches to form the required pattern. Now employed for **jacquard** figured design on machine-woven cloths.

tapping (1) The operation of running molten metal from a furnace into a ladle. (2) Using a tap to thread a hole.

TAPPI standard methods Laboratory test methods conforming to the Technical Association of the Pulp and Paper Industry of the US. Widely used internationally.

tare The weight of a vessel, wrapping, or container, which subtracted from the gross weight gives the net weight.

target In general any surface on which a beam impinges. In sputtering, the surface which is eroded as distinct from the substrate, on which sputtered material is deposited

tawa New Zealand native **hardwood** (*Beilschmedia tawa*), whose **heartwood** is white to yellow, straight-grained, and fine-textured; density about 730 $kg\ m^{-3}$. Not very durable; used for furniture, joinery, fittings and flooring, and in decorative veneers.

Taylor process Making very fine wire by inserting it into a closely fitting glass tube and drawing out the whole during heating.

T_b Temperature of boiling.

TDI Abbrev. for **Tolylene-2,4-DiIsocyanate**.

TD nickel A dispersion strengthened material where high yield and tensile strength are achieved by fine particles of **thoria** which act as reinforcement throughout the nickel matrix and, being essentially insoluble, retain their strengthening effect at high temperature.

teak Valuable and very durable timber from *Tectona grandis*, a **hardwood** tree of the verbena family, native to India and SE Asia. It is a rich brown with darker brown markings, straight- to wavy-grained and coarse-textured; mean density 650 kg m^{-3}; used for shipbuilding, piers, furniture, cabinetmaking, flooring etc, for its acid-resistance in vats, laboratory benches etc, and in plywood and veneers.

tear factor The ratio of the lateral tearing strength of a paper to its grammage.

tearing energy The equivalent of **toughness** when obtained from a **tear test** on eg an elastomer. See **strength measures** p. 301.

tear test See **strength measures** p. 301.

teazle The dried seed head of the thistle-like plant, *Dipsacus fullonum*, fitted on machines to raise the pile of certain fabrics.

technetium Radioactive element not found in ores. First produced as a result of deuteron and neutron bombardments of molybdenum. Symbol Tc. The most common isotope ^{99}Tc has half-life of 2.1×10^6 years. Found among fission products of uranium. See Appendix 1, 2.

Technical Association of the Pulp and Paper Industry US industrial association specifying widely used test procedures known as *TAPPI standard methods*.

teem (1) To pour molten glass from a pot in the rolling process. (2) To fill ingot moulds from a ladle of molten metal.

teeming The operation of filling ingot moulds from a ladle of molten metal.

Teepol TN of liquid anionic detergent based on mixed sodium alkyl sulphates of long-chain alcohols, such as lauryl alcohol, *n*-dodecanol, $CH_3(CH_2)_{11}OH$.

Teflon TN for **polytetrafluoroethylene** (US).

Teklan TN for synthetic textile fibre based on a modified acrylic (modacrylic) base. Chiefly noted for its inherent flameproof properties.

tele- Prefix from Gk *tele*, afar, at a distance.

teleo- Prefix from Gk *teleios*, perfect.

tellurium A semi-metallic element, tin-white in colour, symbol Te. Used in the electrolytic refining of zinc in order to eliminate cobalt; alloyed with lead to increase the strength of pipes and cable sheaths. The

chief sources are the slimes from copper and lead refineries and the fine dusts from telluride gold ores. See Appendix 1, 2.

telo- Prefix from Gk *telos*, end.

TEM Abbrev. for **Transmission Electron Microscope** (or microscopy).

temper (1) Vague term describing the relative condition of the hardness and mechanical properties of a metal. For example, a solid solution alloy which may be extensively cold-worked may exhibit a range of properties from its softest state when annealed to fully work hardened. This range is divided into hardness bands, refered to as *tempers*, eg quarter hard, half hard etc. A given temper can be achieved either by full annealing folowed by careful work-hardening or by full work-hardening followed by careful annealing. The latter is referred to as *temper annealing* or *back annealing*. See **annealing** p. 13. (2) The amount of residual stress in annealed glassware measured by comparison with strain disks.

temper annealing See **temper**.

temperature A measure of whether two systems are relatively hot or cold with respect to one another. Two systems brought into contact will, after sufficient time, be in thermal equilibrium and will have the same *temperature*. A *thermometer* using a temperature scale established with respect to an arbitrary zero (eg **Celsius scale**) or to absolute zero (**Kelvin thermodynamic scale of temperature**) is required to establish the relative temperatures of two systems. See zeroth law of **thermodynamics**.

temperature coefficient The fractional change in any particular physical quantity per degree rise of temperature.

temperature coefficient of resistance In any conductor, if

$$R = R_0 [1 + \alpha(T - T_0)],$$

R is the resistance at temperature T, compared with the resistance R_0 at temperature T_0, the mean coefficient in the range T_0 to T being α. This is useful only for *linear resistances*, ie pure metals.

temper brittleness A type of brittleness that is shown by notched bar tests, but not by the tensile test, in certain types of steel after tempering; influenced by the composition of the steel, the tempering temperature and the subsequent rate of cooling.

temper colour In tempering hardened steel cutting tools etc, the colour of the oxide layer which forms on reheating and which indicates approximately the correct quenching temperature for a particular purpose. See **tempering**.

temper-hardening A term applied to alloys

that increase in hardness when worked or re-heated after rapid cooling; and to the operations of producing this effect. Also *artificial ageing*; distinguished from ageing, which occurs at atmospheric temperature. See **precipitation hardening**.

tempering The reheating of hardened steel at any temperature below the critical range, in order to decrease the hardness. Sometimes *drawing*. May be applied to reheating after rapid cooling, even when this results in increased hardness, eg in the case of steels that exhibit secondary hardening. See **austempering, steels** p. 296.

tenacity Obsolete term for **ultimate tensile stress**. See **strength measures** p. 301.

Tenite TN for cellulose acetate-butyrate (US).

Tensar TN for polymer mesh used for fencing, supporting road foundations etc. Made by punching holes in extruded polyolefin sheet and hot-stretching along one or two axes.

tensile creep test See **creep, tensile test, viscoelasticity** p. 332.

tensile modulus The quantity corresponding to **Young's modulus** in non-linear and/or viscoelastic materials such as polymers. Cf **creep modulus, secant modulus**.

tensile strength See **strength measures** p. 301.

tensile stress See **stress, ultimate tensile stress**.

tensile test Main forms are those in which: (1) a static increasing pull is applied to a specimen until fracture results; from this a **stress-strain curve** may be plotted and the **Young's modulus, proof stress, yield point, ultimate tensile stress** and **elongation** determined; (2) a dynamic load is applied giving data on **fatigue** and **impact**; (3) a constant load is applied and the increasing deformation with time recorded (*tensile creep test*). See **strength measures** p. 301.

tensile testing machine A machine for applying tensile or compressive loads to a test piece, by means of hand- or power-driven screws, or by a hydraulic ram (or actuator).

tension rod A structural member subject to tensile stress only. Also *tie rod*.

tension wood Form of **reaction wood** developed in hardwoods, with a higher cellulose content, hence more rubbery, and with lower compressive but higher tensile strength than normal. Cf **compression wood**.

tensometer A versatile testing machine used for a variety of mechanical tests, including tensile tests.

tera- Prefix, symbol T, denoting 10^{12} times, eg a *terawatt-hour* is 10^{12} watt-hours.

terephthalic acid $C_6H_4(COOH)_2$, *benzene-1,4-dicarboxylic acid*, used in the manufacture of polyethylene terephthalate (PET). See **step-polymerizarion** p. 298.

terbium A metallic element, a member of the rare earth group, symbol Tb. It occurs in the same minerals as dysprosium, europium and gadolinium. See Appendix 1, 2.

termination Ending of polymerization reaction by several possible mechanisms. See **chain polymerization** p. 56.

terminator Chemical compound which ends a chain-polymerization reaction.

ternary diagram Phase diagram or alloy system formed by three components, represented by the ternary constitutional or equilibrium diagram. Also *ternary system*. See **phase diagram** p. 236.

terne metal Lead alloy with up to 18% tin and 1.5–2% antimony, used to coat steel to improve its corrosion resistance and working properties.

terne plate Iron or steel sheet coated by hot-dipping with **terne metal** which acts as a carrier for lubricant used in subsequent drawing operations, as rust protection, as a paint base or to facilitate soldering.

terpolymer Copolymer of three different monomers. See **copolymer** p. 75.

terra-cotta From the Latin for 'baked earth'. A hard, unglazed ceramic ware of characteristic reddish-brown colour and used for decorative tiles and bricks, statues, vases, kitchenware etc. Cf **bone china, earthenware, faience, porcelain, stoneware**.

terry fabric Woven cloth using an additional warp to produce loops on one or both sides of the ground fabric. Often used for towels. See **plush** (weft-knitted).

tertiary alcohols Alcohols containing the group

$$\overset{\textstyle |}{\underset{\textstyle |}{-C}}-OH.$$

When oxidized, the carbon chain is broken, resulting in the formation of two or more oxidation products containing a smaller number of carbon atoms in the molecule than the original compound.

tertiary carbon atom Carbon atom linked to three other carbon atoms and one hydrogen atom. Generally less stable than secondary carbon atoms, and a weak point in polymer chains, making them more susceptible to degradation and oxidation.

tertiary structure The three-dimensional configuration of polymers or biomaterials (eg proteins) which is a stable folding of the sequence of units (groups, bases or peptides) along the polymer, ie of their secondary structure.

Terylene TN for straight-chain polyester fibres derived from condensation of terephthalic acid with ethan-1,2-diol, used widely in the manufacture of fabrics, clothing materials and other textiles. See **step polymerization** p. 298.

tesla SI unit of magnetic flux density or magnetic induction equal to 1 weber m⁻². Equivalent definition; the magnetic induction for which the maximum force it produces on a current of unit strength is 1 newton. Symbol T.

test piece A piece of material accurately turned or shaped, often to specified standard dimensions, for subjecting to a tensile test, shock test etc, in a testing machine.

tetrahydrofuran *Tetramethylene oxide.* Organic solvent used in refining lubricating oils, a solvent for a wide range of polymers, and as source of **polytetramethylene etherglycol**, PTMEG.

tex The basic unit of the tex system, used to express the **linear density** (mass per unit length) of fibres, filaments, yarns or other linear material. The tex of a material is the mass in grams of a 1 km length of material (hence *decitex, kilotex* etc). This system has now replaced the traditional ones based on the **denier** and other units.

texture Physical appearance in terms of roughness and shape of surface features; in microscopical examination it relates to microstructural features such as grain shape, distribution of phases and crystallographic orientation. See **banding, fractography.**

textured yarn Yarn of synthetic continuous filaments that has been treated to make its surface highly irregular by crimping or by introducing loops having a variety of shapes. Widely used methods include: (1) false-twist texturing in which the yarn is continuously twisted, heat-set and untwisted (no overall twist is produced in the filaments), (2) stuffer box crimping; (3) in air-jet texturing the entangled loops are formed by subjecting the yarn to a turbulent air stream. The yarns have increased bulk and are sometimes described as *bulked continuous filaments* (BCF). They are also more easily extended and are used in articles (eg tights) that closely conform to the shape of the body. Was called *bulked yarn.*

T_g Symbol for **glass transition temperature.**

TGA Abbrev. for **ThermoGravimetric Analysis.**

T_g point Obsolete designation of temperature of the glass transition. See **T_g.**

TGS Abbrev. for *TriGlycine Sulphate.* See **dielectrics and ferroelectrics** p. 92.

thallium Metallic element, symbol Tl. White malleable metal like lead. Several thallium isotopes are members of the uranium, actinium, neptunium and thorium radioactive series. Thallium isotopes are used in scintillation crystals. Its compounds are very poisonous. See Appendix 1, 2.

theoretical stiffness Estimate of the stiffness of any material from atomic/molecular data. For example it can be done for completely oriented polymer chains from a knowledge of chain structure, bond lengths and angles, and force constants. For high-density polyethylene, calculation gives a theoretical tensile modulus along the chains of 316 GPa and about 9 GPa perpendicular to the chains. See **theoretical strength, high-performance fibres** p. 158.

theoretical strength   Estimate of the strength of any material from atomic/molecular data. For example it can be done for oriented polymer chains using molecular data such as **Morse equation** for the intermolecular potential energy. Calculation for high-density polyethylene gives a theoretical strength along the chains of 31 GPa. See **theoretical stiffness, high-performance fibres** p. 158.

thermal analysis   The use of cooling or heating curves in the study of physical changes in materials. The freezing points and the temperatures of any polymorphic changes occurring in pure materials may be determined, as well as ranges and temperatures for changes in solid alloys. The data obtained are used in eg constructing **phase diagrams** p. 236. See **differential scanning calorimetry, differential thermal analysis.**

thermal conductivity A measure of the rate of flow of thermal energy through a material in the presence of a temperature gradient. If (dQ/dt) is the rate at which heat is transmitted in a direction normal to a cross sectional area A when a temperature gradient (dT/dx) is applied, then the thermal conductivity is

$$k = -\frac{(dQ/dt)}{A\,(dT/dx)}.$$

SI unit is W m⁻¹ K⁻¹. Materials with high electrical conductivities tend to have high thermal conductivities.

thermal diffusion Process in which a temperature gradient in a mixture of fluids tends to establish a concentration gradient. Has been used for isotope separation. Also known as *Soret effect.*

thermal diffusivity **Thermal conductivity** divided by the product of specific heat capacity and density; more generally applicable than thermal conductivity in most heat transfer problems. Unit is m² s⁻¹. Symbol κ.

thermal expansion coefficient The fractional expansion of matter (ie the expansion of the unit length, area or volume) per degree rise in temperature. If the coefficients of linear, areal and cubical expansion of a substance are α, β, and γ respectively, β is approximately twice, and γ three times, α.

thermal fatigue Fatigue failure resulting from strains caused by expansion and contraction during thermal cycling. See **fatigue**.

thermally activated process Any physical or chemical process whose rate increases with the amount of thermal energy supplied to the system. See **Arrhenius's (rate) equation**.

thermally bonded non-woven fabrics Fabrics formed from a web of fibres that have been partially melted by heat and thereby converted into a coherent sheet.

thermal shock The transient thermal stresses resulting when a body is subjected to sudden changes in temperature. See **thermal shock resistance**.

thermal-shock resistance Merit index of a material's ability to withstand thermal shock without cracking, or otherwise failing; different indices apply to different heat transfer conditions.

thermal spray Spraying molten and finely divided metals or non-metallic particles on to a material to form a coating.

thermionic emission The emission of electrons from a heated material as a consequence of thermal energy. See **Richardson–Dushman equation**.

thermionics Strictly, science dealing with the emission of electrons from hot bodies. Applied to the broader subject of subsequent behaviour and control of such electrons, esp. *in vacuo*.

thermionic work function Thermal energy surrendered by an electron liberated through **thermionic emission** from a hot surface. Cf **photoelectric work function**.

thermistor Abbrev. for *thermal resistor*. A semiconductor, a mixture of cobalt, nickel, and manganese oxides with finely divided copper, whose resistance is very sensitive to temperature. Sometimes incorporated in a waveguide system to absorb and measure all the transmitted power. Also used for temperature compensation and measurement.

thermite A mixture of aluminium powder and half an equivalent amount of iron oxide (or other metal oxides) which gives out a large amount of heat on igniting with magnesium ribbon. The molten metal forms the medium for welding iron and steel (*thermit welding*). See **aluminothermic process**.

Thermit process See **aluminothermic process**.

thermo- Prefix from Gk *therme*, heat. Also *therm-*.

thermochemistry The study of the heat changes accompanying chemical reactions and their relation to other physicochemical phenomena.

thermocouple If two wires of different metals are joined at their ends to form a loop, and a temperature difference between the two junctions unbalances the contact potentials, a current will flow round the loop. If the temperature of one junction is kept constant, that of the other is indicated by measuring the current, or more usually the potential difference set up when the circuit is open. Conversely, a current driven round the loop will cause a temperature difference between the junctions. See **Peltier effect**.

thermodynamics The mathematical treatment of the relation of heat to mechanical and other forms of energy. Its chief applications are to heat engines (steam engines and IC engines) and to chemical reactions in thermochemistry. *Laws of thermodynamics*: *zeroth law*: if two systems are each in thermal equilibrium with a third system then they are in thermal equilibrium with each other. This statement is tacitly assumed in every measurement of temperature. *First law*: the total energy of a thermodynamic system remains constant although it may be transformed from one form to another. This is a statement of the principle of the conservation of energy. *Second law*: heat can never pass spontaneously from a body at a lower temperature to one at a higher temperature (Clausius) *or* no process is possible whose only result is the abstraction of heat from a single heat reservoir and the performance of an equivalent amount of work (Kelvin–Planck). *Third law*: the entropy of a substance approaches zero as its temperature approaches absolute zero.

thermoelectric cooling Abstraction of heat from electronic components by **Peltier effect**, greatly improved and made practicable with solid-state materials, eg Bi_2Te_3. Devices utilizing this effect, eg frigistors, are used for automatic temperature control etc and are energized by d.c.

thermoelectric effect See **Seebeck effect**.

thermoelectricity Interchange of heat and electric energy. See **Peltier effect**, **Seebeck effect**, **Thomson effect**.

thermoelectric materials Any set of materials (metals) which constitute a thermoelectric system, eg binary (bismuth and tellurium), ternary (silver, antimony and tellurium), quaternary (bismuth, tellurium, selenium and antimony, called *Neelium*).

thermoelectroluminescence See **electrothermoluminescence**.

thermoforming Polymer shaping method where extruded sheet is heated to above the glass-transition temperature and pressed into shape by male or female tool.

thermogravimetric analysis Method of thermal analysis of materials, where sample is heated at a controlled rate and the change in mass recorded as a function of time. Used for studying degradation of polymers, oxidation of metals etc. Abbrev. *TGA*.

thermoluminescence Release of light by previously irradiated phosphors upon subsequent heating.

thermoluminescent dosimeter An instrument which registers integrated radiation dose, the read-out being obtained by heating the element and observing the thermoluminescent output with a photomultiplier. It has the advantage of showing very little fading if read-out is delayed for a considerable period, and forms an alternative to the conventional film badge for personnel monitoring.

thermomagnetic effect See **magnetocaloric effect**.

thermometer An instrument for measuring temperature. A thermometer can be based on any property of a substance which varies predictably with change of temperature. For instance, the *constant volume gas thermometer* is based on the pressure change of a fixed mass of gas with temperature, while the *platinum resistance thermometer* is based on a change of electrical resistance. The commonest form relies on the expansion of mercury or other suitable fluid with increase in temperature.

thermometry The measurement of temperature.

thermoplastic Becoming plastic on being heated. Specifically in plastics, any resin which can be melted by heat and then cooled, the process being apparently repeatable in theory any number of times without appreciable change in properties, eg cellulose polymers, vinyl polymers, polystyrenes, polyamides and acrylic polymers. Also *thermosoftening*. Cf **thermosets** p. 318.

thermoplastic elastomer Class of rubbers not needing vulcanization, so relatively easy to injection mould, extrude etc. Includes polyester rubber, some polyurethanes, ethylene vinylacetate and styrene butadiene styrene block copolymers. See **copolymer** p. 75.

thermosets See panel on p. 318.

thermosetting Describes polymer materials in which chemical reactions, including crosslinking, take place while the resins are being moulded; the appearance and chemical and physical properties are entirely changed, and the product is resistant to further applications of heat (up to charring point). See **thermosets** p. 318.

thermosiphon The method of establishing circulation of a cooling-liquid by utilizing the slight difference in density of the hot and the cool portions of the liquid.

thermosoftening Synonym for **thermoplastic**.

thermostat An apparatus which maintains a system at a constant temperature which may be preselected. Frequently incorporates a bimetallic strip.

theta solvent Organic solvent for polymers producing **random coil** conformation of individual chains.

THF Abbrev. for **TetraHydroFuran**.

thick film Film deposited by a screen printing or similar process. For thick film resistors for example, an 'ink' carries a mixture of glass and semiconductor oxide particles. On firing, the ink is driven off and the particles sinter. See **Birox resistor**.

thick-film lubrication The state of fluid-medium lubrication which exists when the lubricant separates the journal and bearing surfaces to prevent metal-to-metal contact. The bearing friction is then dependent largely on the force necessary to shear the lubricant film, and perfect lubrication is said to prevail.

thin film Film made by **chemical** or **physical vapour deposition**, or electrolysis, typically no more than a few micrometres thick.

thin-film lubrication When metal-to-metal contact exists in a journal bearing for all or part of the time, thin-film or imperfect lubrication is said to prevail and the materials and surface characteristics of journal and bearing affect the bearing friction significantly.

thin-film transistor A **metal-oxide-silicon field effect transistor** (MOSFET) in which the semiconductor is not a single crystal but polycrystalline or amorphous. Used as switches for matrix displays and in solar cells, having the advantage that they can be deployed over wider area substrates than feasible in current single-crystal technology.

Thiokol TN for a synthetic rubber of the polysulphide group, derived from sodium tetrasulphide and organic dichlorides, with sulphur as an accelerator.

thiourea resins Thermosetting resins made from thiourea and an aldehyde. They are more water-resistant and more stable than the urea resins, but they cure more slowly, and the sulphur in them causes trouble with the steel moulds and the dyestuffs used to colour them.

thiourea resins Formerly resins made from thiourea and an aldehyde, but now superseded by MF thermoset resins. See **thermosets** p. 318.

thirds A premetrication size of cut card, 38 × 76 mm (1½ × 3 in).

thixotropy Rheological property of a fluid

thermosets

Thermoplastics soften or melt when heated and harden on cooling, a process which can apparently be repeated indefinitely without change in properties (but see **recycling ratio**). Since the T_g or T_m of most common plastics is relatively low (80–120°C), their use in many products is severely limited. Two approaches have been used to overcome the problem. One is the synthesis of much stiffer chains with higher transitions such as polycarbonates or poly-sulphones. Some of these thermally stable polymers show no transitions at all, but rather degrade chemically at high enough temperature (eg **aramids, poly-imides**). The other approach is to tie the chains together with covalent bonds into a network comprising one single molecule. The crosslinking agents needed may be activated by simple mixing with the monomer or prepolymer, by photoinitiation or by the action of heat. It is the last class of material that has given rise to the term *thermoset*, and hence to the group of crosslinked materials so produced. Apart from vulcanized rubbers, conventional thermosets include phenolics, aminoplastics, epoxies, polyesters and some polyurethanes, although it is important to note that most 'thermoplastics' can be crosslinked using an appropriate reagent. See **ion exchange resins**, **Sclair** polyethylene.

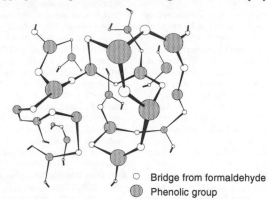

○ Bridge from formaldehyde
◉ Phenolic group

Fig. 1. **Schematic drawing of cross-linked phenol formaldehyde resin**.

Phenolic resins are one of the oldest thermosets (Bakelite) and are formed by reacting phenol and formaldehyde to form *novolak* or *resol* prepolymers in the A stage. Further polymerization gives B-stage resins and final heat cure gives a rigid, non-crystalline network with a very high **crosslink density**. Since the network is permanent, analysis of thermosets is difficult because most methods involve dissolution in a suitable organic fluid. It is possible to identify lightly crosslinked rubbers by swelling tests, but this cannot be used for most thermosets which are usually rigid, infusible materials. They find application in products subjected to severe conditions such as heat-resistant surfaces (laminates etc), electrical insulators (eg printed circuit boards) and composites, where they can be combined with inorganic fibres such as glass or higher performance materials like carbon fibre. See **high-performance fibres** p. 158.

where applied stress lowers viscosity which returns to normal when the stress is released. Property of many polymer fluids and mixtures (eg ketchup), and widely used in paints. Fluids which behave in the opposite

way are known as *rheopectic*.

Thomas resistor A standard manganin resistor which has been annealed in an inert atmosphere and sealed into an envelope.

Thomson effect The electromotive force

produced by temperature differences in a single conductor, and the heat change associated with current flow between temperature differences.

thoriated cathode Tungsten cathode containing a small proportion of **thorium** to reduce the temperature at which copious electronic emission takes place. With heat, the thorium diffuses to the surface, forming a tenuous emitting layer.

thorium A metallic radioactive element, dark-grey in colour, symbol Th. It occurs widely in beach sands (*monazite*). The thorium radioactive series starts with thorium of mass 232 which disintegrates on capture of fast neutrons (ie it is fissile) and ^{233}U (fissile with slow neutrons) being formed from ^{232}Th by neutron capture and subsequent beta decay. See Appendix 1, 2.

thread The general name for a yarn although commonly used specifically for sewing thread.

thread rolling Producing a screw thread by rolling between flat or cylindrical dies an alloy sufficiently plastic to withstand the cold working forces without disintegrating.

three-high mill A rolling mill with three rolls, which are rotated in such a way that the metal is passed in one direction through the bottom pair of rolls and in the opposite direction through the top pair.

threshold voltage The voltage which must be applied between the gate and substrate of a **metal-oxide-silicon** (MOS) structure to cause the formation of an inversion channel.

throat The submerged channel through which glass passes from the melting to the working end of a tank furnace. See **glasses and glassmaking** p. 144.

throwing In silk manufacture, the processes of reeling, doubling and twisting, to bring the silk filaments into the form of a thread. Now extended to include similar processes carried out with continuous filament man-made fibres.

thulium A metallic element, a member of the rare earth group, symbol Tm. One of the rarest elements, occurring in small quantities in euxenite, gadolinite, xenotime etc. Radioactive isotope emits 84 keV gamma-rays, frequently used in radiography. See Appendix 1, 2.

thyristor A semiconductor power switch capable of bistable operation. It can have from two to four terminals and can be triggered from its off state to on at any chosen point within a single 90° quadrant of the applied a.c. voltage. The semiconductor rectifier (SCR) is the most common unidirectional thyristor. See **triac**.

tie A frame member sustaining only a tensile load.

tie bar One of four massive horizontal columns on which the tools, bolster and other parts are mounted on an injection moulding machine. See **injection moulding** p. 169.

tie beam A structural member connecting the lower ends of a pair of principal rafters to prevent them from moving apart. Also applied to a member connecting parts of a structure which may otherwise move apart under load.

tie line See phase diagram p. 236.

tie molecule Single polymer chain connecting adjacent single crystal lamellae, providing bulk material with structural integrity. See **crystallization of polymers** p. 82.

tie rod See **tension rod**.

TIG See **tungsten inert gas welding**.

tight binding model A model from which the electronic band structure of certain transition elements can be calculated. The overlap of wavelengths from atom to atom is regarded as small enough to permit the crystal wavefunction to be calculated by a *linear combination of atomic orbital* (LCAO) approach. See **band theory of solids**.

timber Felled trees or logs suitable for conversion by sawing or otherwise. US *lumber*.

time-temperature equivalence See **viscoelasticity** p. 332.

time-temperature-transformation diagram See **isothermal transformation diagrams** p. 177. Abbrev. *TTT diagram*.

tin A soft, silvery-white metallic element, ductile and malleable, existing in three allotropic forms; symbol Sn (Latin *stannum*, tin). Not affected by air or water at ordinary temperatures. The principal use is as a coating on steel in tinplate; also used as a constituent in alloys and with lead in low melting-point solders for electrical connections. Occurs as tin oxide, SnO_2, or *cassiterite*. See **tin alloys**, Appendix 1, 2.

tin alloys Tin is an essential constituent in soft solders, type metals, fusible alloys and certain bearing metals. These last contain 50–92% of tin alloyed with copper and antimony and sometimes lead. Tin is also a constituent of bronze and pewter.

tinman's solder A tin-lead solder melting at about 183° C, used for tinning. The most fusible solder contains 65% tin, the eutectic composition.

tin-nickel Metal finish which results from the simultaneous electrodeposition of tin and nickel on a polished surface of brass in a carefully controlled bath, resulting in a non-tarnishable and non-corrodible polished surface of low friction.

tin-plate Thin sheet-steel covered with an adherent layer of tin formed by passing the steel though a bath of molten tin or by electrodeposition. Resists atmospheric oxidation

---

titanium alloys

Titanium metal remained a laboratory curiosity until the mid 1950's, but it is now produced in quantities of thousands of tons annually by electrode melting processes in a vacuum or under argon. In commercially pure form it is widely used for corrosion resistant applications; the metal itself is highly reactive but a tenacious oxide film on the surface renders it practically inert to atmospheric attack and resistant to the majority of industrial chemicals. Controlled amounts of oxygen (up to 0.2% by weight) significantly raise the yield and tensile strength, giving rise to a number of commercial grades.

Pure titanium undergoes an allotropic change from *hexagonal close packing* to the *body-centred cubic* form at 885°C and alloying elements are classified by whether they stabilize the alpha (low temperature) or the beta (high temperature) crystal forms. For creep resistance the alpha stabilizers, aluminium and tin, are required while for maximum strength the beta stabilizers, vanadium, molybdenum, zirconium, are added. The heat treatments of the alpha and beta alloys to achieve maximum strength are complex.

As they are only half the density of steel, titanium alloys offer attractive strength/weight ratios, since some of the heat treatable alloys may be produced with tensile strengths of the order of 1200 MN m^{-2}. The major alloying additions are aluminium and vanadium, with lesser amounts of tin, copper, molybdenum and zirconium. These alloys find wide application in aerospace where the requirements justify the high cost. They also used in the rotors of centrifuges which have to withstand very high centrifugal forces.

Commercially pure titanium and the alpha forms are weldable, but not the heat-treated high-strength alloys

Titanium carbide is widely used in industry for cutting tools etc, but the interatomic bonds are strong and this material is not classified as an alloy of titanium.

---

and attack by many organic acids. Used for food-containers etc.

tin-zinc Metal finish which results from the simultaneous electrodeposition of tin and zinc on a clean steel surfaces, giving a non-corrodible finish to chassis for electronic apparatus.

tip The part of a cutting tool containing the cutting edges, if made of a material of superior quality to that of the remainder of the tool, eg tungsten carbide, and securely fastened to the latter by brazing or otherwise.

titania See **titanium dioxide**.

titanium A metallic element resembling iron, symbol Ti. Manufactured commercially since 1948, it is characterized by strength, lightness and corrosion resistance. Widely used in aircraft manufacture, for corrosion resistance in some wet extraction processes, as a deoxidizer for special types of steel, in stainless steel to diminish susceptibility to intercrystalline corrosion and as a carbide in cemented carbides. Sometimes used for the solid horns of magnetostrictive generators. It occurs only in combination, and is widely distributed on the Earth's crust (0.6% by weight). See Appendix 1, 2.

titanium alloys See panel on p. 320.

titanium dioxide *Titanium (IV) oxide*, TiO2. Pure white pigment of great opacity, much used industrially for paints, plastics etc. Also *titania*. See **titanizing**.

titanizing Use of titanium dioxide, TiO$_2$, in the manufacture of heat-resisting and durable glass by incorporating it to replace certain proportions of soda, the viscosity increasing proportionately.

T_m Symbol for **crystal melting point**.

Tobin bronze A type of alpha-beta brass or Muntz metal containing tin. It contains 59–62% copper, 0.5–1.5% tin, the remainder being zinc. Used when resistance to sea water is required. See **copper alloys** p. 76.

tolylene-2,4-diisocyanate Type of isocyanate monomer used for making polyurethanes. They are reacted with polyglycols or polyesters, forming urethane links with terminal hydroxy groups. Abbrev. *TDI*. See **diphenylmethane diisocyanate**.

ton A unit of mass for large quantities. The *long ton*, once commonly used in UK, is 2240 lb. The *short ton*, commonly used in America, is 2000 lb. The *metric ton* or *tonne* (1000 kilograms; symbol t) is 2204.6 lb. In

UK the *short ton* was used in metalliferous mining, the *long ton* in coal-mining. See Appendix 3.

tonne Metric ton, 1000 kg. See **ton**.

tool steel Steel suitable for use in tools, usually for cutting or shaping wood or metals. The main qualities required are hardness, toughness, ability to retain a cutting edge etc. Contain 0.6–1.6% carbon. Many tool steels contain high percentages of alloying metals: tungsten, chromium, molybdenum etc. See **high-speed steel**. Usually quenched and tempered, to obtain the required properties. See **steels** p. 296.

top Originally used to describe a sliver of combed wool ready for spinning into worsted yarns. Now also applied to the material produced by cutting or breaking continuous filament tows of man-made fibres.

torque The turning moment exerted by a *tangential* force acting at a distance from the axis of rotation. It is measured by the product of the magnitude of the force and the distance. Unit newton-metre. See **moment**.

torsion State of strain set up in a body by twisting or applying a **torque**. The external twisting effort is opposed by the shear stresses induced in the material.

torsional braid analysis Method of following the curing reaction of thermosets by coating an absorbent braid with the starting materials in liquid form, and making this the specimen in a **torsion pendulum**. The oscillation characteristics of the system are affected by the degree of cure.

torsion bush Rubber spring comprising a cylinder of rubber enclosing a shaft and enclosed by a steel cover, for resisting torsion, tilt etc. Commonly used for flexible couplings.

torsion meter Device for measuring **torque**.

torsion pendulum A dynamic mechanical test for polymeric materials, where the tape, ribbon or fibre is attached to an inertial mass at one end and a fixed support at the other.The mass is rotated and its amplitude of oscillation measured as a function of time. The ratio of successive amplitudes gives the logarithmic decrement and hence the **loss factor**.

toti- Prefix from L *totus*, all, whole.

touch and close fastener A pair of tapes made from synthetic fibres that stick together on pressing by hand and then separate by peeling when required, eg Velcro. The tapes have pile surfaces, one formed of hooks and the other of loops, which engage on being pressed together.

toughened glass See **safety glass**.

toughness Qualitatively used to denote a condition intermediate between **brittleness** and softness, as indicated in tensile tests by high **ultimate tensile stress** and low to moderate **elongation** and reduction in area, or by high values of energy absorbed by **impact tests**. More precisely it is the value of the critical **strain energy release rate**, G_C, required for propagation in a material. In a fracture test on a cracked specimen of thickness, t, it is evaluated from

$$G_C = \frac{1}{2t} \times \frac{\partial C}{\partial a},$$

where C is the **compliance** of the speciment and a is the crack length. Formally G_C is twice the **fracture energy** of the material. Also *crack driving force*. See **Griffith equation**, **strength measures** p. 301. Cf **fracture toughness**.

tough pitch A term applied to copper in which the oxygen content has been correctly adjusted at about 0.03% by poling. Distinguished from *overpoled* and *underpoled* copper. See **poling**.

tow A collection of a large number of parallel continuous filament fibres, not twisted together. By cutting or breaking (ie converting) a **top** is produced, the process being known as *tow-to-top conversion*.

TPE Abbrev. for **ThermoPlastic Elastomer**.

TPX TN for poly(4-methylpent-1-ene) (Japan).

trace element analysis Type of elemental analysis which seeks to identify origin of materials by finding minor traces of elements accidentally present in samples (eg impurities). The principle can be extended to compound analysis using such hypersensitive methods as nuclear magnetic resonance.

tracheid The principal structural and conduction cells in softwoods and also present to a lesser extent in hardwoods. See **structure of wood** p. 305.

track A conducting path in a circuit, forming interconnections.

tracking Surface breakdown on an electrical insulator eg owing to moisture or other impurities, often leading to waxing and carbon formation.

tracking resistance Ability of an insulator to withstand surface discharge of electricity, esp. over long times. Track formation often occurs at a critical voltage and consists of paths of carbonized material formed by degradation in the case of polymers. Standard tests include measurement of the comparative tracking resistance (CTR).

tramp element (1) Noble metals (eg copper) present in steel scrap which are difficult to remove by refining. (2) General term for any elements present in scrap material which are difficult to remove before the material can be recycled. By extension, *tramp alloys*, *tramp compounds*, *tramp metals*. (3) Metal contaminating a structure, eg *tramp uranium* on nuclear reactor pipework. (4) Stray metal pieces which are accidentally entrained in processed materials or products, eg a piece of steel broken from a machine in a load of rock from a mine.

trans- That geometrical or rotational isomer in which like groups are on opposite sides of a double bond.

transcrystalline failure The normal type of failure observed in metals. The line of fracture passes through the crystals, and not round the boundaries as in intercrystalline failure.

transfer moulding Development of compression moulding, where powder polymer is placed in small chamber adjacent to tool mould and when heated, is forced into tool cavity. Not to be confused with **injection moulding**.

transformation A constitutional change in a solid material, eg the change from gamma to alpha iron, the formation of pearlite, bainite or martensite from austenite in steels, or transformation from alpha to beta quartz in a silica refractory.

transformation temperature The temperature at which phase changes occur in the solid state. Denoted by various symbols: eg for steels, A_c, the temperature on heating; A_e, the temperature at equilibrium; A_r, the temperature on cooling. Subscripts refer to the transformation in question.

transformer An electrical device without any moving parts, which transfers energy of an alternating current in the primary winding to that in one or more secondary windings, through electromagnetic induction. Except in the case of the autotransformer there is no electrical connection between the two windings and usually (excepting the isolating transformer) a change of voltage is involved in the transformation.

transformer core The structure, usually of laminated iron or **ferrite**, forming the magnetic circuit joining the different parts of a transformer, of any degree of coupling. Also *mutual coupling*.

transformer oil A mineral oil of high dielectric strength, forming the cooling and insulating medium of electric power transformers.

transformer stampings The laminations, stamped out of transformer plate, which are assembled to form the transformer core.

transistor See **bipolar transistor**, **field effect transistor**.

transition In atomic and nuclear physics, the terms used to describe the change from one quantum state to another. A transition from a higher to a lower energy state may be accompanied by the *emission* of a *photon*, while a transition from a lower to a higher

state requires the *absorption* of a photon. Transitions are governed by *selection rules* which forbid some transitions.

transition metal One of the group which have an incomplete inner electron shell. They are characterized by large atomic magnetic moments. Also *transition element*.

transition region The part of a doped semiconductor over which the impurity concentration varies.

transition temperature Temperature at which some critical change occurs such as in magnetism (**Curie temperature**) and superconductivity.

transmissibility Ratio of the amplitude of motion of isolated structure to that of its foundation, and a key starting point for analysis of vibration isolation. Main problem is minimizing it at and near the natural frequency of the system, where it rises, often catastrophically. Often solved by using rubber springs.

transmission electron microscopy A method for exploring the microstructure of materials using an electron beam controlled by magnetic lenses. At conventional accelerating voltages (100 kV), samples must be very thin (~20 nm) for electron penetration, causing problems of interpretation if the sample does not represent the bulk. A routine method for determining crystal structure in thin metal foils by electron diffraction and identification of inclusions etc, as well as dislocation observation. Ceramics can be thinned down by ion-beam method, polymers by ultramicrotomy or by casting a thin film from solution. Contrast must usually be enhanced in copolymers by staining with OsO_4. Million volt microscopes allow much thicker samples to be examined. Abbrev. *TEM*.

transmittance Ratio of energy transmitted by a body to that incident on it. If scattered emergent energy is included in the ratio, it is termed *diffuse transmittance*, otherwise *specular transmittance*. Also *transmission*.

trap Crystal lattice defect at which current carriers may be trapped in a semiconductor. This trap can increase recombination and generation or may reduce the mobility of the charge carriers.

TRAPATT diode Abbrev. for *TRApped Plasma Avalanche Transit-Time diode*. A diode which can operate as an oscillator at microwave frequencies, the frequency being determined by the thickness of the active layer in the diode. The avalanche zone moves through the drift region a trapped space-charge plasma within the p-n junction.

trash (1) The unwanted material present in bales of raw cotton. (2) Generic term (US) for post-consumer waste.

tread See **tyre technology** p. 325.

trepanning Producing a hole by removing an annular ring of material, as opposed to drilling out the entire volume corresponding to the hole.

Tresca yield criterion When a triaxial stress system acts in a material, yield will occur when the maximum **shear stress** reaches a critical value τ_c. Since by convention, σ_1 is the largest **principal stress** and σ_3 the smallest, the criterion can be stated as

$$\tau_c = \frac{\sigma_1 - \sigma_3}{2}.$$

In terms of the uniaxial **yield stress**, σ_Y, obtained from a **tensile test**, the criterion becomes

$$\sigma_1 - \sigma_3 = \sigma_Y.$$

tri-. Prefix from L, *tres*, Greek *tria*, three.

triac A bi-directional gate-controlled **thyristor** for full-wave control of a.c. power. By altering the phase of the gate switching signal, load current can be adjusted from a few per cent to nearly 100% of the full load current.

triacetate Textile fibres of cellulose acetate of which nearly all (92%) of the hydroxyl groups are acetylated.

triboelectric series Ranking of insulators esp polymers, on a linear scale of severity with positive charging at one end, negative charging at other. See **static electrification**.

triboelectrification Separation of charges through surface friction. If glass is rubbed with silk, the glass becomes *positive* and the silk *negative*, ie the silk takes electrons. The phenomenon is that of contact potential, made more evident in insulators by rubbing.

tribology The science and technology of interacting surfaces in relative motion (and the practices related thereto), including the subjects of friction, lubrication and wear.

triboluminescence Luminescence generated by friction.

tricot Fabric knitted on a warp-knitting machine.

tricotine (1) A fine-ribbed, plain weave fabric with a silk warp and cotton weft. (2) A weft-face woven fabric with a cotton warp and worsted wool weft containing a fine twill line.

trimming The removal of flash from the edge of a workpiece, eg a plastic moulding or metal casting.

triple bond A covalent bond between two atoms involving the sharing of three pairs of electrons. Shown as $\equiv$. See **alkyne**.

triple point The temperature and pressure at which the three phases of a substance can coexist (see **phase rule**). The triple point of water is the equilibrium point (273.16 K and $610 \, N \, m^{-2}$) between pure ice, air-free water and water vapour obtained in a sealed vac-

uum flask. It is one of the fundamental **fixed points** of the international practical scale of temperature.

triplet The three resonance peaks in spectroscopy, often produced by two unpaired electrons (electron spin resonance) or two hydrogen bonds (nuclear magnetic resonance). Singlet, doublet etc are similarly defined.

trippage Number of times a product is reused as part of its normal product life cycle; thus the UK milk bottle is reused on average eight times before being scrapped for **cullet**.

tritolylphosphate Polyvinyl chloride plasticizer with high-temperature resistance, although sensitive to hydrolysis. Abbrev. *TTP*.

trixylylphosphate Polyvinyl chloride plasticizer with high-temperature resistance, although hydrolysis sensitive. Abbrev. *TXP*.

Trogamid TN for non-crystalline and glassy **nylon** with complex, mixed aliphatic/aromatic backbone chain. Used for mouldings.

trouble-shooting See **fault-finding**.

trouser tear test Test method for measuring tearing energy of various materials of film or sheet form, after shaping the test specimen the legs are grasped and pulled in tension. See **strength measures** p. 301.

Troutonian fluids Fluids whose coefficient of tensile viscosity is unaffected by increasing stress. Equivalent to Newtonian fluids tested in shear. Most polymer melts are Troutonian, but low-density polyethylene is tension stiffening, and thus ideal for the blown film process.

Trouton's rule For most non-associated liquids, the ratio of the latent heat of vaporization per mole, measured in joules, to the boiling-point, on the absolute scale of temperature, is approx. equal to 88 at atmospheric pressure.

true bias The direction in a fabric along which it will show the greatest elongation in response to a given load.

true strain See **strain**.

true stress See **stress**.

TTP Abbrev. for **TriTolylPhosphate**.

TTT diagram See **isothermal transformation diagram** p. 177.

tube-drawing The production of seamless tubes by drawing a large, roughly formed tubular piece of material through dies of progressively decreasing size, usually using a mandrel or plug to control the internal diameter.

tube extrusion A method of producing tubes by direct extrusion from a billet, using a mandrel to shape the inside of the tube.

tube plate End wall of a surface condenser, between which the water tubes are carried; they are bolted between the casing and water-chamber covers. See **condenser tubes**.

tube sinking Drawing an existing tube through a die or rolls to reduce its diameter without an interior plug or mandrel.

Tufnol A proprietary laminated plastic; lightweight, tensile strength approx. 55–110 MN m^{-2}; strong insulation qualities, but increasingly replaced as an engineering plastic for bearings, gear wheels and pulleys by **acetal resin**.

tufted carpet A carpet formed by inserting with needles U-shaped lengths of yarn, or similar, into a strong backing material (eg a hessian or polyolefin fabric or a plastic foam).

tulip tree See **American whitewood**.

tulipwood See **Brazilian whitewood, American whitewood**.

tulle Traditionally, a fine plain-woven silk net. Now also applied to a net with hexagonal holes produced on a warp-knitting machine.

tumbling A method of removing sand, irregularities etc. from castings or forgings by rotating them in a box with abrasives or special metal slugs.

tungsten Element, symbol W. A hard grey metal, resistant to corrosion, used in cemented carbides for drills and grinding tools, and as wire in incandescent electric lamps. Also *wolfram*. See Appendix 1, 2.

tungsten alloy A protective material containing tungsten, copper and nickel, and having a density about 50% greater than that of lead, and thus providing better protection from ionizing radiation.

tungsten bronze Malleable, machinable alloys with high densities based on tungsten alloyed with 6–7% nickel and 3–4% copper or iron.

tungsten inert gas welding Electric welding in which the tungsten electrode is not consumed and a *filler rod* supplies the metal to the joint which is protected from reaction by an inert gas, eg argon. Abbrev. *TIG*.

tunnel diode Junction diode with such a thin depletion layer that electrons bypass the potential barrier. See **tunnel effect**. Negative resistance characteristics can be exhibited and such diodes can be used as low noise amplifiers or as oscillators, up to microwave frequencies. Also *Esaki diode*.

tunnel effect Piercing of a narrow potential barrier by a current carrier which is impossible according to classical physics, but has a finite probability according to wave-mechanics.

tunnel furnace Kiln through which material moves slowly on cars, racks, or suspending gear.

turbulent flow Fluid flow in which the particle motion at any point varies rapidly in magnitude and direction. This irregular eddying motion is characteristic of fluid motion at high Reynold's numbers. Gives rise to high drag, particularly in the *boundary layer* of aircraft. Also *turbulence*. See **laminar flow, Reynolds number**.

turning Producing cylindrical, flat or tapered workpieces in a lathe.

turnings Chips or **swarf** produced as waste in turning.

tussah silk A coarse silk produced by a wild silkworm eg *Antheraea mylitta* (from India), *A. pernyi* (from China) and *A. yamamai* (from Japan). The fibres are pale brown and are usually rather short. They are spun into a yarn which has many irregular slubs.

tussore A fabric woven from **tussah silk**.

Twaron TN for aramid fibre (Europe).

tweed Traditionally a coarse, heavy, rough wool outerwear fabric. Now applies to other wool fabrics having a wide range of weights and weave effects (eg **twills**).

twills Woven fabrics with diagonal lines on the face. Regular twills have continuous lines; zigzag twills have the lines reversed at intervals.

twin One of a pair of two and related entities similar in structure (eg crystals) or function; often synonymous with *double*.

twinned crystal Crystal containing a fault in the atomic stacking which causes the lattice structure on one side of the fault to appear as the mirror image of that on the other. A second fault will correct the effect of the first. Observed in some minerals and commonly in microstructures of copper and nickel alloys after recrystallization, both of which have low stacking fault energies. It does not appear in aluminium alloys which have similar crystal structures but a high stacking fault energy. Twins may also be produced in certain crystals by high strain rate mechanical deformation.

twin-screw extruder Extruder fitted with intermeshing, counter-rotating screws which provide better mixing of polymers and additives, esp. polyvinyl chloride compounds.

twist Fibres in a yarn are held together by the degree of twist introduced in spinning. This may be quantified as the twist level, ie the number of revolutions per unit length. Lively yarns have a tendency to untwist.

twist direction If the direction of twist as viewed on a yarn held vertically goes diagonally from left to right it has *S-twist*; if from right to left it has *Z-twist*.

twist drill A hardened steel drill in which cutting edges, of specific rake, are formed by the intersection of helical flutes with the conical point which is backed off to give clearance; of universal application.

twisting paper A long-fibred paper suitable

tyre technology

Over 60% of all rubber produced today, whether synthetic or natural, is used in car or truck tyres. Although when cut open, a tyre reveals an apparently homogeneous matrix material, it is in fact composed of a mixture of different blends (Fig. 1). The obvious reinforcing radial plies are composed of high-tenacity rayon, polyethylene terephthalate or nylon 6,6 fibre. The tread breaker, which keeps the tread firmly gripping the road, is composed of *bias-belted* steel wire and often aramid fibre, while the steel cord of the bead keeps the tyre firmly attached to the wheel. The elastomers (see p. 106) used in the tread are usually styrenebutadiene and natural rubber blends for high hysteresis and hence good road grip. The requirements of the sidewall are quite the opposite: since they must flex to absorb vibrations caused by adverse road conditions, cornering etc; heat build-up must be reduced. Low hysteresis blends of natural and polybutylene rubbers are therefore used. The tyre must also be impermeable to the pressurized air contained within it, a function of the butyl (often bromo-butyl) rubber inner lining.

Fig. 1. **The structure of a radial-ply tyre.**

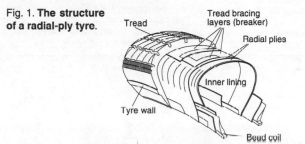

All the rubber components are compounded with carbon black, sulphur accelerator, activator, anti-oxidants, anti-ozonants, oil-extenders etc. Apart from the key crosslinking components, carbon black is the most important modifier of mechanical properties. When used as a pigment in thermoplastics, there is little reinforcing action (rather, the reverse) but in elastomers, extremely fine particles are used, eg *High Abrasion Fine* (HAF) of mean diameter 30 nm. If well dispersed, the rubber chains are adsorbed on the particle surfaces (Fig. 2).

Fig. 2. **The effect of carbon black on the rubber elastomer.**

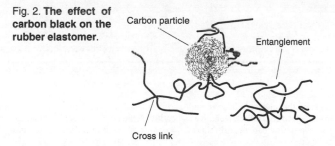

Some chemical reaction occurs between the free radicals present on the carbon black surface and the highly-reactive double bonds of the rubber chains, so that the crosslink density is effectively increased. The maximum elongation is reduced by the rigidity of the graphitic particles together with their high loading. The carbon black does not affect the position of the glass transition (T_g), but smears it out over a wider temperature range rather than increasing it as in **ebonite**. This is important, because tyres must remain elastomeric at the low temperatures encountered in cold climates. An additional benefit is that the aromatic rings of the carbon black help absorb ultraviolet radiation from the Sun and protect against degradation without affecting the rubber molecules.

for waxing and intended for wrapping around sweets, toffee etc.

two-roll mill Mixing process for polymers, esp. rubbers where materials are mixed with additives and fillers on two adjacent rollers. **Dispersive mixing** occurs at the nip, and **distributive mixing** by cutting and overlap of the polymer sheet.

TXP Abbrev. for **TriXylylPhosphate**.

Tylose TN for regenerated cellulose with properties resembling flesh.

type I, II superconductors See **superconductors** p. 308.

type metal A series of alloys of lead, antimony and tin, used for printing type. One composition is antimony 10–20%, tin 2–12% and the remainder lead. In another, tin may reach 26%, and up to 1% of copper may be added.

tyre-building machine Centrepiece of tyre factory, where all the parts are assembled to make the green tyre. See **tyre technology** p. 325.

tyre disposal Problem created by growth in the use of cars and increasingly stringent regulations on tread depth (min. 1.6 mm). As a thermoset, little of the bulk of a tyre is recyclable, but incineration can yield useful energy. Care is needed to limit SO_2 emissions.

tyre lining See **tyre technology** p. 325.

tyre technology See panel on p. 325.

tyre textiles Pneumatic tyres are made from composite materials of rubber reinforced with several layers of textile fabrics or steel wire. The fabrics are mostly made of synthetic fibres and include some, the tyre cord, where the warp threads predominate and the few weft threads are present only to assist processing. Radial and cross-ply denote the way the cords are arranged in forming the carcass. See **tyre technology** p. 325.

Tyvak TN for bonded, non-woven paper made from synthetic polymer fibre such as polyethylene (US).

U – Z

u Symbol for unit of unified scale.

Ubbelohde viscometer Type of viscometer used for measuring dilute solution viscosity of polymers. Consists of glass capillary and reservoir on which are enscribed two fixed marks. Held vertical, the solution is allowed to flow through the capillary from the reservoir, and the time for movement between the fixed marks recorded. This is repeated for several different concentrations, from which data, the intrinsic viscosity of the polymer can be calculated.

UF Abbrev. for *Urea-Formaldehyde plastics* See **urea resins**.

Uganda mahogany See **African mahogany**.

UHMPE Abbrev. for **Ultra-High Molecular Mass PolyEthylene**.

UL Abbrev. for **Underwriters Laboratories**.

UL 94 flammability Burning test for polymers where rod, held in flame under test conditions, is rated on a scale depending on the rate of burning etc. See **NBS smoke test**.

ultimate tensile stress The highest load applied to a material in the course of a **tensile test**, divided by the original cross-sectional area. In brittle or very tough materials it coincides with the point of fracture, but usually extension continues under a decreasing stress, after the ultimate stress has been passed. Also *tenacity* (obs.). See **strength measures** p. 301.

ultra- Prefix from L *ultra*, beyond.

ultra-high molecular mass polyethylene Linear polyethylene with molecular mass of several million, so difficult to process by normal means. Sintered to shape in hip joint socket, and raw material for gel-spinning into high-performance fibre. Abbrev. *UHMPE*. See **molecular mass distribution** p. 212.

ultra large scale integration Grouping of many electronic components on a single chip beyond that of **very large scale integration**. Abbrev. *ULSI*.

ULSI Abbrev. for **Ultra Large Scale Integration**.

ultrasonic cleaning A cleaning process used in conjunction with water or solvents and effective for small crevices, blind holes etc. Ultrasonic frequency vibrations are transferred to the cleaning fluid producing a turbulent penetrating action.

ultrasonic imaging See **acoustic microscope**.

ultrasonic machining Process of removing material by abrasive bombardment and crushing in which a relatively soft tool, matching the shape produced, is made to oscillate at ultrasonic frequencies and drives the abrasive grit, suspended in a liquid, against the work, blasting away fine particles of the material. Operations include drilling round and odd-shaped holes, diesinking and forming wire drawing dies. Primarily for machining hard brittle materials like carbides, ceramics and sintered metals.

ultrasonic soldering Form of soldering in which a specially designed soldering bit replaces the machine tool. Soldering is particularly difficult with aluminium and the application of ultrasonics is supposed to break up the aluminium oxide layer.

ultrasonic testing A method of testing for flaws in which an ultrasonic source is pressed against the part to be tested, using some form of gel to act as a sonic coupling to the surface, and sound is passed into the material. Reflections or echoes occur from the back face and, additionally, any internal discontinuity will reflect the sound wave and generate a signal in the receiver. The time lags of the echoes are measured to determine the thickness of the workpiece and the distance to the discontinuity. The appropriate probe and type of sonic wave permits the size and shape of a flaw to be determined as well as its position within the material.

ultrasonic welding A solid-state process for bonding sheets of similar or dissimilar materials, usually with a lap joint. No heat is applied but vibratory energy at ultrasonic frequencies is applied in a plane parallel to the surface of the weldment.

ultraviolet radiation Electromagnetic radiation in a wavelength range from 400 nm to 10 nm approximately, ie between the visible and X-ray regions of the spectrum. The *near* ultraviolet is from 400 to 300 nm, the *middle* from 300 to 200 nm and the *extreme* from 200 to 10 nm. Abbrev. *UV*.

ultraviolet spectroscopy Method of detecting ultraviolet absorption by aromatic groups or conjugated bonds, esp. in polymers in a suitable solvent. Complementary tool to infrared and nuclear magnetic resonance spectroscopy.

umen The non-mineralized substances of coal etc, and their distillation residues.

Umklapp process A type of collision between phonons, or phonons and electrons, in which crystal momentum is not conserved. Such processes provide the greater part of the thermal resistance in solid dielectrics.

unbleached kraft paper Brown packaging paper made from unbleached sulphate wood pulp, suitable for use as wrapping material or for conversion into sacks, bags, gummed tape etc.

uncertainty principle There is a fundamental limit to the precision with which a position co-ordinate of a particle and its momentum in that direction can be simultaneously known. Also, there is a fundamental limit to the knowledge of the energy of a particle when it is measured for a finite time. In both statements, the product of the uncertainties in the measurements of the two quantities involved must be greater than $h/2\pi$, where h is Planck's constant. See **Planck's law**. The principle follows from the wave nature of particles. Also *Heisenberg uncertainty principle, indeterminacy principle.*

undercure Rubber technology term for state of rubber compound prior to desired cross-link density has been achieved.

undercut (1) A mould or pattern, eg in a foundry, is undercut when it has a re-entrant portion and thus cannot be opened or withdrawn in a straight motion. (2) A short portion of a bolt, stepped shaft or similar part, with a diameter smaller than the peripheries or diameters of the adjacent features. It obviates difficulties of machining sharp corners and facilitates subsequent assembly. (3) A fault appearing as a cutting or washing away of parent plate adjacent to a weld fillet, which reduces the effective section thickness of one of the components joined.

undercutting In semiconductor processing, the etching of material beneath the edge of a mask owing to lateral etching from an adjacent unmasked region; it is an inevitable consequence of isotropic etching.

underpoled copper See **poling**.

underwater cutting A method of cutting iron or steel under water, utilizing a combusted mixture of oxygen and hydrogen (cf **oxy-hydrogen welding**), the flame being protected by an airshield formed by a hood.

Underwriters Laboratories Insurers' organisation which sets standards for testing materials, as well as performing tests (US). Abbrev. *UL.*

undrawn yarn Extruded man-made filament yarn not yet subjected to the drawing process that orients the linear molecules and gives strength to the yarn.

uniaxial, uniaxial crystal A term embracing all those crystalline minerals in which there is only one direction of single refraction (parallel to the principal crystal axis and known as the optic axis). All minerals which crystallize in the tetragonal, trigonal, and hexagonal systems are *uniaxial.* Cf. **biaxial**.

uniform extension The plastic extension produced along the gauge length of a tensile test piece before localised necking commences, ie up to the stage when the maximum load is reached. Thereafter further plastic deformation to fracture is confined to the region of the neck. See **strength measures** p. 301.

union fabric A woven fabric with the warp of one fibre (eg linen) and the weft of another (eg cotton).

union kraft laminated packaging material made of two layers of kraft paper bonded by a layer that is impermeable to water or water vapour, eg bitumen or polyethylene.

unipolar transistor One with one polarity of carrier.

unit A dimension or quantity which is taken as a standard of measurement.

unit cell The smallest group of atoms, ions, or molecules, whose repetition at regular intervals, in three dimensions, produces the lattice of a given crystal.

unsaturated (1) Less concentrated than a saturated solution or vapour. (2) Containing a double or a triple bond, esp. between two carbon atoms; unsaturated molecules can thus add on other atoms or radicals before saturation is reached.

unsoundness The condition of a solid metal which contains blowholes or pinholes due to gases, or cavities resulting from shrinkage during contraction from liquid to solid state, ie contraction cavities.

unstable Subject to spontaneous change.

upsetting Metalworking so as to produce an increase in section of part of a component over and above its starting size, as in the forming of the head of a bolt or rivet from round bar.

UPVC Abbrev. for **Unplasticized PiolyVinyl Chloride**. Also *uPVC*. See **polyvinyl chloride** p. 249.

uranium A hard grey metal with a number of isotopes, symbol U. ^{235}U is the only naturally occurring readily fissile isotope and exists as one part in 140 of natural uranium. See Appendix 1, 2.

urea-formaldehyde plastics See **urea resins**. Abbrev. *UF.*

urea resins The thermosetting resins manufactured by heating together urea and an aldehyde, generally formaldehyde. Pale coloured or water-white and translucent, they can therefore take delicate dyes and tints. They are non-flammable, and are resistant to weathering and fluid attack. Also *aminoaldehydic resins.* See **thermosets** p. 318.

UV Abbrev. for *ultraviolet.* See **ultraviolet radiation**.

UV absorbers Aromatic chemicals added to polymers exposed to sunlight so that ultraviolet degradation can be inhibited.

UV degradation Ultra-violet initiated

breakdown of polymers owing to photon absorption at chain defects, such as carbonyl groups in backbone chain. Commonly caused by effect of sunlight on products like garden furniture, resulting in surface cracking. It is closely related to the oxidation of polymers. Effect is exploited, however, in **photoresists** and **photolithography**.

vacancy Site unnoccupied by atom or ion in a crystal lattice.

vacuum arc melting Using a consumable electrode to melt an ingot with the arc in vacuo, the reduced pressure causing the metal to out-gas.

vacuum bag moulding See **bag moulding**.

vacuum forming Method of thermoforming thermoplastic polymers, where shaping is effected by a vacuum applied through holes in a female tool. The normal process for making polymethyl methacrylate domestic baths.

vacuum furnace One in which material, either solid or liquid, may be outgassed by heating in vacuo. See **vacuum arc melting**.

vacuum induction melting Using induction to melt metal in vacuo preparatory to casting.

valence band Range of energy levels of electrons which bind the atoms of a crystal together.

valence electrons Those in the outer shell of an *atom*, which, by gaining, losing, or sharing such electrons, may combine with other atoms to form *molecules*.

valency The combining power of an atom or group in terms of hydrogen atoms (or equivalent). The valency of an ion is equal to its charge. For *valency bond*, *valence*, see **chemical bond**.

value engineering An approach to engineering design aimed at achieving required performance, reliability and quality at minimum cost by attention to simplicity, avoidance of unnecessary functions and integration of design and manufacturing techniques.

vanadium A very hard, whitish metallic element, symbol V. Its principal use is as a constituent of alloy steel, eg in chromium-vanadium, manganese-vanadium and high-speed steels. See Appendix 1, 2.

vanadium steel Those containing small quantities of vanadium as alloying ingredient though seldom as the sole addition. A typical compostion might be carbon 0.4–0.5%, chromium 1.1–1.5%, vanadium 0.15–0.2%, balance iron. See **steels** p. 296.

van der Waals forces Weak attractive forces between atoms or molecules which vary inversely as the sixth power of the interatomic or intermolecular distance, and are due to momentary dipoles caused by fluctuations in the electronic configuration of the atoms or molecules. Also *London forces*. See **bonding** p. 33.

vapour deposition See **chemical vapour deposition, physical vapour deposition**.

varistor Two-electrode semiconductor with a non-linear resistance dependent on the instantaneous voltage. Used to shortcircuit transient high voltages in delicate electronic devices.

vector A vector or vector quantity is one which has magnitude and direction, eg force or velocity; two such quantities of the same kind obey the parallelogram law of addition. A *localized vector* is one in which the line of action is fixed, as contrasted with a *free vector*, in which only the direction is fixed.

vegetable parchment A dense, grease-resistant, packaging paper with a high wet strength, produced by treating a base paper with sulphuric acid followed by washing and/or neutralizing.

Velcro TN for fabric fastening system, relying on mechanical bond between looped nylon filaments and hooked nylon filament ends. Developed by mimicking natural burrs. It is a **touch and close fastener**.

vellum An early form of writing surface made from the skins of calves, lambs, or goats. In modern usage, a thick writing paper.

velocity (1) The rate of change of displacement of a moving body with time; a vector expressing both magnitude and direction (cf. *speed* which is scalar). (2) For a wave, the distance travelled by a given phase divided by the time taken. Symbol v.

velour A heavy pile or napped woven fabric or felt with the surface fibres all lying in the same direction. Also, a warp-knitted fabric with long raised loops. The velour used for hats is usually a rabbit fur that has been felted and raised.

velour paper Paper made by depositing short wool fibres on an adhesive coated paper. Also *flock paper*.

velvet Woven fabric with a dense short pile, formed in loops which are then cut. In one method of manufacture, two cloths are woven face to face with the pile warp weaving through both and is then sliced in the loom to produce two separate fabrics.

velveteen A woven pile fabric, usually of cotton, that is made by cutting extra floating weft threads after weaving.

veneer Thin layer of decorative timber glued to the surface of a less expensive wood.

Venetian fabric (1) A satin lining fabric made of mercerized cotton. (2) A satin weave, woollen cloth used in overcoats.

venting Special features which allow gases trapped in barrel or tool to escape without causing burn marks etc.

vertical flash tool Injection moulding tool where mating surfaces between moving parts are so designed that tool can open slightly during maximum pressurization. See **mould breathing**.

very large scale integration Integration of the order of 10^4–10^6 electronic components on a single chip. Abbrev. *VLSI*.

Vibram TN for type of vulcanized rubber sole with characteristic tread pattern, used for climbing boots etc.

vibrating reed Method of measuring dynamic mechanical properties of materials, esp. polymers, using resonance vibration of the sample. Contrast forced vibration experiments, where specimen is controlled by apparatus, eg **torsion pendulum**.

vibration isolation Design method for selecting best device for the absorption of unwanted vibrations. See **rubber spring**. **transmissibility**.

Vickers hardness number See **Vickers hardness test** p. 331.

Vickers hardness test See panel on p. 331.

Victrex PEEK TN for a polyether ether ketone polymer (UK).

vicuña (vicunsia) The fine hair from the undercoat of the vicuña (or vicunsìa), *Lama vicugna*, a kind of llama from S America. The fabric woven from it is the softest and finest of any made of wool or hair.

vinhatico Hardwood tree (*Plathymenia reticulata*) principally from Brazil. Sometimes called *Brazilian* or *yellow mahogany*. **Heartwood** is a lustrous, yellow-orange brown with darker and lighter streaks. Used for high-class furniture and cabinetmaking, shipbuilding, veneers etc. Density 600 kg m^{-3}.

vinyl polymers See **polymers** p. 247.

violetwood See **purpleheart**.

virgin metal Metal or alloy first produced by smelting, as distinct from secondary metal containing recirculated scrap.

virola C American **hardwood** tree with rather featureless pinkish-brown wood. The wood has a slight tendency to split. It is used for packing-cases, coffins, matches, matchboxes and roofing shingles and in plywood and veneers. Density 430–590 kg m^{-3} (average 530 kg m^{-3}). Also *banak*.

viscoelasticity See panel on p. 332.

viscometer An instrument for measuring viscosity. Many types of viscometer employ **Poiseuille's formula** for the rate of flow of a viscous fluid through a capillary tube.

viscose process Method of producing regenerated cellulose fibre (viscose rayon) or film (cellophane) by digestion of natural cellulose. Treatment with carbon disulphide

(CS_2) yields cellulose xanthate with ca one xanthate group for every two glucose units. Digestion in caustic soda yields a solution of viscose polymer which is coagulated in acid solution (H_2SO_4) and then spun or extruded.

viscosity The resistance of a fluid to shear forces, and hence to flow. In a **Newtonian fluid**, such shear resistance is proportional to the relative velocity between the two surfaces on either side of a layer of fluid, the area in shear, the viscosity of the fluid and the reciprocal of the thickness of the layer of fluid. The viscosity, or *coefficient of viscosity*, η is defined by:

$$\eta = \tau \left(\frac{d\gamma}{dt} \right)^{-1}$$

where τ is the shear stress and ($d\gamma/dt$) is the shear strain rate. The shear stress equals the shear force, F divided by the area A over which it acts, and the shear strain rate is equivalent to the velocity gradient, dv/dz. For normal ranges of temperature, η for a liquid decreases with increase in temperature and is independent of the pressure. Unit of measurement is N m^{-2} s in SI units (=10 poise or dyne cm^{-2} s in CGS units). More complex fluids are non-Newtonian. Important for polymer melts, whose viscosities decrease with increasing shear rate (*pseudoplastic fluids*). In others (*dilatant fluids*), the viscosity increases with shear rate.

viscosity-average molecular mass The measure of molecular mass derived from dilute solution viscometry, and usually falling between $\overline{M}_n$ and $\overline{M}_w$. Symbol $\overline{M}_v$. See **molecular mass distribution** p. 212.

visible radiation Electromagnetic radiation which falls within the wavelength range of 780 to 380 nm, over which the normal eye is sensitive.

Viton TN for synthetic rubber based on a copolymer of vinylidene fluoride and hexafluoropropene. Maintains its rubber-like properties over a very wide temperature range and is chemically very resistant. Used for O-rings etc.

vitreous enamel Glazed coating fused on to metal surface for protection and/or decoration. Also *enamel*.

vitreous silica Glassy material consisting almost entirely of silica, made in translucent and transparent forms. The former has minute gas bubbles disseminated in it. Compared to ordinary soda-lime-silica glass, it has a much lower thermal expansion coefficient, a higher thermal shock resistance and a higher temperature limit. Also *fused silica*, *quartz glass*, *silica glass*. See **silica and silicates** p. 285.

VLSI Abbrev. for **Very Large Scale Integration**.

Vickers hardness test

This is one of the most commonly used methods of hardness measurement. A 136° diamond pyramid (Fig. 1a) is pushed with a constant force, F, into the surface of a specimen for a specified period during which material flows plastically away from the indenter. After the diamond is withdrawn, the diagonal lengths, d, of the indentation (Fig. 1b) are measured. The *Vickers Hardness Number*, H_V (also *VHN*, *VPN* or *DPH*), is the force divided by the contact surface area of the indentation:

$$H_V = 2F \sin(136°/2)/d^2$$

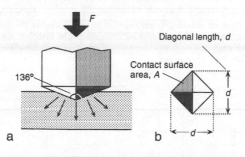

Fig. 1. **The Vickers hardness test.** (a) The diamond pyramid. (b) The indentation.

Most machines have a set of tables for each loading force. The user measures the average diagonal length and reads the hardness from the tables. Loads vary from 30 kg down to 5 g (*microhardness*). The response of materials depends on their ratio of yield stress, σ_Y to Young's modulus, E. For the softer metals, with low σ_Y/E, when they are either non-work-hardening or fully work-hardened, it is found that

$$H_V \approx 3\sigma_Y$$

Note that H_V is conventionally cited in units of kgf mm^{-2}, so that σ_Y must have the same units in the above equation.

For materials with higher σ_Y/E, which include a wide range of metals, glasses and plastics, the mode of deformation changes, and the relationship between H_V and σ_Y becomes more complex . However, there is a useful empirical relationship for steels between the tensile strength (in MN m^{-2}) and H_V (in kgf mm^{-2}), namely

$$\sigma_{TS} \approx 3.2\, H_V$$

Values of H_V for a range of materials are shown in the table below.

Material	H_V	Material	H_V	Material	H_V
tin	5	limestone	250	polypropylene	7
aluminium	25	MgO	500	polycarbonate	14
gold	35	window glass	550	PVC	16
copper	40	fused silica	720	polyacetal	18
iron	80	granite	850	PMMA	20
mild steel	230	quartz	1200	polystyrene	21
full hard steel	1000	Al_2O_3	2500	urea formald.	41
		tungsten carbide	2500	epoxy	45

Finally, it should be borne in mind that, in those materials which exhibit time-dependence of either σ_Y (eg metals above 0.5 T_m) or both σ_Y and E (eg the majority of plastics), the size of the indentation will increase with time, and thus their H_V value will depend on how long the load is applied.

viscoelasticity

The response of viscoelastic materials to an applied load is a combination of viscous and elastomeric components. The viscous component in polymers is caused by chain slippage in the amorphous zones and is time-dependent and irreversible. The elastomeric component arises in the same zones but is reversible, being caused by entanglements trapping chain segments which strive to return to their equilibrium position when unstressed. See **elastomers** p. 106. At low temperatures, below the *glass transition temperature* (T_g), long-range chain motion is impossible and the response is more nearly elastic or *Hookean*. Crystalline regions also respond in an elastic fashion, above or below T_g. The changing mechanical behaviour of polymers is often described by a master curve, a plot of creep- or stress-relaxation modulus versus temperature. The master curve is obtained by superimposing many stress-relaxation curves using the shift factor (a_T), which itself is described by the **Williams–Landel–Ferry equation**. See Fig. 1.

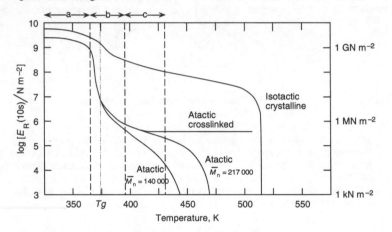

Fig. 1. **Short-term stress relaxation modulus of various polystyrenes**.

The curves show the short-term (10 sec) stress-relaxation modulus of atactic amorphous polystyrene, isotactic crystalline polystyrene in a laboratory specimen and lightly-crosslinked atactic polystyrene of two molecular masses. There are four regions of behaviour as temperature rises:

(a) glassy, brittle behaviour below T_g;

(b) leathery transition region centred on T_g;

(c) elastomeric region above T_g;

(d) viscous flow region.

If the polymer is lightly crosslinked, viscous flow is eliminated; if it is highly crystalline, the polymer shows an enhanced modulus which drops suddenly at the crystal melting point T_m. All four viscoelastic regimes are shown rather dramatically by 'potty-putty', a low-molecular mass silicone rubber cross-linked with boron oxide. By the **time-temperature equivalence principle**, increase of temperature is equivalent to increase of time scale; so when hit with a hammer, the material shatters like glass. When rolled into a ball and dropped, it bounces like rubber but yet can be kneaded like putty with the fingers. At long time scales it flows like a viscous liquid and will cover a wide area without any applied force. Viscoelastic behaviour is characteristic of polymeric materials and is shown by asphalts, bitumens, paints, foods and many biomaterials such as wood and bone.

VMOS A **metal oxide silicon** (MOS) technology in which diffused layers are formed in silicon and V-shaped grooves are precisely etched through the layers. **Inversion** channels form along the arms of the V under gate control (in contrast with other MOS technologies where the channel is parallel to the wafer surface). Higher densities of current and components per chip are possible than with other MOS techniques.

void (1) Hollow formed within a moulding, especially at thick sections, a defect lowering product strength by concentrating stress. (2) In a powder compact or in a powder fluid system, the spaces between the particles.

voidage In powder **compact** or powder fluid system, the volume fraction of voids in the system. For a system containing dense particles, the voidage and the **powder porosity** are numerically equal.

Voigt model Conceptual model for a viscoelastic material consisting of spring and dashpot in parallel. See **Maxwell model**.

voile A light, open, plain-weave fabric made from highly-twisted yarns.

volume fraction Expression of concentration, esp. for a component in **composite materials**; it is the fraction of the total volume occupied by the component, eg the fibres. Cf **weight fraction**.

volume lifetime That of current carriers in the bulk of a semiconductor. Cf **surface lifetime**.

volume resistivity See **resistivity**.

volumetric strain The algebraic sum of three mutually perpendicular principal strains in a material.

von Mises criterion In a polycrystalline material, **slip** can only occur when five independent slip systems (ie sets of planes in the crystal) can operate. *Independent* means that the change in shape of the crystal due to slip in any one system cannot be achieved by slip in any combination of the other systems.

Vulkollan TN for crosslinkable polyurethane rubber which can be cast to shape. Extremely tough, but susceptible to hydrolysis (Europe).

vulcanite Hard vulcanized rubber, in the making of which a relatively high proportion of sulphur is used. Ebonite is one form; coloured varieties are obtained by adding various ingredients, such as the sulphides of antimony and mercury. See **vulcanization of rubber**.

vulcanization The treatment of rubber with sulphur or sulphur compounds to produce crosslinks between the polymer chains, resulting in a change in the physical properties of the rubber. Sulphur is absorbed by the rubber, and the process can be carried out either by heating raw rubber with sulphur at

a temperature between 135°C and 160°C, or by treating rubber sheets in the cold with a solution of S_2Cl_2. To increase the rate of vulcanization, accelerators may be used.

vulcanized fibre Neither a true fibre nor a product of vulcanization. Zinc chloride solution causes absorbent paper to swell and the cellulose fibres to be coated with amyloid gel. With removal of the inorganic salt, laminate is created by pressing the plies together. Electrical insulator.

vulcanizing agents Speciality chemicals for effecting crosslinking of rubbers. May be mixture of sulphur, accelerator, process aids etc. esp. for vulcanizing natural rubber, butadiene rubber. Chloroprene rubber's can be vulcanized with metal oxides alone.

vultex See **latex**.

W Symbol for **watt**.

wale A column of loops running along the length of a knitted fabric. The number of wales per unit length helps to characterize the fabric.

wall energy The energy per unit area stored in the **domain** wall bounding two oppositely magnetized regions of a ferromagnetic material.

Wallner lines Fracture surface markings, usually curved, caused by interaction between a propagating crack front and elastic waves released when pre-existing flaws or cracks are unloaded as the crack passes them. Only observed in brittle materials such as glass, but can be used to determine the speed of the crack. See **fractography**.

walnut European, American and S. American varieties of **hardwood** tree (*Juglans*) with a rich brown **heartwood**, straight- to wavy-grained and coarse-textured; density about 640 kg m^{-3}. Durable and popular for high-class furniture, panelling, gun stocks, musical instruments etc, and in plywood and decorative veneers.

wane A defect in converted timber; some of the original rounded surface of the tree is left along an edge.

warp (1) Permanent distortion of a timber from its true form, due to causes such as exposure to heat or moisture. (2) The threads of a woven fabric that run continuously along the length of the fabric.

warp knitted fabric Fine, knitted fabric formed wholly from warp yarns wound on a beam the width of the machine. The loops from each warp thread run the whole length of the fabric.

waste (1) Textile wastes are frequently recycled usually by degrading them to fibres. Soft waste is most easily processed because it is obtained in the earlier processes before twisting etc has been carried out. Hard waste includes all materials such as thread and rags which have to be pulled before they can be reused. (2) Generic term for materials or products discarded during manufacture or following useful life.

water hydrogen oxide, H_2O; a colourless, odourless, tasteless liquid, mp 0°C, bp 100°C, whose molecules associate extensively through hydrogen bonding, which gives it its unique properties. It forms a large proportion of the Earth's surface, occurs in all living organisms, and combines with many salts as **water of crystallization**. Water has its maximum density of 1000 kg m^{-3} at a temperature of 4°C. This fact has an important bearing on the freezing of ponds and lakes in winter, since the water at 4°C sinks to the bottom and ice at 0°C forms on the surface. Besides being essential for life, water has a unique combination of solvent power, thermal capacity, chemical stability, permittivity and abundance. See **bonding** p. 33.

water finish The high surface finish produced by the machine **calenders** when fitted with water doctors which apply a thin film of water at the **nip**.

waterleaf Paper which has not been sized with rosin prior to tub-sizing.

watermark A device in paper visible by transmitted light as a lighter or darker local area produced by an appropriate relief or intaglio design on a hand mould, cylinder mould or dandy roll.

water of crystallization The water present in hydrated compounds. When crystallized from solution in water these compounds retain a definite amount of water, eg copper (II) sulphate, $CuSO_4.5H_2O$. Also *water of hydration*.

water of hydration See **water of crystallization**.

waterproof paper Packaging paper that has been treated eg by impregnation, coating or lamination to render it resistant to the transmission of liquid water.

water resistor One made by immersing two electrodes in an aqueous solution. The resistance depends on the strength of the solution, and dimensions of the conducting path.

wave function Mathematical equation representing the space and time variations in amplitude for a wave system. The term is used particularly in connection with the Schrödinger equation for particle waves.

wave mechanics The modern form of the *quantum theory* in which events on an atomic or nuclear scale are explained in terms of the interactions between wave systems as expressed by the **Schrödinger equation**. For a bound particle, eg an electron in an atom, standing wave solutions are found for which only certain wavelengths are permitted, and consequently the energy is quantized. See **statistical mechanics**.

wave soldering Method of soldering large preassembled circuit boards by passing through a standing wave of molten solder.

wavy paper A defect of paper showing as undulations, esp. near the edges, due to the moisture content of the paper not being in equilibrium with the surrounding atmosphere.

wawa See **obeche**.

WAXS Abbrev. for **Wide Angle X-ray Scattering**.

wear The process of losing material from two surfaces that have been rubbed against one another. Mechanisms include **abrasive-**, **adhesive-**, **fatigue-**, and **chemical-** (or *corrosive-*) **wear**.

weaving The interlacing of warp and weft threads running at right angles to each other to form a fabric.

weaving machine, loom A machine that produces woven fabrics. In most, a device transports the **weft** threads and interlaces them with **warp** threads. In many looms this is done by a shuttle but in more modern machines other means are used such as projectiles, rapiers, and air or water jets.

web The relatively slender vertical part or parts of an I-beam or built-up girder (such as a box girder) separating the two flanges.

webbing A woven narrow fabric that is strong and able to sustain loads. Used in the manufacture of upholstery and of seatbelts for cars and aeroplanes.

weft The threads that run across the width of woven fabrics. In the weaving loom they are interlaced with the **warp** threads.

weft knitting A method of making a fabric by normal knitting with the loops being formed right across the fabric in straight lines at right angles to the direction in which the fabric is produced.

Weibull's modulus A measure of the degree of scatter in an extreme-value distribution, eg the results of repeated measurements of the tensile strength of a brittle material. Symbol m. Cf **strength measures** p. 301.

weight-average molecular mass See **molecular mass distribution** p. 212.

weight fraction Less frequently used alternative to **volume fraction** to express concentration of components in **composite materials**; It is the weight of the component present divided by the total weight of material.

Weissenberg effect Shown by elastic liquids during rotary stirring. The melt or solution climbs up the stirring rod, in some cases winding itself entirely around the rod. Caused by normal stresses developed in the sheared fluid.

Weiss theory Early theory of ferromagnetism based on the concept of independent molecular magnets.

weld decay Form of pitting corrosion which takes place in heat affected zones adjacent to welds in non-stabilized stainless steels. Overcome by small additions of titanium or niobium which prevent separation of chromium carbides.

welded joint A joint between two metals or plastics, made by fusion or diffusion to create interatomic bonding between the parts joined.

welding (1) Joining pieces of suitable metals or plastics, usually by raising the temperature at the joint so that the pieces may be united by fusing or by forging or under pressure. The welding temperature may be attained by external heating, by passing an electric current through the joint or by friction. (2) Joining pieces of suitable metals by striking an electric arc between an electrode or filler metal rod and the pieces. (3) Joining of thermoplastic parts using solvent, heat (eg hot-plate welding), friction (eg spin welding), adhesives, RF radiation, ultrasound etc.

welding rod Filler metal in the form of a wire or rod; in electric welding the electrode supplies the filler metal to the joint. Also *filler rod*. See **metal inert gas welding**, **tungsten inert gas welding**.

weld line Line on surface of moulding where melt fronts have met, but fused poorly. Kind of mould defect, often inevitable in complex injection mouldings, but may seriously weaken product due to stress concentration.

weldment A welded assembly.

Welvic TN for polyvinyl chloride (UK).

Western red cedar N. American **softwood** (*Thuja plicata*) which is not a true cedar. Reddish-brown **heartwood**, straight-grained and coarse-textured; density about 370 kg m^{-3}. Durable and used in exterior cladding, fencing, sheds etc.

West Indian ebony See **cocuswood**.

Weston standard cadmium cell A practical portable standard of emf, in which the cathode is 121/2% Cd and 871/2% Hg by weight, with anode of amalgamated Pt or highly purified Hg. Saturated sol. aq. Cd_2SO_4 as electrolyte. Electromotive force of cell at 20°C is 1.018 636 V, temperature coefficient only 0.000 04 V °C^{-1}. Used for calibrating potentiometers and hence all other voltage measuring devices. Cells with unsaturated solutions have a lower temperature coefficient of emf, but do not give an equally high absolute standard of reproduc-

ibility. Also *standard cell*.

wet electrolytic capacitor One in which the negative electrode is a solution of a salt, eg aluminium borate, which is suitable for maintaining the aluminium oxide film without spurious corrosion.

wet end The wire and press parts of a papermaking or boardmaking machine where the sheet is formed and water removed by drainage, suction and pressure.

wet etching Etching with wet chemicals as distinct from those processes based on gas discharges. Cf **dry etching**.

wet felt A continuous felt or synthetic fabric used in the press section of the papermaking or boardmaking machine to convey the web, assist in water removal and prevent crushing at the press nip.

wet laying Production of non-woven fabric by using a similar wet laying technique to that employed for making paper.

wet spinning (1) Making filaments from a solution of the polymer by **extrusion** followed by precipitation with a liquid, eg **viscose rayon**. (2) Method for making fine *flax* yarns in which the **roving** passes through hot water to soften it and so assist **drafting**.

wet strength paper Any paper so treated that it retains an appreciable proportion of its dry strength when completely wet and has good wet rub resistance. These properties may be obtained by adding suitable resins to the stock and curing (eg urea or melamine formaldehyde) or by **parchmentizing**.

wetwood Wood with an abnormally high water content and a translucent or water-soaked glassy appearance. This condition develops only in living trees and not through soaking in water.

wheatsheaf Growth form of polymer lamellae. See **crystallization of polymers** p. 82.

whipcord Fabric woven from cotton or worsted with a bold steep warp twill, used chiefly for dresses, suits and coats.

white bombway Timber of the **hardwood** tree, *Terminalia procera*, from the Andaman Islands. Planed surfaces of the wood are mildly silky, the texture moderately open and rather uneven, while the grain is straight. Wood used for joinery, panelling, furniture and interior fittings.

white deal See **whitewood**.

white gold Gold alloyed with nickel or platinum in order to render it silvery white in colour.

white-heart process See **malleable cast-iron**.

white heat As judged visually, temperature exceeding 1000°C.

white iron Pig-iron or cast-iron in which all the carbon is present in the form of cementite

(Fe_3C). White iron has a silvery-white crystalline fracture, and is hard and brittle.

white lead Basic *lead (II) carbonate* or *hydroxycarbonate*. Made by several processes of which the oldest and best known is the *Dutch* (or stack) *process*. Formerly used extensively as a paint pigment and for pottery glazes.

white metal Usually denotes a tin-base alloy (over 50% tin) containing varying amounts of lead, copper and antimony; used for bearings, domestic articles and small castings; sometimes also applied to alloys in which lead is the principal metal. Also *antifriction metal, bearing metal*.

whitewood Timber from the **softwoods**, silver fir (*Abies alba*) and Norway spruce (*Picae abies*). It varies from white to yellowish-brown, with a straight grain and fine texture; mean density 470 kg m^{-3}. Non-durable and susceptible to insect attack. Widely used, general purpose timber, used in plywood and veneers, important for keyboard instruments and the bellies of violins and guitars. Also *European silver pine, white deal*. Not to be confused with **American whitewood**.

whiting Powdered chalk or calcium carbonate used as filler for putty, thermoplastic and thermoset resins etc.

wick A yarn, group of yarns, or a narrow woven fabric or braid with good capillary properties. Used particularly in candles and oil lamps. See **candlewick**.

wide angle X-ray scattering Method of structure determination in materials. See **X-ray crystallography**. Abbrev. *WAXS*. If diffraction occurs, then *WAXD*.

Widmanstätten structure A mesh-like distribution of a precipitating phase in a solid state transformation which occurs along preferred crystal planes. Usually produced by rapid cooling and when the transforming phase has a large grain size. Originally observered in iron meteorites. Also *basket-weave structure*.

Wiedemann effect Tendency to twist in a rod carrying a current when subject to a magnetizing field.

Wien's laws for radiation from a black body (1) *Displacement law*: $\lambda_m T$ = constant (0.0029 metre kelvin). (2) *Emissive power* (E_λ): within the maximum intensity wavelength interval $d\lambda$, $E_\lambda = CT^5 d\lambda$, where $C = 1.288 \times 10^{-5}$ W m^{-2} K^{-5} (*T* is absolute temperature in kelvins). (3) *Emissive power* (*dE*) in the interval $d\lambda$ is

$$dE = A\lambda^{-5} \exp(-B / \lambda T)\, d\lambda.$$

λ_m is wavelength at E_{max}, $A = 4.992$ mJ, B = 0.0144 mK.

Wigner effect Changes in physical properties of graphite resulting from the displacement of lattice atoms by high-energy neutrons and other energetic particles in a reactor. It results in the building up of stored energy (*Wigner energy*) in the change of crystal lattice dimensions and hence in the change of overall bulk size.

wild silk See **tussah**.

Williams–Landel–Ferry equation An equation which relates the shift factor (a_T) to glass-transition temperature of polymer, *Tg*. It is

$$\log(a_T) = -\frac{c_1\,(T - T_g)}{(c_2 + T - T_g)},$$

where *T* is the temperature, c_1 and c_2 are constants. Abbrev. *WLF equation*. See **viscoelasticity** p. 332.

willow Timber from **hardwood** trees of widespread northern hemisphere species (Salix) with pinkish-white **heartwood** and white **sapwood**, straight-grained and fine-textured; density range 340–450 kg m^{-3}. Non-durable and liable to insect attack. Famous for cricket bats, also used in artificial limbs, wickerwork, plywood and decorative veneers.

wincey A light woven flannel made from mixed yarns containing wool.

winceyette A plain or simple twill cotton cloth of light weight, raised slightly on both sides; used for pyjamas, nightgowns, or underwear. May be colour-woven or piece-dyed.

winchester A narrow- or wide-mouthed cylindrical bottle used for the transportation of liquids, capacity 2.5 litres.

window Type of mould defect in partly crystalline polymers where a clear zone is left in opaque or translucent material, due to incomplete crystallization.

wirebar High-purity copper cast to a tapered ingot suitable for processing into wire.

wired glass A form of **flat glass** which incorporates a wire mesh into the sheet to act as a reinforcement by holding the fragments together in the event of the sheet being fractured. This makes it especially useful in fire-resistant glazing. It is not as strong as nonwired glass of the same thickness. Also *Georgian wired glass* after the wire mesh pattern.

wire-drawing (1) The process of reducing the diameter of rod or wire by pulling it through successively smaller holes.

wire gauge Any system of designating the diameter of wires by means of numbers, which originally stood for the number of successive passes through the die-blocks necessary to produce the given diameter.

wire rope Steel rope made by *twisting* or *laying* a number of strands over a central core, the strands themselves being formed by twisting together steel wires. The advantages are flexibility and high tensile strength. A fibre core impregnated with lubricant is incorporated at the centre of ropes which are subjected to repeated flexing.

witness mark (1) A remnant of original surface or scribed line, left during machining or hand working to prove that a minimum quantity of material has been removed or an outline accurately preserved. (2) A mark left on the surface of a moulding by tool parts, eg ejector pin marks.

WLF equation   Abbrev. for **Williams–Landel–Ferry equation**.

wood The universally used structural material derived from hard and softwoods, whether directly by machining etc followed by seasoning or in derived form eg blockboard, clipboard etc. Heartwood contains a greater lignin content and is harder and stiffer than sapwood, the outer layers of the the treetrunk. Softwood is the principal source of cellulose fibre both for processing into cellulose fibres (eg viscose rayon) and for papermaking. Over 100 different species of wood are currently exploited commercially worldwide. See **structure of wood** p. 305.

wood flour Finely ground wood waste used as an inexpensive reinforcing filler for plastics.

wood pulp Pulp produced from wood by any pulping method.

Wood's glass A glass with very low visible and high ultraviolet transmission.

Wood's metal Fusible alloy based on the quarternary eutectic of lead-bismuth-tin-cadmium system. Contains 50% bismuth, 25% lead, 12.5% each of tin and cadmium and melts at 69°C. There are other compositions with different melting ranges but all below the boiling point of water.

wool The fibres obtained from the fleece of sheep. See **hair**.

work One manifestation of energy. The work done by a force is defined as the product of the force and the distance moved by its point of application along the line of action of the force. For example, a tensile force does work in increasing the length of a piece of wire; work is done by a gas when it expands against a hydrostatic pressure. As for all forms of energy, the SI unit of work is the *joule*, performed when a force of 1 newton moves its point of application through 1 metre along the line of action of the force. Alternatively, when 1 watt of power is expended for 1 second, 1 joule of work is done. See **joule, kilowatt-hour**.

work function The minimum energy that must be supplied to remove an electron from a surface so that it can just exist outside a material under vacuum conditions. The energy can be supplied by heating the material (*thermionic* work function) or by illuminating it with radiation of sufficiently high energy (*photoelectric* work function). Also *electron affinity*. See **Einstein photoelectric equation**.

work-hardening The increase in strength and hardness (ie resistance to plastic deformation) produced by plastic deformation of metals at temperatures below about 0.5 *Tm*. It results from increasing numbers of dislocations and their entanglement and is accompanied by reduction in ductility. When extensive cold working is necessary to form a product, it is essential to anneal from time to time to remove the work hardening effects and permit further deformation. However, controlled work hardening may be used to advantage since it allows substantial increases in yield and tensile strength for a given cross section, eg in hard drawn steel wire or temper rolled aluminium alloy sheet.

working point One of the reference temperatures in **glassmaking** p. 144.

working stress The (safe) working stress of a component is the ultimate strength of a material divided by the applicable factor of safety.

work lead See **base bullion**.

worsted Woollen yarns, and fabrics and garments made from them, spun from fibres that have been combed and are fairly parallel.

wove paper Paper which does not exhibit a laid design when viewed by transmitted light.

wrinkling Uneven texture developing on surface during working.

wrought iron Composite material consisting of stringers of slag in an almost pure ferrite matrix, formerly made by the puddling process. Now obsolete.

x Symbol for mole fraction.

X A general symbol for an electro-negative atom or group, esp. a halogen.

xanthates The salts of xanthic acid, $CS(OC_2H_5)SH$. Potassium xanthate, obtained by the action of potassium ethoxide on carbon disulphide. Cellulose xanthate is the basis of the viscose rayon process.

Xe The symbol for **xenon**.

xeno- Prefix from Gk *xenos*, strange, foreign.

xenon A noble gas, present in the atmosphere in the proportion of 6×10^{-9} by volume. Symbol Xe. Forms several compounds, eg XeF_2, XeF_6, XeO_3. See Appendix 1, 2.

Xenoy TN for polycarbonate/polybutylene terephthalate blend or alloy used for car bumpers.

xerography Non-chemical photographic process in which light discharges a charged dielectric surface. It is the basis of most office photocopiers. An image is projected onto a drum whose surface is a charged dielectric, usually a **chalcogenide** film. This is dusted with a dielectric polymer powder, which adheres to the charged areas, rendering the image visible. Permanent images can be obtained by transferring particles to a suitable backing surface (eg paper or plastic) and fixing, usually by heat. Used for document copying, for making lithographic surfaces, usually on paper, for small-offset printing and as a photoresist.

xer-, xero- Prefix from Gk. *xeros*, dry.

X-ray crystallography Study of crystalline structures using diffraction of X-rays. The Debye-Scherrer technique uses diffraction of monochromatic rays from a powdered crystalline sample (giving a range of angles of incidence with respect to crystallographic axes). The arrangement of atoms in space can be inferred from the distribution of intensity in the diffraction pattern. The unit cell determines the geometry of diffraction maxima while the distribution of atoms within the unit cell controls the intensity of each reflection. The Laue method examines polychromatic diffraction from a single crystal. See **Bragg equation**.

X-ray fluorescence spectrometry A method of chemical analysis in which the sample is bombarded by X-rays or gammarays, and secondary radiations, characteristic of the elements present, are studied spectroscopically. Also *X-ray spectrography*. See **energy dispersive analysis of X-rays**, (EDAX).

X-ray protective glass Glass containing up to 50 wt% lead oxide and sometimes also barium oxide, which has a high absobtance to X-rays. See **lead glass**.

X-ray radiography

X-ray spectrography See **X-ray fluorescence spectrometry**.

xylenes $C_6H_4(CH_3)_2$, dimethylbenzenes.

xylenol resin Thermoset resin of the phenolic type. Produced by the condensation of a xylenol with an aldehyde.

Xylonite TN for a thermoplastic of the nitrocellulose type.

xyl-, xylo- Prefix from Gk. *xylon*, wood.

Yankee machine See **machine glazed**.

yard Unit of length in the foot-pound-second system formerly fixed by a line standard (Weights and Measures Act of 1878), redefined in 1963 as 0.9144 m. See Appendix 3.

yarn A thread ie a long thin material made of

fibres or filaments usually twisted together.

yarn count A quantity which designates the size of cross-section of a yarn. Historically this has been the number of lengths per unit mass with many local variations for the units, but the higher the number, the finer the thread. Cf **linear density**.

yew A **softwood** tree (*Taxus baccata*), indigenous to Europe and Asia whose **heartwood** is orange gold-brown streaked with mauve and dark brown, straight- or curly-grained and medium-textured; density about 670 kg m^{-3}. Durable and immune to powder-post beetle attack. The traditional material for the English longbow, it is also used for indoor and outdoor furniture, joinery, fencing, gate posts, etc., and as a very decorative veneer.

yield extension The amount of plastic strain which occurs at constant or decreasing load in metals which exhibit a pronounced **yield point**.

yield point See panel on p. 339, **strength measures** p. 301.

yield stress See **strength measures** p. 301.

YIG Abbrev. for *Yttrium Iron Garnet*. A material which has a lower acoustic attenuation loss than quartz and which has been considered for use in delay lines.

YIG filter One using a **YIG** crystal which is tuned by varying the current in a surrounding solenoid, a permanent magnet being used to provide the main field strength.

Young's modulus Symbol E. One of the most important **elastic constants** of materials, it is the ratio of the **tensile stress**, σ to the **tensile strain**, ε in a linear elastic material at loads less than the elastic limit of the material, ie $E = \sigma/\varepsilon$. It is related to the other elastic constants, G (**shear modulus**), ν (**Poisson's ratio**) and K (**bulk modulus**) by

$$E = 2G(1 + \nu) + 3K(1 - 2\nu) = \frac{9KG}{3K + G} \,.$$

ytterbium A metallic element, a member of the rare-earth group. Symbol Yb, mp of metal about 1800°C. See Appendix 1, 2.

yttrium A metallic element usually classed with the rare earths because of its chemical resemblance to them. Symbol Y, mp of metal 1250°C. See Appendix 1, 2.

Z Symbol for **section modulus**.

z-average molecular mass Symbol $\overline{M}_z$. Molecular mass measure giving greater emphasis to bigger polymer chains, usually obtained by sedimentation analysis in an ultracentrifuge. See **molecular mass distribution** p. 212.

Z-blade mixer Type of mixer for viscous polymer (eg. prepolymer) and filler plus additives etc. Often involves two counter-rotat-

yield point

Essentially, the limit of the elastic region of a stress-strain curve where plastic deformation occurs. However, in low-carbon steel and certain other materials the yield point is associated with a sudden drop in the stress and a small plastic extension. See Fig. 1.

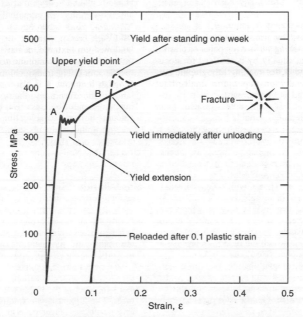

Fig. 1. Stress-strain curve for 0.1% carbon steel showing yield point phenomena.

The effect is attributed to *solute atmospheres* which anchor **dislocations**. When the stress reaches a sufficiently high level a sudden extension movement occurs until the released dislocations tangle so that a higher force is needed to start them moving again. This gives rise to the small amount of plastic deformation which precedes the usual development of the work-hardening part of the stress-strain curve.

In steels the atmospheres are formed by the interstitial carbon atoms which readily diffuse to dislocations because, by so doing, the overall strain energy in the crystal lattice is lowered. The yield point phenomenon is most pronounced in low-carbon steels containing 0.05–0.2% by weight carbon. Typical yield strains may be as high as 2–3% as at A, Fig. 1.

If the stress is taken off and immediately re-applied as at B, Fig. 1, there is no yield drop but, if time is allowed for solute atmospheres to reform, a yield point wll appear again. In steels the yield point begins to re-appear after 2 to 3 days at room temperature and is fully established again in about a week.

The deformation associated with the yield point can give rise to surface irregularities on manufactured products, such as pressings where varying amounts of deformation are required to form the three-dimensional shape from the sheet. These irregularities spoil the visual appearance of the surface and are referred to as *stretcher strains*. Individual bands of yielded material adjacent to unyielded are called *Lüders lines*.

ing Z-shaped blades esp. for DMC mixtures for composite products.

zein The maize protein used for experimental production of thermoset plastics by cross-linking with formaldehyde. See **casein**. Other proteins hitherto unsuccessfully used for plastics include those extracted from ground nuts (Ardil), soyabean, castor bean and blood.

Zener breakdown Temporary and non-destructive increase of current in diode because of critical field emission of holes and electrons in depletion layer at definite voltage.

Zener diode Characteristically showing a sharp increase in reverse current at a certain critical voltage; the current can increase indefinitely at this point, unless limited. This makes such diodes suitable for use as a voltage reference. Diodes which show this effect up to about six volts depend on **Zener breakdown**. At higher voltages, avalanche effects are more prevalent, though the diodes are still called *Zener* diodes.

Zener voltage That at which **Zener breakdown** occurs in certain types of diode. It denotes the negative voltage at which the reverse current increases very rapidly due to Zener breakdown and avalanche effects, and the voltage at which the diode can provide a reference source. See **Zener diode**.

zephyr Lightweight, plain weave, fine cotton fabric ornamented with stripes, checks, cords, etc. and used in dresses, blouses and shirts.

zero-cut crystal Quartz crystal cut at such an angle to the axes as to have a zero frequency/temperature coefficient. Used for accurate frequency and time standards.

Z-glass An alkali-resistant, zirconia glass, composition (wt %): SiO_2 70, Zr_2O 16, Na_2O 12, Li_2O 1, Al_2O_3 0.2, developed for the reinforcement of **Portland cement** (where the equilibrium **pH** is 12.5–13) and concrete.

zibeline Heavy, usually woollen, fabric for overcoats, with a lustrous, satin-type appearance due to a long hairy nap laid in one direction and pressed flat.

Ziegler-Natta catalyst Catalyst, discovered by Ziegler, which induces specific steric orientation. Used for organic polymerization, notably the low-pressure polyethylene process (for high-density polyethylene). Prepared by the reaction of compounds of strongly electropositive transition metals with organometallic compounds, eg titanium trichloride with aluminium alkyls.

zig-zag chain Conformation adopted by polyethylene chain in crystalline state; it is the shape of lowest energy in which the steric hindrance is minimal.

zinc A hard white metallic element with a bluish tinge. Symbol Zn, at.no. 30, r.a.m. 65.37. Because of its good resistance to atmospheric corrosion, zinc is used for protecting steel (see **galvanized iron**, **sheradizing, spelter**). It is also used in the form of sheet and as a constituent in alloys (see **zinc alloys**). Used as an electrode in a Daniell cell and in dry batteries. It is important nutritionally, trace amounts being present in many foods. See Appendix 1, 2.

zinc alloys Alloys based on the crystallographic form of zinc (HCP) The most important are those used for die castings, which contain essentially 3–4% aluminium, sometimes with copper 0–3.5% and magnesium 0.02–0.1%. Zinc is also the principal alloying ingredient of brasses and is used in lesser amounts in heat treatable aluminium alloys, but the amounts do not merit them being classed as zinc alloys.

zinc ferrite Non-magnetic ferrite, having normal as opposed to inverse **spinel** structure.

zinc telluride Semiconductor able to operate at high temperatures (up to about 750°C) without excessive intrinsic conductivity.

zirconium A metallic element, symbol Zr. When purified from hafnium, its low neutron absorption and its retention of mechanical properties at high temperature make it useful for the construction of nuclear reactors. Also used as a refractory, as a lining for jet engines, and as a getter in the manufacture of vacuum tubes. The zirconates are finding application as acoustic transducer materials. Tritium adsorbed in zirconium is a possible target in accelerator neutron sources. See Appendix 1, 2.

ZMC TN for method of injection moulding long-fibre reinforced polymers. High locking force machines are used to produce external car body parts. See **car body materials** p. 46.

zone melting Localized melting of a narrow band or portion of, usually, a column of metal by an induction coil which is moved along the axis. It is possible with certain alloys to arrange the rate of the passage of the molten zone so that the impurities concentrate in the liquid, allowing considerable purification of the remainder.

zone refining The use of **zone melting** for purification. By making several sweeps of the molten zone in the same direction, a bar of metal or semiconductor can be brought to an extremely high degree of purity, by concentrating the impurities at one end which is finally discarded. The very pure silicon required for semiconductors is refined in this way.

Z-twist See **twist direction**.

zyg-, zygo- Prefixes from Gk *zygon*, yoke.

Zytel TN for toughened nylon (US).

Appendix 1. The Periodic Table.

All the naturally occurring elements can be classified in terms of their **atomic structure** and physical or chemical properties. Physical properties include melting and boiling points, atomic volume (relative atomic mass divided by density) and mechanical properties like compressibility (reciprocal of **bulk modulus**). Chemical properties cover such characteristics as oxidation number or **valency**, ionic and **atomic radii** as well as **electronegativity**. These properties which are listed in Appendix 2 all show periodic associations (or *periodicity*) as a function of atomic number. If the elements are arranged in sequence of atomic number starting with hydrogen ($Z=1$) from left to right as in the table below, elements in vertical columns (or groups) tend to show similar properties. The first element hydrogen is ambiguous as it shows both metallic and non-metallic behaviour, forming a cation or an anion depending on its environment.

Table 1. **The periodic table with the elements arranged in 16 groups, 1a — 0.**

1a	2a	3b	4b	5b	6b	7b	8						1b	2b	3a	4a	5a	6a	7a	0
1 H					Metals												Non-metals		1 H	2 He
3 Li	4 Be														5 B	6 C	7 N	8 O	9 F	10 Ne
11 Na	12 Mg														13 Al	14 Si	15 P	16 S	17 Cl	18 Ar
19 K	20 Ca	21 Sc	22 Ti	23 V	24 Cr	25 Mn	26 Fe	27 Co	28 Ni	29 Cu	30 Zn		31 Ga	32 Ge	33 As	34 Se	35 Br	36 Kr
37 Rb	38 Sr	39 Y	40 Zr	41 Nb	42 Mo	43 Tc	44 Ru	45 Rh	46 Pd	47 Ag	48 Cd		49 In	50 Sn	51 Sb	52 Te	53 I	54 Xe
55 Cs	56 Ba	57* to 71	72 Hf	73 Ta	74 W	75 Re	76 Os	77 Ir	78 Pt	79 Au	80 Hg		81 Tl	82 Pb	83 Bi	84 Po	85 At	86 Rn
87 Fr	88 Ra	89+ to 92						Metals									Non-metals			
*Lanthan- ides	57 La	58 Ce	59 Pr	60 Nd	61 Pm	62 Sm	63 Eu	64 Gd	65 Tb	66 Dy	67 Ho	68 Er	69 Tm	70 Yb	71 Lu					
+Actin- ides	89 Ac	90 Th	91 Pa	92 U																

In the periodic table, illustrated above, metals are found on the left of the bold line and non-metals on the right of the dotted line. Between metals and and non-metals is a cluster of six semi-metals, starting with boron, which have intermediate properties such as being semiconductors. There is also a group of 14 lanthanides or rare earth elements with very similar properties and four radioactive heavy metals, the actinides.

Appendix 2. The principal properties of the elements.

The tables on the following three pages list all the naturally occurring elements alphabetically together with some of their properties. The following abbreviations are used.

bc = body centred; bcc = body-centred cubic
cubic (diam) = diamond structure
Electroneg = Electronegativities as assigned by Pauling on a scale of 0 to 4.
fc = face centred; fcc = face-centred cubic
graph = graphite
hcp = hexagonal close-packed
hex = hexagonal
mon = monoclinic
ortho = orthorhombic

Ox. Nos = principal oxidation numbers
r = red and y = yellow
r_a, r_i = atomic, ionic radii calculated from closest distances between atoms in crystal structure
rhombic = rhombohedral
tetra = tetragonal
T_m, T_b = temperatures of melting and boiling, all exc. arsenic, helium at atmospheric pressure
T_{TRANS} = temperature for transformations in crystal structure column

Table 2. **Properties of the elements actinium to helium.** The abbreviations used are listed at the foot of p. 341.

Element, symbol	Z	r.a.m.	Crystal structure	T_{TRANS} K	r_a, pm	Ox. Nos	r_i, pm	Electroneg.	Density, ρ, kg m^{-3}	T_m, K	T_b, K
Actinium, Ac	89	227	fcc		188	3+	118	1.1	10 100	1320	3470
Aluminium, Al	13	26.98	fcc		142	3+	51	1.5	2700	933.2	2740
Antimony, Sb	51	121.75	rhombic		145	3+, 5+	76/62	1.9	6700	903.7	1650
Argon, Ar	18	39.95	fcc		174	0	154	–	1.66	83.7	87.4
Arsenic, As	33	74.92	rhombic		125	3+, 5+	58/46	2.0	5730	1090	886
Astatine, At	85	210	–		–	7+	62	2.2	–	520	623
Barium, Ba	56	137.34	bcc		217	2+	134	0.9	3600	1000	1910
Beryllium, Be	4	9.01	hcp/cubic	1527	112	2+	35	1.5	1800	1550	3243
Bismuth, Bi	83	208.98	rhombic		155	3+, 5+	96/74	1.9	9800	544.4	1830
Boron, B	5	10.81	tetra/rhombic		88	3+	23	2.0	2500	2600	2820
Bromine, Br	35	79.90	ortho		114	1–, 5+	196/47	2.8	3100 (298 K)	265.9	331.9
Cadmium, Cd	48	112.40	hcp		148	2+	97	1.7	8650	594.2	1038
Caesium, Cs	55	132.90	bcc		262	1+	167	0.7	1870	302.6	960
Calcium, Ca	20	40.08	fcc/bcc		196	2+	99	1.0	1540	1120	1760
Carbon, C	6	12.01	hex/cubic	graph/diam 737	71/77	4+, 4–	16/260	2.5	2300	>3800	5100
Cerium, Ce	58	140.12	fcc/hex/fcc/bcc	95/263/998	183	3+, 4+	103/92	1.1	6800	1070	3740
Chlorine, Cl	17	35.45	tetra		91	1–, 5+	181/34	3.0	3.21 (273 K)	172.1	283.5
Chromium, Cr	24	52.00	bcc		125	3+, 6+	63/52	1.6	7200	2160	2755
Cobalt, Co	27	58.93	hcp/fcc	690	125	2+, 3+	72/63	1.8	8900	1765	3170
Copper, Cu	29	63.55	fcc		128	1+, 2+	96/72	1.9	8930	1356	2868
Dysprosium, Dy	66	162.50	rhombic/hcp	86	175	3+	91	1.2	8500	1680	2900
Erbium, Er	68	167.26	hcp		173	3+	88	1.2	9000	1770	3200
Europium, Eu	63	151.96	bcc		198	3+, 2+	95/109	1.1	5200	1100	1712
Fluorine, F	9	19.00	–		60	1–, 7+	133/8	4.0	1.7 (273 K)	35.5	85.01
Francium, Fr	87	223	–		–	1+	187	0.7	–	303	920
Gadolinium, Gd	64	157.25	hcp/bcc	1537	178	3+	94	1.2	7900	1585	3000
Gallium, Ga	31	69.72	fcc or ortho		121	1+, 3+	81/62	1.6	5950	302.9	2676
Germanium, Ge	32	72.59	cubic (diamond)		122	4+	53	1.8	5400	1210.5	3100
Gold, Au	79	196.97	fcc		144	1+, 3+	137/85	2.4	19 300	1336.1	3239
Hafnium, Hf	72	178.49	hcp/bcc	2050	158	4+	78	1.3	133 000	2423	5700
Helium, He	2	4.003	hcp/cubic		176	0	–	–	0.166	0.95	4.21

Table 3. **Properties of the elements holmium to protoactinium.** The abbreviations used are listed at the foot of p. 341.

Element, symbol	Z	r.a.m.	Crystal structure	T_{TRANS}, K	r_a, pm	Ox. Nos	r_i, pm	Electroneg.	Density, ρ, kg m⁻³	T_m, K	T_b, K
Holmium, Ho	67	164.93	hcp		176	3+	89	1.2	8800	1734	2900
Hydrogen, H	1	1.008	hcp/cubic		46	1+	154	2.1	0.08987 (273 K)	14.01	20.4
Indium, In	49	114.82	bc tetragonal		162	3+	81	1.7	7 310	429.8	2300
Iodine, I	53	126.90	ortho		135	1-	216	2.5	4 940	386.6	457.4
Iridium, Ir	77	192.2	fcc		135	4+	68	2.2	22 420	2716	4800
Iron, Fe	26	55.85	bcc/fcc/bcc	1180/1670	123	2+,3+	74	1.8	7 870	1808	3300
Krypton, Kr	36	83.80	fcc		201	0	—	—	3.49	116.5	120.8
Lanthanum, La	57	138.91	hcp/fcc/bcc	583/1137	187	1+,3+	139/102	1.1	6 150	1190	3742
Lead, Pb	82	207.19	fcc		174	2+,4+	120/84	1.8	11 340	600.4	2017
Lithium, Li	3	6.94	hcp/fcc/bcc	74/140	152	1+	68	1.0	534	452	1590
Lutetium, Lu	71	174.97	hcp		173	3+	85	1.2	9 800	1925	3600
Magnesium, Mg	12	24.31	hcp		160	1+,2+	82/66	1.2	1 741	924	1380
Manganese, Mn	25	54.94	cubic		112	2+,3+	80/66	1.5	7 440	1517	2370
Mercury, Hg	80	200.59	rhombic		156	1+,2+	127/110	1.9	13 590 (273K)	234.3	629.7
Molybdenum, Mo	42	95.94	bcc		136	4+,6+	70/62	1.8	10 200	2880	5830
Neodymium, Nd	60	144.24	hcp/bcc	1135	181	3+	100	1.1	6 960	1296	3300
Neon, Ne	10	20.18	fcc		160	0	—	—	0.839	24.5	27.2
Nickel, Ni	28	58.71	fcc		124	2+,3+	69	1.8	8 900	1726	3005
Niobium, Nb	41	92.91	bcc		143	5+	69	1.6	8 570	2741	5200
Nitrogen, N	7	14.01	cubic/hcp	35.4	71	3+,5+	16/13	3.0	1.165	63.3	77.3
Osmium, Os	76	190.2	hcp		135	4+	69	2.2	22 480	3300	4900
Oxygen, O	8	16.00	rhombic		60	2-	132	2.1	1.33	54.7	90.2
Palladium, Pd	46	106.4	fcc		137	2+,4+	80/65	2.2	12 000	1825	3200
Phosphorus, P	15	30.97	cubic		120	3+,4+	44/35	2.1	2 200 (r) 1 800 (y)	317.2	552
Platinum, Pt	78	195.09	fcc		138	2+,4+	80/65	2.2	21 450	2042	4100
Polonium, Po	84	209	monoclinic		168	2+,6+	/67	2.0	9 400	527	1235
Potassium, K	19	39.10	bcc		231	1+	133	0.8	860	336.8	1047
Praseodymium, Pr	59	140.91	hcp/bcc	1065	182	3+	101	1.1	6 800	1208	3400
Promethium, Pm	61	145			—	3+	98	1.1	—	1308	3000
Protactinium, Pa	91	231.04	tetra		160	3+,4+	113/98	1.5	15 400	1500	4300

Table 4. **Properties of the elements radium to zirconium.** The abbreviations used are listed at the foot of p. 341.

Element, symbol	Z	r.a.m.	Crystal structure	T_{TRANS} K	r_a pm	Ox. Nos	r_i pm	Electroneg.	Density, ρ, kg m^{-3}	T_m, K	T_b, K
Radium, Ra	88	226.03	—		—	2+	143	0.9	5 000	970	1410
Radon, Rn	86	222	—		—	0	—	—	9.73 (273K)	202	211.3
Rhenium, Re	75	186.2	hcp		137	4+	72	1.9	20 500	3450	5900
Rhodium, Rh	45	102.91	fcc		134	3+	68	2.2	12 440	2230	4000
Rubidium, Rb	37	85.47	bcc		246	1+	147	0.8	1 530	315.0	961
Ruthenium, Ru	44	101.07	hcp		133	4+	67	2.2	12 400	2520	4200
Samarium, Sm	62	150.35	rhombic/fcc	1190	179	3+	96	1.1	7 500	1345	2200
Scandium, Sc	21	44.96	hcp/fcc	1223	160	3+	73	1.3	3 000	1812	3000
Selenium, Se	34	78.96	hcp		116	2-	191	2.4	4 810	490	958
Silicon, Si	14	28.09	cubic	5	118	4+,4-	42/38	1.8	2 300	1680	2628
Silver, Ag	47	107.87	fcc/hcp		144	1+	126	1.9	10 500	1234	2485
Sodium, Na	11	22.99	bcc		185	1+	97	0.9	970	371	1165
Strontium, Sr	38	87.62	fcc/hcp/bcc	506/813	215	2+	112	1.0	2 600	1042	1657
Sulphur, S	16	32.06	fc ortho		106	2-,4+	184/37	2.5	2 070	386	717.7
Tantalum, Ta	73	180.95	bcc		143	5+	68	1.5	16 600	3269	5698
Technetium, Tc	43	98.91	hcp		135	7+	98	1.9	11 400	2500	4900
Tellurium, Te	52	127.60	hcp		143	2-	211	2.1	6 240	722.6	1260
Terbium, Tb	65	158.92	hcp/rhombic	1590	177	3+	92	1.2	8 300	1629	3100
Thallium, Tl	81	204.37	hcp/fcc	503	171	1+	147	1.8	11 860	576.6	1730
Thorium, Th	90	232.04	fcc/bcc	1673	180	4+	102	1.3	11 500	2000	4500
Thulium, Tm	69	168.93	hcp/bcc	1158	174	3+	87	1.2	9 300	1818	2000
Tin, Sn	50	118.69	cubic(diam)/bcc		140	2+,4+	93/71	1.8	7 300	505.1	2540
Titanium, Ti	22	47.90	hcp/bcc	1158	146	4+	68	1.5	4 540	1948	3530
Tungsten, W	74	183.85	bcc		137	6+	62	1.7	19 320	3650	6200
Uranium, U	92	238.03	rhombic/tetra	941	138	4+,6+	97	1.7	19 050	1405.4	4091
Vanadium, V	23	50.94	bcc		131	3+,5+	74/59	1.6	6 100	2160	3300
Xenon, Xe	54	131.30	fcc		221	0, (3+,5+)	—	—	5.50	161.2	166.0
Ytterbium, Yb	70	173.04	fcc/bcc	1071	193	3+	86	1.2	7 000	1097	1700
Yttrium, Y	39	88.91	hcp/bcc	1763	181	3+	89	1.2	4 600	1768	3200
Zinc, Zn	30	65.37	hcp		133	2+	74	1.6	7 140	692.6	1180
Zirconium, Zr	40	91.22	hcp/bcc	1100	160	4+	79	1.4	6 500	2125	3851

Appendix 3. Units of measurement.

The International System (SI) of units has seven *base units*, two *supplementary units* (the radian and the steradian) and a variety of *derived units*. The seven physical quantities on which the system is based, together with their dimensions, the SI base units and their symbols are listed in the table below.

Basic physical quantity	Dimension	SI base unit	Symbol
length	L	metre	m
mass	M	kilogram	kg
time	T	second	s
electric current	I	ampere	A
temperature	θ	kelvin	K
luminous intensity	J	candela	cd
amount of substance	—	mole	mol

Although the mole is strictly dimensionless, it is usually expressed in the form of g-mole, and represents the amount of substance which contains as many atoms or molecules as there are atoms in 0.012 kg of the carbon isotope carbon-12 ($^{12}_{6}C$).

SI derived units

The SI derived units for other physical quantities are formed from the base units via the equation defining the quantity involved. Thus for example, force = mass × acceleration and the unit of force, the newton, is equivalent to kg m s^{-2}. The two supplementary units and the principal derived units, together with some of the more important equivalents are shown below.

Quantity	Dimension	Unit	Symbol	Equivalent
plane angle	—	radian	rad	$(=180°/\pi)$
solid angle	—	steradian	sr	
density	$M L^{-3}$		kg m^{-3}	
velocity or speed	$L T^{-1}$		m s^{-1}	
acceleration	$L T^{-2}$		m s^{-2}	
momentum	$M L T^{-1}$		kg m s^{-1}	
moment of inertia	$M L^2$		kg m^2	
force	$M L T^{-2}$	newton	N	kg m s^{-2}
pressure, stress	$M L^{-1} T^{-2}$	pascal	Pa	N m^{-2}
energy, work	$M L^2 T^{-2}$	joule	J	N m
power	$M L^2 T^{-3}$	watt	W	J s^{-1}
viscosity	$M L^{-1} T^{-1}$			N s m^{-2} kg m^{-1} s^{-1}
frequency	T^{-1}	hertz	Hz	s^{-1}
electric conductance	$L^{-2} M^{-1} T^3 I^2$	siemens	S	Ω^{-1}
electric charge	$T I$	coulomb	C	A s
electric potential difference	$L^2 M T^{-3} I^{-1}$	volt	V	W A^{-1}
electric capacitance	$T^4 I^2 L^{-2} M^{-1}$	farad	F	C V^{-1}
electric resistance	$L^2 M T^{-3} I^{-2}$	ohm	Ω	V A^{-1}
magnetic flux	$L^2 M T^{-2} I^{-1}$	weber	Wb	V s
magnetic flux density	$M T^{-2} I^{-1}$	tesla	T	V s m^{-2}
inductance	$L^2 M T^{-2} I^{-2}$	henry	H	V s A^{-1}
luminous flux	J	lumen	lm	cd sr
illuminance	$J L^{-2}$	lux	lx	lm m^{-2}
radioactivity	T^{-1}	becquerel	Bq	s^{-1}
absorbed dose, ionizing radiation	$L^2 T^{-2}$	gray	GY	J kg^{-1}

Units of Measurement in Mechanics

In any system of measurement in mechanics three fundamental units are required. In SI these are the metre, kilogram and second. Through common usage certain multiples of the three fundamental units have been given names. Some of these are given below, together with units of pressure and textile line density which are still in frequent use. None of them is a recognized SI unit.

Physical quantity	Unit	Symbol	SI equivalent
length	micron	μ	10^{-6} m = 1 μm
length	ångström	Å	10^{-10} m = 0.1 nm
mass	tonne	t	10^3 kg = 10^6 g
time	minute	min	60 s
time	hour	h	3600 s
time	day	d	86 400 s
time	year	a	3.1578×10^7 s
pressure	atmosphere	atm	1.013×10^5 N m^{-2}
pressure	bar	bar	10^5 N m^{-2}
pressure	torr	Torr	133.3 N m^{-2}
textile line density or	tex		1 g km^{-1} = 10^{-6} kg m^{-1}
yarn count	denier		0.1111 g km^{-1} = 1.111×10^{-7} kg m^{-1}

Conversion Factors

After SI, the two most important systems of units are the *cgs* and *fps systems*, the latter being still very frequently used in the US. Like SI they are both based on the second as unit of time, but differ in their base units for length (centimetre and foot) and for mass (gram and pound). Some of the more important conversion factors between SI units and the other two systems are set out below.

Physical quantity	cgs unit	fps unit
length	*centimetre*, 1 cm = 10^{-2} m	*foot*, 1 ft = 0.3048 m
		inch, 1 in = 2.54×10^{-2} m
mass	*gram*, 1 g = 10^{-3} kg	*pound*, 1 lb = 0.4536 kg
		ounce, 1 oz = 2.835×10^{-2} kg
		ton, 1 ton = 1.016×10^3 kg
area	1 cm^2 = 10^{-4} m^2	1 ft^2 = 9.290×10^{-2} m^2
		1 in^2 = 6.452×10^{-4} m^2
volume	1 cm^3 = 10^{-6} m^3	1 ft^3 = 2.832×10^{-2} m^2
	1 litre = 10^{-3} m^3	1 in^3 = 1.639×10^{-5} m^2
		gallon, 1 gal (UK) = 4.546×10^{-3} m^3
		1 gal (US) = 3.786×10^{-3} m^3
		fluid ounce, 1 fl oz = 2.841×10^{-5} m^3
density	1 g cm^{-3} = 10^{-3} kg m^{-3}	1 lb ft^{-3} = 16.02 kg m^{-3}
		1 lb in^{-3} = 2.786×104 kg m^{-3}
velocity or speed	1 cm s^{-1} = 10^{-2} m s^{-1}	1 ft s^{-1} = 0.3048 m s^{-1}
momentum	1 g cm s^{-1} = 10^{-5} kg m s^{-1}	1 lb ft s^{-1} = 0.1383 kg m s^{-1}
moment of inertia	1 g cm^2 = 10^{-7} kg m^2	1 lb ft^2 = 4.214×10^{-2} kg m^2
force	*dyne*, 1 dyn = 10^{-5} N	*poundal*, 1 pdl = 0.1383 N
		pound force, 1 lbf = 4.448 N
pressure or stress	1 dyn cm^{-2} = 10^{-1} Pa	1 lbf in^{-2} (psi) = 6.895×10^3 Pa
	bar, 1 bar = 10^5 Pa	
energy or work	*erg*, 1 erg = 10^{-7} J	1 ft pdl = 4.214×10^{-2} J
		1 ft lbf = 1.356 J
power	1 erg s^{-1} = 10^{-7} W	1 ft pdl s^{-1} = 4.214×10^{-2} W
		1 ft lbf s^{-1} = 1.356 W
		horsepower, 1 hp = 745.7 W
viscosity	*poise*, 1 P = 10^{-1} N s m^{-2}	1 lbf s ft^{-2} = 47.88 N s m^{-2}
energy (thermal)	*calorie*, 1 cal = 4.187 J	*British thermal unit*, 1 BTU = 1.055×10^3 J